ANALYSIS METHODS
for
ELECTROMAGNETIC WAVE
PROBLEMS

The Artech House Microwave Library

Analysis, Design, and Applications of Fin Lines, Bharathi Bhat and Shiban K. Koul
E-Plane Integrated Circuits, P. Bhartia and P. Pramanick, eds.
Filters with Helical and Folded Helical Resonators, Peter Vizmuller
GaAs MESFET Circuit Design, Robert A. Soares, ed.
Gallium Arsenide Processing Techniques, Ralph Williams
Handbook of Microwave Integrated Circuits, Reinmut K. Hoffmann
Handbook for the Mechanical Tolerancing of Waveguide Components, W.B.W. Alison
Handbook of Microwave Testing, Thomas S. Laverghetta
High Power Microwave Sources, Victor Granatstien and Igor Alexeff, eds.
Introduction to Microwaves, Fred E. Gardiol
LOSLIN: Lossy Line Calculation Software and User's Manual, Fred. E. Gardiol
Lossy Transmission Lines, Fred E. Gardiol
Materials Handbook for Hybrid Microelectronics, J.A. King, ed.
Microstrip Antenna Design, K.C. Gupta and A. Benalla, eds.
Microstrip Lines and Slotlines, K.C. Gupta, R. Garg, and I.J. Bahl
Microwave Engineer's Handbook: 2 volume set, Theodore Saad, ed.
Microwave Filters, Impedance Matching Networks, and Coupling Structures, G.L. Matthaei, L. Young and E.M.T. Jones
Microwave Integrated Circuits, Jeffrey Frey and Kul Bhasin, eds.
Microwaves Made Simple: Principles and Applications, Stephen W. Cheung, Frederick H. Levien, et al.
Microwave and Millimeter Wave Heterostructure Transistors and Applications, F. Ali, ed.
Microwave Mixers, Stephen A. Maas
Microwave Transition Design, Jamal S. Izadian and Shahin M. Izadian
Microwave Transmission Line Filters, J.A.G. Malherbe
Microwave Transmission Line Couplers, J.A.G. Malherbe
Microwave Tubes, A.S. Gilmour, Jr.
MMIC Design: GaAs FETs and HEMTs, Peter H. Ladbrooke
Modern Spectrum Analyzer Theory and Applications, Morris Engelson
Monolithic Microwave Integrated Circuits: Technology and Design, Ravender Goyal, et al.
Nonlinear Microwave Circuits, Stephen A. Maas
Terrestrial Digital Microwave Communications, Ferdo Ivanek, et al.

ANALYSIS METHODS for ELECTROMAGNETIC WAVE PROBLEMS

Eikichi Yamashita
Editor

Artech House
Boston • London

Library of Congress Cataloging-in-Publication Data

[Denjiha mondai no kiso kaiseki ho. English]
Analysis methods for electromagnetic wave problems. / editor Eikichi Yamashita.
p. cm.
Translation of : Denjiha mondai no kiso kaiseki ho.
Rev. ed. of English translation published as: Basic analysis methods for electromagnetic wave problems. 1987.
ISBN 0-89006-364-8
1. Electromagnetic waves. 2. Numerical analysis. I. Yamashita, Eikichi, 1933- . II. Title: Basic analysis methods for electromagnetic wave problems.
QC661.D4613 1990 90-587
539.2'01'515--dc20 CIP

British Library Cataloguing in Publication Data

Analysis methods for electromagnetic wave problems
1. Electromagnetic waves
I. Yamashita, Eikichi
530.141

ISBN 0-89006-364-8

685 Canton Street
Norwood, MA 02062

International Standard Book Number: 0-89006-364-8
Library of Congress Catalog Card Number: 90-587

10 9 8 7 6 5 4 3 2

Contents

Preface

The twentieth century perhaps has been the century of electromagnetic waves. With Maxwell's electromagnetic theory and Hertz's experiments in the nineteenth century as the point of departure, Marconi's "wireless" communication systems have been running and various types of electromagnetic waves in the form of radio and television are now criss-crossing the earth, endlessly carrying information.

The wireless systems using the propagation of electromagnetic waves in space include other applications such as satellite communication systems, automobile telephones, and radar. Antennas are always necessary for establishing wireless systems.

There are also "wire" systems for transmitting electromagnetic waves. Waveguides and striplines use the propagation of the electromagnetic waves over long tubes or strip conductors. Optical fibers guide electromagnetic waves along thin dielectric wires called the *core*.

Other applications of electromagnetic waves have been found recently in noncommunication fields such as particle accelerators in physics experiments, hyperthermia devices for cancer treatment, and microwave ovens at home.

These phenomena or devices are all treated as electromagnetic wave problems in which Maxwell's equations are to be solved under some boundary, media, or excitation conditions. The purpose of this book is to cover basic analysis methods that are now widely used to analyze these problems and to instruct such methods for researchers, engineers, and students, who will use them so that they can be easily understood and applied without reading many publications. This book also could be used as a textbook for engineering graduate students or as a reference book for undergraduate students writing their theses.

Each chapter of this book has been written by an expert in that field as an independent article. Several chapters of the book discuss numerical methods to treat as many problems as possible. The main features of this book are:

1. The historical background of each method is described and important (but not all) references quoted.

2. The characteristics of the method and the range of applications are discussed in a section of each chapter for readers to judge the applicability of the method to a particular problem in an early stage.
3. Some precautions and techniques are described for the programming of computational processes in the chapters on numerical methods that assume the use of computers.

We should mention here that some basic analysis methods have not been included in this book because of limited space but they are expected to be published later on.

The authors of this book thank the members of the Editorial Department of the IEICE who have made great efforts in the publication of this book, especially in the difficult proofreading of many complex mathematical expressions.

EIKICHI YAMASHITA
AUGUST, 1987

This book was originally published by the Institute of Electronics, Information, and Communication Engineers, Japan, in October 1987, to instruct up-to-date analysis methods to solve electromagnetic wave problems for researchers, engineers, and graduate students. For the English version of this book, published by Artech House, Inc., each chapter has been translated from Japanese by the original author of that chapter, who is responsible for the contents.

JANUARY, 1990

The Authors

Prof. Makoto Ando
Dept. of Electrical and Electronic Engineering
Tokyo Institute of Technology
Tokyo, Japan 152

Prof. Kazuhiko Atsuki
Dept. of Electronic Engineering
University of Electro-communications
Tokyo, Japan 182

Prof. Masahiro Hashimoto
Dept. of Applied Electronic Engineering
Osaka Electro-communication University
Osaka, Japan 572

Prof. Tatsuo Itoh
Dept. of Electrical and Computer Engineering
University of Texas
Austin, TX 78712, USA

Ms. Seiko Kitazawa
Dept. of Electronic Engineering
University of Tokyo
Tokyo, Japan 113

Prof. Kazuya Kobayashi
Dept. of Electrical and Electronic Engineering
Chuo University
Tokyo, Japan 112

Prof. Masanori Koshiba
Dept. of Electronic Engineering
Hokkaido University
Sapporo, Japan 060

Dr. Nagayoshi Morita
Dept. of Communication Engineering
Osaka University
Osaka, Japan 565

Prof. Takanori Okoshi
Research Center for Advanced Science and Technology
University of Tokyo
Tokyo, Japan 153

Prof. Yoichi Okuno
Dept. of Electrical Engineering and Computer Science
Kumamoto University
Kumamoto, Japan 860

Prof. Hiroshi Shigesawa
Dept. of Electronics
Doshisha University
Kyoto, Japan 602

Prof. Eikichi Yamashita
Dept. of Electronic Engineering
University of Electro-communications
Tokyo, Japan 182

Prof. Norinobu Yoshida
Dept. of Electrical Engineering
Hokkaido University
Sapporo, Japan 060

Chapter 1
The Finite-Element Method

Masanori Koshiba

1.1 HISTORICAL BACKGROUND

Recently, the finite-element method has been applied to a wide range of problems in electrical and electronic engineering, and many successful results have been obtained as well in the field of electromagnetic-wave engineering, dealing with wave fields. In particular, as the computer becomes larger, faster, and more manifold (from personal and minicomputers to supercomputers), it is spawning a new interdisciplinary field, called *computational mechanics,* in which the methods such as finite-element method, finite-difference method, and boundary-element method play important roles [1]. The new field, in turn, has generated more interest in the finite-element method.

The research on the application of the finite-element method to the electromagnetic-wave engineering has been made in earnest since the latter half of the 1960s, particularly from the 1970s, and is now being made extensively. Among many applications of the finite-element method, the research on its use with electromagnetic waveguide problems, as shown in Figure 1.1, continues incessantly, and it is being established as one of the powerful numerical methods for applying electromagnetic waveguides to various waveguiding structures from microwave to optical wavelength regions. However, the most serious difficulty in applying the finite-element method to such electromagnetic waveguide problems is the appearance of the so-called spurious, nonphysical solutions. Consequently, the development of a method to suppress or eliminate such spurious solutions is pressingly needed, and research on this topic has been extensive in recent years. In the finite-element method, the field region of finite extent is subdivided into elements, and then all the element contributions to the system under consideration are assembled. Thus, the original finite-element scheme cannot be applied straightforwardly to unbounded waveguiding structures, which are frequently used in millimeter-wave and optical wavelength

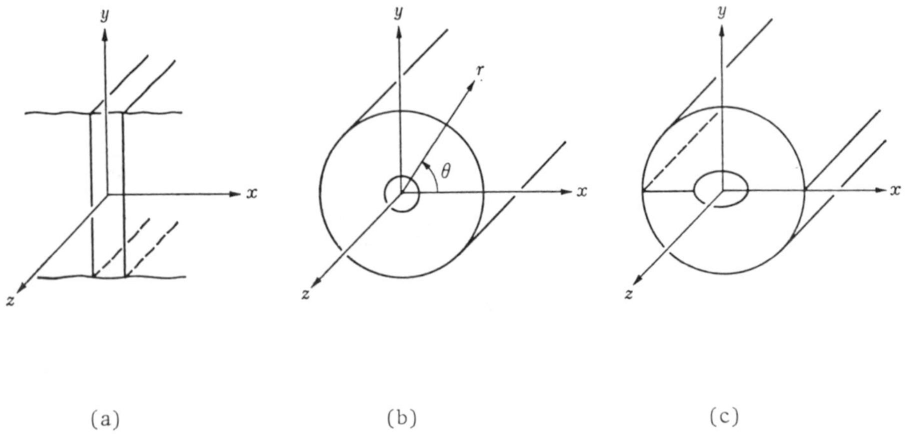

Figure 1.1 Electromagnetic waveguides: (a) planar waveguide; (b) axially symmetrical waveguide; (c) waveguide with arbitrarily shaped cross section.

regions, and consequently various approaches have been developed to deal with such unbounded field problems.

Therefore, the concept of the finite-element method and its fundamentals are discussed in this chapter, and the development of the finite-element method for solving electromagnetic-wave problems, especially waveguide problems, is presented. Although many books on the finite-element method are now available, to save the space, only the books [2–17] written for the readers majoring in the field of electrical and electronic engineering are cited as references in this book. The book by Kagawa *et al.* [15] seems to be particularly helpful for the readers with a special interest in finite-element applications in electromagnetic-wave problems. A large number of related research papers that have already been presented in the previous review papers [18–22] may be omitted from the present reference list.

The waveguide problem in Figure 1.1 is a so-called eigenvalue problem, and the periodic waveguide problem and the cavity problem, shown in Figures 1.2 and 1.3, respectively, also are eigenvalue problems. In addition to these eigenvalue problems are many other practical important problems: for instance, the waveguide-discontinuity problem and the radiation or scattering problem of electromagnetic waves. These problems are called *deterministic problems*. Finite-element applications in various eigenvalue and deterministic problems have been studied extensively.

1.2 THE RANGE OF APPLICATIONS

From various viewpoints, electromagnetic-wave problems may be classified into the following categories:

1. steady and unsteady problems
2. eigenvalue and deterministic problems
3. one-, two-, and three-dimensional problems
4. scalar- and vector-field problems
5. homogeneous and inhomogeneous problems
6. isotropic and anisotropic problems
7. conservative and nonconservative problems
8. bounded- and unbounded-field problems
9. linear and nonlinear problems
10. forward and inverse problems

The waveguide problem discussed in this chapter belongs to the categories of steady problem (1) and eigenvalue problem (2). Because the problem region is the

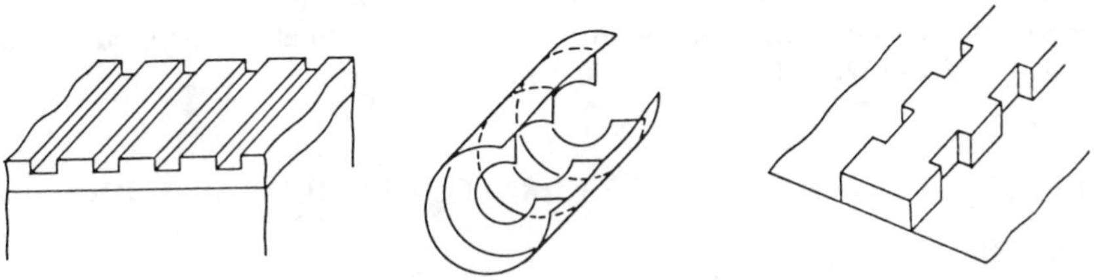

(a) (b) (c)

Figure 1.2 Periodic waveguides: (a) two-dimensional periodic waveguide (dielectric thin-film waveguide grating); (b) axially symmetrical periodic waveguide (waveguide with cylindrically corrugated surface); (c) three-dimensional periodic waveguide (dielectric image guide grating).

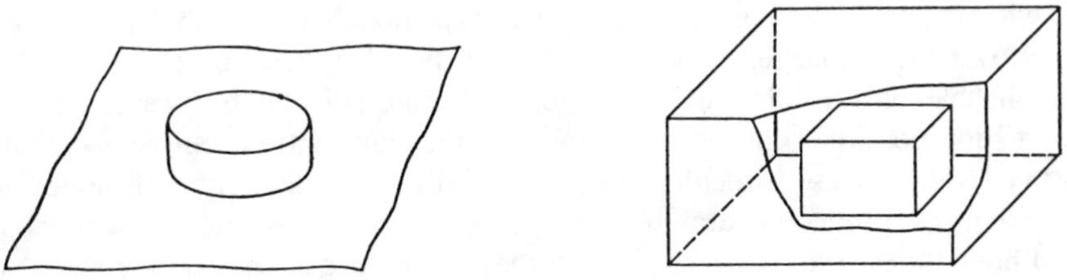

(a) (b)

Figure 1.3 Resonators: (a) axially symmetrical resonator (cylindrical dielectric resonator); (b) three-dimensional resonator (dielectric-loaded resonator).

waveguide cross section, the waveguide problem can be regarded as the one- or two-dimensional problem (3). For waveguides with arbitrarily shaped cross sections, which are treated as two-dimensional problems, different types of finite-element schemes, according to the components of electromagnetic fields, are used in the formulation. In this chapter, various formulations are presented, and emphasis is on the finite-element formulation in terms of the magnetic-field vector. This formulation is quite suitable for a wide range of practical, complicated problems in categories (4) through (8). Therefore, the magnetic-field vector formulation has been used widely for the solution of various waveguiding structures in microwave, millimeter-wave, and optical wavelength regions, and recently has been utilized as the waveguide solver of CAD packages [23].

Finite-element applications are now limited to the linear (9) and forward (10) problems. Nonlinear waveguide problems have been tackled, too [24]. From the viewpoint of waveguide design, it seems that the inverse problem, in which waveguide parameters are determined with the help of information for guided modes, becomes more and more important in the future.

1.3 ELEMENTARY IDEAS OF THE FINITE-ELEMENT METHOD

In the finite-element method, instead of differential equations (governing equations) for the system under consideration, corresponding functionals (variational expressions) to which a variational principle is applied are set up, where the region of interest is divided into the so-called elements; an equivalent discretized model for each element is constructed; and then all the element contributions to the system are assembled. In other words, the finite-element method can be considered a subclass of the Ritz-Galerkin method, in which piecewise defined polynomial functions are used for trial functions and infinite degrees of freedom of the system are discretized or replaced by a finite number of unknown parameters. In classic analytical procedures without subdivision processes, the system is modeled using analytical functions defined over the whole region of interest, and therefore these procedures generally are applicable only to simple geometries and materials. Of the various forms of discretization possible, one of the simplest is the finite-difference method. And its traditional versions use a regular grid; that is, a rectangular grid with nodes at the intersections of orthogonal straight lines. However, a regular grid is not suitable for curved boundaries or interfaces, because they intersect gridlines obliquely at points other than the nodes. Moreover, a regular grid is not suitable for problems with very steep variations of fields. The finite-element method is somewhat similar to the finite-difference method. In the finite-element method, the field region is subdivided into elements; that is, into subregions. Elements can have various shapes, such as triangles and rectangles, allowing the use of an irregular grid. Therefore, the finite-element method is suitable for problems with very steep variations of fields. Furthermore, this approach can be easily adapted to inhomogeneous and anisotropic problems, and it is possible to systematically increase the accuracy of solutions obtained, as necessary. Furthermore, the finite-element scheme can be established not only by the variational method but by the Galerkin method, which is a weighted residual method. Therefore, the finite-element method may be applicable to problems where a variational principle does not exist or cannot be identified and certainly will become more widespread in the near future.

1.4 OUTLINE OF FINITE-ELEMENT CALCULATIONS

To clarify ideas of the finite-element method, consider a particular problem governed by the Helmholtz equation.

The specific governing equation is now written for a domain Ω as

$$(1/p)\nabla^2\phi + \lambda q\phi = 0 \quad \text{in} \quad \Omega \tag{1.1}$$

where ϕ is the electric or magnetic field component, the quantity λ is related to frequency, and the material properties of the medium concerned are represented by p and q. For instance, in a case of a planar waveguide ($\partial/\partial y = 0$) as shown in Figure 1.1(a), $\nabla^2 = \partial^2/\partial^2 x^2 - k_0^2 n_{\text{eff}}^2$ and $\lambda = k_0^2$, where $k_0 = \omega\sqrt{\varepsilon_0\mu_0}$ and $n_{\text{eff}} = \beta/k_0$. Here, k_0 is the wavenumber of free space; ω is the angular frequency; ε_0 and μ_0 are the permittivity and permeability of free space, respectively; n_{eff} is the effective refractive index; and β is the phase constant in the propagation direction. For the analysis of TE modes, $\phi = E_y$, $p = 1$, and $q = \varepsilon_r$. For the analysis of TM modes, $\phi = H_y$, $p = \varepsilon_r$, and $q = 1$. Here, E_y and H_y are the y components of electric and magnetic fields, respectively; and ε_r is the relative permittivity.

Now, we assume that the boundary Γ of the region Ω consists partly of Γ_ϕ on which the value of ϕ is given as $\hat{\phi}$ and partly of Γ_ψ on which the value of $\partial\phi/\partial n \equiv \psi$ is given as $\hat{\psi}$; namely,

$$\phi = \hat{\phi} \quad \text{on} \quad \Gamma_\phi \tag{1.2a}$$

$$\partial\phi/\partial n = \boldsymbol{n}\cdot\nabla\phi = \hat{\psi} \quad \text{on} \quad \Gamma_\psi \tag{1.2b}$$

where $\boldsymbol{n}$ is the outward unit normal vector. The boundary conditions, (1.2a) and (1.2b), are called *Dirichlet* and *Neumann boundary conditions,* respectively. Considering these boundary conditions, the functional equation for (1.1) is given by

$$\mathcal{L} = \frac{1}{2}\int_\Omega \left\{\frac{1}{p}(\nabla\phi)^2 - \lambda q\phi^2\right\} d\Omega - \int_{\Gamma_\psi} \frac{1}{p}\,\phi\hat{\psi}\,d\Gamma \tag{1.3}$$

The first and second terms in the right-hand side of (1.3) denote the integrals over the region Ω and along the boundary Γ_ψ, respectively. The Euler equation derived by the stationary requirement

$$\delta\mathcal{L} = 0 \tag{1.4}$$

coincides with the governing equation, namely, (1.1). The boundary condition in (1.2b) is called the *natural boundary condition,* because this condition is automatically satisfied in the variational procedure. On the other hand, the boundary condition in (1.2a) should be imposed on trial functions; therefore, this condition is called the *forced boundary condition*.

Dividing the region Ω into a number of elements e and considering the functional $\mathcal{L}_e$

$$\mathcal{L}_e = \frac{1}{2}\int_e \left\{\frac{1}{p_e}(\nabla\phi)^2 - \lambda q_e \phi^2\right\} d\Omega - \int_{\Gamma_e} \frac{1}{p_e}\phi\psi d\Gamma \tag{1.5}$$

for each element, and the functional equation for the whole region is given by

$$\mathcal{L} = \sum_e \mathcal{L}_e \tag{1.6}$$

The first and second terms in the right-hand side of (1.5) denote the integrals over each element e and along the element boundary Γ_e, respectively, and the summation Σ_e extends over all different elements.

Arranging m nodes in each element, ϕ can be approximated as

$$\phi = \sum_{i=1}^{m} N_i \phi_i \tag{1.7}$$

where ϕ_i is the ith nodal parameter of the element e and N_i is the interpolation or shape function (see Section 1.5).

We can express (1.7) in matrix form:

$$\phi = \{N\}^{\mathrm{T}}\{\phi\}_e \tag{1.8}$$

where the components of the $\{\phi\}_e$ and $\{N\}$ vectors are ϕ_i and N_i, respectively, and T, $\{\cdot\}$, and $\{\cdot\}^{\mathrm{T}}$ denote a transpose, a column vector, and a row vector, respectively.

Substituting (1.7) into (1.5), we obtain

$$\mathcal{L}_e = \frac{1}{2}\sum_i \sum_j \{\phi_i (K_{ij} - \lambda M_{ij})\phi_j\} - \sum_i \phi_i \psi_i \tag{1.9}$$

$$K_{ij} = \int_e \frac{1}{p_e} (\nabla N_i) \cdot (\nabla N_j) d\Omega \tag{1.10a}$$

$$M_{ij} = \int_e q_e N_i N_j \, d\Omega \tag{1.10b}$$

$$\psi_i = \int_e \frac{1}{p_e} N_i \psi d\Gamma \tag{1.10c}$$

We can express (1.9) in matrix form as

$$\mathcal{L}_e = \frac{1}{2} \{\phi\}_e^{\mathrm{T}} [K]_e \{\phi\}_e - \frac{1}{2} \lambda \{\phi\}_e^{\mathrm{T}} [M]_e \{\phi\}_e - \{\phi\}_e^{\mathrm{T}} \{\psi\}_e \tag{1.11}$$

where the components of the $[K]_e$ and $[M]_e$ matrices are K_{ij} and M_{ij}, respectively.

Assuming that boundary conditions at the interface between the two adjacent elements are $\phi_1 = \phi_2$ and $\psi_1/p_1 = -\psi_2/p_2$, the function for the whole region is given by

$$\mathcal{L} = \sum_e \mathcal{L}_e = \frac{1}{2} \{\phi\}^{\mathrm{T}} ([K] - \lambda [M]) \{\phi\} - \{\phi\}^{\mathrm{T}} \{\psi\} \tag{1.12}$$

where the components of the $\{\phi\}$ vector are the values of ϕ at all nodes, and the global matrices $[K]$ and $[M]$ come from adding the element matrices, $[K]_e$ and $[M]_e$, respectively. The components of the $\{\psi\}$ vector corresponding to nodes on the boundary Γ_ψ are $\hat{\psi}$'s, and the other components become zero.

Using a variational principle

$$\delta \mathcal{L} = \partial \mathcal{L} / \partial \{\phi\}^{\mathrm{T}} = \{0\} \tag{1.13}$$

the discretized algebraic equation is derived as

$$([K] - \lambda [M]) \{\phi\} = \{\psi\} \tag{1.14}$$

where $\{0\}$ is a null vector. For free vibration problems without the excitation term $\{\psi\}$, we obtain

$$[K]\{\phi\} - \lambda [M] \{\phi\} = \{0\} \tag{1.15}$$

The so-called eigenvalue problem, (1.15), can be solved by using computer programs for generalized eigenvalue problems, and thus we obtain eigenvalues for the system under consideration; namely, eigen frequencies.

In this section, (1.14) was derived by applying a variational principle to the function (1.3). It should be noted that the Galerkin method will yield the identical equation to (1.14) derived from a variational principle. The variational expression, (1.3), is not suitable for the study of dissipative systems, and the use of the Galerkin procedure is recommended.

1.5 ELEMENTS

Various elements are available in the finite-element method. However, in this section, we briefly summarize only the elements that are widely used in electromagnetic-wave problems.

Elements are classified as one-, two-, and three-dimensional, as shown in Figures 1.4 to 1.8. Applications of each element are presented in Table 1.1.

As mentioned in Section 1.4, the unknown function ϕ within each element is defined in terms of ϕ_i as in equation (1.7). When the functional value contains first-order derivatives, to guarantee the convergence of solutions, the shape function N_i should satisfy the following two conditions:

1. The variable ϕ and its derivatives must include the constant terms.
2. The variable ϕ must be continuous at the interface between two adjacent elements.

These are called the *completeness* and *compatibility* conditions, respectively. The completeness condition is simple to satisfy if polynomial expressions are used in each element. First-order or linear elements are the most fundamental ones, and use first-order polynomials. High-order elements on the other hand, use higher-order polynomials. The number of nodes within each element, m, coincides with the number of terms in a complete polynomial expansion, and nodes are arranged to satisfy the compatibility condition. Arrangements of nodes are shown in Figures 1.4 to 1.8. Table 1.1 indicates the number of nodes within each first- and second-order element, m, and the shape function N_i. Furthermore, local coordinates, called *line, area,* and *volume coordinates,* respectively, are introduced for one-, two-, and three-dimensional elements. Table 1.1 also gives the coordinate transformation relation between the local and global coordinates. The transformation relations for differentiations and integrations can be derived straightforwardly from the coordinate transformation relation. Therefore, the components of matrices derived from discretization—for instance, the values of K_{ij} and M_{ij} in (1.10)—can be calculated easily using these transformation relations for differentiations and integrations, and the integration formula in the local coordinate presented in Table 1.1.

In this section, only elements with straight sides were discussed. Elements with curved or distorted sides, especially the so-called isoparametric elements in which the polynomial order representing the geometry of an element is coincident with that

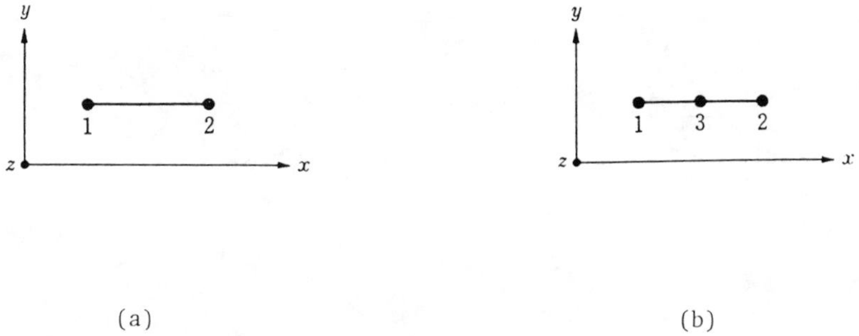

Figure 1.4 Line elements (one dimensional): (a) first order; (b) second order.

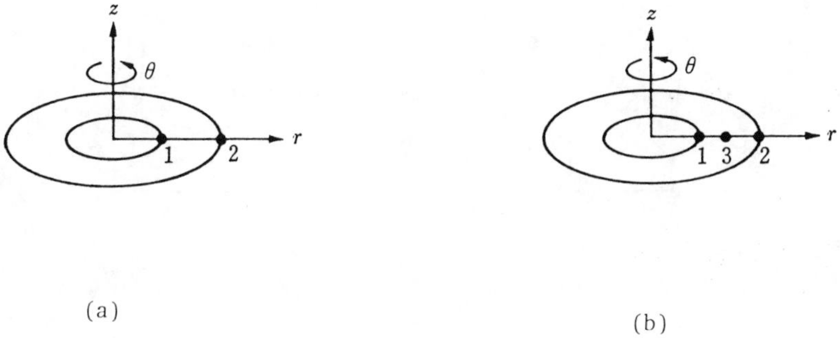

Figure 1.5 Ring elements (one dimensional): (a) first order; (b) second order.

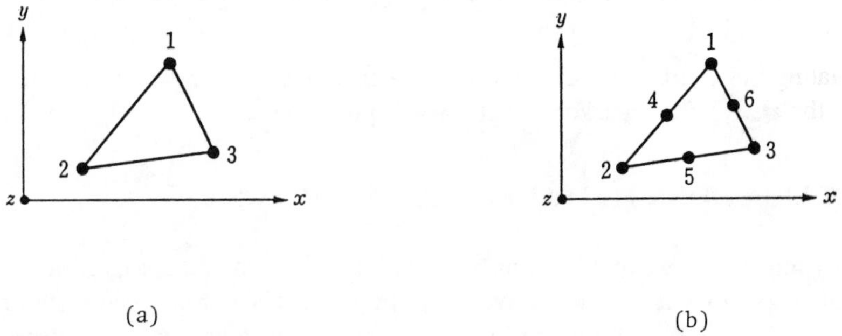

Figure 1.6 Triangular elements (two dimensional): (a) first order; (b) second order.

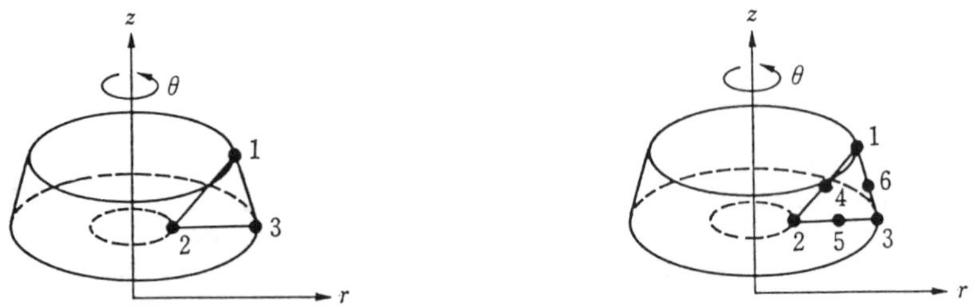

(a) (b)

Figure 1.7 Triangular ring elements (two dimensional): (a) first order; (b) second order.

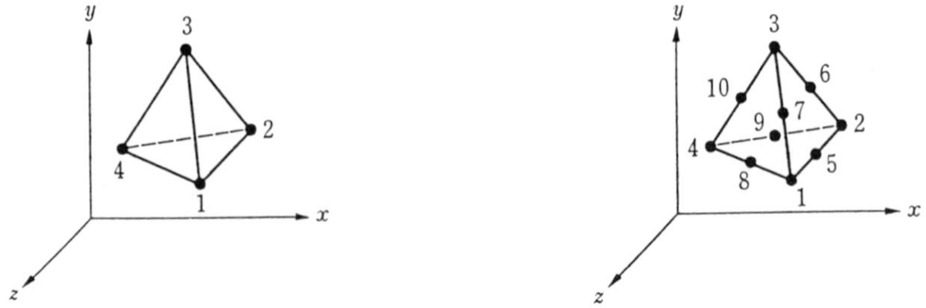

(a) (b)

Figure 1.8 Tetrahedral elements (three dimensional): (a) first order; (b) second order.

interpolating the field in an element, match curvilinear boundaries and are widely used in the analysis of electromagnetic-wave problems.

1.6 APPLICATIONS TO WAVEGUIDE PROBLEMS

The electromagnetic waveguide can be classified into two categories from its cross-sectional shape. One is a planar waveguide (Figure 1.1(a)) or an axially symmetrical waveguide (Figure 1.1(b)), which can be treated as an equivalent one-dimensional problem. The other is the more general arbitrarily shaped waveguide (Figure 1.1(c)), which should be treated as a two-dimensional problem. Furthermore, depending on the eigenmode property in a waveguide, either a scalar wave analysis will be possible,

Table 1.1
Elements

Element	One-Dimensional Element	Two-Dimensional Element	Three-Dimensional Element
(a) Shape (Application)	Fig. 1.4 (Fig. 1.1(a)) Fig. 1.5 (Fig. 1.1(b))	Fig. 1.6 (Fig. 1.1(c) and Fig. 1.2(a)) Fig. 1.7 (Fig. 1.2(b) and Fig. 1.3(a))	Fig. 1.8 (Fig. 1.2(c) and Fig. 1.3(b))
(b) Number of nodes			
first-order element	2	3	4
second-order element	3	6	10
(c) Local coordinates	Line coordinates (L_1, L_2)	Area coordinates (L_1, L_2, L_3)	Volume coordinates (L_1, L_2, L_3, L_4)
(d) Coordinate transformation relation	$\begin{bmatrix} x \\ 1 \end{bmatrix} = \begin{bmatrix} x_1 & x_2 \\ 1 & 1 \end{bmatrix} \begin{bmatrix} L_1 \\ L_2 \end{bmatrix}$ x_i : coordinate of the edge point i (i=1, 2) of line-segment ($x \to r$ for ring element)	$\begin{bmatrix} x \\ y \\ 1 \end{bmatrix} = \begin{bmatrix} x_1 & x_2 & x_3 \\ y_1 & y_2 & y_3 \\ 1 & 1 & 1 \end{bmatrix} \begin{bmatrix} L_1 \\ L_2 \\ L_3 \end{bmatrix}$ x_i, y_i : coordinates of the vertex i (i=1, 2, 3) of triangle ($x \to r$ and $y \to z$ for triangular ring element)	$\begin{bmatrix} x \\ y \\ z \\ 1 \end{bmatrix} = \begin{bmatrix} x_1 & x_2 & x_3 & x_4 \\ y_1 & y_2 & y_3 & y_4 \\ z_1 & z_2 & z_3 & z_4 \\ 1 & 1 & 1 & 1 \end{bmatrix} \begin{bmatrix} L_1 \\ L_2 \\ L_3 \\ L_4 \end{bmatrix}$ x_i, y_i, z_i : coordinates of the vertex i (i=1, 2, 3, 4) of tetrahedron
(e) Shape function			
first-order element	$N_1=L_1$, $N_2=L_2$	$N_1=L_1$, $N_2=L_2$, $N_3=L_3$	$N_1=L_1$, $N_2=L_2$, $N_3=L_3$, $N_4=L_4$
second-order element	$N_1=L_1(2L_1-1)$ $N_2=L_2(2L_2-1)$ $N_3=4L_1L_2$	$N_1=L_1(2L_1-1)$ $N_2=L_2(2L_2-1)$ $N_3=L_3(2L_3-1)$ $N_4=4L_1L_2$ $N_5=4L_2L_3$ $N_6=4L_3L_1$	$N_1=L_1(2L_1-1)$ $N_2=L_2(2L_2-1)$ $N_3=L_3(2L_3-1)$ $N_4=L_4(2L_4-1)$ $N_5=4L_1L_2$, $N_6=4L_2L_3$ $N_7=4L_3L_1$, $N_8=4L_1L_4$ $N_9=4L_2L_4$, $N_{10}=4L_3L_4$
(f) Integration formula	$\int_e L_1^k L_2^l \, dx = L_e \dfrac{k!\,l!}{(k+l+1)!}$ L_e : length of line element	$\iint_e L_1^k L_2^l L_3^m \, dx\,dy = 2A_e \dfrac{k!\,l!\,m!}{(k+l+m+2)!}$ A_e : area of triangular element	$\iiint_e L_1^k L_2^l L_3^m L_4^n \, dx\,dy\,dz = 6V_e \dfrac{k!\,l!\,m!\,n!}{(k+l+m+n+3)!}$ V_e : volume of tetrahedral element

or a vectorial wave analysis will be required. From this point of view, the following classification is adopted here:

1. Scalar wave analysis by the one-dimensional, finite-element method.
2. Vectorial wave analysis by the one-dimensional, finite-element method.
3. Scalar wave analysis by the two-dimensional, finite-element method.
4. Vectorial wave analysis by the two-dimensional, finite-element method.

Here, the one- and two-dimensional, finite-element methods are used to analyze one- and two-dimensional problems, respectively. Furthermore, the scalar wave and vectorial wave analyses indicate guided-mode analyses for electromagnetic waveguides via the scalar wave and vectorial wave equations, respectively.

The spurious solutions do appear in method (4) and, depending on the variational expression used, also in method (2). Here, we review various approaches, laying emphasis on methods (2) and (4), particularly on (4).

1.6.1 Vectorial Wave Analysis by the One-Dimensional, Finite-Element Method

For the finite-element analysis of the hybrid modes in the waveguide of Figure 1.1(a), we use the variational expression coupled between the x-component of electric field and that of magnetic field, and the spurious solutions do not appear.

On the other hand, for the finite-element analysis of the hybrid modes in the waveguide of Figure 1.1(b), we use the variational expression in terms of the z-components (axial components) of the electromagnetic field and analyze the axially symmetrical waveguide with a complicated refractive-index profile like the inhomogeneous-core optical fiber. In applying this approach, pay great attention to how to assign nodes. We write the permittivities at the neighboring nodes i and j as ε_i and ε_j, respectively. When the effective permittivity $\varepsilon_0 n_{\text{eff}}^2$ approaches the average of these permittivities, $(\varepsilon_i + \varepsilon_j)/2$, the matrix derived from discretization is numerically out of condition, and consequently spurious solutions can be involved. If we want to avoid such solutions, it is necessary to assign the node i to the position $r = r_i$ where the permittivity is equal to the effective permittivity; that is, $\varepsilon(r_i)/\varepsilon_0 = n_{\text{eff}}^2$. However, because the term $1/[\varepsilon(r_i)/\varepsilon_0 - n_{\text{eff}}^2]$ is involved in the variational expression in terms of axial electromagnetic field (see equation (1.16)), an integrand included in the variational expression diverges at $r = r_i$. Fortunately, a skillful technique to avoid divergence at such singular points has been developed. In this technique, a special function satisfying nondivergent condition is used for the element with the singular point, whereas the electromagnetic field is expanded using a piecewise polynomial for the element outside the field as in the ordinary finite-element analysis. The spurious solutions do not appear there.

As described earlier, when we use the variational expression in terms of axial electromagnetic-field components, the methods for assigning nodes and dealing with the singular point can be cumbersome. Recently, to avoid this difficulty, a one-dimensional, finite-element method was developed using the variational expression in terms of full vector $\boldsymbol{E}$ field or full vector $\boldsymbol{H}$ field, and the vectorial wave analysis was made of the inhomogeneous-core optical fiber. However, if we use the conventional variational expression in terms of full vector $\boldsymbol{E}$ or $\boldsymbol{H}$ field, many spurious solutions appear that do not satisfy the condition $\nabla \cdot (\varepsilon_r \boldsymbol{E}) = 0$ or $\nabla \cdot \boldsymbol{H} = 0$, respectively. A full detail of this topic is given in Section 1.6.2 because we can suppress or eliminate these spurious solutions in the same way as in the two-dimensional, finite-element method.

1.6.2 Vectorial Wave Analysis by the Two-Dimensional, Finite-Element Method

To rigorously evaluate propagation characteristics of an inhomogeneous waveguide, shown in Figure 1.1(c), vectorial wave analysis is necessary. Although various two-dimensional, finite-element formulations have been developed to date, spurious solutions appear in the vectorial wave analysis using this two-dimensional, finite-element method.

Finite-Element Method Using Axial Electromagnetic-Field Components

The variational expression in terms of axial electromagnetic-field components is given by

$$\begin{aligned} F(E_z, H_z) = & \iint_{\Omega} \frac{1}{\varepsilon/\varepsilon_0 - n_{\text{eff}}^2} \left\{ \frac{\varepsilon}{\varepsilon_0} |\nabla_t E_z|^2 + \frac{\mu_0}{\varepsilon_0} |\nabla_t H_z|^2 \right. \\ & \left. + 2n_{\text{eff}} \sqrt{\mu_0/\varepsilon_0}\, \hat{z} \cdot (\nabla_t E_z \times \nabla_t H_z) \right\} dxdy \qquad (1.16) \\ & - k_0^2 \iint_{\Omega} \left(\frac{\varepsilon}{\varepsilon_0} |E_z|^2 + \frac{\mu_0}{\varepsilon_0} |H_z|^2 \right) dxdy \end{aligned}$$

where $\iint_{\Omega} dxdy$ denotes the integration over the waveguide cross section Ω and $\hat{z}$ is the unit vector for the z-direction.

The two-dimensional, finite-element method using (1.16) is standard for vectorial wave analysis and has been utilized for analyzing various waveguides in microwave and millimeter wave regions. Recently, it has also been utilized for analyzing the waveguide for use in optical wavelength region. Although this approach has been applied to the anisotropic waveguide without off-diagonal elements in a permittivity tensor, no formulation for the arbitrarily anisotropic waveguide with nonzero off-diagonal elements in it yet has been reported.

Application of the finite-element method to (1.16) gives

$$[A]\{X\} - k_0^2\,[B]\{X\} = \{0\} \tag{1.17}$$

where $[A]$ and $[B]$ are matrices obtained from the first and second terms in the right-hand side of (1.16), respectively. The term $\{X\}$ is

$$\{X\} = \begin{bmatrix} \{E_z\} \\ \sqrt{\mu_0/\varepsilon_0}\{H_z\} \end{bmatrix} \tag{1.18}$$

and $\{E_z\}$ and $\{H_z\}$ are the nodal axial electric- and magnetic-field vectors, respectively. Although the matrix $[B]$ is positive definite, the matrix $[A]$ is nonnegative (semipositive definite) or indefinite, depending on the fast-wave ($0 < n_{\text{eff}} < 1$) or slow-wave ($1 < n_{\text{eff}}$) region, respectively. Consequently, spurious solutions appear in the slow-wave region, and what is worse, these are coupled with physical solutions. This results in a great difficulty for the analysis of the dielectric waveguides whose guided modes exist in the slow-wave region. Furthermore, although the natural boundary condition for (1.16) must correspond to the continuity of the tangential electromagnetic-field components at the interface of neighboring media, the spurious solutions, in fact, do not satisfy this condition. Recently, a new variational expression imposing explicitly the continuity of the tangential electromagnetic-field components on (1.16) has been devised by means of Lagrange multipliers, and it has been reported that, although not perfect, most of the spurious solutions can be removed. However, this approach has not yet been applied to various waveguides because of the difficulty in computing the time and accuracy of solutions. Under these circumstances, in the finite-element method in terms of axial electromagnetic-field components, no effective countermeasure has yet been found. Therefore, we must discriminate the physical solutions from the spurious solutions by their field shapes.

*Finite-Element Method Using Full Vector **H** Field*

As already stated, the finite-element method in terms of axial electromagnetic-field components is not applicable to arbitrarily anisotropic waveguides. Thus, the finite-element method using the variational expression in terms of full vector ***H*** or ***E*** field attracts our attention, and it can be used to analyze arbitrarily anisotropic waveguides.

Because the magnetic field is continuous anywhere in dielectric waveguides, it is convenient to use the following variational expression in terms of full vector ***H*** field for the finite-element analysis:

$$F(\boldsymbol{H}) = \iint_{\Omega} (\nabla \times \boldsymbol{H})^* \cdot ([\varepsilon_r]^{-1} \nabla \times \boldsymbol{H}) dxdy - k_0^2 \iint_{\Omega} \boldsymbol{H}^* \cdot \boldsymbol{H} dxdy \qquad (1.19)$$

where $[\varepsilon_r]$ is the relative permittivity tensor and * denotes a complex conjugate.

Application of the standard finite-element method yields

$$[S]\{H\} - k_0^2 [T]\{H\} = \{0\} \qquad (1.20)$$

where $[S]$ and $[T]$ are matrices obtained from the first and second terms in the right-hand side of (1.19), respectively, and $\{H\}$ is the nodal magnetic-field vector.

Although the finite-element method in terms of the variational expression, (1.19), has been applied later than that in terms of (1.16), recent studies have tried to apply it to various waveguiding structures for use in microwave, millimeter-wave, and optical wavelength regions because of its ability to analyze arbitrarily anisotropic waveguide systems. Unfortunately, spurious solutions also appear in this finite-element analysis using full vector $\boldsymbol{H}$ field. Because the matrix $[S]$ in (1.20) is semi-positive definite, it seems that these spurious solutions appear via a different mechanism from those encountered in (1.17). As a result, the spurious solutions encountered in (1.19) do not satisfy the condition $\nabla \cdot \boldsymbol{H} = 0$. These spurious solutions fall into two fairly clear categories. The first one (S_1) can be characterized as follows:

$$\nabla \times \boldsymbol{H} = 0, \quad \nabla \cdot \boldsymbol{H} \neq 0 \quad \text{for} \quad k_0^2 = 0 \qquad (1.21)$$

The second group (S_2) can be characterized as follows:

$$\nabla \times \boldsymbol{H} \neq 0, \quad \nabla \cdot \boldsymbol{H} \neq 0 \quad \text{for} \quad k_0^2 > 0 \qquad (1.22)$$

Now, we discuss the cause of spurious solutions.

Application of a variational principle to (1.19) yields

$$\nabla \times ([\varepsilon_r]^{-1} \nabla \times \boldsymbol{H}) - k_0^2 \boldsymbol{H} = 0 \qquad (1.23)$$

as the Euler equation.

It is obvious that for zero eigenvalues ($k_0^2 = 0$), equation (1.23) does not guarantee the satisfaction of the condition $\nabla \cdot \boldsymbol{H} = 0$.

In the finite-element method, piecewise defined polynomial functions are used for trial functions, and the unknown function and its derivatives are continuous and discontinuous at element interface, respectively. (Note that finite-element solutions are obtained to satisfy the continuity conditions for tangential electric-field components at the interface between elements 1 and 2; namely, $\boldsymbol{n}_1 \times ([\varepsilon_{r1}]^{-1} \nabla \times \boldsymbol{H}_1) =$

$-\boldsymbol{n}_2 \times ([\varepsilon_{r2}]^{-1}\nabla \times \boldsymbol{H})$.) Therefore, for the magnetic field $\boldsymbol{H}$ calculated by the finite-element method, the trivial relation $\nabla \cdot \{\nabla \times ([\varepsilon_r]^{-1}\nabla \times \boldsymbol{H})\} = 0$ is not always satisfied. This means that, for nonzero eigenvalues ($k_0^2 > 0$), the satisfaction of the condition $\nabla \cdot \boldsymbol{H} = 0$ is not guaranteed either. In fact, the derivatives of $\boldsymbol{H}$ do abruptly change at element interfaces, and thus the components of $\boldsymbol{H}$ rapidly oscillate over the whole region. After all, to satisfy the vector formula $\nabla \cdot \{\nabla \times ([\varepsilon_r]^{-1}\nabla \times \boldsymbol{H})\} = 0$, it is necessary to consider the condition $\nabla \cdot \boldsymbol{H} = 0$ as a constraint.

Actually, it has been confirmed in the 1980s that the spurious solutions encountered in (1.19) do not satisfy the condition $\nabla \cdot \boldsymbol{H} = 0$. Soon afterward, it was expected that the spurious solutions could be eliminated if the condition $\nabla \cdot \boldsymbol{H} = 0$ were considered, and several approaches, in which the condition $\nabla \cdot \boldsymbol{H} = 0$ was explicitly imposed, were proposed. The approaches now available are classified into the following:

1. A method choosing a basis function satisfying the condition $\nabla \cdot \boldsymbol{H} = 0$.
2. A method using Lagrange multipliers to enforce the condition $\nabla \cdot \boldsymbol{H} = 0$.
3. A method enforcing the condition $\nabla \cdot \boldsymbol{H} = 0$ in a least-squares way.
4. A method transforming the finite-element equation in terms of full vector $\boldsymbol{H}$ field, (1.20), into that in terms of only transverse magnetic-field component, using the condition $\nabla \cdot \boldsymbol{H} = 0$.
5. A method expressing the nodal magnetic-field vectors with a linear combination of basis vectors for the subspace satisfying the condition $\nabla \cdot \boldsymbol{H} = 0$.

In method (1), it is not necessary to modify the variational expression. However, if we want to consider the condition $\nabla \cdot \boldsymbol{H} = 0$, it is impossible to use the standard finite elements described in Section 1.5. Hence, to date no attempt has been made to suppress or eliminate the spurious solutions by means of this approach.

Although the method (2) is a usual approach to imposing a constraint condition, it has been confirmed to have no effect on suppressing or eliminating the spurious solutions in the finite-element analysis using full vector $\boldsymbol{H}$ field.

The method (3), which is called the *penalty method,* has been studied extensively and applied to various waveguiding problems in recent years because it can suppress or eliminate the spurious solutions. In this approach, the variational expression, (1.19), is modified into

$$\tilde{F}(\boldsymbol{H}) = \iint_{\Omega} (\nabla \times \boldsymbol{H})^* \cdot ([\varepsilon_r]^{-1}\nabla \times \boldsymbol{H})dxdy - k_0^2 \iint_{\Omega} \boldsymbol{H}^* \cdot \boldsymbol{H}dxdy + s \iint_{\Omega} (\nabla \cdot \boldsymbol{H})^*(\nabla \cdot \boldsymbol{H})dxdy \quad (1.24)$$

Equation (1.24) is a variational expression satisfying the condition $\nabla \cdot \boldsymbol{H} = 0$ in a least squares way, where s is a positive constant called the *penalty coefficient.*

Application of the standard finite-element method to (1.24) yields

$$([S] + s[U])\{H\} - k_0^2[T]\{H\} = 0 \tag{1.25}$$

where $[U]$ is a matrix obtained via discretization of the third term on the right-hand side of (1.24).

Although the spurious solutions involved in equation (1.20) are completely eliminated in (1.25), the solutions

$$\nabla \times \boldsymbol{H} = 0, \quad \nabla \cdot \boldsymbol{H} \neq 0 \quad \text{for} \quad k_0^2 > 0 \tag{1.26}$$

are involved because the variational expression, (1.19), has been modified into (1.24). These solutions correspond to a new type of spurious solution. Fortunately, this type of spurious solution can be predicted as a solution of the following equation:

$$\boldsymbol{H} = \nabla \phi \tag{1.27}$$

$$(s\nabla^2 + k_0^2)\phi = 0 \qquad \text{in region } \Omega \tag{1.28a}$$

$$\partial \phi / \partial n = 0 \qquad \text{on the electric wall} \tag{1.28b}$$

$$\phi = 0 \qquad \text{on the magnetic wall} \tag{1.28c}$$

where ϕ is a scalar function. Unlike other spurious solutions encountered in the finite-element method, we need not worry about these solutions because, as found from (1.28), we can suppress them into the following region:

$$n_{\text{eff}} < 1/\sqrt{s} \tag{1.29}$$

Equation (1.29) indicates that the larger is s, the narrower the region for the spurious solutions. For example, if we want to suppress the spurious solutions from the region $n_{\text{eff}} \geq n_{\min}$, we have only to choose the value as $s \geq 1/n_{\min}^2$. In this connection, for $s = 1$, the spurious solutions can be completely eliminated from the slow-wave region. Therefore, this approach is useful, particularly for the analysis of dielectric waveguides. Recently, it has been applied successfully to various waveguiding structures for use in millimeter or optical integrated circuits and single-polarization optical fibers, such as stress-applied and side-tunnel optical fibers.

Method (4), which has been developed only recently, can completely eliminate the spurious solutions. The key point of this approach is that the condition $\nabla \cdot \boldsymbol{H} = 0$; that is,

$$H_z = \frac{1}{j\beta}\left(\frac{\partial H_x}{\partial x} + \frac{\partial H_y}{\partial y}\right) \tag{1.30}$$

is discretized into the following form:

$$[D_z]\{H_z\} = [D_t]\{H_t\} \tag{1.31}$$

via the Galerkin procedure. Here, $\{H_t\}$ is the nodal transverse magnetic-field vector, and the matrices $[D_z]$ and $[D_t]$ are given by

$$[D_z] = \sum_e \iint_e \{N\}\{N\}^{\mathrm{T}} dxdy \tag{1.32a}$$

$$[D_t] = \sum_e \frac{1}{j\beta} \iint_e \left[\{N\} \frac{\partial \{N\}^{\mathrm{T}}}{\partial x} \quad \{N\} \frac{\partial \{N\}^{\mathrm{T}}}{\partial y} \right] dxdy \tag{1.32b}$$

Using (1.31), the nodal magnetic-field vector $\{H\}$ is expressed as

$$\{H\} = [D]\{H_t\} \tag{1.33}$$

$$[D] = \begin{bmatrix} [1] \\ [D_z]^{-1}[D_t] \end{bmatrix} \tag{1.34}$$

where [1] is a unit matrix.

Considering (1.33) into the finite-element equation (1.20) in terms of full vector $\boldsymbol{H}$ field, the following finite-element equation in terms of transverse magnetic-field component can be obtained:

$$[S_{tt}]\{H_t\} - k_0^2[T_{tt}]\{H_t\} = \{0\} \tag{1.35}$$

where

$$[S_{tt}] = [D]^{\dagger}[S][D] \tag{1.36a}$$

$$[T_{tt}] = [D]^{\dagger}[T][D] \tag{1.36b}$$

Here, † denotes a complex conjugate and transposition.

In (1.35), the spurious solutions are completely eliminated, and the matrix size is two-thirds that of (1.20) and (1.25). Moreover, an artificial coefficient such as the one included in the penalty method (the accuracy of the solution depends on the penalty coefficient s in the penalty method) is not included at all. For these reasons, the method (4) is useful for analyzing various electromagnetic waveguides, and more work is expected in the future.

Method (5), which also has been developed only recently, succeeds in completely eliminating the spurious solutions. In this approach, the following variational expression

$$G(\boldsymbol{H}) = \iint_{\Omega} (\nabla \cdot \boldsymbol{H})^*(\nabla \cdot \boldsymbol{H})dxdy - k_0^2 \iint_{\Omega} \boldsymbol{H}^* \cdot \boldsymbol{H}dxdy \qquad (1.37)$$

is first considered to obtain basis vectors for the subspace satisfying the condition $\nabla \cdot \boldsymbol{H} = 0$, and subsequently m eigenvectors $\{D_i\}$ $(i = 1, 2, \ldots, m)$ corresponding to zero eigenvalues ($k_0^2 = 0$) are computed by applying the finite-element method. The nodal magnetic-field vector $\{H\}$ is expressed as a linear combination of these basis vectors:

$$\{H\} = [\tilde{D}]\{\tilde{H}\} \qquad (1.38)$$

$$[\tilde{D}] = [\{D_1\}\,\{D_2\}\ldots\{D_m\}] \qquad (1.39)$$

Considering (1.38) in the finite-element equation (1.20) in terms of the full vector $\boldsymbol{H}$ field, the final finite-element equation can be derived:

$$[\tilde{S}]\{\tilde{H}\} - k_0^2[\tilde{T}]\{\tilde{H}\} = \{0\} \qquad (1.40)$$

where

$$[\tilde{S}] = [\tilde{D}]^\dagger[S][\tilde{D}] \qquad (1.41\text{a})$$

$$[\tilde{T}] = [\tilde{D}]^\dagger[T][\tilde{D}] \qquad (1.41\text{b})$$

The spurious solutions are completely eliminated in (1.40). The problem is how accurately zero eigenvalues of (1.37) and the corresponding eigenvectors can be computed. In particular, as the value of phase constant β is larger, discrimination of zero eigenvalues from nonzero eigenvalues may be difficult. In addition, we must solve the eigenvalue problem twice to compute the k_0 corresponding frequency. For these reasons, future efforts are expected for improving the accuracy of solutions and enhancing efficiency in numerical computations.

To confirm the validity of the preceding discussion, we consider a dielectric square waveguide as shown in Figure 1.9, where n_1 and n_2 are the refractive indices of a core and a cladding, respectively. Because of twofold symmetry of the system, we subdivide only one-quarter of the waveguide cross section into second-order triangular elements. For simplicity, assuming the artificial boundaries AD and CD far from the core region, the original unbounded structure is replaced by a corresponding bounded one. Here, these artificial boundaries are assumed to be perfect electric conductors. Planes of symmetry AB and BC become perfect electric or magnetic

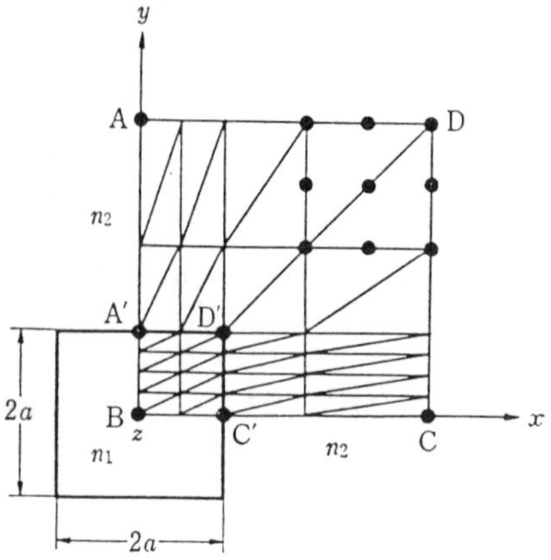

Figure 1.9 Dielectric square waveguide and its element division (second-order elements are used).

conductors according to the kind of mode. In this calculation, we consider the modes such that these planes of symmetry AB and BC become the perfect electric and magnetic conductors, respectively.

Figure 1.10 shows the dispersion characteristics obtained using (1.20), where $n_1 = 1.5$, $n_2 = 1.0$, $v = k_0\, a\sqrt{n_1^2 - n_2^2}/\pi$, and $b = (n_{\rm eff}^2 - n_2^2)/(n_1^2 - n_2^2)$. For com-

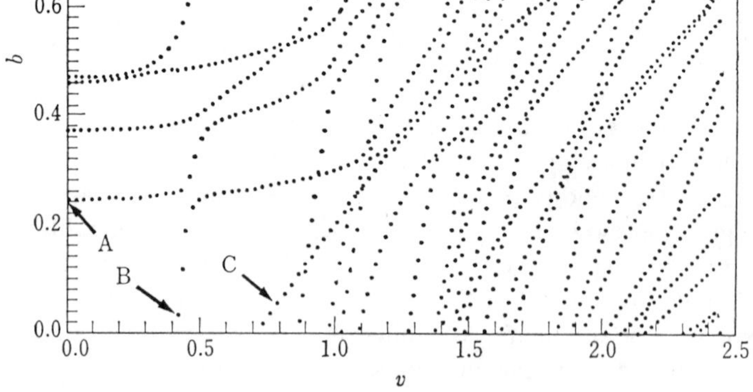

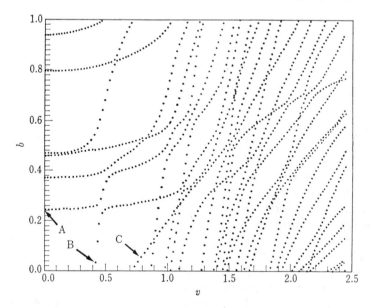

Figure 1.10 Solutions of equation (1.20) for the waveguide in Figure 1.9.

parison, the results of the collocation method [25] are indicated by dots in Figure 1.11. We find from Figures 1.10 and 1.11 that, in addition to physical solutions, numerous spurious solutions appear. Field distributions corresponding to the solutions A, B, and C in Figure 1.10 are shown in Figures 1.12, 1.13, and 1.14, respectively. The magnetic fields for solutions A and B rapidly oscillate over the whole region. On the other hand, the field distribution for physical solution C, corresponding to the so-called E_{11}^{x} mode [25] is indeed reasonable as a whole. However, there are slight ripples in parts, and it seems that, also for the physical solutions, the condition $\nabla \cdot \boldsymbol{H} = 0$ is not satisfied with sufficient accuracy. The solutions obtained by (1.25), (1.35), and (1.40), in which the condition $\nabla \cdot \boldsymbol{H} = 0$ is considered, are indicated by the solid, dashed, and dash-dotted lines in Figure 1.11, respectively. These solutions agree well with the results of the collocation method [25], and spurious solutions do not appear. When using (1.25), a penalty coefficient was set to 1. Figure 1.15 shows the field distribution obtained by (1.25). The frequency in this case is the same as that in Figure 1.14. We find from Figures 1.14 and 1.15 that, when applying the penalty method, the field distribution is dramatically improved in smoothness.

Because the eigenvalues of the finite-element method using full vector $\boldsymbol{H}$ field correspond to k_0^2, in practical calculations, β is first given as an input datum, and subsequently k_0 is obtained as a solution. Therefore, iterations are required if we want to deal with a medium whose material constant varies with k_0 (dispersive me-

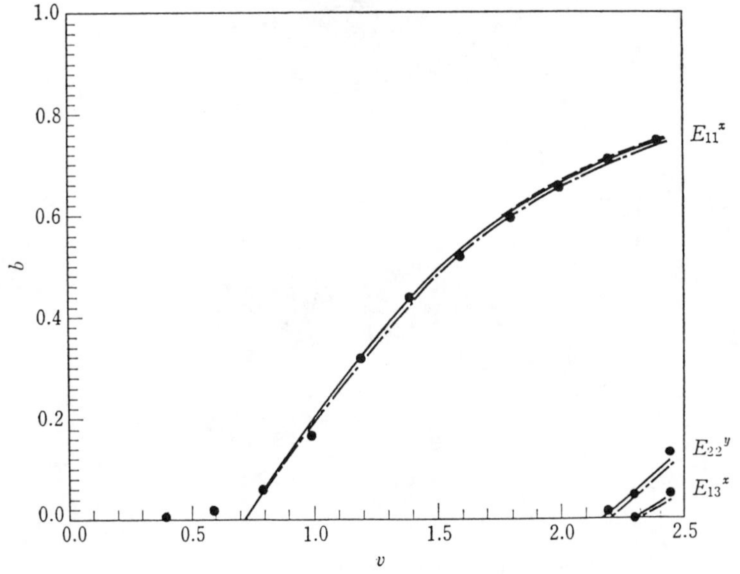

Figure 1.11 Dispersion characteristics of the waveguide in Figure 1.9.

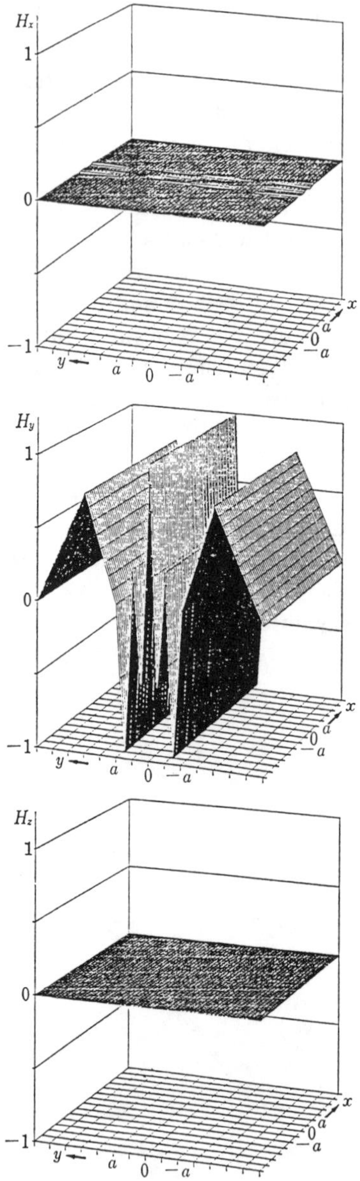

Figure 1.12 Field distributions for the solution of equation (1.20) corresponding to point A in Figure 1.10 (spurious solution).

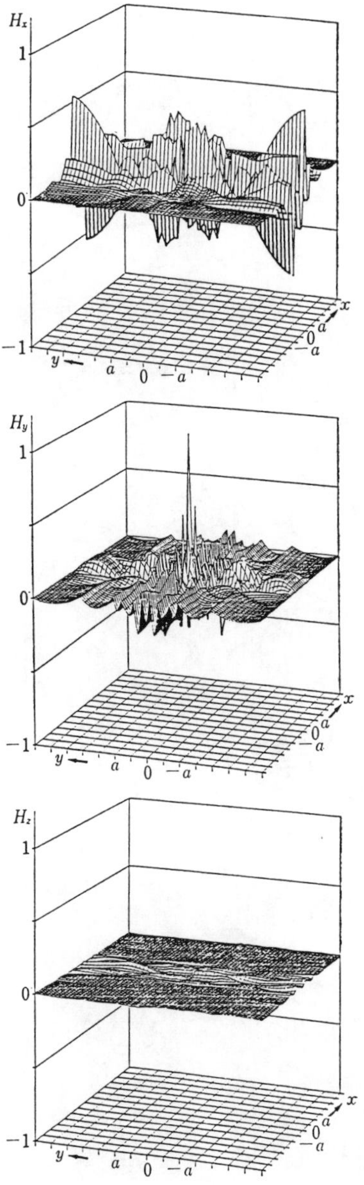

Figure 1.13 Field distributions for the solution of equation (1.20) corresponding to point B in Figure 1.10 (spurious solution).

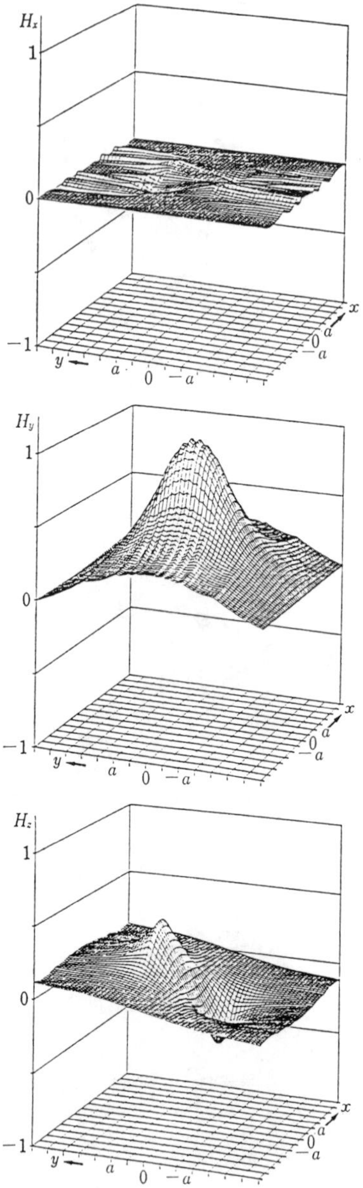

Figure 1.14 Field distributions for the solution of equation (1.20) corresponding to point C in Figure 1.10 (E_{11}^x mode).

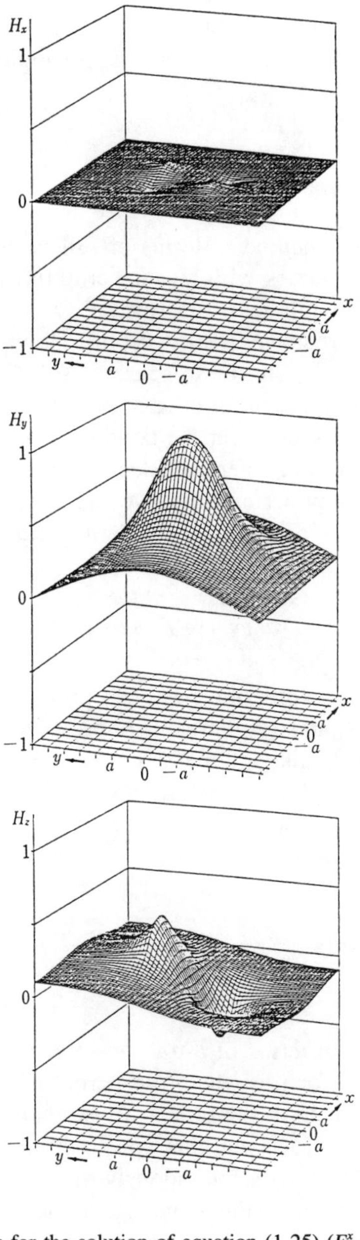

Figure 1.15 Field distributions for the solution of equation (1.25) (E_{11}^x mode).

dium) or a medium with loss or gain (complex medium). To avoid such a difficulty, a method that can obtain β directly by giving k_0 as an input datum recently has been developed in connection with method (4). In this approach, iterations are not necessary. Whereas this approach is applicable in principle to method (3), it is not applicable to method (5).

*Finite-Element Method Using Full Vector **E** Field*

Although the finite-element method in terms of full vector $\boldsymbol{H}$ field is applicable to arbitrarily anisotropic waveguides with tensor permittivity, it is difficult to apply it to anisotropic waveguides with tensor permeability. In this case, the variational expression in terms of full vector $\boldsymbol{E}$ field is useful. However, spurious solutions are involved in this approach, too, and these spurious solutions do not satisfy the condition $\nabla \cdot (\varepsilon_r \boldsymbol{E}) = 0$. To suppress or eliminate such unwanted solutions, methods (3) and (4) have been used. If we consider the continuity of the tangential electric-field component and the normal component of electric-flux density at the interface between the media with different permittivity, we can discretize the variational expression in terms of full vector $\boldsymbol{E}$ field using standard finite elements.

Finite-Element Method Using Transverse Electromagnetic-Field Components

Recently, a new variational expression has been proposed using transverse electromagnetic-field components, and a finite-element procedure has been developed using this variational expression. Unlike the method using conventional variational expressions, this method, in principle, can obtain β from k_0 because the eigenvalue corresponds to β. Furthermore, imposing only the condition that β is a real number, all the spurious solutions can be avoided. More work is expected in the future.

Finite-Element Method Using a Transverse Electric- or Magnetic-Field Component

The finite-element method in terms of a transverse electric- or magnetic-field component has been developed for use where the permittivity varies continuously over the waveguide cross section. In this approach, β is obtained from k_0 and no spurious solution has been reported. However, this approach cannot be applied straightforwardly if the permittivity varies discontinuously in the cross section. For example, we should introduce an extremely thin transition region between the elements with different permittivity to replace the discontinuous change of permittivity by a continuous one. This procedure may be laborious in practical computation.

1.6.3 Treatment of Infinite Regions

Because dielectric waveguides in general, are unbounded, some proper account must be made of infinite regions. These unbounded domains recently have been handled in various ways, and the main methods used in waveguide problems are

1. a method using artificial boundaries
2. a method using infinite elements
3. a method using exterior finite elements
4. a method using the conformal mapping techniques
5. a combination of the finite-element method and the analytical method

The simplest of these techniques is undoubtedly method (1), in which the unbounded structure is replaced by a corresponding bounded one by setting artificial boundaries in the position far from the guided region. This method is used widely in both vectorial and scalar wave analyses. The influence of the artificial boundary depends on the operating frequency and kind of mode. One technique, called the *finite-element iterative method,* involves shifting the artificial boundary to satisfy a given criterion for the maximum field strength at that boundary.

In method (2), the bounded region and the semiinfinite homogeneous region are divided into standard finite elements and infinite elements, respectively. Infinite elements are obtained by moving one edge (for one-dimensional element), one side (for two-dimensional element), or one surface (for three-dimensional element) to infinity. In this process, however, an integrand for computing element matrices diverges. Thus, the shape functions for such an element should realistically represent the fields and should be square integrable over an infinite-element area. To satisfy the radiation condition, decay-type infinite elements are generally used. The decay-type infinite element approach can reduce the finite-element domain compared with the artificial boundary approach, but it presents difficulties in choosing the decay parameters. The best value of decay parameter depends on the operating frequency and the kind of mode.

In method (3), noting that finite-element matrices depend only on the shape of the element but not its size, the exterior region Ω_{ext} is implicitly discretized into geometrically similar elements, called *exterior finite elements,* as shown in Figure 1.16. The finite-element matrices for any annular region can be obtained easily from those for the first annular region enclosing the interior region and adjacent to it. The two annular regions are combined by eliminating the nodes on their common interface, and new matrices for the combined region are constructed. When high-order elements are used, internal nodes exist in the annular region, but these nodes also can be eliminated. Thus, the finite-element matrices for the whole exterior region can be obtained from the first annular region in a recursive manner. After the radiation condition is considered, the resultant matrix equation relates only nodes on the boundary Γ between the interior region Ω_{in} and the exterior region Ω_{ext}.

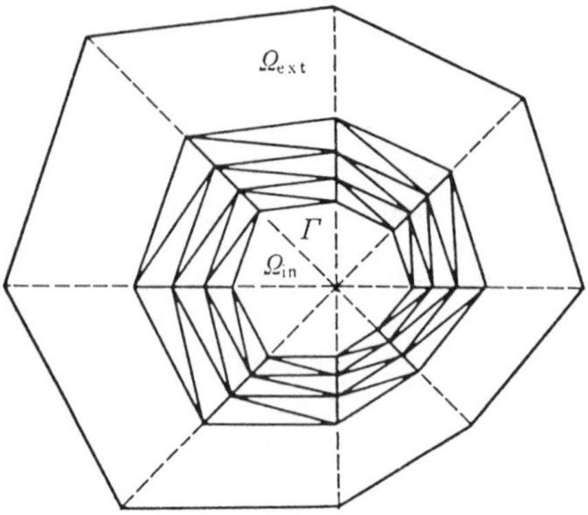

Figure 1.16 Exterior finite elements.

In the method (4), an infinite domain is converted into a finite domain by using the conformal mapping technique, and the linear fractional transformation

$$w = u + iv = f(z) = (z - 1)/(z + 1) \tag{1.42}$$

is used. Here, $z = \bar{x} + j\bar{y}$ and $(\bar{x}, \bar{y})$ is the normalized coordinate. For example, one-quarter of the dielectric square waveguide in Figure 1.9 can be conformally mapped into a unit circle in the w plane, as shown in Figure 1.17, where $x' = x/a$, $y' = y/a$, the boundaries AD and CD are assumed to be located at infinity, and the region surrounded by solid lines corresponds to the core. It is interesting to note that this figure is the same as the so-called Smith chart, which is obtained by conformally mapping the impedance plane into the reflection-coefficient plane.

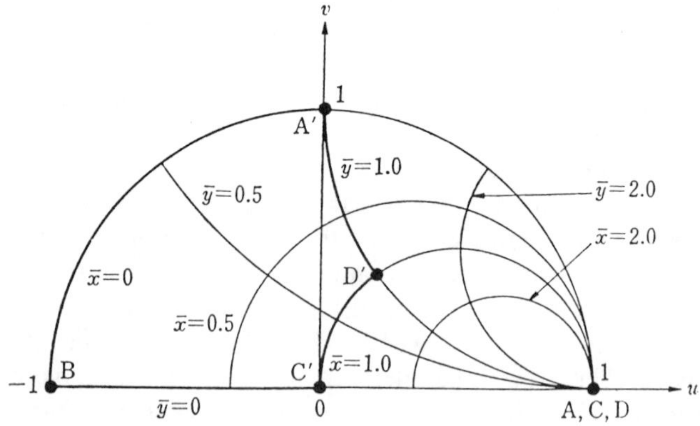

Figure 1.17 Conformal mapping of the waveguide in Figure 1.9.

In method (5), the whole region is divided into interior and exterior regions. The ordinary finite-element approach and the analytical approach are applied to the interior and exterior regions, respectively, and then these two approaches are combined. The function expansion method or the integral equation method is used for the exterior region. Although the formulation is somewhat complex, this combined method can give accurate results even for near-cutoff modes compared with methods (1) to (4). Recently, a boundary-element method has been proposed. This uses a combination of the boundary integral equation and a discretization technique similar to the finite-element algorithm as applied to the boundary. It is expected that a numerical approach based on a combination of finite and boundary elements will be applied to waveguide problems.

For the semiinfinite homogeneous regions of waveguides, as shown in Figures 1.1(a) and (b), which can be treated as an equivalent one-dimensional problem, the closed-form analytical expressions are available, and thus these waveguide problems can be solved by method (5) utilizing the exact analytical solutions.

1.7 SOME PRECAUTIONS FOR PROGRAMMING

In the finite-element method analysis of electromagnetic-wave problems, more than hundreds of nodes are usually employed. Hence, if we input these data (the numbers of nodes and elements, sequential ordering and coordinates of nodes, order of joints between nodes in an element, material constant of each element, boundary conditions, *et cetera*) manually, not only is considerable labor required but input errors may increase. In processing computed results, displays for a variety of figures are needed. However, this process is very cumbersome provided we perform it without relying on on-line processing by computer. For these reasons, we should reduce the labor necessary to input and output data as much as possible by utilizing pre- and postprocessors; and avoiding the troublesome effort in pre- and postprocessings will also be an important topic for investigation [23]. Furthermore, in the finite-element method, problems often arise in the way one treats large-scale matrices. In general, an algorithm should be chosen to save storage regions and use them efficiently, and to shorten computing time and increase accuracy. Here, adaptive use of the optimal technique is preferable, depending on the specific problem.

Domestic supercomputers have recently become available in universities, laboratories, and corporations. In contrast to several mega (10^6) FLOPS (this unit is an abbreviation for *floating-point operations per second* and indicates computing speed; 1 mega FLOPS means 1 million multiplications per second) observed in large-scale general-purpose machines, pipeline-type supercomputers exhibit nearly one giga (10^9) FLOPS, which allows a number of problems to be solved within practical computing time. Indeed, this powerful tool is now attracting researchers in all modern scientific and engineering areas. As is well known, such extremely high-speed computation is attained by either the large-scale integration of computing elements or the vectorial processing. The former approach now seems to be approaching its upper limit, and then substituting GaAs, Josephson, and optical devices for conventional Si-based devices becomes a topical interest. In the latter, superparallel architecture and specialization for specific problems allow this breakthrough. Although the former problem concerning integration of elements is beyond the user's ability, the latter problem concerning vectorial processing is intimately related to the software, indicating that, to effectively derive the performance of the supercomputers, we should make an effort during programming to enhance, as much as possible, the vectorization ratio. When constructing a new program, we should keep our minds on utilizing mathematical libraries for matrix manipulation, which are supplied by hardware makers, or constructing a program according to the text [26] concerning numerical solution methods and programming techniques suitable for the supercomputing architecture. On the other hand, because there are strong needs for transforming existing programs into vectorized ones for use in supercomputers, hardware makers provide the utility programs such as VECTIZER or VECTLIZER as a preprocessing program of FORTRAN compilers. By implementing existing FORTRAN programs under the operation of these preprocessing programs, useful information is given on transforming original programs into vectorizable ones suitable for vector processors. We can improve our programs using such information.

REFERENCES

[1] Kawai, T., ed., *Computrol: Special Issue on Computational Mechanics,* Vol. 8. Corona Pub. Co., Tokyo, 1984 [in Japanese].

[2] Okoshi, T., and T. Miyoshi, *Planar Circuit*. Ohm Pub. Co., Tokyo, 1975 [in Japanese].

[3] Kagawa, Y., *An Introduction to Finite-Element Method for Electrical and Electronic Engineers*. Ohm Pub. Co., Tokyo, 1977 [in Japanese].

[4] Chari, M.V.K., and Silvester, P.P., ed., *Finite Elements in Electrical and Magnetic Field Problems*. John Wiley and Sons, Chichester, England, 1980.

[5] Hara, T., *Fundamentals of Finite-Element Method for the Fields of Nonstructural Engineerings*. Shokodo Pub. Co., Tokyo, 1981 [in Japanese].

[6] Japan Society for Simulation Technology, ed., *Simulation Technology (III): Finite-Element Method*. Baifukan Pub. Co., Tokyo, 1981 [in Japanese].

[7] Kagawa, Y., *Vibration and Acoustic Engineering by the Finite-Element Method: Fundamentals and Applications*. Baifukan Pub. Co., Tokyo, 1981 [in Japanese].

[8] Kagawa, Y., *Practical Finite-Element Method for Electrical and Electronic Engineers*. Ohm Pub. Co., Tokyo, 1982 [in Japanese].

[9] Nakata, T., and T. Takahashi, *Finite-Element Method for Electrical Engineers*. Morikita Pub. Co., Tokyo, 1982 [in Japanese].

[10] Onoe, M., ed., *Fundamentals of Vibrations in Solids for Electrical and Electronics Engineers*. Ohm Pub. Co., Tokyo, 1982 [in Japanese].

[11] Silvester, P.P., and R.L. Ferrari, *Finite Elements for Electrical Engineers*. Cambridge University Press, Cambridge, 1983.

[12] Onuki, T., T. Nakata, and S. Nemoto, *Modern Finite-Element Method for Electrical Engineers*. Ohm Pub. Co., Tokyo, 1983 [in Japanese].

[13] Okoshi, T., K. Okamoto, and K. Hotate, *Optical Fibers*. Ohm Pub. Co., Tokyo, 1983 [in Japanese].

[14] Kagawa, Y., *Finite/Boundary Elements for Unbounded Field Problems*. Science Pub. Co., Tokyo, 1983 [in Japanese].

[15] Kagawa, Y., M. Koshiba, M. Ikeuchi, and S. Kagami, *Finite/Boundary Element Methods for Electrical and Electronic Engineers: Applications to Wave Problems*. Ohm Pub. Co., Tokyo, 1984 [in Japanese].

[16] Aoki, Y., *Numerical Computation by BASIC*. Corona Pub. Co., Tokyo, 1984 [in Japanese].

[17] Kagawa, Y., and K. Murayama, *Finite-Element Method Using BASIC for Electrical and Electronic Engineers*. Kagakugijutsu Pub. Co., Tokyo, 1986 [in Japanese].

[18] Koshiba, M., "Analysis of Electromagnetic Field Problems," *J. Inst. Television Eng. Japan,* Vol. 38, No. 12, December 1984, pp. 1045–1052 [in Japanese].

[19] Koshiba, M., and M. Suzuki, "On the Finite-Element Analysis of Electromagnetic and Optical Waveguides," *Tech. Res. Rep. Inst. Electron. Commun. Eng. Japan,* MW85-39, July 1985 [in Japanese].

[20] Koshiba, M., and M. Suzuki, "On the Finite-Element Analysis of Electromagnetic Eigenvalue Problems," *Tech. Res. Rep. Inst. Electr. Eng. Japan,* EMT-86-3, January 1986 [in Japanese].

[21] Adachi, S., and K. Sawaya, "Analysis of Electromagnetic Fields in Inhomogeneous Media," *J. Electr. Eng. Japan,* Vol. 106, No. 9, September 1986, pp. 887–894 [in Japanese].

[22] Koshiba, M., K. Hayata, and M. Suzuki, "Finite-Element Method Analysis of Microwave and Optical Waveguides—Trends in Countermeasures to Spurious Solutions," *Trans. Inst. Electron. Commun. Eng. Japan,* Vol. J69-C, No. 12, December 1986, pp. 1477–1486 [in Japanese].

[23] Young, T.P., and P. Smith, "Finite Element Modelling of Integrated Optical Waveguides," *GEC J. Res.,* Vol. 4, No. 4, 1986, pp. 249–255.

[24] Hayata, K., M. Koshiba, and M. Suzuki, "Finite-Element Solution of Arbitrarily Nonlinear, Graded-Index Slab Waveguides," *Electron. Lett.,* Vol. 23, No. 8, April 1987, pp. 429–431.

[25] Goell, J.E., "A Circular-Harmonic Computer Analysis of Rectangular Dielectric Waveguides," *Bell Syst. Tech. J.,* Vol. 48, No. 7, September 1969, pp. 2133–2160.

[26] Murata, K., T. Oguni, T. Miyoshi, and Y. Oyanagi, *Numerical Simulation in Engineering Science: Application of Supercomputer*. Maruzen Pub. Co., Tokyo, 1988 [in Japanese].

Chapter 2
The Boundary-Element Method

Nagayoski Morita

2.1 INTRODUCTION

The boundary-element method is a computer technique for choosing, as the basic equations, boundary integral equations and solving these equations numerically by dividing integration domains into a set of elements. The most distinctive feature of this method, compared with the finite-element method, is that, whereas the latter method needs to take unknowns throughout the region, the former requires unknowns only on boundaries, demanding therefore far less computation time. An excellent quality of the boundary-element method, similar to the finite element method, is the flexibility that enables us to cope with various types of complex boundary shapes in a like manner, without fundamental changes in the steps of solution procedures, and that allows solving a wide range of different classes of differential equations on almost the same principle. This is the major reason that the boundary-element method is a highly practical computer technique, applicable to the analysis of almost all sorts of physical phenomena, with only minor devices needed for solving specific problems.

2.2 HISTORICAL BACKGROUND

The numerical solution based on the integral equation formulation, or the method of first discretizing boundary integral equations (the integral equations with unknowns only on boundaries) to get linear equations and then obtaining their numerical solutions, has a relatively long history. This method was applied, for example, to the evaluation of the electrical capacitance of a cube [1] as early as 1951. However, only in the early 1960s, when electronic computers began to be put to practical use, did the integral equation method come to be clearly recognized as a practical technique in the field of electromagnetics [2–4]. From the 1960s to the 1970s, this technique developed rapidly, closely linked with the so-called method of moments [5], mainly in the field of electromagnetic wave scattering; and now it has become widely accepted as a particularly useful method.

In the fields of structural and continuum mechanics and the like, the integral equation method has attracted much attention since the late 1970s [6]. In those fields, however, the finite-element method already was widespread, together with a variety of computer programs based on the method. Because of this, the integral-equation method was contrasted with the finite-element method, and hence the following advantages came to be emphasized [6, 7]:

1. As the unknowns need be taken only on boundaries of regions being investigated, the number of unknowns as well as the computation time can be greatly reduced, in comparison with finite-element methods, in which unknowns are required all through the regions.
2. Unlike the finite-element method, the boundary conditions at infinity can be automatically incorporated.
3. The same boundary discretization schemes and quadrature formulas as for the finite-element method can be used, enabling us to hold the merit of the finite-element method that is suitable for universal usage.

In the fields of structural and continuum mechanics, people used to almost exclusively employ the method of dividing boundaries into a set of elements for discretizing integral equations. Consequently, the name *boundary element method* came to be used for the method using integral equations. This name began to spread into the field of electromagnetics as well after 1980 [8, 9]. However, this method is essentially the same as a particular case of the method called the *integral-equation method* or the *moment method* in the field of electromagnetics: that is, the case of expanding unknown functions in terms of subdomain functions [10].

2.3 LIMITATIONS OF THE METHOD

The major limitation of this method is that it is not suited for dealing with inhomogeneous regions. This limitation results from an essential characteristic of the method: unknowns are taken only on boundaries. The method has another drawback, a possibility of involvement of unphysical solutions, that is, what are called *resonant solutions*. This drawback, however, can be surmounted by taking pertinent means, such as constructing integral equations having no possibility of involvement in resonant solutions. We may be able to enumerate as other flaws of this method several points, such as certain pertinent mathematical steps required for each specified problem, some analytical treatment demanded with respect to singularities of Green's functions associated with integral equations, coefficient matrices becoming dense in contrast with the finite-element method or the finite-difference method in which the matrices remain sparse (band form), increased computation time and hence decreased efficiency when functional values within the regions need to be obtained, and so on. However, these are rather minor flaws.

Overall, the boundary-element method can be said to have no remarkable defects as compared with other various computer methods, except for being unsuited to the problems with inhomogeneous regions.

2.4 INTEGRAL REPRESENTATIONS

To solve a problem by the present method means to accomplish a series of steps of, first, deriving integral equations with respect to unknowns taken on boundaries, next reducing these equations to linear equations by an appropriate discretization method to get the numerical solutions, and last obtaining the values of various physical quantities from these solutions by means of integral representations. Integral representations are normally used during this process not only in deriving integral equations but also in evaluating physical quantities; and hence they play a key role in the present method. In other words, determining which integral representations on which to base the analysis or which integral representations to use for the analysis is a very important point in this solution method. Varieties of integral representations are available depending on the problems to be solved and the detailed methods employed. This section presents several examples of such integral representations for some typical fundamental problems. For Green's functions associated with the integral representations, the free-space Green's functions are used exclusively in this chapter. However, this usage should not be a general rule, and we recommend using the specialized Green's functions that satisfy boundary conditions on some boundaries for the problem configuration and formulation method chosen. For the sake of simplicity, however, only the case using free-space Green's functions is examined in this chapter.

2.4.1 Two-Dimensional Problems

Consider the problem of two-medium regions for which both the geometry of regions and the electromagnetic fields are uniform in the z direction. Let the interior of the boundary C be the region V of the medium with wavenumber k_1 and let the exterior be the region $\bar{V}$ of the medium with wavenumber k_2 as in Figure 2.1. In this problem, the case of the electric field having the z component alone (TM wave or E wave) and the case of the magnetic field having the z component alone (TE wave or H wave) are completely separable, and the general solution can be expressed as a superposition of the solutions for the two cases.

For both TM and TE waves, the electromagnetic fields can be obtained by means of the solutions ϕ_i to the scalar Helmholtz equation

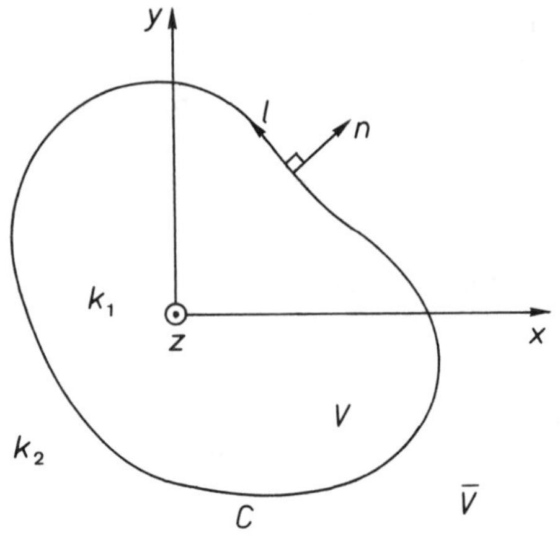

Figure 2.1 Two-dimensional regions and relevant coordinates.

$$(\nabla_t^2 + k_i^2)\phi_i = 0, \quad (i = 1, 2) \tag{2.1}$$

as follows. TM wave:

$$\boldsymbol{E}_i = \boldsymbol{i}_z E_{zi} = \boldsymbol{i}_z \phi_i \tag{2.2}$$

$$\boldsymbol{H}_i = -\frac{1}{j\omega\mu_i} \boldsymbol{\nabla} \times \boldsymbol{E}_i = \frac{1}{j\omega\mu_i} \boldsymbol{i}_z \times \boldsymbol{\nabla}_t \phi_i \tag{2.3}$$

TE wave:

$$\boldsymbol{H}_i = \boldsymbol{i}_z H_{zi} = \boldsymbol{i}_z \phi_i \tag{2.4}$$

$$\boldsymbol{E}_i = \frac{1}{j\omega\varepsilon_i} \boldsymbol{\nabla} \times \boldsymbol{H}_i = -\frac{1}{j\omega\varepsilon_i} \boldsymbol{i}_z \times \boldsymbol{\nabla}_t \phi_i \tag{2.5}$$

where the time factor $\exp(j\omega t)$ is understood and

$$\boldsymbol{\nabla}_t^2 = \boldsymbol{\nabla}^2 - \frac{\partial^2}{\partial z^2}, \quad \boldsymbol{\nabla}_t = \boldsymbol{\nabla} - \boldsymbol{i}_z \frac{\partial}{\partial z} \tag{2.6}$$

and where $i = 1$ and 2 stand for the regions V and $\bar{V}$, respectively; the note $i = 1, 2$ will be omitted hereafter. The Green's function $\hat{G}_i$ associated with the equation (2.1) is a solution to the equation

$$(\nabla_t^2 + k_i^2)\hat{G}_i = -\delta(\boldsymbol{\rho} - \boldsymbol{\rho}') \tag{2.7}$$

and the $\hat{G}_i$ subject to the radiation condition is given by

$$\hat{G}_i = \frac{1}{4j} H_0^{(2)}(k_i R) \tag{2.8}$$

where $H_0^{(2)}$ is the Hankel function of the second kind and order zero, $\delta(\boldsymbol{\rho} - \boldsymbol{\rho}')$ denotes the Dirac delta function, $\boldsymbol{\rho}$ denotes the two-dimensional coordinate vector, and R is the distance between the observation and the integration points:

$$R = |\boldsymbol{\rho} - \boldsymbol{\rho}'| \tag{2.9}$$

Applying the Green's theorem with respect to functions ϕ_1 and $\hat{G}_1$ to the region V and using (2.1) and (2.7), we have the following two different relations, depending upon the position of the observation point P [11]. For $P \in V$:

$$\phi_1(\boldsymbol{\rho}) = -\int_C \left\{ \phi_1(\boldsymbol{\rho}') \frac{\partial \hat{G}_1}{\partial n'} - \hat{G}_1 \frac{\partial \phi_1(\boldsymbol{\rho}')}{\partial n'} \right\} dl' \tag{2.10}$$

For $P \in \bar{V}$:

$$0 = -\int_C \left\{ \phi_1(\boldsymbol{\rho}') \frac{\partial \hat{G}_1}{\partial n'} - \hat{G}_1 \frac{\partial \phi_1(\boldsymbol{\rho}')}{\partial n'} \right\} dl' \tag{2.11}$$

where n and l denote the local coordinates in the normal and tangential directions, respectively, at a point on C, with the prime indicating the coordinates with respect to the integration point. Equation (2.10) is the integral representation for the interior region, and (2.11) is the relationship that expresses an important property inherent in the integral representation (see Figure 2.2). Assuming that the incident wave comes from infinity and applying similarly the Green's theorem to the region $\bar{V}$ bounded by the closed contour C inside and the closed contour C_∞ at far distances, we get the term of the incident wave ϕ^{inc} from the integration over C_∞, resulting in the following relationships (see Figure 2.3). For $P \in \bar{V}$:

$$\phi_2(\boldsymbol{\rho}) = \phi^{\text{inc}}(\boldsymbol{\rho}) + \int_C \left\{ \phi_2(\boldsymbol{\rho}') \frac{\partial \hat{G}_2}{\partial n'} - \hat{G}_2 \frac{\partial \phi_2(\boldsymbol{\rho}')}{\partial n'} \right\} dl' \tag{2.12}$$

For $P \in V$:

$$0 = \phi^{\text{inc}}(\boldsymbol{\rho}) + \int_C \left\{ \phi_2(\boldsymbol{\rho}') \frac{\partial \hat{G}_2}{\partial n'} - \hat{G}_2 \frac{\partial \phi_2(\boldsymbol{\rho}')}{\partial n'} \right\} dl' \tag{2.13}$$

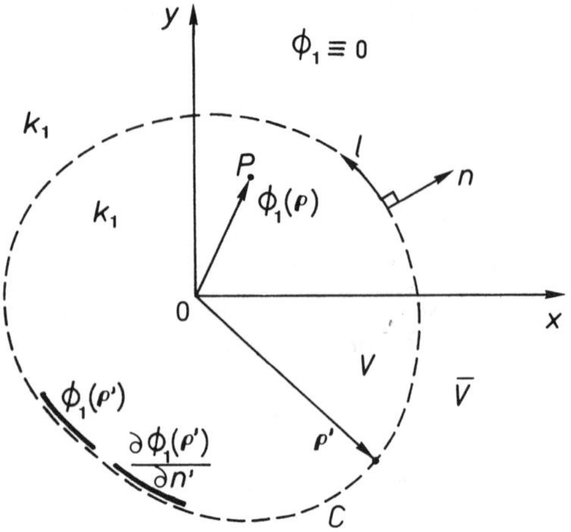

Figure 2.2 Illustration of the integral relationships (2.10) and (2.11).

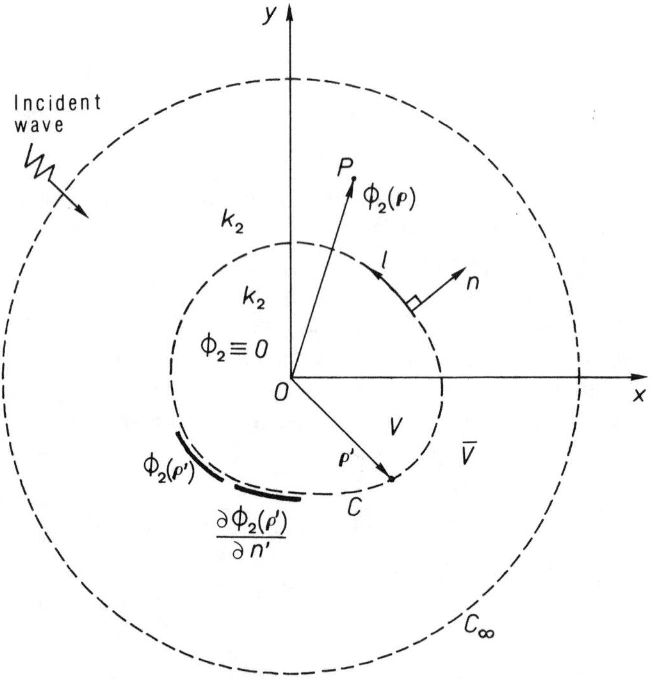

Figure 2.3 Illustration of the integral relationships (2.12) and (2.13).

The signs in front of the integrals of (2.12) and (2.13) are opposite to those of (2.10) and (2.11), which is because the positive direction of the coordinate n is inward to the region we are now considering. In (2.10) to (2.13), the subscripts 1 and 2 are purposely attached to ϕ to clarify the true meaning of the equations. Equation (2.12) is the integral representation for the exterior region and (2.13) is the relationship that expresses an important property inherent in the integral representation. Equation (2.12) is rewritten, on use of the scattered wave ϕ^s, as follows. For $P \in \bar{V}$:

$$\phi = \phi^{\text{inc}} + \phi^{\text{s}} \tag{2.14}$$

$$\phi^{\text{s}} = -\frac{j}{4}\int_C \left\{\phi \frac{\partial H_0^{(2)}(k_2R)}{\partial n'} - H_0^{(2)}(k_2R)\frac{\partial \phi}{\partial n'}\right\} dl' \tag{2.15}$$

2.4.2 Fields Due to Source Distributions

Electromagnetic fields $\boldsymbol{E}$ and $\boldsymbol{H}$ in the medium of wavenumber k, generated by the electric current of density $\boldsymbol{J}$ and the electric charge of density ρ, are expressible in terms of the vector potential $\boldsymbol{A}$ and the scalar potential ϕ, the solutions to

$$\nabla^2 \boldsymbol{A} + k^2 \boldsymbol{A} = -\mu \boldsymbol{J} \tag{2.16}$$

$$\nabla^2 \phi + k^2 \phi = -\frac{\rho}{\varepsilon} \tag{2.17}$$

in the form

$$\boldsymbol{E} = -j\omega \boldsymbol{A} - \nabla \phi \tag{2.18}$$

$$\boldsymbol{H} = \frac{1}{\mu} \nabla \times \boldsymbol{A} \tag{2.19}$$

No confusion should arise from the symbol ρ used for the electric charge density and the radius ρ of the two-dimensional coordinates. Equation (2.18) is tantamount to the following equation expressed without ϕ:

$$\boldsymbol{E} = -j\omega \boldsymbol{A} + \frac{1}{j\omega\varepsilon\mu} \nabla(\nabla \cdot \boldsymbol{A}) \tag{2.20}$$

Solutions of (2.16) and (2.17) are written in terms of the Green's function G, the solution of

$$\nabla^2 G + k^2 G = -\delta(\boldsymbol{r} - \boldsymbol{r}') \tag{2.21}$$

in the form

$$\boldsymbol{A} = \int_{V_0} \mu \boldsymbol{J} G dv' \tag{2.22}$$

and

$$\phi = \int_{V_0} \frac{\rho}{\varepsilon} G dv' \tag{2.23}$$

respectively; where V_0 denotes the region where $\boldsymbol{J}$ or ρ is distributed, and $\boldsymbol{r}$ is the three-dimensional coordinate vector. The integral representation for the electric field $\boldsymbol{E}$ produced by the surface electric current of density $\boldsymbol{K}$ flowing on a surface S is obtained by replacing $\boldsymbol{J}$ in V_0 in (2.22) with $\boldsymbol{K}$ on S and using (2.20), yielding

$$\boldsymbol{E} = \int_S \left\{ -j\omega\mu \boldsymbol{K} G + \frac{1}{j\omega\varepsilon} \nabla(\boldsymbol{K} \cdot \nabla G) \right\} ds' \tag{2.24}$$

In the static limit, (2.17) reduces to

$$\nabla^2 \phi = -\frac{\rho}{\varepsilon} \tag{2.25}$$

the solution of which is given by means of the Green's function satisfying

$$\nabla^2 G = -\delta(\boldsymbol{r} - \boldsymbol{r}') \tag{2.26}$$

as

$$\phi = \int_{V_0} \frac{\rho}{\varepsilon} G dv' \tag{2.27}$$

Hence, the integral representation for the electric field produced by the surface electric charge of density η on a surface S is obtained, using (2.18), as follows:

$$\boldsymbol{E} = -\int_S \frac{\eta}{\varepsilon} \nabla G ds' \tag{2.28}$$

2.4.3 General Vector-Field Problems

When three-dimensional problems are treated by means of the integral equation formulation, three-dimensional integral representations are required. This section summarizes general three-dimensional integral representations for electromagnetic fields, omitting the detailed derivation processes. In principle, integral representations for the general three-dimensional fields can be derived through use of the same procedures as the integral representations (2.10) and (2.12) for two-dimensional problems. Consider first the representations of electric fields in a region with neither current nor charge. The electric field obeys the following form of the vector equation:

$$\nabla \times \nabla \times \boldsymbol{E}_i - k_i^2 \boldsymbol{E}_i = 0 \tag{2.29}$$

Accordingly, we choose two functions, the electric field $\boldsymbol{E}_i$ and $\boldsymbol{a}G_i$, where G_i is the solution to (2.21) and $\boldsymbol{a}$ is an arbitrary constant vector, to apply the vector analogue of Green's theorem[1] to a region V, resulting in the following integral representation and relation (see references [12, 13] for the details of the derivation):

$$\left.\begin{array}{ll} P \in V, & \boldsymbol{E}_1 \\ P \in \bar{V}, & 0 \end{array}\right\} = \int_S \{j\omega\mu_1 G_1(\boldsymbol{i}_n' \times \boldsymbol{H}_1) - (\boldsymbol{i}_n' \times \boldsymbol{E}_1) \times \nabla' G_1 - (\boldsymbol{i}_n' \cdot \boldsymbol{E}_1)\nabla' G_1\} ds' \tag{2.30}$$

The meanings of regions V and $\bar{V}$ and subscripts 1 and 2 are the same as in the two-dimensional problems in Section 2.4.1. The S is the closed surface separating V and $\bar{V}$, and $\boldsymbol{i}_n$ is the unit normal directed outward (from V to $\bar{V}$). Likewise, we have, for the magnetic field

$$\left.\begin{array}{ll} P \in V, & \boldsymbol{H}_1 \\ P \in \bar{V}, & 0 \end{array}\right\} = -\int_S \{j\omega\varepsilon_1 G_1(\boldsymbol{i}_n' \times \boldsymbol{E}_1) + (\boldsymbol{i}_n' \times \boldsymbol{H}_1) \times \nabla' G_1 + (\boldsymbol{i}_n' \cdot \boldsymbol{H}_1)\nabla' G_1\} ds' \tag{2.31}$$

Following similar steps for the part of region $\bar{V}$ within the closed surface S_∞ at far distances, we obtain

$$\left.\begin{array}{ll} P \in V, & 0 \\ P \in \bar{V}, & \boldsymbol{E}_2 \end{array}\right\} = \boldsymbol{E}^{\text{inc}} - \int_S \{j\omega\mu_2 G_2(\boldsymbol{i}_n' \times \boldsymbol{H}_2) - (\boldsymbol{i}_n' \times \boldsymbol{E}_2) \times \nabla' G_2 - (\boldsymbol{i}_n' \cdot \boldsymbol{E}_2)\nabla' G_2\} ds' \tag{2.32}$$

$$\left.\begin{array}{ll} P \in V, & 0 \\ P \in \bar{V}, & \boldsymbol{H}_2 \end{array}\right\} = \boldsymbol{H}^{\text{inc}} + \int_S \{j\omega\varepsilon_2 G_2(\boldsymbol{i}_n' \times \boldsymbol{E}_2) + (\boldsymbol{i}_n' \times \boldsymbol{H}_2) \times \nabla' G_2 + (\boldsymbol{i}_n' \cdot \boldsymbol{H}_2)\nabla' G_2\} ds' \tag{2.33}$$

Equations (2.30) to (2.33) express quite general integral representations and relationships from which we can derive specific integral representations for almost all sorts of electromagnetic problems. We will now show a few examples, starting with the equation for $P \in \bar{V}$ in (2.32).

Example 1. Expressing electromagnetic fields on a surface S in terms of surface electric current $\boldsymbol{K}$, surface magnetic current $\boldsymbol{K}_m$, surface electric charge η, and surface magnetic charge η_m, which are equivalent to fields on S and related by

$$\begin{aligned} \boldsymbol{i}_n \times \boldsymbol{H} &= \boldsymbol{K}, \quad \boldsymbol{i}_n \times \boldsymbol{E} = -\boldsymbol{K}_m \\ \boldsymbol{i}_n \cdot \boldsymbol{E} &= \frac{\eta}{\varepsilon_2}, \quad \boldsymbol{i}_n \cdot \boldsymbol{H} = \frac{\eta_m}{\mu_2} \end{aligned} \tag{2.34}$$

when the incident wave is absent, we get

$$\boldsymbol{E}_2 = -\int_S \left\{ j\omega\mu_2 G_2 \boldsymbol{K} + \boldsymbol{K}_m \times \nabla' G_2 - \frac{\eta}{\varepsilon_2} \nabla' G_2 \right\} ds' \tag{2.35}$$

which reduces, by the formal replacement of surface sources $\boldsymbol{K}$, $\boldsymbol{K}_m$, η, and η_m with volume sources $\boldsymbol{J}$, $\boldsymbol{J}_m$, ρ, and ρ_m, respectively, in volume region V_0, to

$$\boldsymbol{E}_2 = -\int_{V_0} \left\{ j\omega\mu_2 G_2 \boldsymbol{J} + \boldsymbol{J}_m \times \nabla' G_2 - \frac{\rho}{\varepsilon_2} \nabla' G_2 \right\} dv' \tag{2.36}$$

This is just the integral representation for the electric field radiated from sources $\boldsymbol{J}$, $\boldsymbol{J}_m$, ρ, and ρ_m. On use of potentials $\boldsymbol{A}$ and ϕ given by (2.22) and (2.23), expression (2.36) is found to become (2.18) when $\boldsymbol{J}_m$ is absent. (Note that $\nabla G = -\nabla' G$.)

Example 2. Consider the problem of a two-dimensional TM wave. Using, in place of boundary surface S, a closed line C and, in place of three-dimensional Green's function G_2, the two-dimensional counterpart $\hat{G}_2$, and expressing the electromagnetic fields just outside C as[2]

$$\boldsymbol{E}_2 = \boldsymbol{i}_z E_{2z}, \quad \boldsymbol{H}_2 = \boldsymbol{i}_n H_{2n} + \boldsymbol{i}_l H_{2l} \tag{2.37}$$

to substitute into (2.32), and then making use of the following relation obtained from Maxwell's equations

$$j\omega\mu_2 H_{2l} = \frac{\partial E_{2z}}{\partial n} \tag{2.38}$$

we have

$$E_{2z} = E_z^{\text{inc}} - \frac{j}{4}\int_C \left\{ E_{2z} \frac{\partial H_0^{(2)}(k_2 R)}{\partial n'} - H_0^{(2)}(k_2 R) \frac{\partial E_{2z}}{\partial n'} \right\} dl' \tag{2.39}$$

This is just the equation obtained by replacing ϕ_2 with E_{2z} in (2.12).

Example 3. Consider a problem whose configuration is two-dimensional but whose fields are not two-dimensional and are propagated with an angle θ with respect to the positive z-axis, as illustrated in Figure 2.4. Typical examples of such problems are those of scattering of plane waves obliquely incident on cylinders and those of guided modes of waveguides. Although several methods are available for obtaining integral representations for this type of problem, we show here a comparatively simple method in which the representations are derived from the general three-dimensional representations by carrying out the z coordinate integration.

In the present problem, all field components $A(\boldsymbol{r})$ have the form of product of the cross-sectional expression $\hat{A}(\boldsymbol{\rho})$ and the common factor of z dependence $F(z)$:

$$A(\boldsymbol{r}) = \hat{A}(\boldsymbol{\rho})F(z) \tag{2.40}$$

with

$$F(z) = \exp(jk_2 z \cos\theta) \tag{2.41}$$

When the form of (2.40) is used in (2.32), the z' integration $\int dz'$ contained in $\int ds'$ can be performed analytically, making use of the following relation [11]:

$$\int_{-\infty}^{\infty} F(z')G_i dz' = \frac{1}{4j} F(z)H_0^{(2)}(h_i R) \tag{2.42}$$

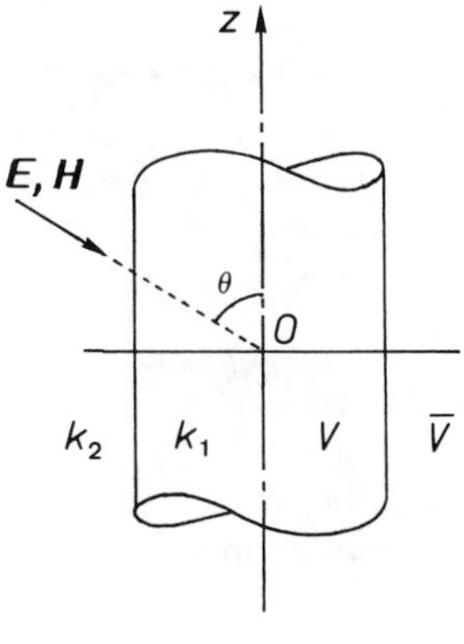

Figure 2.4 The problem of two-dimensional configurations with wave propagation in an oblique direction.

which is obtained from the integral formula

$$\int_{-\infty}^{\infty} \frac{2j}{\sqrt{a^2 + x^2}} \mathrm{e}^{-jb\sqrt{a^2+x^2}-jxy} dx = 2\pi H_0^{(2)}(a\sqrt{b^2 - y^2}) \tag{2.43}$$

where

$$h_i = \sqrt{k_i^2 - k_2^2 \cos^2\theta} \tag{2.44}$$

Although the details are omitted here, by performing z' integration, we obtain from (2.32) the following integral representation for the electric field in the cross-sectional plane perpendicular to the z-axis [11, 14]:

$$\begin{aligned} \boldsymbol{E}_2 = \boldsymbol{E}^{\text{inc}} + \frac{j}{4}\int_C &[j\omega\mu_2(\boldsymbol{i}_n' \times \boldsymbol{H}_2) - (\boldsymbol{i}_n' \times \boldsymbol{E}_2) \times (\boldsymbol{\nabla}_t' - jk_2 \cos\theta\, \boldsymbol{i}_z) \\ &- (\boldsymbol{i}_n' \cdot \boldsymbol{E}_2)(\boldsymbol{\nabla}_t' - jk_2 \cos\theta\, \boldsymbol{i}_z)]H_0^{(2)}(h_2 R)dl' \end{aligned} \tag{2.45}$$

where

$$\boldsymbol{\nabla}_t = \boldsymbol{\nabla} - \boldsymbol{i}_z \frac{\partial}{\partial z} = \boldsymbol{i}_n \frac{\partial}{\partial n} + \boldsymbol{i}_l \frac{\partial}{\partial l} \tag{2.46}$$

In (2.45), the hat symbol or caret on the cross-sectional expressions $\hat{A}$ is omitted for simplicity, and so it will be hereafter.

Consider next the problem of guided modes. All field components, in this case, have the common factor

$$F(z) = \exp(-j\beta z) \tag{2.47}$$

with β being the propagation constant. Compared with (2.41), we see that the correspondence

$$\cos\theta = -\frac{\beta}{k_2} \tag{2.48}$$

holds. This means that $|\cos\theta| > 1$ because $k_2 < \beta < k_1$. Noting that the incident wave is absent in the problem of guided modes, and also that the relations

$$H_0^{(2)}(h_2 R) = \frac{2j}{\pi} K_0(\gamma R) \tag{2.49}$$

$$\gamma = \sqrt{\beta^2 - k_2^2} = jh_2 \tag{2.50}$$

hold, instead of (2.45), we obtain the following integral representation:

$$\begin{aligned} \boldsymbol{E}_2 = -\frac{1}{2\pi} \int_C & [j\omega\mu_2(\boldsymbol{i}_n' \times \boldsymbol{H}_2) - (\boldsymbol{i}_n' \times \boldsymbol{E}_2) \times (\boldsymbol{\nabla}_t' + j\beta \boldsymbol{i}_z) \\ & - (\boldsymbol{i}_n' \cdot \boldsymbol{E}_2)(\boldsymbol{\nabla}_t' + j\beta \boldsymbol{i}_z)] K_0(\gamma R) dl' \end{aligned} \tag{2.51}$$

where $K_0(\cdot)$ is the modified Bessel function of the second kind and order zero. The n component of $\boldsymbol{E}_2$ in the last term within brackets in (2.51) can be given, from Maxwell's equations, as

$$\boldsymbol{i}_n \cdot \boldsymbol{E}_2 = -\frac{j}{\omega\varepsilon_2}\left(\frac{\partial H_{2z}}{\partial l} + j\beta H_{2l}\right) \tag{2.52}$$

which is transformable through use of integration by parts, provided that it is in the integrand, into

$$\boldsymbol{i}_n \cdot \boldsymbol{E}_2 = \frac{j}{\omega\varepsilon_2}\left(H_{2z}\frac{\partial}{\partial l} - j\beta H_{2l}\right) \tag{2.53}$$

or with the aid of the first equation of (2.34), into

$$\boldsymbol{i}_n \cdot \boldsymbol{E}_2 = -\frac{j}{\omega\varepsilon_2}\left(K_l\frac{\partial}{\partial l} + j\beta K_z\right) \tag{2.54}$$

Of these expressions, we should choose the suitable one according to the method of choice of unknowns in constructing integral equations and the method of discretization of the integral equations.

2.5 INTEGRAL EQUATIONS

In deriving integral equations, we must first get the integral representations with the observation point taken on the boundaries of regions, for the electromagnetic quantities (occasionally auxiliary quantities such as scalar and vector potentials). Once these representations are prepared, the integral equations are immediately obtained by applying boundary conditions in accord with the problem under consideration. This section briefly describes these integral representations obtained as the observation point is taken on the boundary, for the solutions to the two-dimensional and three-dimensional scalar Helmholtz equations and the solutions to the vector wave equations. Next, we summarize the fundamental integral equations derived by imposing boundary conditions, for the problems of scattering from conducting bodies and dielectric bodies and for the problems of eigenvalues of hollow waveguides and dielectric waveguides. These fundamental integral equations for the problems of the exterior region may have unphysical solutions, that is, internal-resonance solutions, mingled with the true solutions. Therefore, it would be desirable in some cases to construct, from the outset, those kinds of integral equations that have no possibility of internal-resonance solutions. However, we will discuss the subject only briefly and omit the specific expressions of those equations.

2.5.1 Expressions with the Observation Point Taken on the Boundary

If the observation point of integral representations is located on a boundary, there appears a point where the integration point coincides with the observation point. Because of the singularity of Green's functions at this point, special consideration is required for the boundary integration in the neighborhood of this point. More specifically, the boundary integration in this case must be handled by taking the limit as the observation point approaches the boundary. The result obtained with this limiting procedure, however, is generally very simple. That is, it can simply be interpreted such that there appears a jumping-out term equivalent to half of the quantity for which the integral representation is given,[3] provided the boundary shape is smooth in the vicinity of the observation point. The processes by which this result is drawn will now be shown briefly, for some specific integral representations.

(1) Two-Dimensional Problems

Let us first consider the limit as the observation point P approaches the boundary for the integral representation (2.10); that is, the following one, which holds when $P \in V$:

$$\phi_1 = -\int_C \left(\phi_1 \frac{\partial \hat{G}_1}{\partial n'} - \hat{G}_1 \frac{\partial \phi_1}{\partial n'} \right) dl' \tag{2.55}$$

Let the point O be a point on the boundary where P approaches and let the shape of the boundary C be smooth in the neighborhood of O. There are two different ways to take the limit; one way, as illustrated in Figure 2.5(a), is to place P at the position of O from the outset, deform the shape of C into a shape with a hollow of a semicircle of infinitesimal radius a, and take the limit as $a \to 0$ after calculating the integral over the deformed shape [15]; the other way, as in Figure 2.5(b), is to place P a finite length y away from S, carry out the integral over a segment of finite width a,

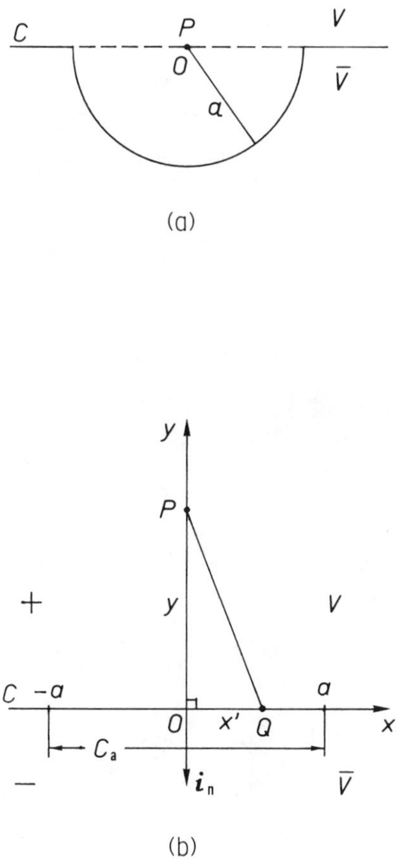

Figure 2.5 Two ways of limiting procedures.

and take the limit as $a \to 0$ after letting $y \to 0$ [4]. If our purpose was only to get the analytical expression for the point P placed just on a boundary, the former approach would be preferable to the latter on account of its simplicity. However, this limit procedure is also used when integral equations are discretized, if we want to evaluate the integrals over a subdomain of finite width $2a$ with relatively high accuracy. In such a case, the former approach is inappropriate because the boundary shape is different from the real one. For this reason, we employ the latter approach consistently in this section.

Let C_a be a segment of infinitesimal width $2a$ centered at O. The functions ϕ_1 and $\partial\phi_1/\partial n$ can be assumed to be continuous in the vicinity of O insofar as the boundary shape is smooth there. Then, we have the following approximate relations:

$$\int_{C_a} \hat{G}_1 \frac{\partial \phi_1}{\partial n'} dl' \approx \frac{\partial \phi_1}{\partial n} (O) \int_{C_a} \hat{G}_1 dl' \tag{2.56}$$

$$\int_{C_a} \phi_1 \frac{\partial \hat{G}_1}{\partial n'} dl' \approx \phi_1(O) \int_{C_a} \frac{\partial \hat{G}_1}{\partial n'} dl' \tag{2.57}$$

Therefore, it follows that the essential properties can be found by examining the following two types of integrals:

$$\hat{I}_1 = \int_{C_a} \hat{G}_1 dl', \quad \hat{I}_2 = \int_{C_a} \frac{\partial \hat{G}_1}{\partial n'} dl' \tag{2.58}$$

In the present case, $\hat{G}_1$ is the free-space Green's function for the two-dimensional scalar Helmholtz equation (the case of $i = 1$ in (2.8)). However, the principal term of this function reduces, as $|\boldsymbol{\rho} - \boldsymbol{\rho}'| \to 0$, to the Green's function for the equation obtained by putting $k = 0$ in the two-dimensional scalar Helmholtz equation; that is, the Green's function for Laplace's equation in two dimensions:

$$\hat{G}_1 = -\frac{1}{2\pi} \ln |\boldsymbol{\rho} - \boldsymbol{\rho}'| \tag{2.59}$$

Therefore, we can use (2.59) for $\hat{G}_1$ in examining the integrals of (2.58). In terms of the coordinates shown in Figure 2.5(b), we have

$$n' = -y', \quad \boldsymbol{\rho} = (0, y), \quad \boldsymbol{\rho}' = (x', y') \tag{2.60}$$

Hence $\hat{I}_1$ and $\hat{I}_2$ are calculated as

$$\hat{I}_1 = 2 \int_0^a -\frac{1}{2\pi} \ln \sqrt{y^2 + x'^2} dx' = -\frac{1}{2\pi} \left\{ a \ln (a^2 + y^2) - 2a + 2y \tan^{-1} \frac{a}{y} \right\} \tag{2.61}$$

$$\hat{I}_2 = 2 \int_0^a \left[\frac{1}{2\pi} \frac{\partial}{\partial y'} \ln \sqrt{(y' - y)^2 + x'^2} \right]_{y'=0} dx' = -\frac{1}{\pi} \tan^{-1} \frac{a}{y} \tag{2.62}$$

In the limit as $P \to O$; that is, $y \to 0$

$$\hat{I}_1 \to -\frac{a}{\pi} (\ln a - 1), \quad \hat{I}_2 \to -\frac{1}{2} \tag{2.63}$$

In the further limit as $a \to 0$

$$\hat{I}_1 \to 0, \quad \hat{I}_2 \to -\frac{1}{2} \tag{2.64}$$

We assume in the preceding that $P \in V$. However, the limit of P approaching O from the $\bar{V}$ side can also be dealt with in exactly the same fashion. Noting that the unit normal $\boldsymbol{i}_n$ is directed, in this case, to the region of P existing, we find that the result with respect to $\hat{I}_2$ is obtained with the sign reversed; that is,

$$\hat{I}_2 \to \frac{1}{2} \tag{2.65}$$

The integral over $C - C_a$ reduces to the integral over C at the limit as $C_a \to 0$ and remains to be integrable. As a result, we obtain the following general formulas[4] (illustrated in Figure 2.6):

$$\int_{C_\pm} \hat{G}_1 dl' = \int_C \hat{G}_1 dl' \tag{2.66}$$

$$\int_{C_\pm} \frac{\partial \hat{G}_1}{\partial n'} dl' = \mp \frac{1}{2} + \int_C \frac{\partial \hat{G}_1}{\partial n'} dl' \tag{2.67}$$

where C^+ and C^- signify the limits as P approaches the boundary C from the V side and the $\bar{V}$ side, respectively. This is the reason why the signs + and − are written in the regions V and $\bar{V}$, respectively, in Figure 2.5(b). The upper (or lower) signs in (2.67) go with each other. If $\partial/\partial n'$ is replaced by $\partial/\partial n$ in (2.67), the term $\mp 1/2$ must be replaced by $\pm 1/2$. Applying the results of (2.66) and (2.67) together with (2.56) and (2.57) to (2.55), we get, as the expression on the boundary,

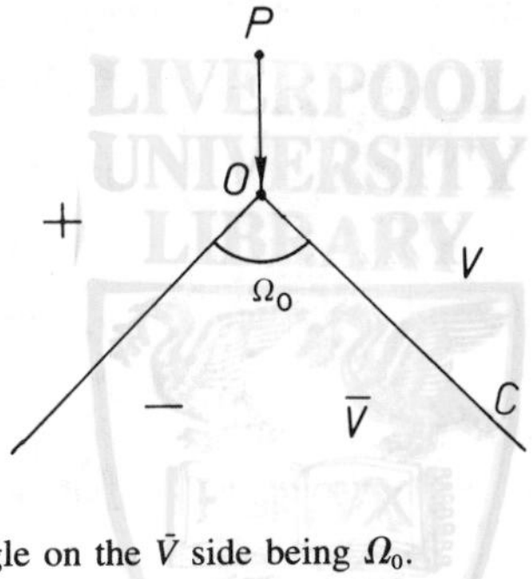

Figure 2.6 A wedge with the angle on the $\bar{V}$ side being Ω_0.

$$\frac{1}{2}\phi_1 = -\int_C \left(\phi_1 \frac{\partial \hat{G}_1}{\partial n'} - \hat{G}_1 \frac{\partial \phi_1}{\partial n'}\right) dl' \tag{2.68}$$

Writing (2.68) in combination with (2.10) and (2.11) yields[5]

$$-\int_C \left(\phi_1 \frac{\partial \hat{G}_1}{\partial n'} - \hat{G}_1 \frac{\partial \phi_1}{\partial n'}\right) dl' = \begin{cases} \phi_1, & P \in V \\ \frac{1}{2}\phi_1, & P \in C \\ 0, & P \in \bar{V} \end{cases} \tag{2.69}$$

We see from this that the value of integral representation becomes, on the boundary, the arithmetic mean of those in both neighboring regions.

(2) Three-Dimensional Scalar-Field Problems

We consider here the case of ϕ_i, which satisfies the three-dimensional scalar Helmholtz equation

$$(\boldsymbol{\nabla}^2 + k_i^2)\phi_i = 0 \tag{2.70}$$

leaving aside the problem of how this scalar function ϕ_i should be related to the fields $\boldsymbol{E}$ and $\boldsymbol{H}$. The corresponding free-space Green's function is

$$G_i = \frac{1}{4\pi|\boldsymbol{r} - \boldsymbol{r}'|} e^{-jk_i|\boldsymbol{r}-\boldsymbol{r}'|} \tag{2.71}$$

We can easily confirm that the same relationships as (2.10) to (2.13) hold for the solution ϕ_i to the three-dimensional problem also, if only $\hat{G}_i$, $\int_C dl'$, and $\boldsymbol{\rho}$ are replaced by G_i, $\int_S ds'$, and $\boldsymbol{r}$, respectively. Thus, let us consider the limiting procedure analogous to the two-dimensional case for the integral representation obtained from (2.10) with these replacements. We examine the limiting values of integrals I_1 and I_2 as P approaches the point O on the boundary:

$$I_1 = \int_{S_a} G_1 ds', \quad I_2 = \int_{S_a} \frac{\partial G_1}{\partial n'} ds' \tag{2.72}$$

where S_a is an infinitesimal circular region of radius a centered at O. Only the principal term is important, so that we use, for G_i, that with $k_i = 0$ in (2.71); that is, the Green's function for Laplace's equation:

$$G_i = \frac{1}{4\pi|\boldsymbol{r} - \boldsymbol{r}'|} \tag{2.73}$$

We can easily confirm, by simple calculation, that the resulting formulas become the equations obtained from equations (2.64) to (2.69) by simple replacements of $\hat{I}_1$, $\hat{I}_2$, $\hat{G}_1$, C, and dl' with I_1, I_2, G_1, S, and ds', respectively.

(3) Vector-Field Problems

Consider, as an example, the integral representation for $P \in V$ of (2.30); that is,

$$\boldsymbol{E}_1 = \int_S \{j\omega\mu_1 G_1(\boldsymbol{i}_n' \times \boldsymbol{H}_1) - (\boldsymbol{i}_n' \times \boldsymbol{E}_1) \times \nabla' G_1 - (\boldsymbol{i}_n' \cdot \boldsymbol{E}_1)\nabla' G_1\} ds' \tag{2.74}$$

As the field values can be approximated to be constant within the infinitesimal region S_a, it follows that only the integrals of G_1 and $\nabla' G_1$ on S_a need to be examined. In the case of $ds' = dx'dy'$, we can write $\nabla' G_1$ as

$$\nabla' G_1 = \left(\boldsymbol{i}_x' \frac{\partial}{\partial x'} + \boldsymbol{i}_y' \frac{\partial}{\partial y'} + \boldsymbol{i}_n' \frac{\partial}{\partial n'}\right) G_1 \tag{2.75}$$

from which we see that the integrals

$$I_3 = \int_{S_a} \frac{\partial G_1}{\partial x'} ds', \quad I_4 = \int_{S_a} \frac{\partial G_1}{\partial y'} ds' \tag{2.76}$$

must be examined besides I_1 and I_2 of (2.72). However, G_1 is considered to have rotational symmetry within the infinitesimal smooth region S_a, so that I_3 and I_4 can approximately be regarded as

$$I_3 \approx I_4 \approx \frac{1}{4\pi} \int_{S_a} \frac{\partial}{\partial x'} \frac{1}{\sqrt{x'^2 + y'^2 + z^2}} dx'dy' \approx 0 \tag{2.77}$$

Consequently, we get, from the limiting procedure,

$$\int_{S_a} \{(\boldsymbol{i}_n' \times \boldsymbol{E}_1) \times \nabla' G_1 + (\boldsymbol{i}_n' \cdot \boldsymbol{E}_1)\nabla' G_1\} ds'$$

$$\to \{[\boldsymbol{i}_n \times \boldsymbol{E}_1(\boldsymbol{r})] \times \boldsymbol{i}_n + [\boldsymbol{i}_n \cdot \boldsymbol{E}_1(\boldsymbol{r})]\boldsymbol{i}_n\} \int_{S_a} \frac{\partial G_1}{\partial n'} ds' \to -\frac{1}{2}\boldsymbol{E}_1(\boldsymbol{r}) \tag{2.78}$$

Thus, the relationships corresponding to (2.69) result in [11][6]

$$\int_S \{j\omega\mu_1 G_1(\boldsymbol{i}'_n \times \boldsymbol{H}_1) - (\boldsymbol{i}'_n \times \boldsymbol{E}_1) \times \boldsymbol{\nabla}' G_1 - (\boldsymbol{i}'_n \cdot \boldsymbol{E}_1)\boldsymbol{\nabla}' G_1\} ds'$$

$$= \begin{cases} \boldsymbol{E}_1, & P \in V \\ \dfrac{1}{2}\boldsymbol{E}_1, & P \in S \\ 0, & P \in \bar{V} \end{cases} \quad (2.79)$$

For the representation of $\boldsymbol{H}$, too, we have the same results; that is, the jumping-out term $(1/2)\boldsymbol{H}$ is extracted as a result of the limiting procedure and the value on the boundary becomes the arithmetic mean of those in both adjoining regions.

2.5.2 Fundamental Integral Equations

This section summarizes fundamental integral equations for several particular problems. It always is possible to construct both types of integral equations, those expressing boundary conditions of the electric field (electric-field integral equations) and those expressing boundary conditions of the magnetic field (magnetic-field integral equations). Nevertheless, only one type of integral equation is presented in most cases for each specific problem, because generally one type clearly is more convenient than the other. However, the other type is not meaningless and may play a certain role in the involvement of resonant solutions and the removal of these solutions [11]. Thus, here, we try to show both types of integral equations in parallel.

(1) Scattering from Conducting Cylinders

Consider the problem of scattering from a conducting cylinder infinite in the z direction with the peripheral boundary line being C, immersed in the medium of wavenumber k_2, that is, the case where region V in Figure 2.7 is a perfect conductor.

In the problem of a TM wave, it is convenient to choose E_z for ϕ as in (2.2). Thus, let

$$\phi_2 = E_z, \quad \phi^{\text{inc}} = E_z^{\text{inc}} \quad (2.80)$$

Substituting the boundary condition on C

$$E_z = 0 \quad (2.81)$$

and (2.8) together with (2.68) into (2.12) yields the following electric field integral

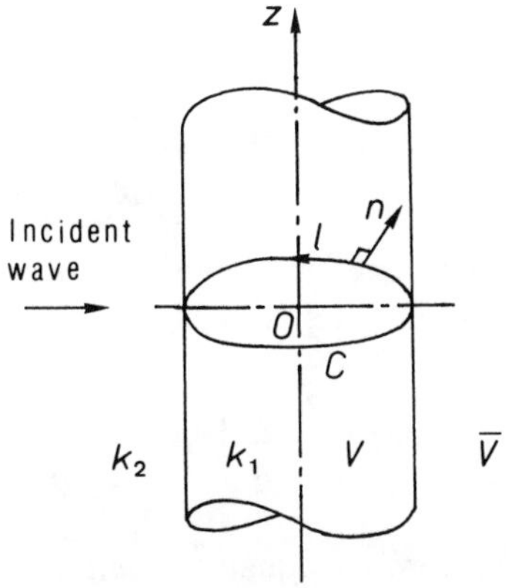

Figure 2.7 The problem of scattering by a cylinder.

equation for the unknown function $\partial E_z/\partial n$:

$$E_z^{\text{inc}} = -\frac{j}{4}\int_C H_0^{(2)}(k_2 R)\frac{\partial E_z}{\partial n'}\,dl' \tag{2.82}$$

On the other hand, taking the limit as the observation point P approaches the boundary C for the equation operated on (2.12) with the n-derivative and using the boundary condition (2.81) and the relation

$$\frac{\partial E_z}{\partial n} = j\omega\mu_i H_l \tag{2.83}$$

along with (2.67), we obtain the following magnetic-field integral equation for the unknown function H_l:

$$\frac{1}{2}H_l = H_l^{\text{inc}} + \frac{j}{4}\int_C \frac{\partial H_0^{(2)}(k_2 R)}{\partial n} H_l\,dl' \tag{2.84}$$

In the problem of a TE wave, we let

$$\phi_2 = H_z, \quad \phi^{\text{inc}} = H_z^{\text{inc}} \tag{2.85}$$

Using this, in (2.12), together with the boundary condition on C

$$E_l = 0 \tag{2.86}$$

and the relation

$$\frac{\partial H_z}{\partial n} = -j\omega\varepsilon_i E_l \tag{2.87}$$

and taking the limit as P approaches the boundary, due to the formula (2.67), we have the following magnetic-field integral equation:

$$\frac{1}{2} H_z = H_z^{\text{inc}} - \frac{j}{4} \int_C H_z \frac{\partial H_0^{(2)}(k_2 R)}{\partial n'} \, dl' \tag{2.88}$$

In a similar fashion, we get, from the equation operated on (2.12) with the n-derivative, the following electric field integral equation:

$$E_l^{\text{inc}} = -\frac{1}{4\omega\varepsilon_2} \text{FP} \int_C H_z \frac{\partial^2 H_0^{(2)}(k_2 R)}{\partial n \partial n'} \, dl' \tag{2.89}$$

The symbol FP, signifying the finite part of Hadamard, is used to show clearly that the integral term on the right-hand side of (2.89), which has strong singularity associated with the quadratic differential, can be made meaningful with the understanding that [16, 13, 11]

$$\text{FP} \int_C f(\boldsymbol{\rho}') \frac{\partial H_0^{(2)}(k_2 R)}{\partial n \partial n'} \, dl' = \lim_{a \to 0} \left\{ \int_{C-C_a} f(\boldsymbol{\rho}') \frac{\partial^2 H_0^{(2)}(k_2 R)}{\partial n \partial n'} \, dl' - \frac{4j}{a\pi} f(\boldsymbol{\rho}) \right\} \tag{2.90}$$

where C_a is a segment of infinitesimal width $2a$ as in Figure 2.5(b). It follows that a finite value of a is used when this term is numerically evaluated practically. The term also can be evaluated by transforming the part of derivative $\partial^2/\partial n \partial n'$ into the form with derivative $\partial^2/\partial l \partial l'$ and then using integration by parts (see the next section). Rather seldom is the integral equation (2.84) or (2.89) used solely, because the quantity expressed by the original integral representation does not correspond to the basic quantity of the problem.

(2) *Scattering from Dielectric Cylinders*

Let V and $\bar{V}$ in Figure 2.7 be the regions of the media of wavenumbers k_1 and k_2, respectively. In the TM wave problem, the tangential components of fields are only E_z and H_l. As in the problem of a conductor, we choose E_z for ϕ, take the limit as the observation point P approaches the boundary, and make use of (2.68). Then we obtain a set of simultaneous electric field integral equations with E_z and $\partial E_z/\partial n$ as unknowns.

$$\frac{1}{2}E_z = \frac{j}{4}\int_C \left\{ E_z \frac{\partial H_0^{(2)}(k_1R)}{\partial n'} - \frac{\mu_1 \partial E_z}{\mu_2 \partial n'} H_0^{(2)}(k_1R) \right\} dl' \tag{2.91}$$

$$\frac{1}{2}E_z = E_z^{\text{inc}} - \frac{j}{4}\int_C \left\{ E_z \frac{\partial H_0^{(2)}(k_2R)}{\partial n'} - \frac{\partial E_z}{\partial n'} H_0^{(2)}(k_2R) \right\} dl' \tag{2.92}$$

The E_z and $\partial E_z/\partial n$ in these equations are those for the exterior region. In the derivation the boundary conditions $E_{1z} = E_{2z} = E_z$ and $\partial E_{2z}/\partial n = (\mu_2/\mu_1)\partial E_{1z}/\partial n = \partial E_z/\partial n$ (cf. (2.83)) are utilized. If the set of equations was derived on the basis of the interior-region quantities E_{1z} and $\partial E_{1z}/\partial n$, the coefficient μ_1/μ_2 would disappear from (2.91) and instead μ_2/μ_1 would appear as the coefficient to $\partial E_z/\partial n$ in (2.92). From the equations operated on (2.10) and (2.12) with the n-derivative, the following set of magnetic-field integral equations results:

$$\frac{1}{2}j\omega\mu_1 H_l = \frac{j}{4}\int_C \left\{ E_z \frac{\partial^2 H_0^{(2)}(k_1R)}{\partial n \partial n'} - j\omega\mu_1 H_l \frac{\partial H_0^{(2)}(k_1R)}{\partial n} \right\} dl' \tag{2.93}$$

$$\frac{1}{2}j\omega\mu_2 H_l = j\omega\mu_2 H_l^{\text{inc}} - \frac{j}{4}\int_C \left\{ E_z \frac{\partial^2 H_0^{(2)}(k_2R)}{\partial n \partial n'} - j\omega\mu_2 H_l \frac{\partial H_0^{(2)}(k_2R)}{\partial n} \right\} dl' \tag{2.94}$$

The unknown functions E_z and H_l may be regarded as those of either interior or exterior region because those functions are continuous across the boundary. The terms of quadratic differential $\partial^2/\partial n \partial n'$ in both equations should be considered to have the meaning of FP defined by (2.90). These terms can be transformed into other forms as the following. First, define φ_n as the angle between $\boldsymbol{i}_n$ and $\boldsymbol{i}_x$ as in Figure 2.8. Then we have

$$\frac{\partial H_0^{(2)}(kR)}{\partial n} = -\boldsymbol{i}_n \cdot \nabla_t' H_0^{(2)}(kR)$$

$$= -\left\{ \cos(\varphi_n - \varphi_n') \frac{\partial}{\partial n'} + \sin(\varphi_n - \varphi_n') \frac{\partial}{\partial l'} \right\} H_0^{(2)}(kR) \tag{2.95}$$

$$\frac{\partial H_0^{(2)}(kR)}{\partial l} = -\boldsymbol{i}_l \cdot \nabla_t' H_0^{(2)}(kR)$$

$$= -\left\{ \cos(\varphi_n - \varphi_n') \frac{\partial}{\partial l'} - \sin(\varphi_n - \varphi_n') \frac{\partial}{\partial n'} \right\} H_0^{(2)}(kR) \tag{2.96}$$

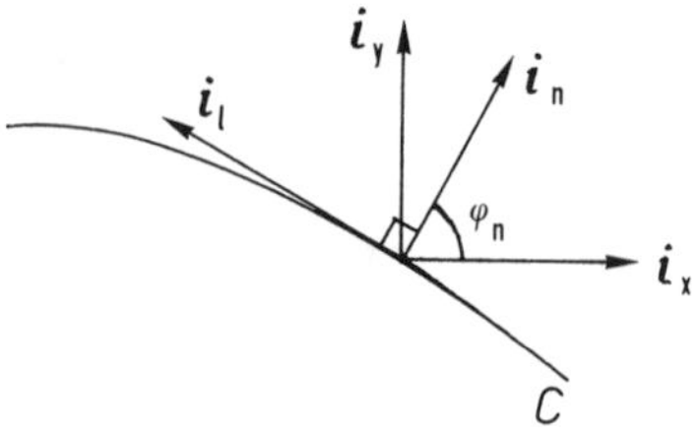

Figure 2.8 Unit vectors at a point on C.

Using these two relationships in conjunction with

$$\left(\frac{\partial^2}{\partial n'^2} + \frac{\partial^2}{\partial l'^2} + k^2\right) H_0^{(2)}(kR) = 0 \tag{2.97}$$

leads us to

$$\frac{\partial^2}{\partial n \partial n'} H_0^{(2)}(kR) = \left\{ k^2 \cos(\varphi_n - \varphi_n') - \frac{\partial^2}{\partial l \partial l'} \right\} H_0^{(2)}(kR) \tag{2.98}$$

Substituting (2.98) into (2.93) and (2.94) and using integration by parts, we obtain

$$\frac{1}{2} H_l = \frac{1}{4} \int_C \left\{ \omega\varepsilon_1 E_z \cos(\varphi_n - \varphi_n') + \frac{1}{\omega\mu_1} \frac{\partial E_z}{\partial l'} \frac{\partial}{\partial l} - jH_l \frac{\partial}{\partial n} \right\} H_0^{(2)}(k_1 R) dl' \tag{2.99}$$

$$\frac{1}{2} H_l = H_l^{\text{inc}} - \frac{1}{4} \int_C \left\{ \omega\varepsilon_2 E_z \cos(\varphi_n - \varphi_n') + \frac{1}{\omega\mu_2} \frac{\partial E_z}{\partial l'} \frac{\partial}{\partial l} - jH_l \frac{\partial}{\partial n} \right\} H_0^{(2)}(k_2 R) dl' \tag{2.100}$$

We see that these equations have no quadratic differentials. Instead, there appear l derivatives of E_z, so that differentiable functions must be chosen as expansion functions for E_z when the set of (2.99) and (2.100) is used for practical calculation; it follows, for instance, that step functions cannot be used.

In the problem of a TE wave, the tangential components of fields are only H_z and E_l. Letting ϕ correspond to H_z, we have, by using the steps similar to those for the TM wave case, the following set of magnetic-field integral equations for unknowns H_z and $\partial H_z/\partial n$:

$$\frac{1}{2}H_z = \frac{j}{4}\int_C \left\{ H_z \frac{\partial H_0^{(2)}(k_1R)}{\partial n'} - \frac{\varepsilon_1}{\varepsilon_2}\frac{\partial H_z}{\partial n'} H_0^{(2)}(k_1R) \right\} dl' \tag{2.101}$$

$$\frac{1}{2}H_z = H_z^{\text{inc}} - \frac{j}{4}\int_C \left\{ H_z \frac{\partial H_0^{(2)}(k_2R)}{\partial n'} - \frac{\partial H_z}{\partial n'} H_0^{(2)}(k_2R) \right\} dl' \tag{2.102}$$

In deriving these, the relations $H_{2z} = H_{1z} = H_z$ and $\partial H_{2z}/\partial n = (\varepsilon_2/\varepsilon_1)\partial H_{1z}/\partial n = \partial H_z/\partial n$ are used (cf. (2.87)). If we employed the relation $\partial H_{1z}/\partial n = \partial H_z/\partial n$ instead, the coefficients to $\partial H_z/\partial n'$ in the preceding equations would become a little different. From the equations operated on (2.10) and (2.12) with the n-derivative, we get the following set of electric-field integral equations for unknowns H_z and E_l:

$$\frac{1}{2}E_l = -\frac{1}{4}\int_C \left\{ \frac{1}{\omega\varepsilon_1} H_z \frac{\partial^2 H_0^{(2)}(k_1R)}{\partial n \partial n'} + jE_l \frac{\partial H_0^{(2)}(k_1R)}{\partial n} \right\} dl' \tag{2.103}$$

$$\frac{1}{2}E_l = E_l^{\text{inc}} + \frac{1}{4}\int_C \left\{ \frac{1}{\omega\varepsilon_2} H_z \frac{\partial^2 H_0^{(2)}(k_2R)}{\partial n \partial n'} + jE_l \frac{\partial H_0^{(2)}(k_2R)}{\partial n} \right\} dl' \tag{2.104}$$

Here also, the terms associated with the quadratic differential should be considered to have the meaning of FP defined by (2.90). Through use of formal procedure of integration by parts with the aid of (2.98), equations (2.103) and (2.104) also are reducible to the following forms:

$$\frac{1}{2}E_l = -\frac{1}{4}\int_C \left\{ \omega\mu_1 H_z \cos(\varphi_n - \varphi_n') + \frac{1}{\omega\varepsilon_1}\frac{\partial H_z}{\partial l'}\frac{\partial}{\partial l} + jE_l \frac{\partial}{\partial n} \right\} H_0^{(2)}(k_1R) dl' \tag{2.105}$$

$$\frac{1}{2}E_l = E_l^{\text{inc}} + \frac{1}{4}\int_C \left\{ \omega\mu_2 H_z \cos(\varphi_n - \varphi_n') + \frac{1}{\omega\varepsilon_2}\frac{\partial H_z}{\partial l'}\frac{\partial}{\partial l} + jE_l \frac{\partial}{\partial n} \right\} H_0^{(2)}(k_2R) dl' \tag{2.106}$$

(3) Guided Modes of Hollow Waveguides

The problem of guided modes of hollow waveguides is analogous to that of scattering from conducting cylinders in the sense that boundaries are those of the conductor, whereas the basic differential equation is different from that for the latter problem (2.1) and is given by

$$(\nabla_t^2 + h_1^2)\phi_1 = 0 \tag{2.107}$$

where

$$h_1 = \sqrt{k_1^2 - \beta^2} \tag{2.108}$$

In the present problem, all electromagnetic quantities can be expressed in the form with common factor $F(z)$ of (2.47) eliminated. From the fundamental integral equations for the problem of scattering from conducting cylinders, we obtain the integral equations for determining eigenvalues of modes, by reversing the direction of $\boldsymbol{i}_n$, removing the terms of the incident wave, and letting $k_2 \to h_1$, as follows. TM mode:

$$\int_C \frac{\partial E_z}{\partial n'} H_0^{(2)}(h_1 R) dl' = 0 \tag{2.109}$$

TE mode:

$$\frac{1}{2} H_z = \frac{j}{4} \int_C H_z \frac{\partial H_0^{(2)}(h_1 R)}{\partial n'} dl' \tag{2.110}$$

The former is an integral equation of the electric-field type and the latter is of the magnetic-field type. The equations of the magnetic-field type for TM modes and the electric-field type for TE modes are omitted here. Noting that the problem of waveguide modes refers to the interior region, we find that we may employ, as the Green's function, the singular part of the conventional Green's function satisfying the radiation condition:

$$\mathrm{Im}\{H_0^{(2)}(h_1 R)\} = -N_0(h_1 R) \tag{2.111}$$

in place of using the original Green's function as a whole. Computation time, of course, is remarkably reduced in this case. In return, however, a possibility arises that the numerical solutions would be mingled with certain spurious solutions, although these solutions can be distinguished comparatively easily from true solutions, because the spurious ones will change greatly as the discretization methods are changed [5].

(4) Guided Modes of Dielectric Waveguides

The integral equations derived from the integral representations for the interior region differ from those derived from the representations for the exterior region, and those of the electric-field type differ from those of the magnetic-field type. Therefore, we have at least eight different kinds of fundamental integral equations. Moreover, by transforming these into various forms, we could derive many other different equations. However, the number of independent tangential components of fields on C is only four, indicating that four independent integral equations would be sufficient. Although several choices from among many different integral equations would be available for constructing four independent equations, only one particular method will be shown here. In this method, we choose only the equations obtained from the integral representations for the z components of fields to form a characteristic equation [11, 14].

When (2.53) is employed for the expression of $(\boldsymbol{i}'_n \cdot \boldsymbol{E}_2)$, the z component of (2.51) reduces, in the limit as P approaches the boundary, to

$$\frac{1}{2}E_{2z} = \frac{1}{2\pi}\int_C \left(j\frac{\gamma^2}{\omega\varepsilon_2}H_{2l} - \frac{\beta}{\omega\varepsilon_2}H_{2z}\frac{\partial}{\partial l'} + E_{2z}\frac{\partial}{\partial n'}\right) K_0(\gamma R)dl' \quad (2.112)$$

Deriving the equation for the magnetic field counterpart of (2.112), on making use of the duality relationship between (2.32) and (2.33) with respect to the electric and magnetic fields, we have

$$\frac{1}{2}H_{2z} = \frac{1}{2\pi}\int_C \left(-j\frac{\gamma^2}{\omega\mu_2}E_{2l} + \frac{\beta}{\omega\mu_2}E_{2z}\frac{\partial}{\partial l'} + H_{2z}\frac{\partial}{\partial n'}\right) K_0(\gamma R)dl' \quad (2.113)$$

Replacing, in (2.45), subscript 2 with subscript 1 and $k_2 \cos\theta$ with $-\beta$, reducing $\boldsymbol{E}^{\text{inc}}$ to 0, and then reversing the sign in front of the integral, we obtain the integral representation for the interior region. As P approaches the boundary, its z component becomes

$$\frac{1}{2}E_{1z} = -\frac{j}{4}\int_C \left(j\frac{h_1^2}{\omega\varepsilon_1}H_{1l} + \frac{\beta}{\omega\varepsilon_1}H_{1z}\frac{\partial}{\partial l'} - E_{1z}\frac{\partial}{\partial n'}\right) H_0^{(2)}(h_1R)dl' \quad (2.114)$$

Likewise, the magnetic field counterpart of (2.114) is derived to become

$$\frac{1}{2}H_{1z} = \frac{j}{4}\int_C \left(j\frac{h_1^2}{\omega\mu_1}E_{1l} + \frac{\beta}{\omega\mu_1}E_{1z}\frac{\partial}{\partial l'} + H_{1z}\frac{\partial}{\partial n'}\right) H_0^{(2)}(h_1R)dl' \quad (2.115)$$

Recalling that the tangential field components are continuous across the boundary, in other words, the field components of subscripts 1 and 2 are equal, we find that the four equations (2.112) to (2.115) have only four unknowns E_z, H_z, E_l, and H_l. It follows, therefore, that these four equations form a set of simultaneous equations for determining eigenvalues of guided modes. Using the relations expressing E_{2l} and H_{2l} in terms of E_{2z} and H_{2z}:

$$E_{2l} = \frac{j}{\gamma^2}\left(\beta\frac{\partial E_{2z}}{\partial l} - \omega\mu_2\frac{\partial H_{2z}}{\partial n}\right) \tag{2.116}$$

$$H_{2l} = \frac{j}{\gamma^2}\left(\beta\frac{\partial H_{2z}}{\partial l} + \omega\varepsilon_2\frac{\partial E_{2z}}{\partial n}\right) \tag{2.117}$$

together with $E_{1l} = E_{2l}$ and $H_{1l} = H_{2l}$, we can change equations (2.112) to (2.115) into the simultaneous integral equations having E_z, H_z, $\partial E_z/\partial n$, and $\partial H_z/\partial n$ as unknowns.

$$\frac{1}{2}E_z = -\frac{1}{2\pi}\int_C\left(\frac{\partial E_z}{\partial n'} - E_z\frac{\partial}{\partial n'}\right)K_0(\gamma R)dl' \tag{2.118}$$

$$\frac{1}{2}H_z = -\frac{1}{2\pi}\int_C\left(\frac{\partial H_z}{\partial n'} - H_z\frac{\partial}{\partial n'}\right)K_0(\gamma R)dl' \tag{2.119}$$

$$\frac{1}{2}E_z = \frac{j}{4}\int_C\left\{\frac{h_1^2\varepsilon_2}{\gamma^2\varepsilon_1}\frac{\partial E_z}{\partial n'} + E_z\frac{\partial}{\partial n'} - \frac{\beta(k_1^2 - k_2^2)}{\gamma^2\omega\varepsilon_1}H_z\frac{\partial}{\partial l'}\right\}H_0^{(2)}(h_1R)dl' \tag{2.120}$$

$$\frac{1}{2}H_z = \frac{j}{4}\int_C\left\{\frac{h_1^2\mu_2}{\gamma^2\mu_1}\frac{\partial H_z}{\partial n'} + H_z\frac{\partial}{\partial n'} + \frac{\beta(k_1^2 - k_2^2)}{\gamma^2\omega\mu_1}E_z\frac{\partial}{\partial l'}\right\}H_0^{(2)}(h_1R)dl' \tag{2.121}$$

Although these four equations are based on the fields in the exterior region, the subscript 2 is eliminated here for notational simplicity. When the set of equations is based on the fields in the exterior region, the equations corresponding to the exterior region take a simpler form, as seen in (2.118) and (2.119); whereas if the set of equations based on the fields in the interior region is derived [17], the equations corresponding to the interior region take simpler form. In the present problem, too, we could invoke the method using, for the interior region, the Neumann function part (or singular part) in place of the free-space Green's function. In this simpler method, however, certain spurious solutions will also arise. Therefore, we must be sufficiently cautious, when using the method, to locate and remove these spurious solutions by some means, such as additional calculation by means of different discretization methods.

2.5.3 Notes on the Involvement of Resonant Solutions

All of the fundamental integral equations presented in the previous section for the problem of the exterior region have a possibility of involving unphysical solutions, that is, internal-resonance solutions. These resonant solutions are the eigensolutions (what is called *internal-resonance mode solutions*) in the problem of the interior region filled with the exterior medium, and these unphysical solutions are different in the nature from the spurious solutions that often appear in the problems of waveguides. Although theoretically these resonant solutions arise only at particular, discrete frequencies, in practical numerical calculations, they increase the errors of solution over a finite range of the frequency spectrum near these discrete frequencies. The degree of the error increase differs depending upon the type of integral equation and the accuracy of calculation; the typical features can be described, for example, as follows. (1) In the problem of scattering from conducting bodies, the integral equations of the electric-field type tend to be affected by resonant solutions much less than those of the magnetic-field type. This is because, in the former equations, the current associated with internal-resonance modes does not contribute theoretically to radiation of fields. (2) In the equations for scattering from dielectric bodies, the frequency ranges where errors are involved are much narrower for both electric-field type and magnetic-field type than in the case of equations for scattering from conducting bodies—and so on. Generally, the frequency range affected by resonant solutions narrows as the accuracy of calculation increases. Also, we oftentimes need not calculate solutions for those frequency ranges influenced by resonant solutions. Taking these points into consideration, we may not need to worry excessively about the internal-resonance solutions. However, the problem of resonant solutions pertains theoretically to essential properties, so that it is very important to have sufficient knowledge of these circumstances as long as we use the integral equation methods. Needless to say, we must resort to some means of removing these resonant solutions whenever occasion demands.

An expedient method removing internal-resonance solutions utilizes, in addition to the fundamental integral equations, what we call the extended boundary condition equations derived from the application of the integral representation for the exterior region to the interior region (e.g., (2.13) or the upper equations of (2.32) and (2.33)) [18, 11].

Several other methods are also available in which we construct and use the special integral equations with uniqueness being assured. We will mention now two typical examples of these methods; one is for the problem of scattering from conducting bodies and the other is for the problem of scattering from dielectric bodies. An integral equation for the former example is constructed by combining those of the magnetic field type and the electric field type at the ratio of 1 to $-(p/\zeta_2)\boldsymbol{i}_n\times$ where p is a real constant and ζ_2 is the wave impedance. In the numerical examples for a conducting sphere, the p values in the range $0.2 \sim 1.0$ seem to be appropriate [19]. A set of integral equations for the latter example is constructed by combining, for the electric-field type, the equation derived from the integral representation for the interior region and that derived from the integral representation for the exterior region at the ratio of $-\alpha$ to 1, and, for the magnetic field type, at the ratio of $-\beta$

to 1 where $\alpha\beta^*$(* denotes the complex conjugate) must be positive and real for the uniqueness to be assured. This set of equations reduces, when $\alpha = \beta = 1$, to the set Poggio and Miller [15] gave, and when $\alpha = -\varepsilon_1/\varepsilon_2$ and $\beta = -\mu_1/\mu_2$, to the Müller's integral equation set [20]. In addition, several other integral equations are proposed that have no possibility of involvement of resonant solutions. More detailed discussions of resonant solutions are given, for example, in [11].

2.6 NUMERICAL CALCULATION OF INTEGRAL EQUATIONS

For solving boundary integral equations numerically, we must first discretize the equations by some methods to reduce them to a set of linear equations (or a matrix equation). A variety of discretization methods are available. In the field of electromagnetics, however, a way of gathering all these discretization methods under a unified concept of "moment method" [5, 10, 11, 21] is widespread. To be brief, the moment method is a general process of first expanding unknown functions in terms of some expansion functions and then taking the inner products with some weighting functions to derive linear equations. Various discretization methods result from various choices of expansion and weighting functions. Each discretization method has its own merits and drawbacks, and which method is best depends on the problem considered and the viewpoint taken. Three-dimensional problems and two-dimensional problems are considerably different in the specific ways of discretization. As regards the application of the integral equation method to general three-dimensional electromagnetic wave problems, a lot of examples, mainly for conductor problems, have been analyzed thus far by means of the methods approximating boundary surfaces with a wire-grid model or a polygonal patch model. All of these, however, are for bodies of special, simple shapes, and there still is room for improvement in discretization methods. Therefore, here, we concentrate our discussions on two-dimensional problems and describe the basic steps in numerical calculation of integral equations, taking up specific simple examples.

2.6.1 Discretization Methods of the Boundary-Element Method

The expansion functions can be divided typically into two classes: the subdomain functions (subdivision of the domain into elements) and the entire-domain functions. In the boundary-element method, the subdomain functions are used almost exclusively. The expansion functions most frequently used are step functions (or pulse functions), triangle functions, and higher-order functions (mainly quadratic functions); the methods expanding unknown functions in terms of these three functions correspond to the method of boundary subdivision by means of constant elements, linear elements, and higher-order (mainly quadratic) elements, respectively. The use of a step-function expansion is probably the easiest in view of its simplicity and tractability, when just simple calculation is attempted. If higher accuracy is desired, the use of functions with higher-order degrees, such as triangle functions and higher-order functions, will be required.

Two choices of weighting functions most often are used: one is the Dirac delta function and the other is the same function as the expansion function. The former is called the *point-matching method* and the latter is called *Galerkin's method*. In the boundary-element method, the former is used almost exclusively.

When unknown functions are expanded in terms of subdomain functions and a point-matching method is applied to convert integral equations into a matrix equation, the elements of the resulting coefficient matrix are usually given in the form of the integration of functions containing Green's functions over a finite range. The approximate calculation of this integration is carried out, in the boundary-element method, by borrowing methods of numerical integration generalized for use for the finite-element method. A typical feature of this method is that the integral contour is deformed from the actual boundary shape into a shape conforming with the expansion functions. This implies that the boundary shape is approximated by a polygon when a step or triangle function expansion is used, whereas the boundary shape is approximated by a higher-order function shape when a higher-order function expansion is used. This would help enhance the universality of the method in the sense that any complicated shape can automatically be approximated by entering only the coordinates of mesh points of subdivided elements. However, at least in the two-dimensional problem, this technique of boundary deformation does not necessarily look desirable. In the case of a circular-arc shape, for example, the approximation by a polygon would be unnecessary because the method using circular coordinates would be more natural and could give more accurate results as well. Also, in the case of a long, straight section, it would be unnecessary to enter the coordinates of all mesh points. Therefore, we will assume in this section that the boundary shape is not being approximated by any shape and keeps its original shape at least in the first stage. However, as will be found later, all the resulting equations can be applied, in substantially the same manner, when the boundary deformation is introduced from the beginning. Details of handling a circular-arc portion of a boundary by means of circular coordinates are found in [11].

When we put matching points just on tips of a wedge shape, such as vertexes of a polygon, we should be very careful whether accuracy is not being degraded, because field values happen to be infinite or zero at such points.

2.6.2 Discretization of Integral Equations and Derivation of Matrix Equations

This section briefly describes basic procedures needed in numerical calculation of integral equations, taking up, as a specific simple example, the scattering of a plane wave from a conducting cylinder embedded in the medium of wavenumber k_2. Let the cylinder axis be chosen parallel to the z-axis and let the boundary periphery line of the cylinder cross section be C, as shown in Figure 2.9. As the basis of the principles and procedures is common to every problem, we explain here only the cases of the simplest fundamental integral equations. Needless to say, these fundamental equations intrinsically have a possibility of involving resonant solutions. Therefore, in applying these practically, we must examine beforehand the problems of involvement of resonant solutions and, if necessary, devise suitable measures for removing those resonant solutions.

Let us consider the two problems in parallel. One is the TM wave case where the electric field of the incident plane wave is given by

$$\boldsymbol{E}^{\text{inc}} = \boldsymbol{i}_z E_z^{\text{inc}} = \boldsymbol{i}_z E_0 \, \mathrm{e}^{-jk_2 x} \tag{2.122}$$

with the fundamental integral equation being (2.82); that is,

$$E_z^{\text{inc}} = -\frac{j}{4} \int_C H_0^{(2)}(k_2 R) \frac{\partial E_z}{\partial n'} \, dl' \tag{2.123}$$

The other is the TE wave case, where the magnetic field of the incident plane wave is given by

$$\boldsymbol{H}^{\text{inc}} = \boldsymbol{i}_z H_z^{\text{inc}} = \boldsymbol{i}_z H_0 \, \mathrm{e}^{-jk_2 x} \tag{2.124}$$

with the fundamental integral equation being (2.88); that is,

$$\frac{1}{2} H_z = H_z^{\text{inc}} - \frac{j}{4} \int_C H_z \frac{\partial H_0^{(2)}(k_2 R)}{\partial n'} \, dl' \tag{2.125}$$

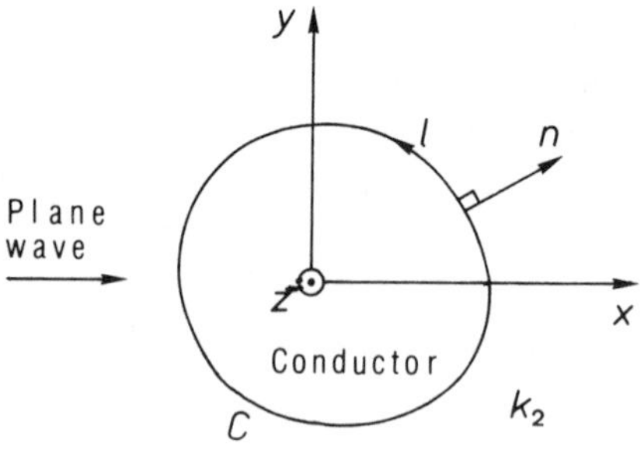

Figure 2.9 Problems of scattering by conducting cylinders.

(1) The Method Using Step-Function Expansion (Constant-Element Approximation) and Point Matching

Consider the case where unknown functions are expanded in terms of N step functions and equations are point matched at N points. Let the boundary C be divided into N small sections $\Delta C_1, \Delta C_2, \ldots, \Delta C_N$, as illustrated in Figure 2.10, within which the unknown function takes constant values $\partial E_z/\partial n(1)$, $\partial E_z/\partial n(2)$, $\ldots$, $\partial E_z/\partial n(N)$ (for the TM wave case) or $H_z(1), H_z(2), \ldots, H_z(N)$ (for the TE wave case). Matching points usually are located at the midpoints of the small sections, so that the values of the incident wave are used only at the midpoints of the sections. Let these values of the incident wave be expressed as $E_z^{\rm inc}(1), E_z^{\rm inc}(2), \ldots, E_z^{\rm inc}(N)$ (for the TM wave case), or $H_z^{\rm inc}(1), H_z^{\rm inc}(2), \ldots, H_z^{\rm inc}(N)$ (for the TE wave case). Then, (2.123) and (2.125) reduce to the following N equations, respectively:

$$E_2^{\rm inc}(p) = \sum_{q=1}^{N} \left(-\frac{j}{4} \int_{\Delta Cq} H_0^{(2)}(k_2 R_p) dl' \right) \frac{\partial E_z}{\partial n}(q), p = 1, 2, \ldots, N \quad (2.126)$$

$$\frac{1}{2} H_z(p) = H_z^{\rm inc}(p) + \sum_{q=1}^{N} \left(-\frac{j}{4} \int_{\Delta C_q} \frac{\partial H_0^{(2)}(k_2 R_p)}{\partial n'} dl' \right) H_z(q), p = 1, 2, \ldots, N \quad (2.127)$$

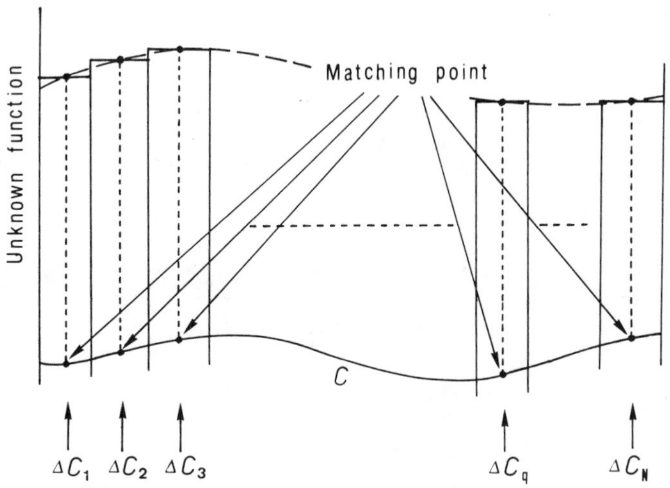

Figure 2.10 Step-function expansion of unknown functions.

with R_p being the distance from the pth observation point (matching point) to the integration point. Both (2.126) and (2.127) can be rewritten in matrix notation as

$$\boldsymbol{AX} = \boldsymbol{B} \tag{2.128}$$

where A is the square matrix of dimension N, and $\boldsymbol{X}$ and $\boldsymbol{B}$ are the column vectors of dimension N: that is,

$$A = \begin{pmatrix} a_{11} & a_{12} & a_{13} \ldots & a_{1N} \\ a_{21} & a_{22} & & \vdots \\ \vdots & & & \\ a_{N1} & \ldots & & a_{NN} \end{pmatrix},$$

$$\boldsymbol{X} = \begin{pmatrix} \alpha_1 \\ \alpha_2 \\ \vdots \\ \alpha_N \end{pmatrix}, \ \boldsymbol{B} = \begin{pmatrix} b_1 \\ b_2 \\ \vdots \\ b_N \end{pmatrix} \tag{2.129}$$

Each element a_{pq}, α_p, and b_p of A, $\boldsymbol{X}$, and $\boldsymbol{B}$ is given for the TM wave case, as seen from (2.126), as

$$a_{pq} = -\frac{j}{4}\int_{\Delta C_q} H_0^{(2)}(k_2R_p)dl' \tag{2.130}$$

$$\alpha_p = \frac{\partial E_z}{\partial n}(p),$$

$$b_p = E_z^{\text{inc}}(p) = E_0\,\mathrm{e}^{-jk_2x_p} \tag{2.131}$$

and for the TE wave case, as seen from (2.127), as

$$a_{pq} = \frac{1}{2}\,\delta_{pq} + \frac{j}{4}\int_{\Delta C_q} \frac{\partial H_0^{(2)}(k_2R_p)}{\partial n'}\,dl' \tag{2.132}$$

$$\alpha_p = H_z(p),$$

$$b_p = H_z^{\text{inc}}(p) = H_0\,\mathrm{e}^{-jk_2x_p} \tag{2.133}$$

where x_p is the x coordinate of the midpoint of section ΔC_p, and δ_{pq} denotes the Kronecker delta ($\delta_{pq} = 0$ unless $p = q$, in which case $\delta_{pq} = 1$).

(2) The Method Using Triangle-Function Expansion (Linear-Element Approximation) and Point Matching

Let the boundary C be divided into N small sections ΔC_1, ΔC_2, ..., ΔC_N, with the width of small section ΔC_q being W_q, and let l_{q-1} and l_q be the l coordinates of nodes of both ends of ΔC_q. We expand the unknown functions in terms of the triangle functions represented in Figure 2.11 and written as

$$T_q = \begin{cases} 1 - \dfrac{l' - l_q}{W_{q+1}}, & l_q \le l' \le l_{q+1} \\ 1 - \dfrac{l_q - l'}{W_q}, & l_{q-1} \le l' \le l_q \\ 0, & \text{otherwise} \end{cases} \tag{2.134}$$

as follows:

$$\frac{\partial E_z}{\partial n} = \sum_{q=1}^{N} T_q \frac{\partial E_z}{\partial n}(q) \tag{2.135}$$

$$H_z = \sum_{q=1}^{N} T_q H_z(q) \tag{2.136}$$

where $\partial E_z/\partial n(q)$ and $H_z(q)$ are the unknowns, being the values of $\partial E_z/\partial n$ and H_z,

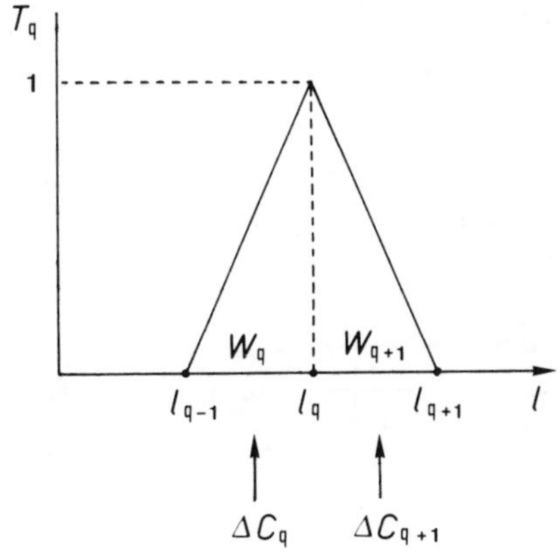

Figure 2.11 A triangle function.

respectively, at the node l_q. With the expansion of (2.135) and (2.136), $\partial E_z/\partial n$ and H_z are approximated by straight-line segments connected (we call this *line shape* hereafter for brevity) as illustrated in Figure 2.12. Inserting (2.135) and (2.136) into (2.123) and (2.125), and taking N points for observation points, we have a matrix equation in the form of (2.128).

If the observation points are placed on the nodes (the positions of l_1, l_2, ...), the elements a_{pq}, α_p, and b_p of $\boldsymbol{A}$, $\boldsymbol{X}$, and $\boldsymbol{B}$ are given as follows. TM wave case:

$$a_{pq} = -\frac{j}{4}\int_{\Delta C_q+\Delta C_{q+1}} H_0^{(2)}(k_2R_p)T_q dl' \tag{2.137}$$

$$\alpha_p = \frac{\partial E_z}{\partial n}(p),\ b_p = E_z^{\text{inc}}(p) = E_0\,\mathrm{e}^{-jk_2x_p} \tag{2.138}$$

TE wave case:

$$a_{pq} = \frac{1}{2}\,\delta_{pq} + \frac{j}{4}\int_{\Delta C_q+\Delta C_{q+1}} \frac{\partial H_0^{(2)}(k_2R_p)}{\partial n'}\,T_q dl' \tag{2.139}$$

$$\alpha_p = H_z(p),\ b_p = H_z^{\text{inc}}(p) = H_0\,\mathrm{e}^{-jk_2x_p} \tag{2.140}$$

where x_p is the x coordinate of the point l_p.

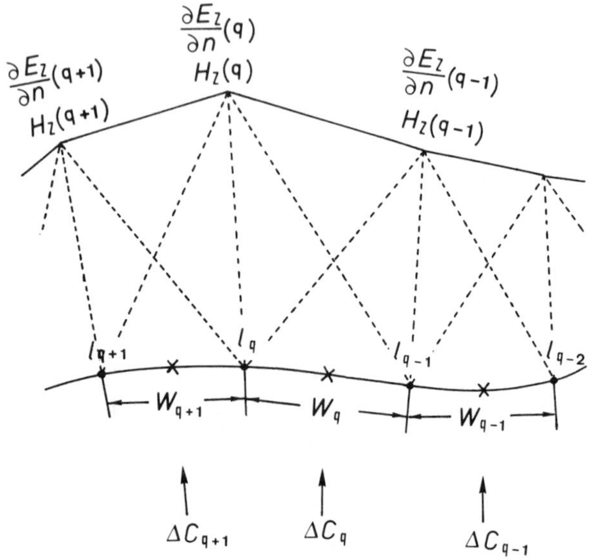

Figure 2.12 Approximation of $\partial E_z/\partial n$ and H_z by connected line segments.

It is also possible to place observation points at the midpoints (the points marked × in Figure 2.12) of sections ΔC_q. In this case, a_{pq} and b_p become as follows. TM wave case:

$$a_{pq} \rightarrow \text{the same as (2.137) in functional expression} \tag{2.141}$$

$$b_q = E_0\, \mathrm{e}^{-jk_2\bar{x}_p} \tag{2.142}$$

TE wave case:

$$a_{pq} = \frac{1}{4}(\delta_{pq} + \delta_{p,q+1}) + \frac{j}{4}\int_{\Delta C_q + \Delta C_{q+1}} \frac{\partial H_0^{(2)}(k_2 R_p)}{\partial n'} T_q dl' \tag{2.143}$$

$$b_p = H_0\, \mathrm{e}^{-jk_2\bar{x}_p} \tag{2.144}$$

where $\bar{x}_p$ in (2.142) and (2.144) is the x coordinate of the midpoint of section ΔC_p.

2.6.3 Numerical Calculation of Elements of Coefficient Matrices

The element a_{pq} of coefficient matrix A results in the form of integration over a finite range depending on the discretization method employed, as shown in the previous section. Thus, only the way to perform these integrals over a finite range remains. First, when written in detailed expression, the n'-derivative of $H_0^{(2)}(k_2R_p)$ becomes

$$\frac{\partial H_0^{(2)}(k_2 R_p)}{\partial n'} = -\,k_2 H_1^{(2)}(k_2 R_p)\frac{\partial R_p}{\partial n'} \tag{2.145}$$

$$\frac{\partial R_p}{\partial n'} = \cos\varphi_R' = \cos\varphi_n' \cos\varphi_x' + \sin\varphi_n' \sin\varphi_x' \tag{2.146}$$

where φ_R' and φ_n' are the angles that the unit vector $\boldsymbol{i}_n'$ makes with vector $\boldsymbol{R}_p$ and the positive x direction, respectively, and φ_x' is the angle that the vector $\boldsymbol{R}_p$ makes with the positive x direction (see Figure 2.13). With Cartesian coordinates of the observation and integration points being (x, y) and (x', y'), respectively, we have

$$\cos\varphi_x' = \frac{x' - x}{R_p}$$

$$\sin\varphi_x' = \frac{y' - y}{R_p} \tag{2.147}$$

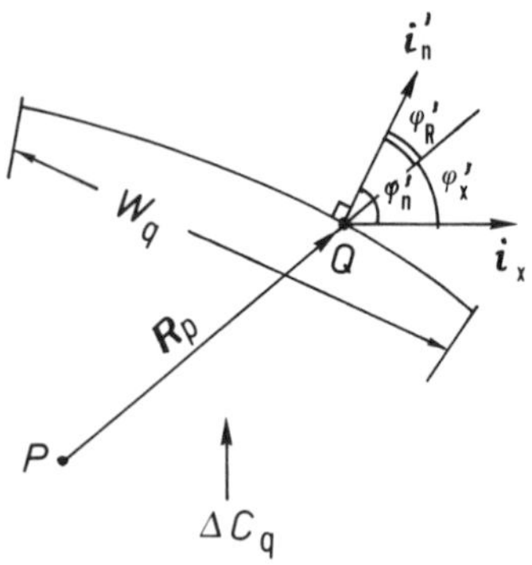

Figure 2.13 Section ΔC_q, integration point Q, and the relevant coordinates.

If the position of the observation point P is not included within the integration range, the integral of a_{pq} can be managed, for any a_{pq}, through simple numerical integration. For handling this simply, we may use a step-function approximation, trapezoidal rule, or a Simpson's rule. If more accurate calculation is required, we may use pertinent Gaussian quadrature formulas. For this integration, we often use a calculation technique in which the l'-integration over each section is transformed into the integration with respect to a new variable ξ defined for the region $-1 \leq \xi \leq 1$. Then, a certain interpolation function is frequently introduced. This technique can be said to be nothing but a coordinate transformation, however. Details of such numerical integration techniques can be found in standard texts on boundary element methods [6, 7].

On the other hand, if point P is included in the integration range, a different approach is required, because singularity of Hankel functions arising at P prevents the application of simple numerical integration. Thus, we must manage the integration with the help of an analytical treatment. For both case (1) and case (2), we will next describe methods of dealing with the integrals over the sections within which P is included.

(1) The Case of Step-Function Expansion and Point Matching

In this case, the specific integrals to be considered are those of (2.130) and (2.132), and the integration range is the section of $p = q(\Delta C_p)$. Let ΔC_{pp} be a smaller section of width w_p taken around point P within section ΔC_p, as represented in Figure 2.14. Let the width w_p be sufficiently small so that ΔC_{pp} can be regarded as a straight segment. Then, the integrands never become singular in the region ΔC_p–ΔC_{pp}. Hence, the integrals over this region can be carried out by means of suitable numerical methods; for example, a method using step function approximation. Thus, only the

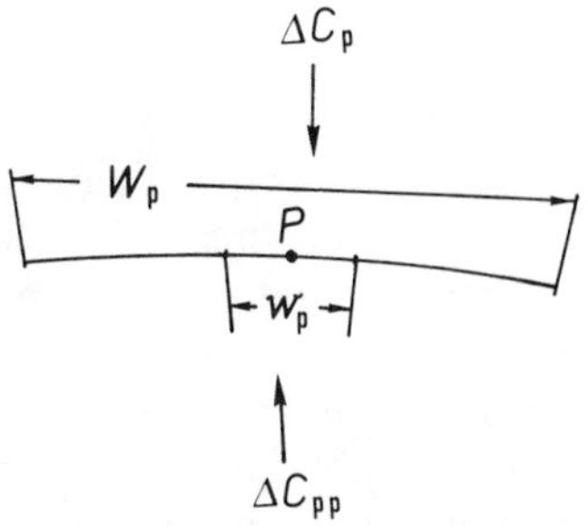

Figure 2.14 Smaller section ΔC_{pp} within section ΔC_p.

integrals over ΔC_{pp} are the object of analytical treatment. If the width W_p of ΔC_p itself is small enough or the boundary shape is approximated by a line shape, ΔC_{pp} can be regarded as ΔC_p itself. Under the assumption that ΔC_{pp} is a straight segment, we have $\cos\varphi_R' = 0$. Consequently, from (2.145) and (2.146), the integral of (2.132) over ΔC_{pp} results in

$$\frac{j}{4}\int_{\Delta C_{pp}} \frac{\partial H_0^{(2)}(k_2 R_p)}{\partial n'}\, dl' \approx 0 \tag{2.148}$$

whereas that of (2.130) over ΔC_{pp} can be obtained, using the approximation of the Hankel function for a small argument, for example,

$$H_0^{(2)}(z) \approx \frac{j2}{\pi} \ln \frac{2}{\gamma_0 z} + 1 \tag{2.149}$$
$$(\gamma_0 = e^{\gamma} = 1.781072\ldots,\ \gamma;\ \text{Euler's constant})$$

as follows:

$$\begin{aligned} -\frac{j}{4}\int_{\Delta C_{pp}} H_0^{(2)}(k_2 R_p)dl' &= -\frac{j}{2}\int_0^{w_p/2} H_0^{(2)}(k_2 l')dl' \\ &\approx -\frac{w_p}{2\pi}\left(\ln\frac{\gamma_0 k_2 w_p}{4} + j\frac{\pi}{2} - 1\right) \end{aligned} \tag{2.150}$$

If a more accurate integral is desired, any higher-order approximation of the Hankel function than (2.149) could be used. The preceding results are equally applicable when the boundary shape is approximated by a line shape.

(2) The Case of Triangle-Function Expansion and Point Matching

Consider first the case of P being taken on nodes. In this case, the integrals of (2.137) and (2.139) over the sections corresponding to $q = p$, $q = p + 1$, and $q = p - 1$ become the object of analytical treatment. Let us begin with the case $q = p$. We again set the smaller sections ΔC_{pp} and ΔC_{pp+1}, small enough to be regarded as straight, in the neighborhood of point P within ΔC_p and ΔC_{p+1}, respectively (see Figure 2.15). Then, we can perform numerical integration for the integrals over the regions ΔC_p–ΔC_{pp} and ΔC_{p+1}–ΔC_{pp+1}. If ΔC_p and ΔC_{p+1} are sufficiently small or their boundary shape is approximated by a line shape, we may regard ΔC_{pp} and ΔC_{pp+1} as ΔC_p and ΔC_{p+1}, respectively. As long as we can consider ΔC_{pp} and ΔC_{pp+1} to be straight, we have the approximate relation

$$\frac{j}{4}\int_{\Delta C_{pp}+\Delta C_{pp+1}} \frac{\partial H_0^{(2)}(k_2 R_p)}{\partial n'} T_p dl' \approx 0 \tag{2.151}$$

whereas, using the approximation (2.149) together with (2.134), we get

$$\begin{aligned} -\frac{j}{4}\int_{\Delta C_{pp}} H_0^{(2)}(k_2 R_p) T_p dl' &= -\frac{j}{4}\int_{l_p - w_p}^{l_p} H_0^{(2)}(k_2(l_p - l'))\left(1 - \frac{l_p - l'}{W_p}\right) dl' \\ &= -\frac{j}{4}\int_0^{w_p} H_0^{(2)}(k_2 l)\left(1 - \frac{l}{W_p}\right) dl \\ &\approx -\frac{w_p}{2\pi}\left\{\left(1 - \frac{w_p}{2W_p}\right)\left(\ln\frac{\gamma_0 k_2 w_p}{2} + j\frac{\pi}{2} - 1\right) - \frac{w_p}{4W_p}\right\} \end{aligned} \tag{2.152}$$

The same integral, with integration range ΔC_{pp} replaced by ΔC_{pp+1}, results in the same expression as the right-hand side of (2.152) except that p is replaced by $p + 1$.

Consider next the cases of $q = p + 1$ and $q = p - 1$. As regards the integrals having the n'-derivative of the Hankel function in the integrand, both integrals over ΔC_{pp} for $q = p - 1$ and that over ΔC_{pp+1} for the case of $q = p + 1$ can be approximated to be zero. On the other hand, the integral without the n'-derivative term is calculated, using (2.149) and (2.134) and with the aid of Figure 2.16, for the case of $q = p - 1$, as

$$\begin{aligned} -\frac{j}{4}\int_{\Delta C_{pp}} H_0^{(2)}(k_2 R_p) T_{p-1} dl' &= -\frac{j}{4}\int_0^{w_p} H_0^{(2)}(k_2 l)\frac{l}{W_p} dl \\ &\approx -\frac{w_p^2}{4\pi W_p}\left\{\ln\frac{\gamma_0 k_2 w_p}{2} + \frac{1}{2}(j\pi - 1)\right\} \end{aligned} \tag{2.153}$$

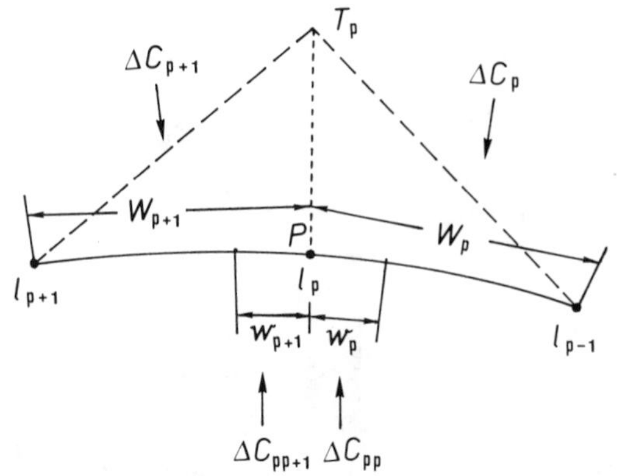

Figure 2.15 Smaller section ΔC_{pp} within ΔC_p, smaller section ΔC_{pp+1} within ΔC_{p+1}, and triangle function T_p, where P is taken on the node.

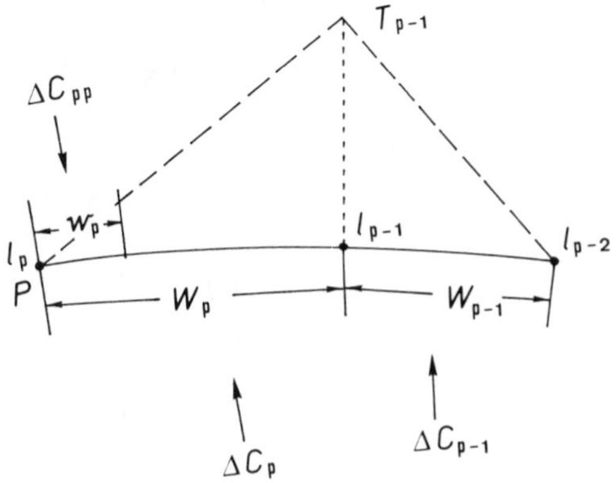

Figure 2.16 Smaller section ΔC_{pp} within ΔC_p and triangle function T_{p-1}, where P is taken on the node.

whereas, in the case of $q = p + 1$, the integral becomes that over ΔC_{pp+1}. This expression turns out to be the same as the right-hand side of (2.153) except for replacing p with $p + 1$. Even when the boundary shape is approximated by a line shape, the results of these integrals remain the same, except that the first term of the right-hand side of (2.139) is modified from $\delta_{pq}/2$ to $(\Omega/2\pi)\delta_{pq}$ (see note 5), where Ω is the angle on the scatterer side of the wedge shape.

Consider next the case of P taken at the midpoint of divided sections. In this case, the integrals of (2.141) and (2.143) are to be considered, and in the two cases $q = p - 1$ and $q = p$, point P is included within the integration range. In the case of $q = p - 1$, the integration range becomes $\Delta C_{p-1} + \Delta C_p$ (see Figure 2.17). Here, too, the integral over the part with ΔC_{pp} removed can be done numerically, and only the integrals over ΔC_{pp} must be dealt with analytically. We again get

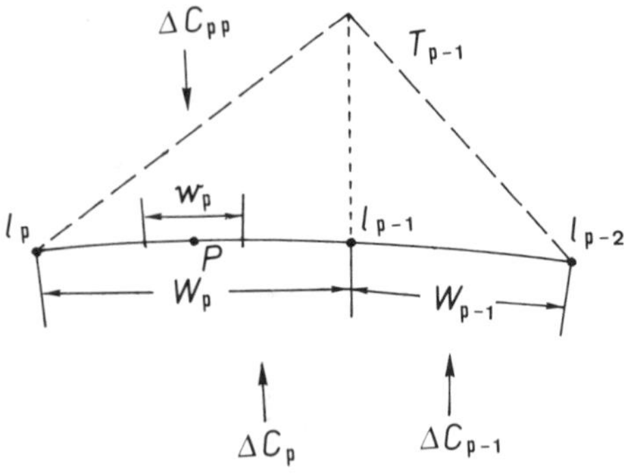

Figure 2.17 Smaller section ΔC_{pp} within ΔC_p and triangle function T_{p-1}, where P is taken at the midpoint of the section.

$$\frac{j}{4}\int_{\Delta C_{pp}} \frac{\partial H_0^{(2)}(k_2R_p)}{\partial n'} T_{p-1}dl' \approx 0 \tag{2.154}$$

and under the approximation of (2.149) we get

$$-\frac{j}{4}\int_{\Delta C_{pp}} H_0^{(2)}(k_2R_p)T_{p-1}dl' = -\frac{j}{4}\int_0^{w_p/2} H_0^{(2)}(k_2l)\left\{\left(\frac{1}{2}+\frac{l}{W_p}\right)\right.$$
$$\left. + \left(\frac{1}{2}-\frac{l}{W_p}\right)\right\}dl \approx -\frac{w_p}{4\pi}\left(\ln\frac{\gamma_0 k_2 w_p}{4} + j\frac{\pi}{2} - 1\right) \tag{2.155}$$

For the case of $q = p$ also, the results of both integrals remain just the same as (2.154) and (2.155). These results hold as well when the boundary shape is approximated by a line shape.

2.6.4 Some Precautions for Programming and Numerical Calculation

Input data include size (or frequency), the coordinates determining boundary shapes, the number of elements to be divided (or division width), the number of divisions for numerical integration over sections, *et cetera*. The criterion for the number of elements to be divided differs considerably depending on specific shapes and medium constants of the problems considered, the expected magnitudes and manners of variation of unknown functions, and so on. However, the division of at least three or four per wavelength is needed. In some cases, it will be necessary to increase the number of divisions in the regions where remarkable fluctuation of unknown functions is expected or where there are sharp changes of boundary shapes.

Major efforts in the calculation pertaining to the present method are devoted to calculating elements a_{pq} of coefficient matrices. In the case of two-dimensional problems, most of the computation time is used to calculate Hankel functions $H_0^{(2)}(\cdot)$ contained in a_{pq}. Therefore, when the computation time must be shortened, the number of calculations of Hankel functions should be decreased as much as possible. This means that we should try not to recalculate the values of Hankel functions when the distance between the observation and the integration points becomes equal to the distance for which Hankel functions are already calculated. For instance, the number of calculations of a_{pq} is practically halved if we utilize the fact that the values of the Hankel functions considered do not change by the interchange of the positions of observation and integration points. Moreover, if symmetry of the geometry modeled is taken into consideration, the amount of numerical calculation can be decreased further, often becoming less than a quarter of the amount.

For the case $\boldsymbol{B} \neq 0$, calculation of numerical solution $\boldsymbol{X}$ of a matrix equation $A\boldsymbol{X} = \boldsymbol{B}$ is generally straightforward and very easy, compared with the calculation of elements of coefficient matrices. We can leave it to a subroutine program for solving linear equations based on a solution such as Gaussian elimination. When the dimension of matrices is large, however, the computation time may increase surprisingly and the capacity of computer memory comes short. For this case, iterative methods such as the conjugate gradient method are very effective [11].

For the case $\boldsymbol{B} = 0$, that is, the case of eigenvalue problems, the problems reduce to obtaining the roots of $\det A = 0$, and the standard solution method for this will be to get the roots one by one via Muller's method, regula falsi, or the Newton-Raphson method, depending on the value of the roots estimated roughly, taking account of physical conditions of the problems. Occasionally, it may be more effective to search for minimal points of $|\det A|$ or $|\det A|^2$.

Once unknowns are obtained, the next step is to evaluate various physical quantities using integral representations. This calculation usually is very similar to the calculation of a_{pq}, so that the steps in calculating a_{pq} often can be utilized. When the field values within the regions are evaluated by means of integral representations, the accuracy can degrade extremely if the observation point is very close to the boundary. This is caused by the inaccuracy of the integration over a part on the boundary near the observation point. In this case, we must improve accuracy of the integration over this part [22].

For examining whether the values obtained are valid, we usually attempt to execute some of the following procedures: (1) solve simplified problems for which rigorous solutions can be obtained; (2) compare the solutions with those of the past literature; (3) numerically check on the degree of satisfaction of physical laws, such as energy theorem, optical theorem, or reciprocity theorem; (4) examine whether the solutions converge pertinently with the increase in the division number or the decrease in the width of divided elements, and so on. When comparing the results with those of the past literature, we should keep in mind that some results of the literature may have lacked accuracy because of the involvement of resonant solutions.

NOTES

1. In some books, this formula is called *Stratton's theorem*.
2. The triad (n, l, z) forms a local "rectangular coordinate system" with the origin at a point on C.
3. This principle applies equally for the case of the Green's function that satisfies boundary conditions on other boundaries.
4. When P approaches the vertex of a wedge the angle of which is Ω_0 on the side of $\bar{V}$, as illustrated in Figure 2.6, the first term $\mp 1/2$ on the right-hand side of (2.67) becomes $-\Omega_0/2\pi$ and $(2\pi - \Omega_0)/2\pi$, respectively. In the three-dimensional case, that is, the case of the point O being at the apex of a conical shape the solid angle of which is Ω_0, those terms become $-\Omega_0/4\pi$ and $(4\pi - \Omega_0)/4\pi$, respectively [11].
5. When P is at the tip of a wedge as in Figure 2.6, the term $(1/2)\phi_1$ on the right-hand side of (2.69) becomes $[1 - (\Omega_0/2\pi)]\phi_1$.
6. When the observation point approaches a tip of a pointed surface, the jumping-out term should be modified as mentioned in note 5. In this case, $(1/2)\,\boldsymbol{E}_1$ on the right-hand side of (2.79) should become $(\Omega/4\pi)\boldsymbol{E}_1$, where Ω is the solid angle on the side of region V. In other words, the coefficient of $\boldsymbol{E}_1$ for the case of $P \in S$ becomes equal to the solid angle on the side of region V divided by the whole solid angle. This also applies to (2.69). Special notes like this for the case of the pointed surface will be omitted hereafter.
7. To be exact, the rigorous steps should be as follows: first, use (2.98) in the stage of integral representations; and then, reconstruct the integral equations using the resulting representations. However, the same results can be obtained by following the steps, as mentioned here, such that (2.98) is applied formally in the integral equations directly and then integration by parts is made use of formally, although this expedient method is not exact mathematically.

REFERENCES

[1] Reitan, D.K., and T.J. Higgins, "Calculations of the Electrical Capacitance of a Cube," *J. Appl. Phys.*, Vol. 22, February 1951, pp. 223–226.

[2] Edwards, T.W., and J. Van Bladel, "Electrostatic Dipole Moment of a Dielectric Cube," *Appl. Sci. Res.*, Sec. B, Vol. 9, 1961–1962, pp. 151–155.

[3] Mei, K.K., and J. Van Bladel, "Low Frequency Scattering by Rectangular Cylinders," *IEEE Trans. Antennas and Propagation*, Vol. AP-11, Jan. 1963, pp. 52–56.

[4] Mei, K.K., and J. Van Bladel, "Scattering by Perfectly-Conducting Rectangular Cylinders," *IEEE Trans. Antennas and Propagation*, Vol. AP-11, March 1963, pp. 185–192.

[5] Harrington, R.F., *Field Computation by Moment Methods*, Macmillan, New York, 1968.

[6] Brebbia, C.A., *The Boundary Element Method for Engineers*, Pentech Press, London, 1978.

[7] For example, C.A. Brebbia, and S. Walker, *Boundary Element Techniques in Engineering*, Newnes-Butterworths, London, 1979; P.K. Banerjee, and R. Butterfield, *Boundary Element Methods in Engineering Science*, McGraw-Hill, London, 1981.

[8] Washisu, S., and I. Fukai, "An Analysis of Electromagnetic Unbounded Field Problems by Boundary Element Method," *Trans. Inst. Electron. and Commun. Eng. Japan,* Sec. E, Vol. E64, Dec. 1981, pp. 819–828.

[9] For example, Y. Kagawa, M. Koshiba, I. Ikeuchi, and S. Kagami, *Finite-Boundary Element Method in Electrical and Electronic Engineering—Applications to Wave Problems.* Ohmsha, Tokyo, 1984 [in Japanese]; Technical Society of Boundary Element Method, ed., *Theory and Application of Boundary Element Method,* Corona, Tokyo, 1986 [in Japanese].

10] Thiele, G.A., "Wire Antennas," in *Computer Techniques for Electromagnetics,* R. Mittra, ed. Chapter 2, Pergamon, Oxford, 1973.

11] Kumagai, N., and N. Morita, *Electromagnetic Waves and Boundary Element Methods,* Morikita Shuppan, Tokyo, 1987 [in Japanese].

12] Stratton, J.A., *Electromagnetic Theory,* McGraw-Hill, New York, 1941.

13] Iijima, T., ed., *Modern Analytical Methods in Electromagnetics,* Inst. Electron. Inf. and Commun. Eng. of Japan, Tokyo, 1979, Chapter 1, [in Japanese].

14] Morita, N., "Surface Integral Representations for Electromagnetic Scattering from Dielectric Cylinders," *IEEE Trans. Antennas and Propagation,* Vol. AP-26, March 1978, pp. 261–266.

15] Poggio, A.J., and E.K. Miller, "Integral Equation Solutions of three-Dimensional Scattering," in *Computer Techniques for Electromagnetics,* R. Mittra, ed., Pergamon, Oxford, 1973, Chapter 4.

16] Bolomey, J.-C., and W. Tabbara, "Numerical Aspects on Coupling between Complementary Boundary Value Problems," *IEEE Trans. Antennas and Propagation,* Vol. AP-21, May 1973, pp. 356–363.

17] Matsuhara, M., "Boundary Element Analysis of Polarization Holding Fibers," *Trans. Inst. Electron. and Commun. Eng. Japan,* Vol. J67-B, Sept. 1984, pp. 968–973, [in Japanese]; H. Sano and S. Kurazono, "Analysis of Dielectric Waveguide Problem by Boundary Element Method," *Trans. Inst. Electron. and Commun. Eng. Japan,* Vol. J68-B, Dec. 1985, pp. 1419–1427, [in Japanese].

18] Morita, N., "Resonant Solutions Involved in the Integral Equation Approach to Scattering from Conducting and Dielectric Cylinders," *IEEE Trans. Antennas and Propagation,* Vol. AP-27, December 1985, pp. 869–871.

19] Mautz, J.R., and R.F. Harrington, "H-Field, E-Field and Combined-Field Solutions for Conducting Bodies of Revolution," *Arch. Elektron Übertragungstech.,* Vol. 32, 1978, pp. 157–164.

20] Müller, C., *Foundations of the Mathematical Theory of Electromagnetic Waves,* Springer, Berlin, 1969.

21] For example, R. Mittra, ed., *Numerical and Asymptotic Techniques in Electromagnetics,* Springer, Berlin, 1975; E.K. Miller and A.J. Poggio, "Moment Method Techniques in Electromagnetics from an Applications Viewpoint," in *Electromagnetic Scattering,* ed. P.L.E. Uslenghi, Academic Press, New York, 1978; E.K. Miller, "A Selective Survey of Computational Electromagnetics," *IEEE Trans. Antennas and Propagation,* Vol. AP-36, Sept. 1988, pp. 1281–1305.

22] Enokizono, M., and T. Todaka, "Boundary Element Analysis Using Revised Integration," *Trans. Inst. Elect. Eng. Japan,* Vol. 106-A, April 1986, pp. 149–156, [in Japanese]; T. Takeda, "On the Nature of Double Node in Boundary Element Method," *Trans. Inst. Elect. Eng. Japan,* Vol. 104-A, June 1984, pp. 291–298, [in Japanese].

Chapter 3
The Point-Matching Method

Eikichi Yamashita and Kazuhio Atsuki

3.1 HISTORICAL BACKGROUND

The name *point-matching method* has been used frequently in the literature, but it also may be called the *collocation method* [1] or the *moment method* [2]. In this method, electromagnetic fields in a region to be analyzed are expanded with a linear combination of appropriate basis functions to satisfy given conditions at only a finite number of points in the region. If appropriate basis functions to approximate electromagnetic fields in the region are found, this method could be applied to electromagnetic field problems of arbitrary boundary geometry.

The early studies of electromagnetic wave problems as the application of the point-matching method are the analysis of the cutoff frequency characteristics of waveguides with arbitrary cross sections by Yee and Auden [3] and two-dimensional scattering problems by infinite cylinders with arbitrary geometrical cross sections by Mullin, Sandburg, and Velline [4].

Though the convergence property of electromagnetic fields with regard to the number of particular basis functions has not been rigorously inspected in these studies, it has been shown that approximate solutions with satisfactory numerical convergence could be attained at least from the engineering point of view. Bates theoretically investigated the treatment of basis functions for perfectly conducting boundary and proved that the cutoff characteristics could be estimated with the point-matching method [5]. He also has shown that the characteristic impedance of the TEM waves along two parallel conductor lines could be calculated with the point-matching method.

Bates pointed out further that the point-matching method could be applicable to the scattering problems for structures with slightly modified conductor cross sections from the circular shape, and that there would be some items to be considered when applying the point-matching method to general structures [6].

A new concept of optical waveguides has been generally accepted since semiconductor lasers and low-loss optical fibers have undergone significant development. Analysis methods for electromagnetic fields of such optical waveguides or dielectric waveguides have been actively studied with this development. During these years, electronic computers have also shown significant progress in computation time and memory capacity, and various numerical methods have been proposed to use these features. The complexity of boundary conditions for dielectric waveguides due to their open structure has been overcome with numerical methods that strongly depend on electronic computers.

Goell showed that the point-matching method could be applied to the analysis of wave propagation characteristics along square dielectric waveguides [7]. Some theoretical problems have also been pointed out by James and Gallet [8] and Bates *et al.* [9] for electromagnetic field analysis based on the point-matching method.

Dielectric waveguides of various types have been proposed with the progress of optical fibers, optical waveguide circuits, and low-loss dielectric waveguides in the millimeter-wave region. The point-matching method also has been found to be effective in the analysis of transmission properties of these waveguides, as discussed in the literature [10–17].

3.2 CHARACTERISTICS OF THE METHOD AND RANGE OF APPLICATION

The point-matching method has thus been applied to the analysis of conductor transmission lines, scattering bodies, and dielectric waveguides. For simplifying explanations to learners, however, we will limit our discussion of the point-matching method to the range of propagation characteristics of dielectric waveguides because most recent works using the point-matching method have been on these topics and these applications are relatively easy to understand.

The main features of the point-matching method are

1. It can be conveniently applied to numerical analyses using the electronic computers available in laboratories.
2. It can be applied to dielectric waveguides having arbitrary cross sections. Because the point-matching method uses electromagnetic fields only at a small number of points on the cross-sectional boundary lines in algebraic equations, numerical analyses are possible with short computation time on a computer with a small memory capacity. Though other numerical methods such as the finite element method can treat these problems, they require electromagnetic fields on many selected points in the cross-sectional area; namely, a large memory capacity. The point-matching method can be applied not only to single-dielectric waveguides with an arbitrary cross section [10] but also to composite dielectric waveguides having multiple dielectric materials [11–14] and coupled dielectric waveguides composed of multiple waveguides [15].

3. It is difficult, however, to apply the point-matching method as described in this chapter to structures with the three-dimensional boundary surface over a large area, such as three-dimensional optical waveguides made on the surface of a large substrate, or to structures having index distributions within a waveguide, such as graded index optical fibers. Good computation accuracy could not be expected for flat structures or waveguides with sharp corners in the cross section, unless appropriate basis functions can be found to express electromagnetic fields in the waveguide.

.3 HOMOGENEOUS DIELECTRIC WAVEGUIDES HAVING 'HE CROSS SECTION OF ARBITRARY BOUNDARY

.3.1 Basic Equations

,et us first consider a homogeneous dielectric waveguide having a cross section of rbitrary boundary geometry as shown in Figure 3.1 to explain fundamental ideas f electromagnetic field analysis with the point-matching method. The waveguide is ssumed to have a uniform cross section along the direction of propagation, the -axis. The interior region denoted with 1 and the exterior region denoted with e are nade of homogeneous, isotropic, and lossless materials whose dielectric constants re ε_1 and ε_e, respectively. When $\varepsilon_1 > \varepsilon_e$, electromagnetic energy is confined near nd in the interior region and some eigenmode waves can propagate. Then, the propgation constant, β, is defined in the range

$$k_1 \geqslant \beta \geqslant k_e \tag{3.1}$$

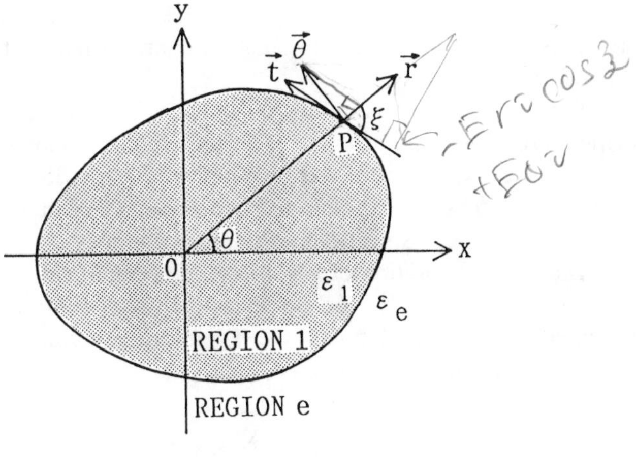

'igure 3.1 Dielectric waveguide with the cross section of an arbitrary boundary.

where these wavenumbers are defined as

$$k_1 = \omega(\varepsilon_1\mu_0)^{1/2} \tag{3.2a}$$

$$k_e = \omega(\varepsilon_e\mu_0)^{1/2} \tag{3.2b}$$

for the interior and the exterior region, respectively. The term ω is the angular fre quency, and μ_0 is permeability in a vacuum.

The longitudinal components of electromagnetic fields of the propagation mod in each region, E_{zi} and H_{zi} ($i = 1, e$), satisfy the following wave equation for th circular coordinate system (r, θ)

$$\left[\frac{\partial^2}{\partial r^2} + \frac{1}{r}\frac{\partial}{\partial r} + \frac{1}{r^2}\frac{\partial^2}{\partial \theta^2} + (k_i^2 - \beta^2)\right]\begin{Bmatrix} E_{zi} \\ H_{zi} \end{Bmatrix} = 0 \tag{3.3}$$

Therefore, our problem is an eigenvalue problem, to seek the propagation constan β satisfying equation (3.3) under the continuity conditions of electromagnetic tan gential fields at the boundary surface between the interior and exterior region [7, 10].

3.3.2 Electromagnetic Fields of Dielectric Waveguides

Electromagnetic fields either inside or outside of dielectric waveguides can be es timated by applying the method of the separation of variables in the case of the cros section of a circular boundary. These fields are written in the form of the Bessel' functions multiplied by sinusoidal functions. The eigenvalue problem given by (3.3 is formulated and solved by connecting these electromagnetic fields as the boundar conditions. As a result, the propagation characteristics of these waveguides can b estimated exactly.

The separation of variables no longer can be applied to dielectric waveguide having a cross section deformed from the circular boundary as shown in Figure 3.1 However, we expect the electromagnetic field distribution of the dielectric wave guides with a cross section close to circular boundary to be similar to that of dielectri waveguides with cross sections of circular boundary.

The general solution of (3.3) is given by the series of circular cylindrical har monic functions; namely, the Bessel's functions or the modified Bessel's function multiplied by sinusoidal functions. Therefore, longitudinal components of electro magnetic fields can be written as

$$E_{z1} = \sum_{n=0}^{\infty} a_n J_n(hr) \sin(n\theta + \phi_n) \tag{3.4a}$$

$$H_{z1} = \sum_{n=0}^{\infty} b_n J_n(hr) \cos(n\theta + \psi_n) \tag{3.4b}$$

for the inside region 1, and

$$E_{ze} = \sum_{n=0}^{\infty} c_n K_n(pr) \sin(n\theta + \phi_n) \tag{3.4c}$$

$$H_{ze} = \sum_{n=0}^{\infty} d_n K_n(pr) \cos(n\theta + \psi_n) \tag{3.4d}$$

for the outside region e, where the phase factor $\exp[j(\omega t - \beta z)]$ has been dropped. The terms h and p are the parameters for the transverse direction defined by

$$h = (k_1^2 - \beta^2)^{1/2} \tag{3.5a}$$

$$p = (\beta^2 - k_e^2)^{1/2} \tag{3.5b}$$

$J_n(hr)$ and $K_n(pr)$ are the Bessel's function of the first kind of the nth order and the modified Bessel's function of the second kind of the nth order, respectively. The former function is finite but the latter is singular at the origin. The terms a_n, b_n, c_n, and d_n are undetermined constants; ϕ_n and ψ_n are undetermined phase angles. The constants are determined later according to the boundary conditions.

The longitudinal components of the electromagnetic fields combined with Maxwell's equations in the circular cylindrical coordinate system lead to the transverse components of the electromagnetic fields as

$$E_{ri} = \frac{-j\beta}{k_i^2 - \beta^2}\left[\frac{\partial E_{zi}}{\partial r} + \frac{\omega\mu_0}{\beta r}\frac{\partial H_{zi}}{\partial \theta}\right] \tag{3.6a}$$

$$E_{\theta i} = \frac{-j\beta}{k_i^2 - \beta^2}\left[\frac{1}{r}\frac{\partial E_{zi}}{\partial \theta} - \frac{\omega\mu_0}{\beta}\frac{\partial H_{zi}}{\partial r}\right] \tag{3.6b}$$

$$H_{ri} = \frac{-j\beta}{k_i^2 - \beta^2}\left[-\frac{\omega\varepsilon_i}{\beta r}\frac{\partial E_{zi}}{\partial \theta} + \frac{\partial H_{zi}}{\partial r}\right] \tag{3.6c}$$

$$H_{\theta i} = \frac{-j\beta}{k_i^2 - \beta^2}\left[\frac{\omega\varepsilon_i}{\beta}\frac{\partial E_{zi}}{\partial r} + \frac{1}{r}\frac{\partial H_{zi}}{\partial \theta}\right] \tag{3.6d}$$

Boundary conditions are the continuity equations of the longitudinal components, E_{zi} and H_{zi}, in addition to the transverse components tangential to the boundary surface, E_{ti} and H_{ti}, of the electromagnetic fields in the inside region 1 and the outside region e as

$$E_{z1} = E_e \quad H_{z1} = H_e \tag{3.7a}$$

$$E_{t1} = E_e \quad H_{t1} = H_e \tag{3.7b}$$

For example, the tangential components of the electromagnetic fields at point P on the boundary in the transverse plane in Figure 3.1 are given by

$$E_{ti} = -E_{ri} \cos\xi + E_{\theta i} \sin\xi \tag{3.8a}$$

$$H_{ti} = -H_{ri} \cos\xi + H_{\theta i} \sin\xi \tag{3.8b}$$

where ξ denotes the angle made by the radial direction vector and the tangential direction vector in the transverse plane.

3.3.3 Symmetry in the Waveguide Cross Section

The symmetry in the waveguide cross section is precious information to reduce computation time and memory requirement in computer calculations. When a waveguide cross section is symmetric with regard to the x-axis, electromagnetic fields in it also should be symmetric or antisymmetric with regard to the x-axis. This fact, associated with (3.4) and (3.6), enables us to expect two possible types of propagation mode. One is the symmetrical mode in which the longitudinal component of the electric field, E_z, is symmetric with regard to the x-axis; that is, the x-axis is regarded as a magnetic wall. The phase angles in (3.4) ard $\phi_n = \psi_n = \pi/2$ in this case.

The other is the antisymmetric mode, in which the x-axis is an electric wall and the phase angles are $\phi_n = \psi_n = 0$. Consequently, the symmetry of electromagnetic fields are expressed in terms of the phase angles as

$$\begin{aligned} \phi_n = \psi_n = \pi/2 \quad &\text{(a magnetic wall on the } x\text{-axis)} \\ \phi_n = \psi_n = 0 \quad &\text{(an electric wall on the } x\text{-axis)} \end{aligned} \tag{3.9}$$

for waveguides having the cross-sectional symmetry with regard to the x-axis. When a waveguide has cross-sectional symmetry with regard to both the x-axis and the y-axis, electromagnetic fields also must be symmetric or antisymmetric with regard to the y-axis as well as the x-axis.

To handle the y-axis symmetry for the mode having an x-axis as an electric wall, the two conditions, $\phi_n = 0$ and $\theta = \alpha + \pi/2$, are substituted in (3.4a) as

$$E_{z1} = \sum_{n=0}^{\infty} a_n J_n(hr) \left[\sin(n\alpha) \cos\left(\frac{n\pi}{2}\right) + \cos(n\alpha) \sin\left(\frac{n\pi}{2}\right) \right] \tag{3.10}$$

We find from (3.10) that the mode having the symmetric distribution of E_z with regard to the y-axis or the y-axis as a magnetic wall should have an odd number of

n because the symmetry condition for α with regard to the y-axis is given by $\cos(n\pi/2) = 0$. Then the electromagnetic fields are expressed by series with odd numbers of n. On the other hand, the condition $\sin(n\pi/2) = 0$ is given for the antisymmetric mode or the mode where the y-axis is an electric wall. Hence, n is an even number, and the electromagnetic fields are expressed by series with even numbers of n.

When a mode has the x-axis as a magnetic wall, the mode is symmetric with regard to the y-axis for even numbers of n and antisymmetric for odd numbers of n. That is, the symmetry of electromagnetic fields for the waveguide having a symmetrical cross section with regard to the y-axis can be expressed by series with even numbers or odd numbers of n.

After all, electromagnetic fields in only a half region of the waveguide cross section must be considered when the structure is symmetric with regard to the x-axis. When the structure is symmetric with regard to the x-axis and y-axis, electromagnetic fields in only a quarter region of the total waveguide cross section have to be considered. Moreover, the number of terms in the expansion series is reduced to half of the original terms because only even or odd numbers of n have to be taken.

The use of symmetry in the point-matching method is thus effective in reducing the required computation time and memory capacity of a computer, and yet it improves accuracy. It also helps us classify propagation modes systematically.

3.3.4 Application of the Point-Matching Method

When the boundary conditions given by equation (3.7) are imposed on the electromagnetic field components of the dielectric waveguide given by (3.4) and (3.6), a set of linear equations concerning the unknown constants, a_n, b_n, c_n, and d_n ($n = 0, 1, 2, \ldots$) is attained. To determine all these constants, we have to solve the simultaneous linear equations with an infinite number of unknown variables subject to boundary conditions. Obviously, this is impossible.

In the point-matching method, we truncate all the infinite expansion series with finite numbers of terms and approximate electromagnetic fields with finite numbers of basis functions. As the next step, we select a finite number of representative points on given boundary lines and impose the boundary conditions to the electromagnetic fields at these points. As a result, we obtain a finite number of linear simultaneous equations by which we can solve given electromagnetic field problems approximately. We call these representative points *matching points*. Carefully consider that the basis functions are selected to accurately approximate electromagnetic fields in a waveguide with the smallest number of terms and that a finite number of the matching points needed to represent electromagnetic fields on boundaries are effectively selected. When waveguides have slightly deformed cross sections from a circular boundary, it is convenient to use basis functions that are the solutions for waveguides having a circular cross section based on the method of the separation of variables as shown in (3.4).

On the other hand, we can expect that the accuracy of computation will be improved by taking many matching points on the boundary where electromagnetic fields change rapidly. The distribution of the matching points must be formed for each propagation mode because the electromagnetic field distribution depends on the kind of mode. The electromagnetic field distribution in waveguides having arbitrary cross sections can be estimated approximately, for example, from the field distribution in waveguides having a circular cross section. However, this approach to matching-point selection is not recommended, as it is not convenient and needs additional computation time. Fortunately, the electromagnetic fields of the lower-order mode propagating in dielectric waveguides generally are distributed all over the cross section so that a uniform distribution of matching points turns out to be effective. Though the uniform distribution of the matching points appears to be too simplified, the convergence of numerical solutions seems satisfactory in our experience.

3.3.5 Boundary Condition Matrices

When the longitudinal components of the electromagnetic fields given by the series in (3.4) are expanded with the first N terms of the basis functions, $4N$ coefficients, a_n, b_n, c_n, and d_n ($n = 0, 1, 2, \ldots, N - 1$), remain unknown.

On the other hand, the boundary conditions in (3.7) yield four equations per matching point. Consequently, M matching points, $P_m(r_m, \theta_m)$ ($m = 1, 2, \ldots, M$), on the boundary require $4M$ boundary conditions. When the number of the basis functions, N, is equal to the number of the matching points, M, $4N$ linear simultaneous equations are obtained to determine the unknown coefficients.

When the form of matrices is used to express these simultaneous linear equations in order to easily understand and simplify this analysis method, the following expressions are given:

$$\boldsymbol{E}^{LA}\boldsymbol{A} = \boldsymbol{E}^{LC}\boldsymbol{C} \tag{3.11a}$$

for the continuation condition of the longitudinal electric fields,

$$\boldsymbol{H}^{LB}\boldsymbol{B} = \boldsymbol{H}^{LD}\boldsymbol{D} \tag{3.11b}$$

for that of the longitudinal magnetic fields,

$$\boldsymbol{E}^{TA}\boldsymbol{A} + \boldsymbol{E}^{TB}\boldsymbol{B} = \boldsymbol{E}^{TC}\boldsymbol{C} + \boldsymbol{E}^{TD}\boldsymbol{D} \tag{3.11c}$$

for the continuation of tangential electric fields on the boundary in the cross section, and

$$\boldsymbol{H}^{TA}\boldsymbol{A} + \boldsymbol{H}^{TB}\boldsymbol{B} = \boldsymbol{H}^{TC}\boldsymbol{C} + \boldsymbol{H}^{TD}\boldsymbol{D} \tag{3.11d}$$

for that of tangential magnetic fields on the boundary in the cross-section, where the matrices, $\boldsymbol{A}$, $\boldsymbol{B}$, $\boldsymbol{C}$, and $\boldsymbol{D}$, are column vectors including the elements, a_n, b_n, c_n, and d_n, respectively.

The matrices $\boldsymbol{E}^{LA}$, $\boldsymbol{E}^{LC}$, $\boldsymbol{H}^{LB}$, $\boldsymbol{H}^{LD}$, ... are the $(M \times N)$ matrices consisting of the elements e_{mn}^{LA}, e_{mn}^{LC}, h_{mn}^{LB}, h_{mn}^{LC}, ... given as follows, respectively:

$$e_{mn}^{LA} = J_n(hr_m)S \tag{3.12a}$$

$$e_{mn}^{LC} = K_n(pr_m)S \tag{3.12b}$$

$$h_{mn}^{LB} = J_n(hr_m)C \tag{3.12c}$$

$$h_{mn}^{LD} = K_n(pr_m)C \tag{3.12d}$$

$$e_{mn}^{TA} = -\beta(J'S\cos\xi_m - JC\sin\xi_m) \tag{3.12e}$$

$$e_{mn}^{TB} = \omega\mu_0(JS\cos\xi_m - J'C\sin\xi_m) \tag{3.12f}$$

$$e_{mn}^{TC} = \beta(K'S\cos\xi_m - KC\sin\xi_m) \tag{3.12g}$$

$$e_{mn}^{TD} = -\omega\mu_0(KS\cos\xi_m - K'C\sin\xi_m) \tag{3.12h}$$

$$h_{mn}^{TA} = \omega\varepsilon_1(JC\cos\xi_m + J'S\sin\xi_m) \tag{3.12i}$$

$$h_{mn}^{TB} = -\beta(J'C\cos\xi_m + JS\sin\xi_m) \tag{3.12j}$$

$$h_{mn}^{TC} = -\omega\varepsilon_e(KC\cos\xi_m + K'S\sin\xi_m) \tag{3.12k}$$

$$h_{mn}^{TD} = \beta(K'C\cos\xi_m + KS\sin\xi_m) \tag{3.12l}$$

where

$$J = \frac{n}{h^2 r_m} J_n(hr_m) \qquad J' = \frac{1}{h} J_n'(hr_m) \tag{3.13a}$$

$$K = \frac{n}{p^2 r_m} K_n(pr_m) \qquad K' = \frac{1}{p} K_n'(pr_m) \tag{3.13b}$$

$$S = \sin(n\theta_m + \phi_n) \quad C = \cos(n\theta_m + \psi_n) \tag{3.13c}$$

Equation (3.11) can be rewritten in the form of matrices as follows:

$$[\boldsymbol{Q}][\boldsymbol{T}] = [0] \tag{3.14}$$

$$[\boldsymbol{Q}] = \begin{bmatrix} \boldsymbol{E}^{LA} & 0 & -\boldsymbol{E}^{LC} & 0 \\ 0 & \boldsymbol{H}^{LB} & 0 & -\boldsymbol{H}^{LD} \\ \boldsymbol{E}^{TA} & \boldsymbol{E}^{TB} & -\boldsymbol{E}^{TC} & -\boldsymbol{E}^{TD} \\ \boldsymbol{H}^{TA} & \boldsymbol{H}^{TB} & -\boldsymbol{H}^{TC} & -\boldsymbol{H}^{TD} \end{bmatrix}, \quad [\boldsymbol{T}] = \begin{bmatrix} \boldsymbol{A} \\ \boldsymbol{B} \\ \boldsymbol{C} \\ \boldsymbol{D} \end{bmatrix} \tag{3.15}$$

The condition for the preceding homogeneous linear simultaneous equations to have nontrivial solutions is that the determinant of the matrix $\boldsymbol{Q}$ vanishes as

$$\det[\boldsymbol{Q}] = 0 \tag{3.16}$$

This equation is the eigenvalue equation for dielectric waveguides. The eigenvalues in this case are the propagation constants, β, which can be estimated by specifying ω and waveguide dimensions as parameters and is used to know the propagation characteristics of the dielectric waveguides.

3.3.6 Designation of Propagation Modes

The TE and TM modes, which have axially circularly symmetrical components of electromagnetic fields, can propagate along dielectric waveguides having circular cross sections in addition to the hybrid modes, the HE mode and EH mode, which have longitudinal components of both electric and magnetic fields. However, only hybrid modes can propagate in some cases of dielectric waveguides having arbitrary cross sections. Because the electromagnetic fields in these cases are affected by the dielectric constant, frequency, and cross-sectional shape and, therefore, there is no standard designation method for propagation modes, we designate these modes according to Goell's convention [7].

When the electric field vector of a hybrid mode is in the direction of the x-axis at the high frequency limit and the homogeneous medium limit where the dielectric constant in the inside region tends to that in the outside region, this mode is designated the E^x_{pq} *mode*. On the contrary, when the electric field vector is in the y-axis it is designated as the E^y_{pq} *mode*. The suffixes, p and q, physically express the variation numbers of the electromagnetic fields in the x- and y-axis. For example, the fundamental HE_{11} mode, whose electric fields are polarized in the x-axis or the y-axis, can propagate along the dielectric waveguides having circular cross sections. This mode is designated the E^x_{11} mode or the E^y_{11} mode according to the convention. When dielectric waveguides have two axes of symmetry in the cross section, propagation modes are designated as shown in Table 3.1 and discussed in Section 3.3.3.

Table 3.1
Mode Designation for the Structures with Two Symmetric Axes

x-AXIS	y-AXIS	PHASE ANGLE	EXPANSION TERMS	E^x_{pq} MODE		E^y_{pq} MODE	
		$\phi_n = \phi_n$	n	p	q	p	q
MAGNETIC WALL	MAGNETIC WALL	$\pi/2$	EVEN	EVEN	ODD	ODD	EVEN
MAGNETIC WALL	ELECTRIC WALL	$\pi/2$	ODD	ODD	ODD	EVEN	EVEN
ELECTRIC WALL	MAGNETIC WALL	0	ODD	EVEN	EVEN	ODD	ODD
ELECTRIC WALL	ELECTRIC WALL	0	EVEN	ODD	EVEN	EVEN	ODD

3.3.7 Numerical Analysis of Dielectric Waveguides Having the Cross Section of a Chipped Circle Boundary

A concrete example of the numerical analysis for the case of a dielectric waveguide is described in this section where the cross section has two symmetrical axes and of a chipped circle with the dimensions as indicated in Figure 3.2 [10].

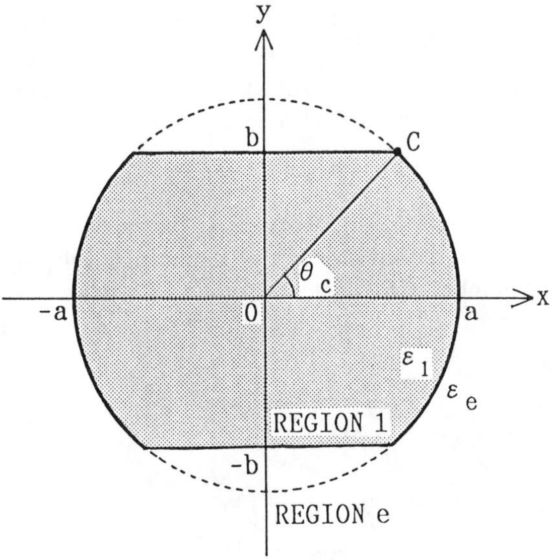

Figure 3.2 Dielectric waveguide with the cross section of a chipped circle boundary.

(1) Selection of Matching Points

Because this structure has two symmetrical axes, we have to consider only the first quadrant. Hence, the x-axis and the y-axis are treated as an electric wall or magnetic wall, as shown in Table 3.1. The electromagnetic fields of propagation modes are expanded in a series of even number or odd number terms. M matching points, $P_m(r_m, \theta_m)$ $(m = 1, 2, \ldots, M)$, are selected on the boundary in the first quadrant to make equiangles, θ_m, defined by

$$\theta_m = \left(m - \frac{1}{2}\right)\frac{\pi}{2M} \quad (m = 1, 2, \ldots, M) \tag{3.17}$$

The term r_m and the angle of tangential lines, ξ_m, are given as

$$r_m = a \quad \xi_m = \frac{\pi}{2} \tag{3.18a}$$

for $\theta_c > \theta_m > 0$ and

$$r_m = \frac{b}{\sin\theta_m} \quad \xi_m = \theta_m \tag{3.18b}$$

for $\pi/2 > \theta_m > \theta_c$ where θ_c is the coordinate of the sharp corner of the chipped circle boundary given by $\theta_c = \sin^{-1}(b/a)$. That is, the matching points are on the circular boundary when $\theta_c > \theta_m$ and on the straight-line boundary when $\theta_c < \theta_m$.

(2) Convergence

Table 3.2 shows the convergence property of the normalized propagation constant, P, for increasing the number of the matching points, M. The normalized propagation constant, P, and the normalized frequency, F, are defined as follows:

$$P = \frac{\beta^2 - k_e^2}{k_1^2 - k_e^2} \tag{3.19}$$

$$F = k_e \frac{2a}{\pi}\left(\frac{\varepsilon_1}{\varepsilon_e} - 1\right)^{1/2} \tag{3.20}$$

It is understood from (3.1) that P exists in the range between 0 and 1. The lower limit value $P = 0$ leads to the relation $\beta = k_e$. This is the cutoff state in which

Table 3.2
The Convergence Property of the Normalized Propagation Constant P ($b = 0.7a$; $\varepsilon_1 = 2.25\varepsilon_e$)

MODE		E^y_{11}	E^y_{31}	E^x_{22}	E^x_{13}
F		2.0	3.2	4.0	4.4
M	5	0.513	0.462	0.400	0.235
	6	0.512	0.460	0.388	0.234
	7	0.513	0.461	0.395	0.236
	8	0.512	0.460	0.389	0.235
	9	0.512	0.461	0.393	0.235
	10	0.512	0.460	0.389	0.235

the electromagnetic energy confined in the waveguide begins to radiate to the outside region. The upper limit value $P = 1$, on the other hand, leads to $\beta = k_1$. This is the state in which the electromagnetic energy is completely confined to inside region 1. The effective number of numerical values in Table 3.2 for all modes are of three digits at least. Especially for the E^y_{11} mode, which is the fundamental mode, these numerical values seem to converge well for $M = 5$ or 6.

The fundamental and the higher-order mode values converge well when M is even, but not when M is odd. This is perhaps due to the effect of the corner point C on the boundary of the chipped circle where the tangential line can not be defined. When M is even, the corner point C is surely a good point in this sense at the middle of two matching points. When M is odd, the corner point is too close to a matching point. Especially when the electromagnetic energy of the propagation mode is concentrated around the corner point C, the existence of this corner badly affects the convergence property.

(3) The Series Expansion of Mode Fields with Odd or Even Number Terms

In the case of the structures with two symmetrical axes, electromagnetic fields are expressed by the expansion series with odd number terms ($n = 1, 3, \ldots, 2N - 1$) or even number terms ($n = 0, 2, \ldots, 2(N - 1)$). For the modes with odd number terms, $4N$ linear simultaneous homogeneous equations on the unknown coefficients, a_n, b_n, c_n, and d_n, are obtained by making the number of basis functions, N, equal to the number of the matching points, M. For the modes with even number terms, however, we have

$$a_0 = c_0 = 0 \tag{3.21a}$$

for the x-axis as an electric wall ($\phi_n = \psi_n = 0$), or

$$b_0 = d_0 = 0 \tag{3.21b}$$

for the x-axis as a magnetic wall ($\phi_n = \psi_n = \pi/2$). Therefore, two unknown coefficients are zero in the beginning. When $N = M$, we have surplus equations as boundary conditions. So, we have to delete two equations from the boundary conditions. This means that there are matching points at which boundary conditions are not applied.

We now investigate ways to select the two boundary conditions to be deleted so as to obtain good numerical results on the E^y_{21} mode as an example of such expansion series with even number terms. Table 3.3 shows the convergence property of the normalized propagation constant, P, for increasing the number of the matching points, M, where the two cases of the deletion of the boundary conditions are considered.

Deletion condition (a) is the case where the boundary conditions for H_z at the two matching points near the x-axis are deleted. Deletion condition (b) is the case where the boundary conditions for H_z at the two matching points near the y-axis are deleted. We clearly can see from this result that convergence is fast in the condition (b) because the electromagnetic energy of the E^y_{21} mode does not exist near the y-axis.

It is thus important to select the matching points so that the electromagnetic energy is not much confined around matching points when two equations of the boundary conditions are to be deleted and to select the two equations of the boundary conditions associated with the weakest field strength out of the four components of the electromagnetic fields.

(4) The Effects of the Cross-Sectional Dimensions and the Dielectric Constants

Table 3.4 shows the convergence property of the normalized propagation constant, P, of the E^x_{22} mode against the dimensional ratio b/a. We can see in this table that

Table 3.3
The Effects of the Two Deleted Boundary Conditions on the Convergence Property ($b = 0.7a$; $\varepsilon_1 = 2.25\varepsilon_e$; $F = 3.2$)

DELETION CONDITION		(a)	(b)
M	5	0.585	0.547
	6	0.523	0.549
	7	0.545	0.551
	8	0.560	0.553
	9	0.556	0.555

Table 3.4
The Effects of Cross-Sectional Dimensions on the Convergence Property ($\varepsilon_1 = 2.25\varepsilon_e$; $F = 4.0$)

	b/a	0.9	0.7	0.5
	5	0.442	0.440	0.180
	6	0.441	0.388	0.218
	7	0.440	0.395	0.226
M	8	0.441	0.389	0.207
	9	0.441	0.393	0.221
	10	0.440	0.389	0.221

he convergence is not good for large values of the dimensional ratio. The reason or this is that as the cross-sectional shape becomes more deformed the electromag- ietic fields become more complicated. This situation requires many basis functions o express the field distribution.

Table 3.5 shows the convergence property of the normalized propagation con- stant, P, of the E_{11}^y mode against the ratio of the dielectric constant for the inside region to that for the outside region, $\varepsilon_1/\varepsilon_e$. Relatively good convergence is seen for small values of the ratio $\varepsilon_1/\varepsilon_e$. This is because, for a large value of $\varepsilon_1/\varepsilon_e$, more electromagnetic energy is confined in region 1 and, therefore, more basis functions are required to express such rapidly changing field distributions.

(5) Dispersion Property

The fundamental mode of the dielectric waveguides having circular cross sections is the HE_{11} mode, which is considered to be the degenerate state of two orthogonally polarized fields. The dashed line in Figure 3.3. shows the dispersion property of the

Table 3.5
The Effect of the Dielectric Constants on the Convergence Property ($b = 0.7a$; $F = 2.0$)

	$\varepsilon_1/\varepsilon_e$	1.02	2.25	4.0
	4	0.607	0.510	0.443
	5	0.609	0.513	0.447
	6	0.608	0.512	0.447
M	7	0.609	0.513	0.448
	8	0.609	0.512	0.447
	10	0.609	0.512	0.446

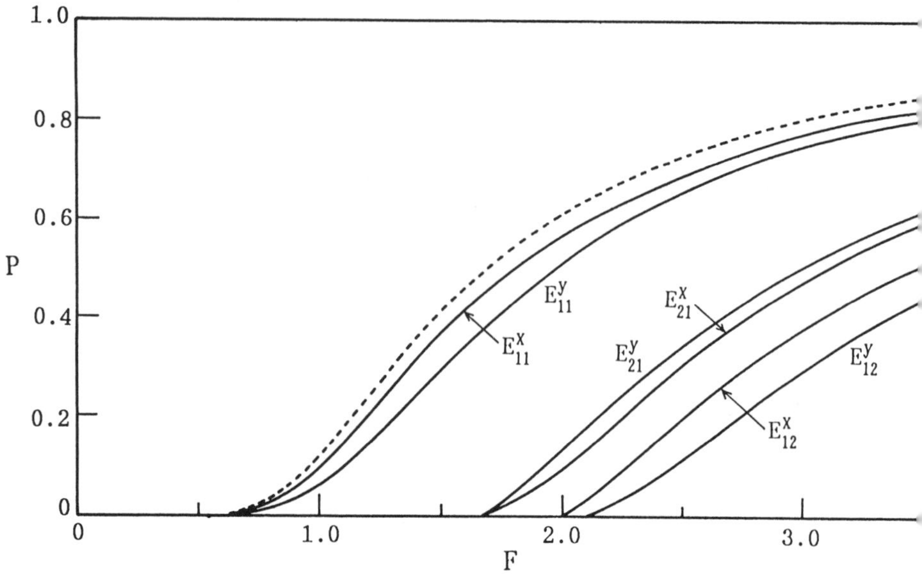

Figure 3.3 Dispersion property of the dielectric waveguide with a chipped-circle boundary. The dashed line shows the HE_{11} mode of the waveguide with a circular boundary cross section (b = $0.7a$; $\varepsilon_1 = 2.25\varepsilon_e$).

fundamental HE_{11} mode, which has been estimated by the separation of variables. The solid lines in Figure 3.3 show the dispersion property of the fundamental and higher-order modes of chipped-circle boundary structures estimated by the point-matching method. The degeneracy of the HE_{11} mode due to the circular boundary has been separated by adopting chipped-circle boundary structure and has produced two propagation modes, E^x_{11} and E^y_{11}.

Because the electric field of the E^x_{11} mode is polarized in the x-axis direction, the electric field energy is concentrated around the x-axis. Therefore, the electric field energy in the chipped part is not strong enough to affect the propagation mode even though this part has been removed from the original circle. This means that the difference between the propagation constant of the HE_{11} mode and the E^x_{11} mode is smaller.

In contrast, the electric field of the E^y_{11} mode is polarized in the y-axis direction so that the electric field energy is concentrated around the y-axis. The propagation constant is strongly affected when the upper and lower parts of the circle have been removed and much of the electric-field energy spread out from the waveguide. The propagation constant of the E^y_{11} mode, therefore, is decreased more than that of the E^x_{11} mode. The electromagnetic energy distribution of higher-order modes, because it spreads out from the boundary, is more affected by chipping the circle boundary than that of the fundamental mode.

3.4 COMPOSITE DIELECTRIC WAVEGUIDES

A feature of the point-matching method is that the electromagnetic-field analysis method based on point matching is well suited to dielectric waveguides having various complex cross sections. Hence, the point-matching method is now applied to two composite dielectric waveguides composed of multiple dielectric media: a dielectric waveguide composed of fan cross sections and a dielectric waveguide composed of elliptical cross sections. Such structures could be utilized as waveguides to transmit millimeter waves or as optical waveguides to maintain stable polarization directions.

3.4.1 Composite Dielectric Waveguides with Cross Sections Composed of Fan-Shaped Boundaries

The point-matching method can also be applied to the composite dielectric waveguides, as shown in Figure 3.4, in which the inside dielectric region is divided by four radial lines and has cross sections composed of fan-shaped boundaries [11, 12]. The structure is assumed to have symmetry with regard to the *x*-axis and the *y*-axis, to simplify explanation of the analysis method, and only the first quadrant is treated here.

The inside dielectric region is composed of two media: region 1 with the dielectric constant ε_1 and region 2 with the dielectric constant ε_2. The outside dielectric

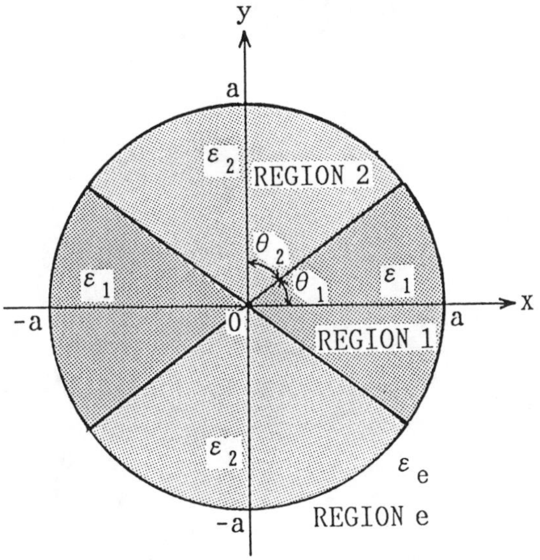

Figure 3.4 Composite dielectric waveguides with the cross sections having fan-shaped boundaries.

region is specified by the region e with the dielectric constant ε_e. We consider the two cases of the dielectric constant distributions: $\varepsilon_1 > \varepsilon_2 > \varepsilon_e$ and $\varepsilon_1 > \varepsilon_e > \varepsilon_2$. The propagation constant, β, of such a waveguide is in the range defined by (3.1). Therefore, the longitudinal components of the electromagnetic fields of the waveguide can be expressed as follows:

$$E_{z1} = \sum_{n=0}^{\infty} a_n J_n(h_1 r) \sin(n\theta + \phi_n) \tag{3.22a}$$

$$H_{z1} = \sum_{n=0}^{\infty} b_n J_n(h_1 r) \cos(n\theta + \psi_n) \tag{3.22b}$$

for inside region 1;

$$E_{z2} = \sum_{n=0}^{\infty} c_n J_n(h_2 r) \sin(n\theta + \phi_n) \tag{3.22c}$$

$$H_{z2} = \sum_{n=0}^{\infty} d_n J_n(h_2 r) \cos(n\theta + \psi_n) \tag{3.22d}$$

for inside region 2 when $k_2 > \beta$;

$$E_{z2} = \sum_{n=0}^{\infty} c_n I_n(p_2 r) \sin(n\theta + \phi_n) \tag{3.22e}$$

$$H_{z2} = \sum_{n=0}^{\infty} d_n I_n(p_2 r) \cos(n\theta + \psi_n) \tag{3.22f}$$

for inside region 2 when $k_2 < \beta$; and

$$E_{ze} = \sum_{n=0}^{\infty} e_n K_n(p_e r) \sin(n\theta + \phi_n) \tag{3.22g}$$

$$H_{ze} = \sum_{n=0}^{\infty} f_n K_n(p_e r) \cos(n\theta + \psi_n) \tag{3.22h}$$

for outside region e, where

$$h_i = (k_i^2 - \beta^2)^{1/2} \quad (i = 1, 2) \tag{3.23a}$$

$$p_i = (\beta^2 - k_i^2)^{1/2} \quad (i = 2, e) \tag{3.23b}$$

Note that the basis function in region 2 is expressed with the Bessel's functions of the first kind or the modified Bessel's function of the first kind, depending on the magnitude of the propagation constant β. The phase angles and the necessary terms of the basis functions are determined as listed in Table 3.1 according to the propagation mode to be analyzed.

When the longitudinal components of the electromagnetic fields are given, the boundary conditions for tangential electromagnetic field components can be formulated in a similar fashion to those of single-dielectric waveguides.

Unlike the case of single-dielectric waveguides, when the electromagnetic fields are approximated with the basis functions of N terms, there are $6N$ unknown coefficients because the waveguide cross section is made of the three different dielectric regions. Therefore, we have a total number of matching points $M = 6N/4$ and must distribute these points in such a way that M_{12} points are allocated for the boundary between regions 1 and 2, M_{1e} points for the boundary between regions 1 and e, and M_{2e} points for the boundary between regions 2 and e, satisfying the relation $M = M_{12} + M_{1e} + M_{2e}$.

Figure 3.5 shows some numerical results. In this composite waveguide, composed of fan-shaped cross sections, the degeneracy of the fundamental mode of a dielectric waveguide with a circular cross section, HE_{11}, is separated and the two polarization modes propagate independently.

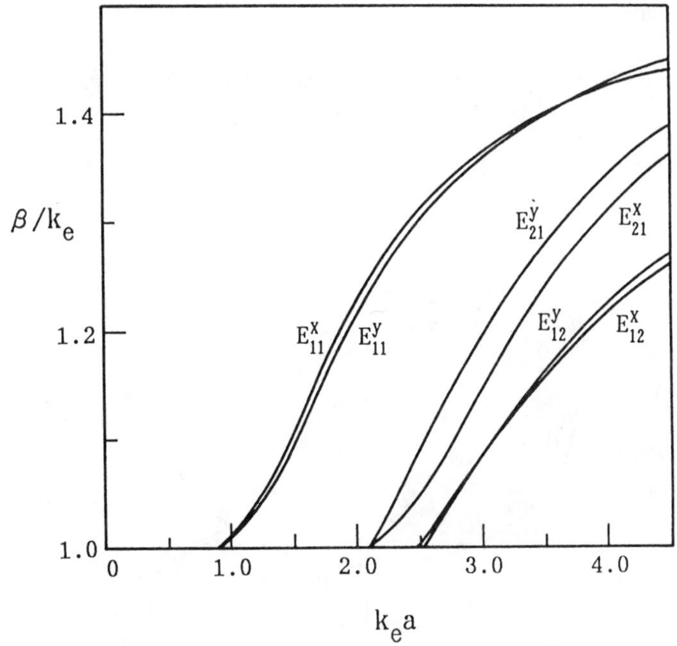

Figure 3.5 Dispersion property of the composite dielectric waveguides with cross sections having fan-shaped boundaries ($\varepsilon_1 = 2.53\varepsilon_e$; $\varepsilon_2 = 2.04\varepsilon_e$; $\theta_1 = \theta_2 = \pi/4$).

3.4.2 Composite Dielectric Waveguides with Cross Sections Composed of Elliptical Boundaries

Figure 3.6 shows a composite dielectric waveguide that has a cross section composed of two elliptical boundaries. This structure is similar to that of optical fibers, called the *double-clad type* or the *W-type,* and that of polarization-maintaining optical fibers.

Here, we consider a symmetrical structure with regard to the x-axis and y-axis, as shown in Figure 3.6, in which the inside dielectric region is composed of two media. These regions have the dielectric constant ε_1, ε_2, and ε_e for regions 1, 2, and e, respectively. There are two cases for the combination of dielectric constants: $\varepsilon_1 > \varepsilon_2 > \varepsilon_e$ and $\varepsilon_1 < \varepsilon_e < \varepsilon_2$. The propagation constant, β, is in the range given by (3.1). The longitudinal components of electromagnetic fields in this waveguide structure are expressed as follows:

$$E_{z1} = \sum_{n=0}^{\infty} a_n J_n(h_1 r) \sin(n\theta + \phi_n) \tag{3.24a}$$

$$H_{z1} = \sum_{n=0}^{\infty} b_n J_n(h_1 r) \cos(n\theta + \psi_n) \tag{3.24b}$$

for inside region 1;

$$E_{z2} = \sum_{n=0}^{\infty} [c_n J_n(h_2 r) + d_n Y_n(h_2 r)] \sin(n\theta + \phi_n) \tag{3.24c}$$

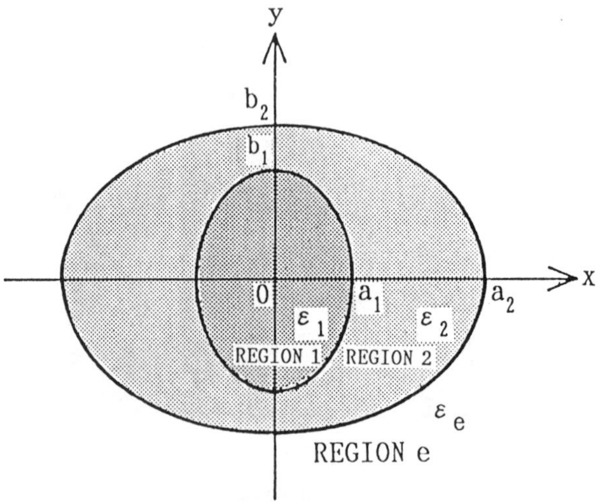

Figure 3.6 Composite dielectric waveguides with the cross sections having elliptical boundaries.

$$H_{z2} = \sum_{n=0}^{\infty} [e_n J_n(h_2 r) + f_n Y_n(h_2 r)] \cos(n\theta + \psi_n) \tag{3.24d}$$

for inside region 2 when $k_2 > \beta$;

$$E_{z2} = \sum_{n=0}^{\infty} [c_n I_n(p_2 r) + d_n K_n(p_2 r)] \sin(n\theta + \phi_n) \tag{3.24e}$$

$$H_{z2} = \sum_{n=0}^{\infty} [e_n I_n(p_2 r) + f_n K_n(p_2 r)] \cos(n\theta + \psi_n) \tag{3.24f}$$

for inside region 2 when $k_2 < \beta$; and

$$E_{ze} = \sum_{n=0}^{\infty} g_n K_n(p_e r) \sin(n\theta + \phi_n) \tag{3.24g}$$

$$H_{ze} = \sum_{n=0}^{\infty} h_n K_n(p_e r) \cos(n\theta + \psi_n) \tag{3.24h}$$

for the outside region e.

Because the origin and infinity are not included in inside region 2, the combinations of the modified Bessel's functions of the first and the second kinds are used as basis functions. The Bessel's functions or the modified Bessel's functions are used depending on the magnitude of β. Because the electromagnetic fields in inside region 2 include twice the number of unknown coefficients compared with inside region 1 and outside region e, the total number of unknown coefficients is $8N$ when the basis functions with N terms are adopted. Consequently, when we put M_{12} matching points on the boundary between inside regions 1 and 2, and M_{2e} matching points on the boundary between inside region 2 and outside region e, then the total number of the matching points is given by $M = 2N = M_{12} + M_{2e}$.

Figure 3.7 shows some numerical results. Though inside region 1 is assumed to have a circular boundary in this particular example, the degeneracy of the fundamental HE_{11} mode is separated in a fashion similar to that of the chipped-circle structure because inside region 2 has an elliptical boundary. The elliptical shape of inside region 2 is long in the x-axis and $\varepsilon_2 > \varepsilon_e$; therefore, the propagation constant of the E^x_{11} mode with the electric field polarized in the x-axis is larger than that of the E^y_{11} mode.

The point-matching method thus can be effectively applied to the analysis of composite dielectric waveguides with cross sections composed of multiple elliptical boundaries and also to those with cross sections composed of intersecting elliptical boundaries [13, 14].

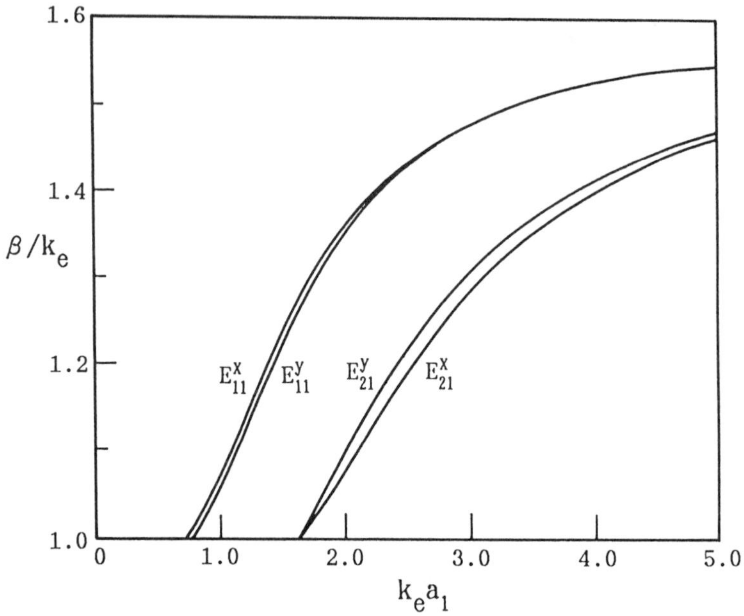

Figure 3.7 Dispersion property of composite dielectric waveguides with the cross sections having elliptical boundaries ($a_1 = b_1$; $a_2 = 1.5a_1$; $b_2 = 1.2a_1$; $\varepsilon_1 = 2.56\varepsilon_e$; $\varepsilon_2 = 1.69\varepsilon_e$).

3.5 COUPLED DIELECTRIC WAVEGUIDES

3.5.1 Coupled Dielectric Waveguides Composed of Two Waveguides

When two dielectric waveguides are located parallel and close to each other, the system acts as a coupled waveguide, which is important as a basic optical circuit element like an optical coupler or an optical multiplexer. In this section, the point-matching method is applied to the analysis of such coupled dielectric waveguides. For simplicity, we consider coupled dielectric waveguides with two waveguides having cross sections with elliptical boundaries and located in parallel, as shown in Figure 3.8. The two waveguides have waveguiding regions, 1 and 2, with the same dielectric constant ε_1. Outside is region e with the dielectric constant ε_e. Because the two waveguides are electromagnetically coupled, the longitudinal components of the electromagnetic fields in each region are given as follows:

$$E_{z1} = \sum_{n=0}^{\infty} a_{n1} J_n(hr_1) \sin(n\theta_1 + \phi_n) \tag{3.25a}$$

$$H_{z1} = \sum_{n=0}^{\infty} b_{n1} J_n(hr_1) \cos(n\theta_1 + \psi_n) \tag{3.25b}$$

or waveguiding region 1;

$$E_{z2} = \sum_{n=0}^{\infty} a_{n2} J_n(hr_2) \sin(n\theta_2 + \phi_n) \tag{3.25c}$$

$$H_{z2} = \sum_{n=0}^{\infty} b_{n2} J_n(hr_2) \cos(n\theta_2 + \psi_n) \tag{3.25d}$$

or waveguiding region 2; and

$$E_{ze} = \sum_{n=0}^{\infty} [c_{n1} K_n(pr_1) \sin(n\theta_1 + \phi_n) + c_{n2} K_n(pr_2) \sin(n\theta_2 + \phi_n)] \tag{3.25e}$$

$$H_{ze} = \sum_{n=0}^{\infty} [d_{n1} K_n(pr_1) \cos(n\theta_1 + \psi_n) + d_{n2} K_n(pr_2) \cos(n\theta_2 + \psi_n)] \tag{3.25f}$$

or outside region e where (r_i, θ_i) $(i = 1, 2)$ denotes the coordinate with the origin t the center of the waveguiding region, i, as shown in Figure 3.8. The electro-nagnetic fields in waveguiding regions 1 and 2 are given by the circular cylindrical armonic functions of this coordinate system as the basis functions, and those in utside region e are given by the superposition of outside fields originated from vaveguides 1 and 2.

When the total structure of the coupled dielectric waveguides is symmetrical vith regard to the x-axis and the y-axis, as shown in Figure 3.8, the propagation node fields exist with regard to these two axes as either an electric wall or a magnetic vall. We take $\phi_n = \psi_n = \pi/2$ for the modes with the x-axis as a magnetic wall, nd $\phi_n = \psi_n = 0$ for the modes with x-axis as an electric wall.

When the x-axis is a magnetic wall, we have the following relations among nknown coefficients:

$$\begin{aligned} a_{n1} &= a_{n2}, \quad b_{n1} = b_{n2}, \\ c_{n1} &= c_{n2}, \quad d_{n1} = d_{n2} \end{aligned} \tag{3.26a}$$

or the y-axis as a magnetic wall; and

$$\begin{aligned} a_{n1} &= -a_{n2}, \quad b_{n1} = -b_{n2}, \\ c_{n1} &= -c_{n2}, \quad d_{n1} = -d_{n2} \end{aligned} \tag{3.26b}$$

or the y-axis as an electric wall.

When the x-axis is an electric wall, on the other hand, we have opposite re-tions among unknown coefficients and the symmetry of the y-axis. Consequently, vhen we select basis functions with N terms, the total number of unknown coeffi-ients is $4N$. Hence, when we select the matching points at the boundary in the first

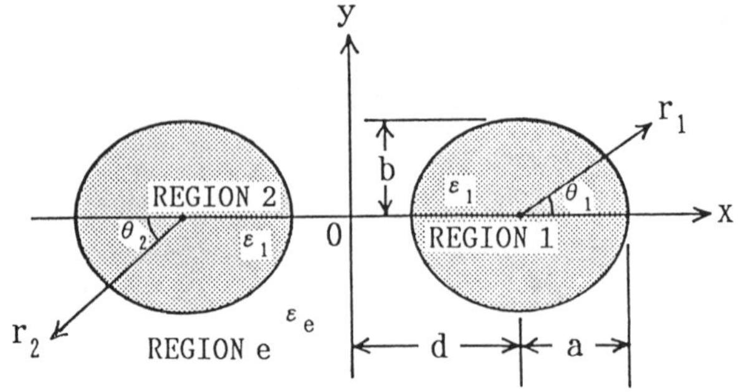

Figure 3.8 Coupled dielectric waveguides composed of two waveguides.

quadrant of waveguiding region, 1, the total number of the matching points, M, i equal to the number of the basis functions, N.

Figure 3.9 shows some numerical results. The fundamental mode for a circul boundary structure, HE_{11}, is changed to the E^x_{11} mode and the E^y_{11} mode with th polarization in the x-axis and the y-axis, respectively, after the circular boundary changed to an elliptical boundary. Moreover, these modes are coupled because c the coupling structure. Thus, two even modes, $e^{E^x_{11}}$ and $e^{E^y_{11}}$ and two odd mode $o^{E^x_{11}}$ and $o^{E^y_{11}}$, are produced and propagate along this structure.

3.5.2 Coupled Dielectric Waveguides Composed of Multiple Waveguides

Coupled dielectric waveguides composed of multiple waveguides closely located i parallel could be used, in the future, as components to distribute the electromagneti

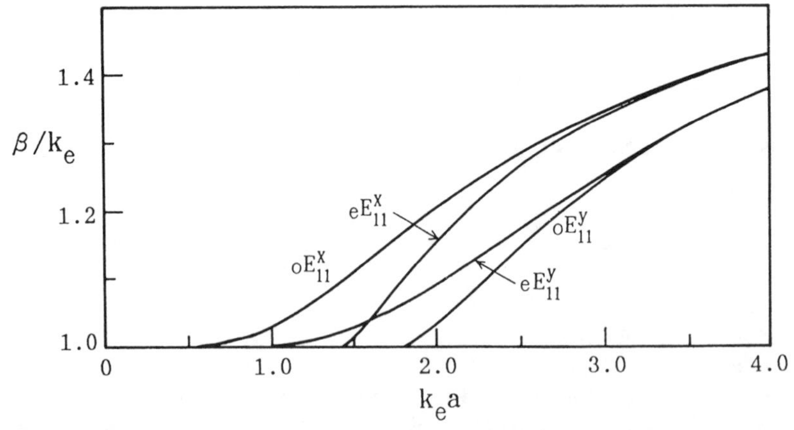

Figure 3.9 Dispersion property of coupled dielectric waveguides composed of two waveguides (b $0.5a$; d = $1.25a$; ε_1 = $2.53\varepsilon_e$).

nergy of a dielectric waveguide to many other waveguides or to realize inverse erformance. In this section, we consider a structure in which waveguides are located t each top of a polygon, as shown in Figure 3.10. This structure is symmetrical ith regard to the x-axis and y-axis; therefore, the point-matching method can be pplied by considering only the first quadrant, as explained earlier. However, we ill need more matching points for more complicated boundaries, even in only the rst quadrant, because of the multiple waveguide structure.

Note that the structure has rotation symmetry in addition to its mirror symetry. When we use group theory and pay attention to the various types of symmetry this structure, we can analyze all its propagation characteristics by distributing the atching points in only a half region of one waveguide [15]. We refer readers to is reference for details of the analysis of electromagnetic fields based on the group eory.

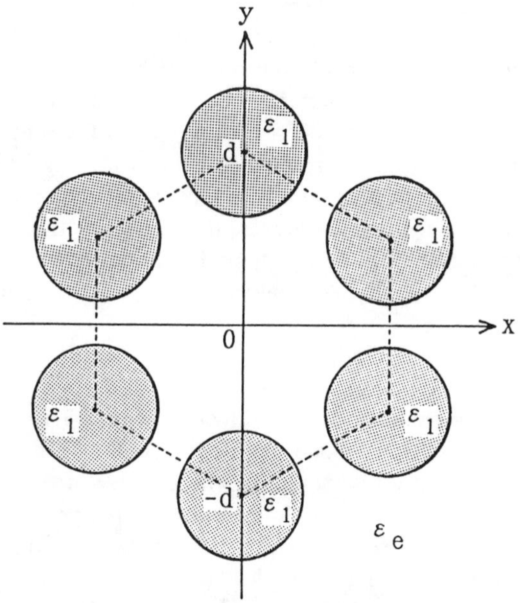

igure 3.10 Coupled dielectric waveguides composed of multiple waveguides.

.6 SOME PRECAUTIONS FOR PROGRAMMING

he wide range of values for each element in the coefficient matrix for unknown onstants in this analysis method calls for caution regarding overflow, underflow, nd round-off errors in computer calculations. The propagation constant, β, as the igenvalue is obtained by finding zeros of the coefficient determinant given by (3.16).

In this case, the values of the determinant are not important but the precise value of β to inverse the sign of the determinant are.

For this purpose, we suggest some ways to minimize variation in matrix element values. We recommend readers employ a method in which we multiply row (or columns) including the terms of the Bessel's functions of the first kind or those of the second kind by $|h_i^2 r_0/J_n(h_i r_0)|$ or $|h_i^2 r_0/Y_n(h_i r_0)|$, respectively. Also, we multiply rows (or columns) including the terms of the modified Bessel's functions of the first kind or those of the second kind by $|p_i^2 r_0/I_n(p_i r_0)|$ or $|p_i^2 r_0/K_n(p_i r_0)|$, respectively. Here, r_0 is defined as the average value of the distances between the origin and each matching point.

When we find a determinant whose dimension is more than 40 (or the number of the matching point is more than 10), we have to use a double precision procedure. When $n = 0$ for the number of terms of the basis functions, the unknown coefficient become zero for the mode with the even number terms as described in Section 3.3.7. Therefore, it is necessary to delete boundary condition equations at the matching point on the boundary where electromagnetic field intensity is minimal to adjust the size of the coefficient matrix to match the total number of unknown coefficients.

Another method to match the number of unknown coefficients and that of the boundary condition equations is to add one more term of the basis function to the longitudinal components of the electromagnetic fields lacking in coefficients. This procedure seems to make the preceding deletion of the boundary condition equation unnecessary. With such a method, however, the basis functions are added only to electric-field components or only to magnetic-field components without satisfying Maxwell's equations. As a result, physically meaningless solutions (called *spurious modes*) are usually generated.

When the point-matching method is applied to the analysis of a circular boundary structure and the basis function with one term is added, a spurious mode appears in addition to the fundamental mode, as shown in Figure 3.11. Because the propagation constant of this spurious mode changes very slowly against the change of frequency and changes rapidly for increasing the number of the matching points, it is simple to distinguish the spurious mode from the normal propagation mode. When a propagation mode and a spurious mode are degenerate, the propagation constant changes according to the coupling strength. The solution of the propagation mode near the degenerate point could not be obtained with good accuracy in this case.

Note that the accuracy of the point-matching method explained in this chapter is not good near cutoff frequencies; namely, where the propagation constant, β, is close to the wavenumber k_e for the outside region. Cutoff frequencies themselves cannot be obtained with this method either. These difficulties are because the transverse components of electromagnetic fields in outside region e given by (3.6) have the denominators including the term $k_e^2 - \beta_2^2$, which is singular at the cutoff frequencies. An improved point-matching method that removes the singular points ha

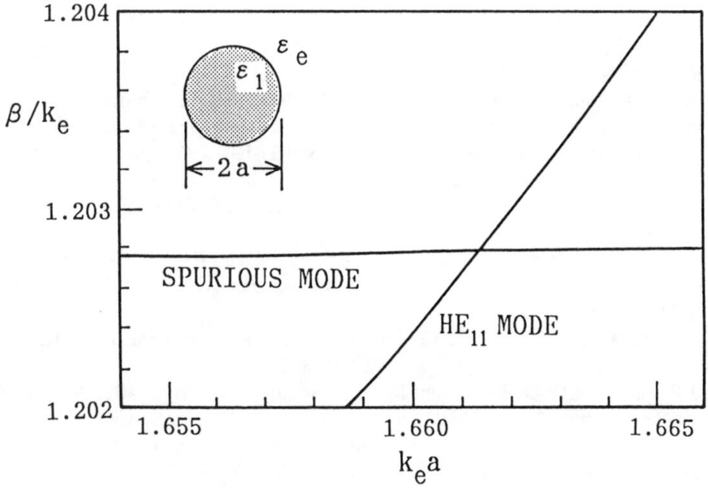

gure 3.11 Appearance of a spurious mode in the numerical results of the point-matching analysis for the dielectric waveguide with the cross section of a circular boundary ($\varepsilon_1 = 2.53\varepsilon_e$).

een proposed to accurately calculate the cutoff frequencies of dielectric waveguides 6, 17].

.7 CONCLUSION

/e introduced the point-matching as an analysis method suited to a class of elec-omagnetic field problems and showed the analysis of wave propagation parameters ong the dielectric waveguides as example. We noticed, in particular, the generality f applicable structures and the reasonable accuracy of the point-matching method rough these calculations.

The point-matching method can be applied easily to single-dielectric wave-uides, composite dielectric waveguides, and coupled dieletric waveguides, as ex-lained in this chapter. Because the calculation time on a computer with this method relatively short, the method could be applied to the analysis of various waveguide ructures to be used in millimeter-wave and optical-wave regions in the future.

REFERENCES

1] Pipes, L.A., and L.R. Harvill, *Applied Mathematics for Engineers and Physicists,* McGraw-Hill, New York, 1958, p. 583.

2] Harrington, R.F., *Field Computation by Moment Methods,* Macmillan, New York, 1968.

[3] Yee, H.Y., and N.F. Auden, "Uniform Waveguides with Arbitrary Cross-Section Considered b the Point-Matching Method," *IEEE Trans. Microwave Theory and Tech.*, Vol. MTT-13, No. 1 Nov. 1965, pp. 847–851.

[4] Mullin, C.R., R. Sandburg, and C.O. Velline, "A Numerical Technique for the Determinati of Scattering Cross Sections of Infinite Cylinders of Arbitrary Geometrical Cross Section," *IEE Trans. Antennas and Propagation,* Vol. AP-13, No. 1, Jan. 1965, pp. 141–149.

[5] Bates, R.H.T., "The Theory of the Point-Matching Method for Perfectly Conducting Waveguid and Transmission Lines," *IEEE Trans. Microwave Theory and Tech.*, Vol. MTT-17, No. 6, Ju 1969, pp. 294–301.

[6] Bates, R.H.T., "The Point-Matching Method for Interior and Exterior Two-Dimensional Boun ary Value Problems," *IEEE Trans. Microwave Theory and Tech.,* Vol. MTT-15, No. 3, Mar 1967, pp. 185–187.

[7] Goell, J.E., "A Circular-Harmonic Computer Analysis of Rectangular Dielectric Waveguides *Bell Syst. Tech. J.,* Vol. 48, Sept. 1969, pp. 2133–2160.

[8] James, J.R., and I.N.L. Gallet, "Point-Matched Solutions for Propagating Modes on Arbitrari Shaped Dielectric Rods," *Radio and Electron. Eng.*, Vol. 42, No. 3, March 1972, pp. 103–11

[9] Bates, R.H.T., J.R. James, I.N.L. Gallett, and R.F. Millar, "An Overview of Point Matching *Radio and Electron. Eng.,* Vol. 43, No. 3, March 1973, pp. 193–200.

[10] Yamashita, E., K. Atsuki, O. Hashimoto, and K. Kamijo, "Modal Analysis of Homogeneo Optical Fibers with Deformed Boundaries," *IEEE Trans. Microwave Theory and Tech.,* V MTT-27, No. 4, April 1979, pp. 352–356.

[11] Yamashita, E., K. Atsuki, and R. Kuzuya, "Composite Dielectric Waveguides," *IEEE Tran Microwave Theory and Tech.*, Vol. MTT-28, No. 9, Sept. 1980, pp. 986–990.

[12] Atsuki, K., T. Aslam, and E. Yamashita, "Modal Analysis of Composite Dielectric Waveguid with Multi-Boundaries in Radial Directions," *Trans. IECE Japan,* Vol. E65, No. 4, April 198 pp. 189–193.

[13] Yamashita, E., K. Atsuki, and Y. Nishino, "Composite Dielectric Waveguides with Two Ellipti Cylinder Boundaries," *IEEE Trans. Microwave Theory and Tech.,* Vol. MTT-29, No. 9, Se 1981, pp. 987–990.

[14] Yoshikawa, H., I. Hinata, T. Hosono, and S. Takano, "Analysis of Polarization Maintaini Single-Mode Optical Fibers with Vacuum Circular Pits across Core-Clad Interface," *Trans. IEC Japan,* Vol. J67-B, No. 1, Jan. 1984, pp. 70–77.

[15] Yamashita, E., S. Ozeki, and K. Atsuki, "Modal Analysis Method for Optical Fibers with Sy metrically Distributed Multiple Cores," *J. Lightwave Tech.*, Vol. LT-3, No. 2, April 1985, p 341–346.

[16] Su, C.C., "Cutoff Frequency of a Homogeneous Optical Fiber with Arbitrary Cross Section *IEEE Trans. Microwave Theory and Tech.*, Vol. MTT-33, No. 11, Nov. 1985, pp. 1101–110

[17] Atsuki, K., S. Kaewsuriyathumrong, and E. Yamashita, "Cutoff Characteristics of Composi Optical Fibers," *Tech. Digest, 1st Optoelectronics Conf.* (OEC '86), Tokyo, July 29–31, 198 pp. 120–121.

Chapter 4
The Mode-Matching Method

Yoichi Okuno

4.1 INTRODUCTION

The scattering of a monochromatic plane wave by an obstacle is formulated as a boundary value problem of the Helmholtz equation. One of the classic methods for solving the problem is the separation of variables [1–3]. In this method we expand the solution in terms of the particular solutions of the Helmholtz equation and determine the expansion coefficients so that the solution satisfies the boundary condition. This is a strong method for problems where the boundaries correspond to coordinate surfaces of some appropriate system of coordinates; for example, plane, sphere or circular cylinder boundaries.

Unfortunately, the shape of the boundaries of practical interest seldom coincide with the coordinate surfaces. Here arises the problem of convergence: Does the infinite series converge everywhere outside and on the obstacle? Actually, the series converges in some cases and diverges in others. This is expected in theory and is not a problem. The problem is that the range where the convergence occurs is so restricted [4–6] that we cannot recognize the separation of variables as a general method of solution. Note that this question is closely related to the Rayleigh assumption.

In this chapter we discuss a *mode-matching method* (MMM) proposed by Yasuura and others [7–21] as a standard method for problems with arbitrarily shaped boundaries. In the context of Yasuura's MMM, we introduce an approximate solution in terms of a finite summation of modal functions instead of the infinite expansion in the method of separation of variables. Here, the modal functions mean a set of outgoing wave functions whose boundary values form a complete set in the mean squares sense. The concept of the modal functions is defined in [7] to include the set of particular solutions. Replacing the series by a sequence of truncated modal expansions, we can remove the restriction of the convergence area.

The MMM consists of the three methods below:

1. The *conventional mode-matching method* (CMMM) [7–12]. This is a metho described in [7]. The coefficients in the truncated expansion are decided so tha the expansion fits the boundary condition in the sense of least squares. Al though some preceding literature makes use of the least-squares method [13] we stress that the CMMM was presented for the first time in [7] because Yasuur and Itakura introduced the concept of the sequence and proved the convergence
2. The *mode-matching method with a smoothing procedure* (MMM with SP) [8 14–17]. Although the CMMM is perfect in theory, it often fails to yield rapidl converging sequence of solutions. This causes a problem for which we can fin no precise solution. In the two-dimensional case (i.e., the field does not var in one direction) and when the surface of the obstacle is smooth, we can emplo the SP to accelerate the convergence.
3. The *mode-matching method with a singular-smoothing procedure* (MMM wit SSP) [18–21]. For a two-dimensional problem in which the obstacle has edg points, we apply the SSP instead of the SP to obtain rapidly converging se quence of solutions.

We explain the three methods taking the scattering by a cylindrical obstacle a an example. All the methods will be reduced to a set of linear equations that decide the coefficients in the truncated expansion. Note that, by changing the forced term of the set of equations, we can find the current density on the surface of the cylinde [7, 22–24].

4.2 FORMULATION OF SCATTERING BY CYLINDRICAL OBSTACLES

We now formulate the problem of plane-wave scattering by cylindrical obstacles The discussion in this section is primarily to provide some formulas that will be use in later sections.

4.2.1 Two-Dimensional Scattering Problems

Figure 4.1 shows the cross section of an infinite cylindrical obstacle made of a per fect electric conductor. In this figure C denotes the contour of the cylinder in the x plane. We assume that C is sufficiently smooth unless otherwise stated. P is a poin in the exterior infinite domain S. A point on C is represented by an arc length measured from a fixed point s_0. We can assume, without any loss of generality, tha the total length of C is equal to unity because of the principle of similitude [25] Under this assumption, the range of s is $0 \leqslant s \leqslant 1$; both $s = 0$ and $s = 1$ correspon to s_0.

In the problem of plane-wave scattering by the cylinder, the incident wave i given by

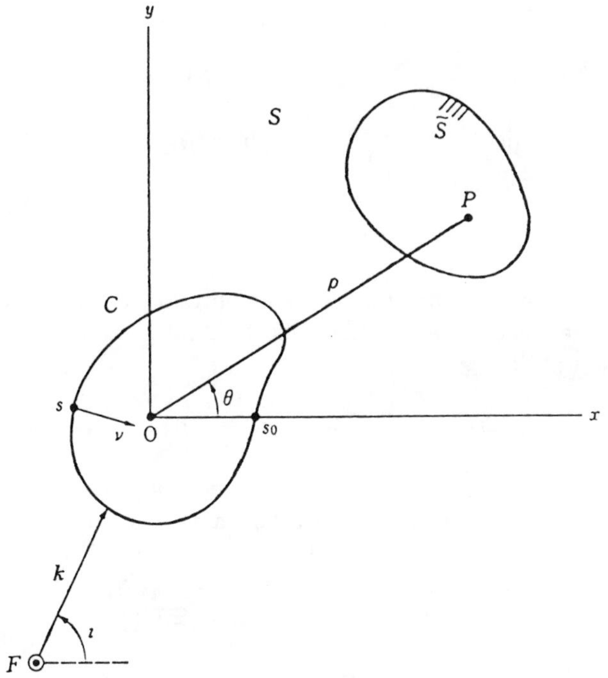

Figure 4.1 Cross section of a cylindrical obstacle with a smooth contour. P is inside an arbitrary closed subdomain $\tilde{S}$ (see Section 4.3.2).

$$F(P) = \exp[-jk\rho\cos(\theta - \iota)] \tag{4.1}$$

where k $(= 2\pi/\lambda)$ is the wavenumber and ι is the angle of incidence. The time factor $\exp(j\omega t)$ is suppressed throughout.

In this problem, there are no field variations in the z direction and we can put

$$\partial/\partial z \equiv 0 \tag{4.2}$$

in the Maxwell equation

$$\begin{aligned} \text{rot}\, \boldsymbol{E} &= -j\omega\mu \boldsymbol{H} \\ \text{rot}\, \boldsymbol{H} &= j\omega\varepsilon \boldsymbol{E} \end{aligned} \tag{4.3}$$

This is a two-dimensional scattering problem, and we can assume that we are in one or the other of two fundamental cases of polarization: the E-wave case and the H-wave case. In the following, we will illustrate that the E- or H-wave scattering can be reduced to the exterior boundary value problem of the Helmholtz equation.

1. E-Wave Case. When the incident wave is polarized in the z direction

$$\boldsymbol{E}^i(P) = \boldsymbol{i}_z F(P) \tag{4.4}$$

the scattered electric field has only a z component $E_z^s(P)$. From (4.2) and (4.3) we have the Helmholtz equation for the $E_z^s(P)$:

$$\nabla^2 E_z^s(P) + k^2 E_z^s(P) = 0 \quad (P \in S) \tag{4.5}$$

where ∇ stands for the two-dimensional nabla operator. Because the scattered field is radiated from a finite surface current, the field must satisfy the Sommerfeld radiation condition [26]:

$$\partial E_z^s(\rho, \theta)/\partial \rho + jkE_z^s(\rho, \theta) = o(\rho^{-1/2}) \quad (\rho \rightarrow \infty) \tag{4.6}$$

Here, $o(\alpha)$ denotes the Landau omicron. On the contour C the total field $E_z^t(s)$ $(= F(s) + E_z^s(s))$ must vanish, as this is a tangential component on a perfect conductor. Hence we have a boundary condition for the $E_z^s(P)$:

$$E_z^s(s) = -F(s) \quad (s \in C) \tag{4.7}$$

If we find the $E_z^s(P)$, we can calculate the scattered magnetic field by

$$\boldsymbol{H}^s(P) = (j\omega\mu)^{-1} \boldsymbol{i}_z \times \nabla E_z^s(P) \tag{4.8}$$

Consequently, the E-wave scattering is reduced to the exterior Dirichlet boundary value problem of the Helmholtz equation.

2. H-Wave Case. If the incident field is polarized in the xy plane and

$$\boldsymbol{H}^i(P) = \boldsymbol{i}_z F(P) \tag{4.9}$$

the scattered magnetic field has only a z component $H_z^s(P)$. The $H_z^s(P)$ again satisfies the Helmholtz equation and the radiation condition. The boundary condition for the $H_z^s(P)$ is given by

$$\partial H_z^s(s)/\partial \nu = -\partial F(s)/\partial \nu \quad (s \in C) \tag{4.10}$$

owing to the condition on a perfect conductor ($\boldsymbol{\nu} \times \mathbf{E} = 0$). The scattered electric field is given by

$$\boldsymbol{E}^s(P) = (j/\omega\varepsilon) \boldsymbol{i}_z \times \nabla H_z^s(P) \tag{4.11}$$

This is an exterior Neumann boundary value problem of the Helmholtz equation.

4.2.2 Scattered Far Fields

When point P is far from the obstacle ($k\rho \gg 1$), the scattered fields behave locally like plane waves. Called *scattered far fields,* they play important roles in many practical situations. In this subsection, we interpret the scattered far field for an E-wave case.

In disregard of terms that fall off more rapidly than $\rho^{-1/2}$ (i.e., $o(\rho^{-1/2})$ terms) as $\rho \to \infty$, the $E_z^s(P)$ satisfying (4.5) and (4.6) can be represented in such a form that [27]

$$E_z^s(P) = D(\iota, \theta)\exp(-jk\rho)/\sqrt{\rho} \quad (k\rho \gg 1) \tag{4.12}$$

The $D(\iota, \theta)$ function is called the *scattering amplitude* when the incident wave is a plane wave with a unit amplitude.

Inserting (4.12) into (4.8), we have the magnetic components of the scattered far field. Because $H_\rho^s(P) = o(\rho^{-1/2})$, the components that contribute to the scattered power at infinity are $E_z^s(P)$ and

$$H_\theta^s(P) = -E_z^s(P)/\zeta \quad (k\rho \gg 1) \tag{4.13}$$

Here, ζ $(= \sqrt{\mu/\varepsilon})$ is the intrinsic impedance of the free space. The Poynting vector has only a ρ component which is given by

$$P_\rho(P) = |E_z^s(P)|^2/\zeta \quad (k\rho \gg 1) \tag{4.14}$$

Let us consider a fictitious line source that isotropically radiates a constant power density $P_\rho(\rho, \theta)$. The total power (per unit length in the z direction) of the line source is $2\pi\rho P_\rho(\rho, \theta) = 2\pi|D(\iota, \theta)|^2/\zeta$. We define a bistatic scattering cross section as the ratio of the total power to the power density of the incident plane wave:

$$\sigma(\iota, \theta) = 2\pi\rho\zeta P_\rho(\rho, \theta) = 2\pi|D(\iota, \theta)|^2 \tag{4.15}$$

We also define a total cross section by

$$\sigma(\iota) = \int_0^{2\pi} |D(\iota, \theta)|^2 d\theta \tag{4.16}$$

This means the total scattered power normalized by the incident power density.

Because the power of the scattered field comes from a part of the incident power, the total field in the forward direction ($\theta = \iota$) must be weaker than the incident field. This suggests a relationship between the total cross section and the forward scattering amplitude. Starting from the energy conservation law (Poynting's theorem), we find that [28]

$$\sqrt{k/8\pi}\sigma(\iota) = -\text{Re}\,[D(\iota, \iota)/\sqrt{j}] \tag{4.17}$$

This is called the optical theorem.

4.3 A CONVENTIONAL MODE-MATCHING METHOD

In this section we explain the CMMM for solving the problems in Section 4.2. An approximate solution is defined in terms of a finite linear combination of the modal functions. The unknown coefficients in the linear combination are determined so that the solution approximately satisfies the boundary condition in the least-squares sense. Note that there is an ambiguity in defining the modal functions. In the following sections, we employ the multipole functions from among various possible selections. Some comments on the definition and examples of the modal functions will be found in Appendix A.

4.3.1 Modal Functions and Approximate Wave Function

For simplicity, let us denote by $E(P)$ the solution of the E-wave problem. We define an approximate wave function for $E(P)$ by

$$E_N(P) = \sum_{m=-N}^{N} A_m(N)\varphi_m(P) \tag{4.18}$$

Here, $\varphi_m(P)$'s are the modal functions and the notation $A_m(N)$ means that the A_m coefficient depends on the number of truncations N.

The scattered field $E(P)$ is a radiative solution of the Helmholtz equation and periodic in θ with period 2π. Therefore we choose

$$\varphi_m(P) = H_m^{(2)}(k\rho)\exp(jm\theta) \quad m = 0, \pm 1, \pm 2, \ldots \tag{4.19}$$

as the modal functions, where $H_m^{(2)}(k\rho)$ stands for the second type Hankel function of order m.

The E_N function defined by (4.18) satisfies the Helmholtz equation and the radiation condition. It will be the rigorous solution if the function agrees with the boundary condition and

$$E_N(s) = -F(s) \quad (s \in C) \tag{4.20}$$

ecause of the uniqueness of the wave functions [29]. Unfortunately, however, a inite summation of the modal functions cannot fulfill the boundary condition strictly. n these circumstances we must make an approximation that

$$E_N(s) \approx -F(s) \quad (s \in C) \tag{4.21}$$

n actual computations. Hence, our present issue is to decide the meaning of the pproximation, which subjects the method of numerical solution.

Keep in mind that the meaning should meet the following conditions:

1. Reliability. The sequence of the approximate solutions converges to the true solution $E(P)$ as N tends to infinity.
2. Practicality. The algorithm for finding the coefficients is suited to numerical computation with a computer.

The CMMM method of solution satisfies these conditions.

4.3.2 Method of Solution: *E*-Wave Case

Considering (4.21), we define the mean square error on the boundary by

$$\Omega^0(N) = \|E_N + F\|^2 \tag{4.22}$$

where $\|f\|$ denotes a norm of a square integrable function $f(s)$:

$$\|f\| = \left[\int_0^1 |f(s)|^2 ds\right]^{1/2} \tag{4.23}$$

And the superscript 0 means the CMMM.

Let point P be inside an arbitrary closed subdomain $\tilde{S}$ (Figure 4.1). Then, according to [8], we have an estimation that

$$|E_N(P) - E(P)|^2 \leq M^0(\tilde{S})\Omega^0(N) \quad (P \in \tilde{S} \subset S) \tag{4.24}$$

where $M^0(\tilde{S})$ is a positive constant independent of N and P. This inequality means that the maximum error in $\tilde{S}$ does not exceed the mean square error on the boundary multiplied by $M^0(\tilde{S})$.

In view of this, we decide the A_m coefficients to minimize the $\Omega^0(N)$. Letting

$$\partial\Omega^0(N)/\partial\overline{A_m(N)} = 0 \quad |m| \leq N \tag{4.25}$$

we have a set of linear equations for the A_m coefficients:

$$\sum_{n=-N}^{N} (\varphi_m, \varphi_n)A_n^0(N) = -(\varphi_m, F) \quad |m| \leq N \tag{4.26}$$

Here, (φ_m, φ_n) denotes an inner product:

$$(\varphi_m, \varphi_n) = \int_0^1 \overline{\varphi_m(s)}\varphi_n(s)ds \tag{4.27}$$

defined on the boundary. Note that the coefficient matrix in (4.26) is positive definit Hermitian.

This is a computer-aided method and meets the practicality condition of Sectio 4.3.1. The method also satisfies the reliability condition, as following theorem wa proved in [7].

Theorem 1. The set of the boundary values of the modal functions $\{\varphi_m(s) : m = 0, \pm 1, \pm 2, \ldots\}$ is complete in the functional space $\mathfrak{H}$, which consists of all squar integrable functions on the boundary.

Because of the completeness, the sequence $\{E_N^0(s) : N = 0, 1, 2, \ldots\}$ with th coefficients obtained by (4.26) converges to $-F(s)$ [30]:

$$\Omega^0(N) = \|E_N^0 + F\|^2 \rightarrow 0 \quad (N \rightarrow \infty) \tag{4.28}$$

At the same time, by virtue of (4.24), the sequence $\{E_N^0(P) : N = 0, 1, 2, \ldots$ converges to $E(P)$ uniformly in $\tilde{S}$. We emphasize that the conventional least-square method guarantees the convergence of solution.

Additionally, the application of the CMMM means a relaxation in the signif icance of the boundary condition. The true solution fitting (4.8) of course satisfie the condition

$$\|E + F\| = 0 \tag{4.29}$$

On the other hand, the sequence $\{E_N^0(s) : N = 0, 1, 2, \ldots\}$ does not always satisf [31] the condition

$$E_N^0(s) + F(s) \rightarrow 0 \quad (N \rightarrow \infty) \tag{4.30}$$

We made this relaxation to develop a computer-aided algorithm for the A_m coeffi cients while ensuring convergence.

4.3.3 Definition of Errors

In this section we define some criteria for the accuracy of the CMMM solutions. I most cases, it will be sufficient to check a mean square error and an energy error Sometimes we observe a relative error of the A_m coefficients, too.

1) Mean Square Error

he mean square error on the boundary condition is defined by (4.22). Considering 4.26), we modify (4.22) to obtain a simplified form where

$$\Omega^0(N) = \sum_{m=-N}^{N} (F, \varphi_m) A_m^0(N) + \|F\|^2 \tag{4.31}$$

2) Energy Error

nserting the asymptotic forms of the Hankel functions for $k\rho \to \infty$ [32] into (4.18) nd comparing the result with (4.12), we have approximations for the forward scatering amplitude and the total cross section. In view of (4.17), we define the energy rror or the error on the optical theorem in terms of the approximation that

$$\varepsilon^0(N) = 2\frac{\left|\sum_{m=-N}^{N} |A_m^0(N)|^2 + \operatorname{Re}\sum_{m=-N}^{N} A_m^0(N)\exp\left[jm\left(\iota + \frac{\pi}{2}\right)\right]\right|}{\left|\sum_{m=-N}^{N} |A_m^0(N)|^2 - \operatorname{Re}\sum_{m=-N}^{N} A_m^0(N)\exp\left[jm\left(\iota + \frac{\pi}{2}\right)\right]\right|} \tag{4.32}$$

3) Relative Error of Coefficients

et d_0 be the radius of the circumscribed circle of C with the center at origin. Then we can show that the scattered field is expanded in a series that converges uniformly n wider sense outside the circle [33]:

$$E(P) = \sum_m A_m \varphi_m(P) \quad (\rho > d_0) \tag{4.33}$$

On the other hand, the sequence $\{E_N^0(P) : N = 0, 1, 2, \ldots\}$ converges uniformly on a circle with a radius d being greater than d_0:

$$E_N^0(d, \theta) \to E(d, \theta) \quad (N \to \infty) \tag{4.34}$$

Multiplying both sides of (4.34) by $\exp(-jm\theta)$ and integrating over $(0, 2\pi)$, we have

$$A_m^0(N) \to A_m \quad (N \to \infty) \tag{4.35}$$

with the aid of (4.33). Therefore, we define the relative error of the A_m coefficients by

$$d_m^0(N) = |A_m^0(N) - A_m^0(\infty)|/|A_m^0(\infty)| \quad (4.36$$

Here, $A_m^0(\infty)$ is an estimated value obtained by extrapolation [34].

4.3.4 Some Precautions for Numerical Computation

(1) Computing the Inner Products

The elements of the coefficient matrix of (4.26) are given by (4.27). The integr in (4.27) cannot be accomplished analytically in general and hence must be evaluate numerically. The integral is of the following form:

$$(\varphi_m, \varphi_n) = \int_0^1 \psi(\rho, \theta) ds \quad (4.37$$

We put L evenly spaced sampling points on C ($s_0, s_1, \ldots, s_L(= s_0)$) and denote b ψ_i the value of the integrand at s_i. The trapezoidal rule approximates (4.37) and th result is

$$[\varphi_m, \varphi_n] = \frac{1}{L} \sum_{i=1}^{L} \psi_i \quad (4.38$$

as $\psi_0 = \psi_L$.

After this approximation the A_m coefficients depend also on $L : A_m(N, L)$ e *cetera*. Increasing the L under a fixed N, we can expect that $A_m(N, L) \to A_m(N)$. I numerical computations, however, choice of L is an important issue because of com putation costs. It is verified numerically that the selection

$$L = 2(2N + 1) \quad (4.39$$

is sufficient for finding the scattered fields [8]. Figure 4.6 in Section 4.6 is an ex ample showing the validity of (4.39).

(2) Monotonic Property of $\Omega^0(N)$

Obviously, $\Omega^0(N)$ should decrease monotonically with increasing N's. Moreover th quantity

$$\tilde{\Omega}^0(N) = \Omega^0(N)/\|F\|^2 \quad (4.40$$

annot exceed unity. If $\tilde{\Omega}^0(N)$ goes beyond unity or if the $\Omega^0(N)$ increases with N, here must be something wrong with the numerical computations. In most cases, the rror is caused by numerical instability in solving (4.26), and there is a maximum alue of N (let us say N_0) that can be used. Assume that $\Omega^0(N_0)$ or $\varepsilon^0(N_0)$ are not ufficiently small. This means that the CMMM cannot solve the problem. In such case, we recommend using the SP we introduce later.

.3.5 Method of Solution: *H*-Wave Case

n this polarization we approximate the magnetic field of the scattered wave $H(P)$ $= H_z^s(P))$ by

$$H_N(P) = \sum_{m=-N}^{N} B_m(N)\varphi_m(P) \tag{4.41}$$

he B_m coefficients are determined in such a manner that $\partial H_N(s)/\partial\nu$ is the best approximation of $-\partial F(s)/\partial\nu$ in the mean squares sense. Hence, the set of linear equations

$$\sum_{n=-N}^{N} \left(\frac{\partial\varphi_m}{\partial\nu}, \frac{\partial\varphi_n}{\partial\nu}\right) B_n^0(N) = -\left(\frac{\partial\varphi_m}{\partial\nu}, \frac{\partial F}{\partial\nu}\right) \quad |m| \leqslant N \tag{4.42}$$

ives the coefficients. The sequence $\{H_N^0(P) : N = 0, 1, 2, \ldots\}$ is proved to converge o the $H(P)$ uniformly in $\tilde{S}$ [7]. Definition of the errors in this polarization are the ame as in Section 4.3.3, and precautions for numerical computations are similar to hose in Section 4.3.4. Note that the errors in H-wave problems are usually ten times s great as the errors in E-wave problems. This means that the H-wave problems re more difficult to solve and application of the SP is strongly recommended.

.4 A SMOOTHING PROCEDURE

Ve can accelerate the convergence of the MMM solutions by employing the SP on condition that the contour is sufficiently smooth. The basic idea of the SP is relaxation of the boundary condition through an indefinite integration. We have already oosened the significance of the boundary condition by using the least squares method. Application of the SP means further relaxation: A sequence of functions does not always converge even though the sequence of their indefinite integrals converge in he mean squares sense. That is, the MMM with the SP yields a rapidly converging equence of solutions $\{E_N(P)\}$ or $\{H_N(P)\}$ at the sacrifice of their mean square convergence on the boundary. Again in this section we deal mainly with the E-wave case as we can handle the H-wave problem similarly.

4.4.1 Approximation in $\mathfrak{H}_\perp$

We denote by $h(s)$ a member of $\mathfrak{H}$ orthogonal to constants: $\|h\| < \infty$ and

$$(1, h) = 0 \tag{4.43}$$

Such $h(s)$'s form a functional space $\mathfrak{H}_\perp$, which is a subspace of $\mathfrak{H}$. Any element of $\mathfrak{H}$ can be decomposed into a constant and an $\mathfrak{H}_\perp$ component:

$$f(s) = (1, f) + f_\perp(s) \tag{4.44}$$

Note that this is the definition of the $\mathfrak{H}_\perp$ component $f_\perp(s)$.

Because a constant 1 is a member of $\mathfrak{H}$, Theorem 1 assures the existence of an element $\varphi_{n_0}(s)$ satisfying the condition

$$(1, \varphi_{n_0}) \neq 0 \tag{4.45}$$

Usually $\varphi_0(s)$ plays the role of $\varphi_{n_0}(s)$.

Letting

$$N \geqslant N_0 = |n_0| \tag{4.46}$$

we can impose the constraint

$$(1, E_N) = -(1, F) \tag{4.47}$$

on the $E_N(s)$. This is a natural restriction; that is, the average value of $E_N(s)$ is equal to that of $-F(s)$. Decomposing the $E_N(s)$ and $-F(s)$ into constants ((1, E_N), $-(1, F)$) and $\mathfrak{H}_\perp$ components ($E_{N\perp}(s)$, $-F_\perp(s)$), we find that (4.47) implies a coincidence of the constants. For this reason, under the constraint, $E_{N\perp}(s)$ should approximate $-F_\perp(s)$ so that $E_N(s)$ approximates $-F(s)$. In fact, minimization of $\|E_{N\perp} + F_\perp\|$ under (4.47) is almost equivalent to the CMMM and perfect in theory; that is, the convergence of the solution by this means can be proved.

4.4.2 Smoothing Procedure

We define an operator $\boldsymbol{K}$ by

$$\boldsymbol{K}f(s) = \int_0^1 K_1(s, t) f(t) ds \tag{4.48}$$

where $K_1(s, t)$ is a kernel function given by

$$K_1(s, t) = u(s - t) - (s - t) - 1/2 \tag{4.49}$$

with $u(s)$ being the Heaviside unit step function. Direct calculation shows that $\boldsymbol{K}f(s)$ is an indefinite integral of $f_\perp(s)$ and a member of $\mathfrak{H}_\perp$: $d\boldsymbol{K}f(s)/ds = f_\perp(s)$ and $(1, \boldsymbol{K}f) = 0$. Next, letting

$$\boldsymbol{K}^p f(s) = \boldsymbol{K}(\boldsymbol{K}^{p-1} f)(s) \quad p = 2, 3, \dots \tag{4.50}$$

we find that $\boldsymbol{K}^p f(s)$ is a p times indefinite integral of $f_\perp(s)$ and again a member of $\mathfrak{H}_\perp$. We call the $\boldsymbol{K}^p$ ($p = 1, 2, \dots$) the *smoothing operator,* as an indefinite integration works to increase the smoothness of a function.

We consider the mean square error:

$$\Omega^p(N) = \|\boldsymbol{K}^p(E_N + F)\|^2 \tag{4.51}$$

instead of $\|E_{N\perp} + F_\perp\|^2$ for constructing the algorithm of the MMM with the SP. When P is inside the closed subdomain $\tilde{S}$, we can prove that [8]

$$|E_N(P) - E(P)|^2 \leqslant M^p(\tilde{S})\Omega^p(N) \quad (P \in \tilde{S} \subset S) \tag{4.52}$$

where $M^p(\tilde{S})$ is a positive constant independent of P and N.

Considering this, let us decide the A_m coefficients so that the $\boldsymbol{K}^p E_N(s)$ is the best approximation of the $-\boldsymbol{K}^p F(s)$ under the constraint of (4.47). For this purpose, we define a quadratic form so that

$$I_N^p = \Omega^p(N) + \overline{\lambda_N^p}(1, E_N + F) + (E_N + F, 1)\lambda_N^p \tag{4.53}$$

with λ_N^p being the Lagrange multiplier, and minimize it with respect to $A_m(N)$'s and λ_N^p [34]. Letting

$$\left.\begin{aligned} &\partial I_N^p/\partial \overline{A_m^p(N)} = 0 \quad |m| \leqslant N \\ &\partial I_N^p/\partial \overline{\lambda_N^p} = 0 \end{aligned}\right\} \tag{4.54}$$

we have the following set of linear equations (Appendix B):

$$\left.\begin{aligned} &\sum_{n=-N}^{N} ((\varphi_m, \varphi_n))_p A_n^p(N) + (\varphi_m, 1)\lambda_N^p = -((\varphi_m, F))_p \quad |m| \leqslant N \\ &\sum_{n=-N}^{N} (1, \varphi_n)A_n^p(N) = -(1, F) \end{aligned}\right\} \tag{4.55}$$

Here, $((\varphi_m, \varphi_n))_p$ denotes an inner product defined by

$$((\varphi_m, \varphi_n))_p = (-1)^p \int_0^1 \int_0^1 \overline{\varphi_m(s)} K_{2p}(s, t) \varphi_n(t) ds dt \tag{4.56}$$

and $K_{2p}(s, t)$ is the $2p$ order iterated kernel of $K_1(s, t)$ (Appendix C).

As for completeness, we can prove Theorem 2 [14].

Theorem 2. The sequence $\{E_N^p(s) : N = N_0, N_0 + 1, \ldots\}$ with the coefficients determined by (4.55) satisfies (4.47) and

$$\Omega^p(N) \to 0 \quad (N \to \infty) \tag{4.57}$$

With the aid of (4.52) we can conclude that the sequence $\{E_N^p(P) : N = N_0, N_0 + 1, \ldots\}$ converges to $E(P)$ uniformly in $\tilde{S}$.

As mentioned earlier SP is a technique to accelerate the convergence of the MMM solutions. Theoretical considerations on the speed of convergence can be found in [8] and [14]. We will verify the effectiveness of the SP in Section 4.6.

4.4.3 Definition of Errors

As in the case of the CMMM, we usually check the mean square error and the energy error. The definition of the energy error is the same as in Section 4.3.3 (the superscript 0 should be replaced by p).

The mean square error is defined by (4.51). Inserting (4.18) into (4.92) in Appendix B, we see that

$$\Omega^p(N) = (F, 1)\lambda_N^p + \sum_{n=-N}^{N} ((F, \varphi_n))_p A_n^p(N) + ((F, F))_p \tag{4.58}$$

because $A_m^p(N)$'s are the solutions to (4.55). A normalized mean square error is defined by

$$\Omega^{p\prime}(N) = \Omega^p(N)/((F, F))_p \tag{4.59}$$

or alternatively by

$$\tilde{\Omega}^p(N) = \Omega^p(N)/\Omega^p(\tilde{N}) \quad (\tilde{N} : \text{arbitrarily fixed}) \tag{4.60}$$

4.4.4 Some Precautions on Using the MMM with the SP

The cautions for numerical computations in Section 4.3.4 apply as well to the present case except in the numerical method for the inner products. We indicate the following two points:

1. Order of the SP. As we will observe later, the convergence of solutions becomes faster as we increase p, the order of the SP. However, we do not expect to have good solutions by applying the SP with p greater than some p_0, say $p_0 = 4$, because a higher-order SP may neglect an important contribution of some space harmonics. If this is the case, the $E_N^p(P)$ cannot be a precise approximation even if $\Omega^p(N)$ falls off rapidly.
2. Computing the Inner Products. The coefficients in (4.55) are evaluated numerically by the two-dimensional trapezoidal rule. The number of divisions again is twice the number of modal functions ($L = 2(2N + 1)$).

4.4.5 Computing the Near Field

As stated earlier, MMM solutions converge in any closed subdomain $\tilde{S}$. We therefore can trust the value of $E_N^p(P)$ calculated at a point away from the scatterer provided that the errors are small. Even if this is the case, the boundary value $E_N^p(s)$ may not approximate the $-F(s)$ in the usual sense: mean squares sense or uniform sense. The norm of the boundary value may even diverge ($\|E_N^p\| \to \infty$ while $\Omega^p(N) \to 0$ as $N \to \infty$) if circumstances require. This suggests that accuracy can be lower near the scatterer as the $E_N^p(P)$ is continuous from infinity up to the scatterer surface. Results of numerical computations show that the $E_N^p(P)$ can be a good solution if the distance between P and the surface is no less than $\lambda/100$ to $\lambda/10$.

Note that the preceding discussion does not apply to the CMMM ($p = 0$). This is because on a smooth contour the $E_N^0(s)$ converges to $-F(s)$ uniformly owing to the fact that $-F(s)$ is continuously differentiable with respect to s. Consequently, $E_N^0(P)$ can be a good approximation up to the surface if it has sufficiently small errors.

4.4.6 Method of solution: *H*-Wave Case [35]

Replacing $\varphi_m(s)$, $-F(s)$, and $A_m(N)$ by $\partial\varphi_m(s)/\partial\nu$, $-\partial F(s)/\partial\nu$, and $B_m(N)$, respectively, we have the algorithm for the H-wave problems.

4.5 A SINGULAR-SMOOTHING PROCEDURE

We use the SSP to accelerate the convergence of solutions for problems with edge scatterers. By an edged scatterer we mean a cylindrical obstacle with sharp wedges In this section we apply the first-order SSP to the problem of an E-wave scatterin by an edged scatterer and show the method of solution. If the cylinder has depresse wedges or if the incidence is an H-wave, the SP (not the SSP) can be used to obtai rapidly converging solutions [15]. Theoretical details including some generalizations for example, a higher-order SSP, can be found in [18–21].

4.5.1 *E*-Wave Scattering by an Edged Scatterer

Figure 4.2 shows the geometry of the problem. In this figure Θ denotes the ape angle; other notations are the same as in Figure 4.1 except that both $s = 0$ and $s =$ 1 correspond to the edge point.

We consider the problem to seek the scattered electric field $E(P)$ when a E-wave $\boldsymbol{E}^i(P) = \boldsymbol{i}_z F(P)$ is incident. The solution $E(P)$ satisfies the Helmholtz equa tion (4.6), the radiation condition (4.7), and the boundary condition (4.8), as state in Section 4.2.1. The only difference is that the dominant part of the normal deriv ative of the total field $E^t(P)$ $(= E(P) + F(P))$ has the form

$$\frac{\partial E^t}{\partial \nu}(s) \propto \begin{cases} s^{\sigma-1} & (s \approx 0) \\ (1-s)^{\sigma-1} & (s \approx 1) \end{cases} \tag{4.61}$$

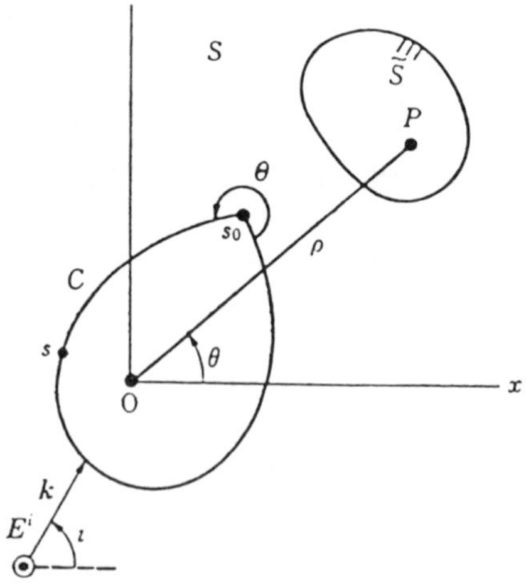

Figure 4.2 Cross section of a cylindrical obstacle with an edged contour.

near the edge point. Here, σ is a parameter defined by

$$\sigma = \pi/\Theta \tag{4.62}$$

Because $1/2 < \sigma < 1$ in our present problem, (4.61) means that the induced current density diverges at the edge point, which suggests the importance of the boundary values near the edge point.

4.5.2 Singular-Smoothing Procedure

Again in this context we approximate an indefinite integral of $-F_{\perp}(s)$ by that of $E_{N\perp}(s)$ under the constraint (4.47). In the present method, however, we decide the A_m coefficients to minimize the weighted mean square error:

$$\Omega^s(N) = \|w\boldsymbol{H}(E_N + F)\|^2 \tag{4.63}$$

Here, the weighting function

$$w(s) = [s(1 - s)]^{-1} \tag{4.64}$$

is used to make a close account of the boundary values near the edge point. $\boldsymbol{H}$ denotes an operator of indefinite integration, and the superscript s means the SSP.

We define the operator $\boldsymbol{H}$ considering following two conditions. First, $\boldsymbol{H}f(s)$ must be an indefinite integral of $f_{\perp}(s)$; that is,

$$d\boldsymbol{H}f(s)/ds = f_{\perp}(s) \tag{4.65}$$

To state the second condition, we rewrite (4.64) following the modification in Appendix B and obtain

$$\Omega^s(N) = \int_0^1 \int_0^1 \overline{[E_N(s) + F(s)]} W(s, t)[E_N(t) + F(t)]dsdt \tag{4.66}$$

Here, $W(s, t)$ denotes a symmetric kernel function defined by

$$W(s, t) = \int_0^1 [w(v)]^2 H(v, s)H(v, t)dv \tag{4.67}$$

with $H(s, t)$ being the unknown kernel of $\boldsymbol{H}$. Note that $W(s, t)$ corresponds to $(-1)^p K_{2p}$ (s, t) in (4.92) or (4.56). We should have an explicit functional form of $W(s, t)$ in order to develop a numerical algorithm based on the minimization of $\Omega^s(N)$. This makes the second condition clear: The kernel $H(s, t)$ must have zeros of an order no less than 1/2 at $s = 0$ and $s = 1$. We employ the kernel defined by

$$H(s, t) = u(s - t) - s \tag{4.68}$$

which satisfies these conditions. The $W(s, t)$ defined by (4.67) can be represented in the form

$$\begin{aligned} W(s, t) = \log[t(1 - s)/s(1 - t)] + 1/s \\ + 1/(1 - t) - 2 \; (0 < t < s < 1) \end{aligned} \tag{4.69}$$

When P is inside $\tilde{S}$, we can prove that

$$|E_N(P) - E(P)|^2 \leqslant M^s(\tilde{S})\Omega^s(N) \quad (P \in \tilde{S} \subset S) \tag{4.70}$$

under (4.47), where $M^s(\tilde{S})$ is a positive constant. We therefore determine the A_m coefficients that minimize $\Omega^s(N)$ under (4.47). Following the discussion in the case of the MMM with the SP, we obtain the set of linear equations:

$$\left.\begin{aligned} \sum_{n=-N}^{N} ((\varphi_m, \varphi_n))_s A_n^s(N) + (\varphi_m, 1)\lambda_N^s = -((\varphi_m, F))_s \quad |m| \leqslant N \\ \sum_{n=-N}^{N} (1, \varphi_n) A_n^s(N) = -(1, F) \end{aligned}\right\} \tag{4.71}$$

Here, $((\varphi_m, \varphi_n))_s$ denotes an inner product defined by

$$((\varphi_m, \varphi_n))_s = \int_0^1 \int_0^1 \overline{\varphi_m(s)} W(s, t) \varphi_n(t) ds dt \tag{4.72}$$

We may assume, though this is not proved in [7], that the set $\{\varphi_m(s)\}$ is complete in $\mathfrak{H}$. Then, we can prove Theorem 3.

Theorem 3. The sequence $\{E_N^S(s) : N = N_0, N_0 + 1, \ldots\}$ with the coefficients obtained by (4.71) satisfies (4.74) and

$$\Omega^s(N) \to 0 \quad (N \to \infty) \tag{4.73}$$

Consequently, the sequence $\{E_N^S(P) : N = N_0, N_0 + 1, \ldots\}$ converges to $E(P)$ in the closed subdomain $\tilde{S}$ uniformly. We expect that the sequence converges more rapidly than the $\{E_N^0(P) : N = 0, 1, \ldots\}$ obtained by the CMMM.

4.5.3 Some Precautions on Numerical Computations

We must evaluate numerically the inner products defined by (4.72) with the singular kernel $W(s, t)$. Because the usual trapezoid rule does not obtain a good result for a singular integration, we use an alternative means. Although a variety of half analytic methods evaluate a singular integral precisely, we employ the following method for simplicity. First, we divide the interval [0, 1] into $2L$ subdivisions with the nodes $s_i (i = 0, 1, \ldots, 2L)$. Then, we part the integral so that

$$\int_0^1 = \int_0^{s_1} + \int_{s_1}^{s_{2L-1}} + \int_{s_{2L-1}}^1$$

and make an approximation of the middle term by the trapezoid rule with $1/L$ being the width of a subdivision. As for the first and the last term, we put

$$\int_0^{s_1} \approx \alpha\psi_1/2L \quad \text{and} \quad \int_{s_{2L-1}}^1 \approx \alpha\psi_{2L-1}/2L$$

where α is a parameter no less than 1. Hence, we have an approximation of the integral

$$\frac{1}{L}\left[\frac{1+\alpha}{2}\psi_1 + \sum_{i=1}^{L-2}\psi_{2i+1} + \frac{1+\alpha}{2}\psi_{2L-1}\right] \tag{4.74}$$

Results of sample calculations show that $L = 3(2N + 1)$ is sufficient for stability of $A_m(N, L)$ and that a variation of α in the range $1 \leqslant \alpha \leqslant 2$ does not affect the value of $A_m(N, L)$ seriously.

4.6 NUMERICAL EXAMPLES

This section illustrates some numerical examples obtained by the three methods: the CMMM, the MMM with the SP, and the MMM with the SSP. Numerical computations are carried out using double precision on a middle-sized computer (Fujitsu M340). The linear equations are solved by the Gauss-Jordan algorithm. A note on the solution of the least squares problem is attached in Appendix D.

4.6.1 Scattering by a Periodic Deformed Cylinder

Figure 4.3 shows the cross section of a periodic deformed cylinder whose contour is governed by

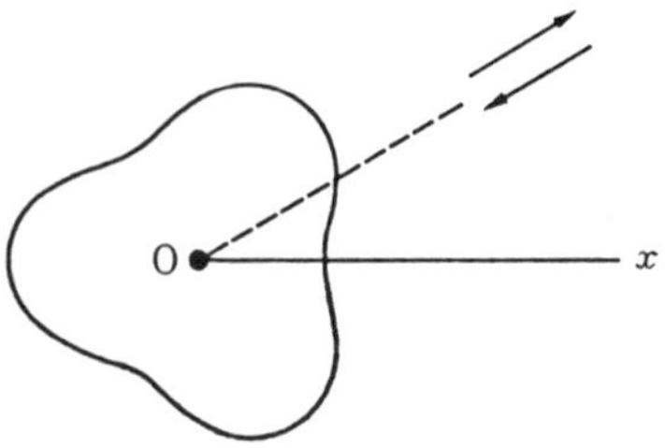

Figure 4.3 A periodic deformed cylinder.

$$\rho_s = a(1 - \delta \cos 3\theta_s) \tag{4.75}$$

We analyze the problem of an E- or H-wave scattering using the CMMM.

In this problem we introduce polyphase wave functions [16, 19] to save a great deal of numerical computation. That is, a wave function is divided into three independent phase components: -1st, 0th, and 1st phase. Here, a μth phase component of a wave function $f(\rho, \theta)$ is defined by the relation that $f^{\mu}(\rho, \theta + 2\pi/3) = \exp(2\mu\pi j/3) f^{\mu}(\rho, \theta)$. The modal functions can be classified by the number m: the -1st, 0th, and the 1st phase include $m = 3n - 1$, $3n$, and $3n + 1$, respectively. The boundary values, $-F(s)$ and $-\partial F(s)/\partial \nu$, also can be decomposed according to the definition. After this, we match the boundary condition phase by phase and determine the A_m or B_m coefficients for each phase. Let the total number of modal functions be $6N + 1$. Then, we must solve a set of linear equations with $6N + 1$ unknowns if we apply (4.26) or (4.42) without introducing the polyphase wave functions. While using the polyphase wave functions, we can find the same solutions by solving one set with $2N + 1$ unknowns for the 0th phase and two sets with $2N$ unknowns for the -1st and the 1st phase. Moreover, we can calculate the inner products on only one-third of the contour.

Figure 4.4 illustrates a backscattering transfer function [36]

$$H(ka) = \sqrt{2/a} D(\iota, \iota - \pi) \tag{4.76}$$

as a function of a normalized frequency, ka. The solid and dashed curves correspond to the E- and H-wave cases. In this figure the energy error is less than 1.0×10^{-5} percent for both polarizations so that three significant digits of $H(ka)$ are verified. The number of truncations required to achieve this accuracy is in the range $10 \leq N \leq 20$ $(61 \leq 6N + 1 \leq 121)$, and CPU time is between 15 to 30 seconds per point.

This example shows that the CMMM has a wide range of applicability provided that we can employ a sufficient number of modal functions.

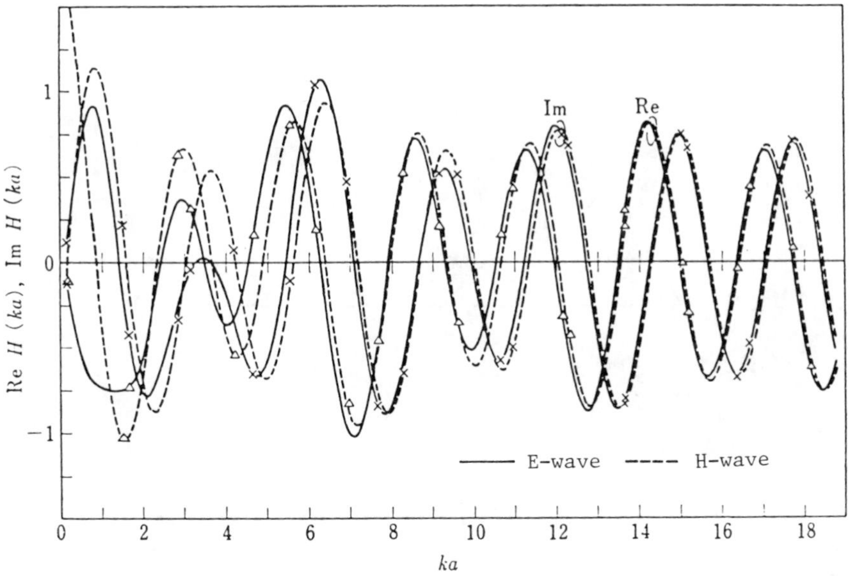

Figure 4.4 Backscattering transfer function of the cylinder shown in Figure 4.3 ($\delta = 0.2$; $\iota = 7\pi/6$; $\theta = \pi/6$).

4.6.2 Diffraction by a Fourier Grating

A typical Fourier grating [37] has a profile given by

$$y = h[\sin(2\pi x/D) + \gamma \sin(2l\pi x/D + \delta)] \quad l = 2, 3 \tag{4.77}$$

In this section we analyze the problem of plane-wave diffraction by a Fourier grating using the MMM with the SP. The incident wave is given by

$$F(P) = \exp(-j\alpha_0 x + j\beta_0 y) \tag{4.78}$$

where

$$\alpha_0 = k \sin\iota, \qquad \beta_0 = k \cos\iota \tag{4.79}$$

with ι being the incident angle shown in Figure 4.5.

Consider the E-wave case. The diffracted field $E(P)$ satisfies the Helmholtz equation in a semiinfinite domain above the grating surface and the radiation condition in the y direction. Also, because of the periodic nature of the problem, $E(P)$ satisfies the condition

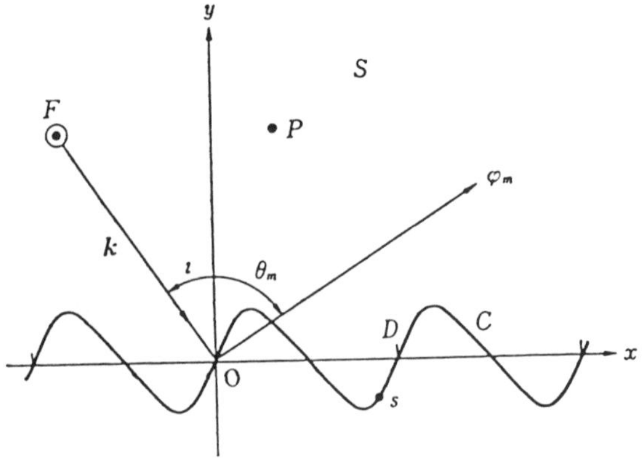

Figure 4.5 A typical profile of a Fourier grating.

$$E(x + D, y) = \exp(-j\alpha_0 D)E(x, y) \tag{4.80}$$

Therefore, we employ

$$\varphi_m(P) = \exp(-j\alpha_m x - j\beta_m y) \quad m = 0, \pm 1, \pm 2, \ldots \tag{4.81}$$

$$\alpha_m = \alpha_0 + 2m\pi/D, \quad \beta_m = \sqrt{k^2 - \alpha_m^2}, \quad \mathrm{Re}\beta_m \geqslant 0, \quad \mathrm{Im}\beta_m \leqslant 0 \tag{4.82}$$

as the modal functions for grating problems. Note that the φ_m function represents a plane wave traveling along $\theta_m = \tan^{-1}(\alpha_m/\beta_m)$ when β_m is positive. This is called the $-m$ *order diffracted wave* in optics.

We define an approximate wave function in terms of a finite linear combination of the φ_m functions. The CMMM in Section 4.3.2 can be applied directly to find the A_m coefficients. On the other hand, we should make a slight modification of the boundary values before using the MMM with the SP: The $\varphi_m(s)$ and $-F(s)$ must be replaced by $\varphi_m(s)\exp(i\alpha_0 x)$ and $-F(s)\exp(i\alpha_0 x)$, respectively. For details refer to [15] and [17].

Figure 4.6 shows the convergence of $A_m^p(N, L)$ with respect to the number of divisions L for a fixed N. Here, the second-order SP ($p = 2$) is combined with the MMM and the number of truncations is 8 ($N = 8$, $M = 2N + 1 = 17$). The L is on the abscissa and the ordinate shows a relative error from $A_m^2(8)$, which is an estimated value for $L \to \infty$. We observe that the error falls off rapidly with an increasing L. Putting $L = 2(2N + 1) = 34$, we have approximate coefficients with 10^{-4} relative errors. We accept these coefficients because the accuracy of the solution in this case is about 10^{-3} in energy error (which we will define later). Such a tendency is strongly in evidence, although the value and the speed of decrement of the

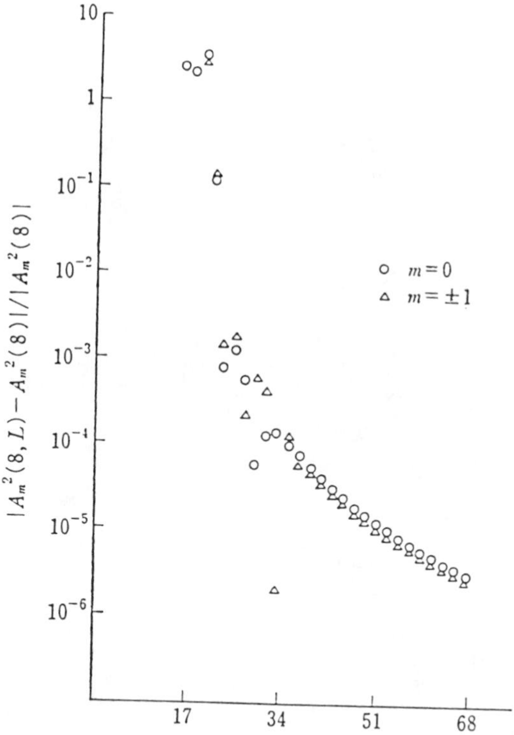

Figure 4.6 Convergence of the A_m coefficients with respect to L ($h/D = 0.2$; $\gamma = 0$ (sinusoidal); $\iota = 0$; $kD = 10$; $p = 2$; $N = 8$).

relative error depends on the order of the SP and other parameters of the problem. Especially for some dominant modes with relatively small $|m|$, the selection $L = 2(2N + 1)$ is usually sufficient to obtain coefficients with high accuracy.

A problem of diffraction by a sinusoidal grating is solved by the CMMM ($p = 0$) and the MMM with the SP ($p = 1, 2, 3$), and the errors of the solutions are plotted in Figure 4.7: Panels (a) and (b) show the mean square errors and the energy errors, respectively. Here, the energy error in grating problems is defined by

$$\varepsilon^p(N) = \left| 1 - \sum_{\beta_m > 0} \frac{\beta_m}{\beta_0} |A_m^p(N)|^2 \right| \tag{4.83}$$

In the figure we observe that a higher-order SP solution converges more rapidly than a lower-order one. This proves the effectiveness of the SP numerically. Further, in Figure 4.7(b), the $\varepsilon^3(N)$ does not decrease when N is greater than 10. We understand that this is caused by the fact described in Section 4.4.4 (1).

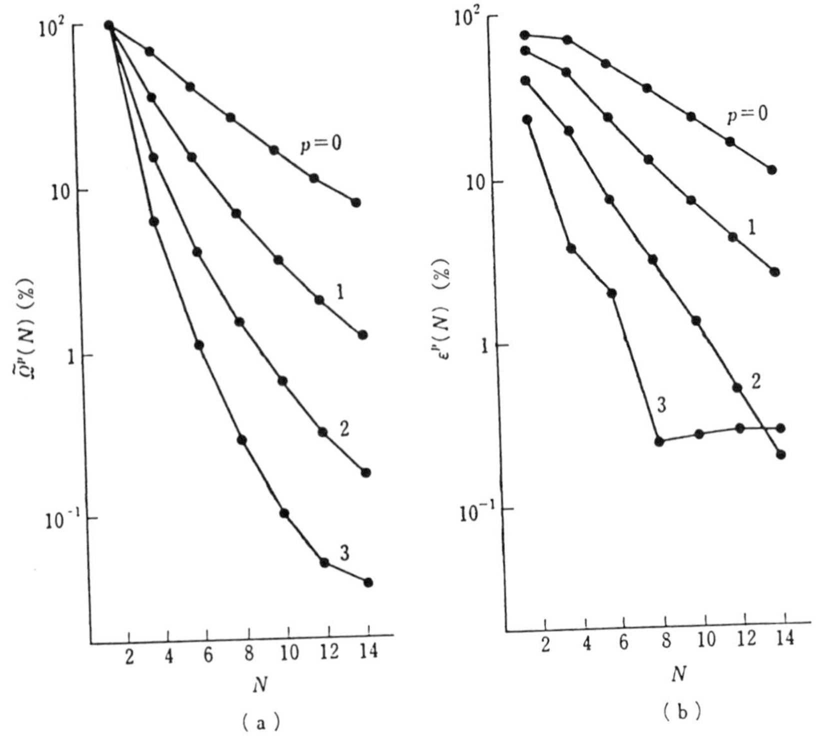

Figure 4.7 Comparison of rate of convergence: (a) mean square error; (b) energy error ($h/D = 0.25$; $\gamma = 0$ (sinusoidal); $\iota = 0$; $kD = 14.451$).

Figure 4.8 is an example of -1 order diffraction efficiency defined by

$$\eta_{-1}^E = \beta_{-1}|A_{-1}|^2/\beta_0, \; \eta_{-1}^H = \beta_{-1}|B_{-1}|^2/\beta_0 \tag{4.84}$$

The maximum number of truncation is 14 and the $p = 3$ method (MMM with third-order SP) is utilized. The accuracy is less than 1 percent in energy error, and the CPU time is between 2 seconds ($N = 10$) and 4 seconds ($N = 14$) per point [17].

In Section 4.4.5 we made a note on calculating the near scattered/diffracted field. Figure 4.9 illustrates an example of calculated total field $|F(3D/4, y) + E_{12}^p(3D/4, y)|$ as a function of y. The "exact" curve shows a precise total field obtained from the surface current density [22]. The MMM with the SP also yields accurate solutions at a point away from the grating surface: By using $p = 2$ or $p = 3$ method, we find a solution with $\varepsilon^2(12) = 0.04$ percent or $\varepsilon^3(12) = 0.02$ percent. These three curves coincide with each other for $y \geqslant 0.05\lambda$. However, we observe that the last two curves have serious errors near the surface. Moreover, the

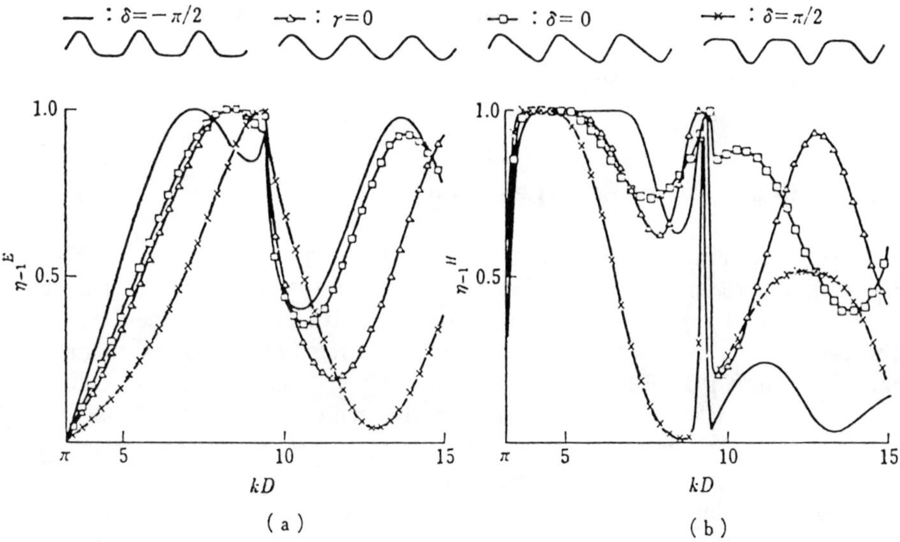

Figure 4.8 Example of -1 order diffraction efficiency in Littrow mounting ($\theta_{-1} = -\iota$): (a) E-wave; (b) H-wave ($h/D = 0.2$; $l = 2$, $\gamma = 0.3$).

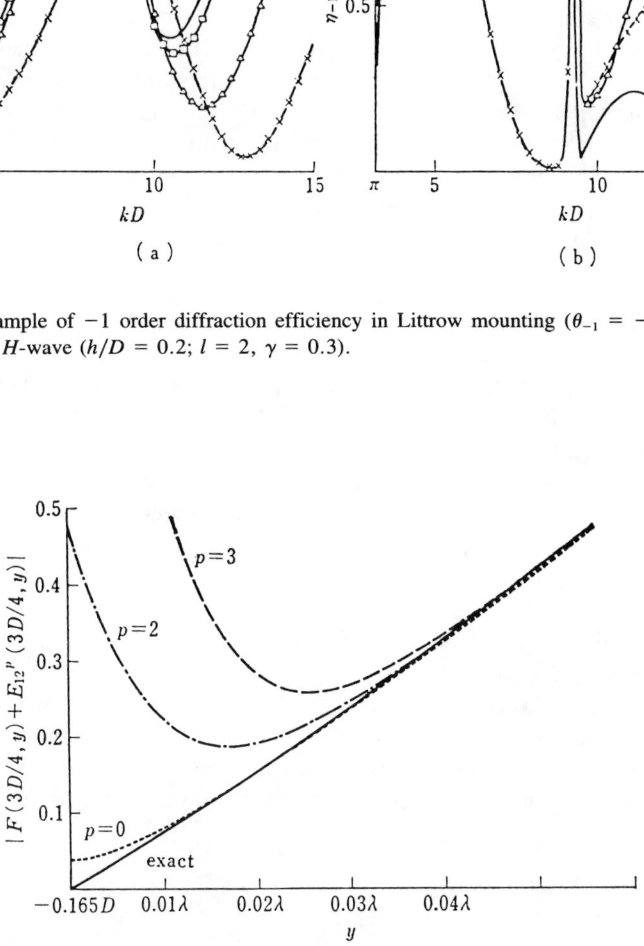

Figure 4.9 A sample plot of the total field near the bottom of a groove. The abscissa shows the distance from the bottom ($0.75D$, $-0.165D$) ($h/D = 0.15$; $l = 3$; $\gamma = 0.1$; $\delta = \pi$; $\iota = 0$, $kD = 10$).

$p = 3$ solution, which is more accurate in the far-field region, contains more errors than the $p = 2$ solution. On the other hand, the $p = 0$ solution has fewer errors in the near field although its level of errors in the far-field region ($\varepsilon^0(12) = 0.6$ percent) is not very small.

4.6.3 Diffraction by an Echelette Grating

An echelette grating [38] has a triangular profile with 90° apex angle, as shown in Figure 4.10. We analyze the problem of an E-wave diffraction by an echelette grating. Figure 4.11 shows the -1 order diffraction efficiency for some blaze angles. The energy error is less than 1 percent with the maximum number of truncations being 14. The CPU time is 7.5 seconds for $N = 10$ and 27 seconds for $N = 14$. Note that the CMMM cannot obtain a solution with the errors less than 1 percent except for a shallow grating case; for example, $\varphi_B \leqslant 20°$. Also by the present method the convergence becomes slow for gratings with $\varphi_B > 40°$. In such a case we can obtain precise solutions by employing a higher-order SSP [18] or the method in [39].

4.7 CONCLUSION

In this chapter we concentrated our discussion on directions for using the MMM because we already have a detailed introduction to the theory [8] written by Yasuura. We recommend that interested readers consult his interpretation.

Although we described the method taking the scattering by a perfectly conducting cylindrical obstacle as an example, this is a general method for solving two-dimensional boundary value problems in homogeneous media: The problem of a dielectric cylinder [40, 41] or a dielectric waveguide [42] and the problems with infinite boundaries [43, 44] also are solved by the MMM. For a three-dimensional problem, a method equivalent to the CMMM is proved to obtain a sequence that converges to the solution [45].

However, we should not employ the MMM for problems with large dimensions; that is, dimensions of the obstacle are more than ten times the wavelength. This is because such problems require a huge number of modal functions and the GTD approximation obtains a precise solution for them. Consequently, the recom-

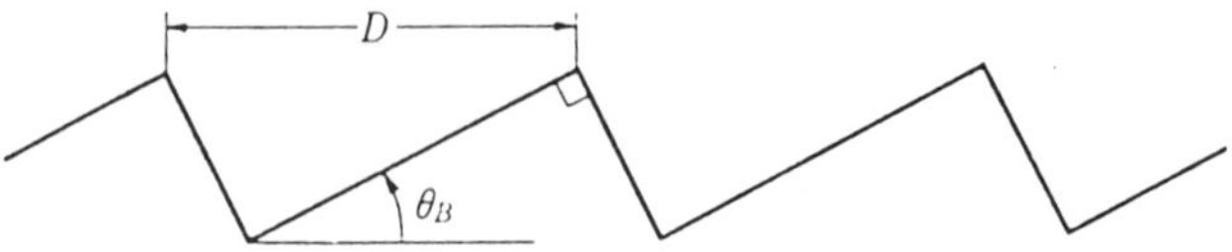

Figure 4.10 An echelette grating with a θ_B blaze angle.

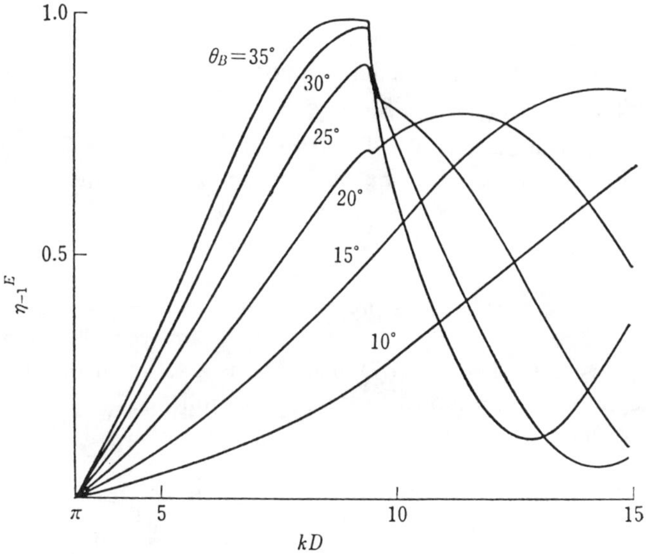

Figure 4.11 A − 1 order diffraction efficiency in Littrow mounting ($\theta_{-1} = -\iota$).

range of applications of the MMM is the so-called resonance region. Note that an appropriate choice of the modal functions according to the geometry of the problem is important to increase the effectiveness of the method [46].

APPENDIX 4A. A GENERAL EXPRESSION AND SOME EXAMPLES OF MODAL FUNCTIONS

We show a general expression of the modal functions for the exterior problems and give some examples for exterior and interior problems. A rigorous definition can be found in [7].

Let L be an arbitrary contour inside C and s_L be a point on L. We denote by $\{f_m(s_L) : m = 1, 2, \ldots\}$ a set of functions complete in $\mathfrak{H}_L$, which consists of all the square integrable functions on L. Then, both of the set of functions

$$\varphi_m(P) = \frac{1}{4j} \int_L H_0^{(2)}(kr) f_m(s_L) ds_L \quad m = 1, 2, \ldots \tag{4.85}$$

and

$$\varphi_m(P) = \frac{j}{4} \int_L \frac{\partial}{\partial \nu} H_0^{(2)}(kr) f_m(s_L) ds_L \quad m = 1, 2, \ldots \tag{4.86}$$

are the modal functions in S. Here, r is the distance between s_L and P. Equations (4.85) and (4.86) are the general expressions.

Some examples of exterior modal functions are as follows:

1. The set $\{\varphi_m(P)\}$ is defined by (4.19).
2. Divide the interior domain of C by a mesh with $P_m (m = 1, 2, \ldots)$ as the nodal points. Let r_m be the distance between P_m and P, and define

$$\varphi_m(P) = H_0^{(2)}(kr_m) \quad m = 1, 2, \ldots \tag{4.87}$$

Then, we have a set of modal functions $\varphi_m(P)$ as $M \to \infty$.

3. Denote by $\{P_m : m = 1, 2, \ldots\}$ a sequence of points inside C that accumulate to a point inside C. Then, (4.87) again yields a set of modal functions.

An example of interior modal functions is as follows:

4. The set of functions given by

$$\varphi_m(P) = J_m(k\rho)\exp(jm\theta) \quad m = 0, \pm 1, \pm 2, \ldots \tag{4.88}$$

where the origin 0 is arbitrarily fixed.

5. As another example, divide an arbitrary bounded domain by a mesh with nodal points P_m $(m = 1, 2, \ldots, M)$. Let the distance between P_m and P be r_m, and we have the set of modal functions

$$\varphi_m(P) = J_0(kr_m) \quad m = 1, 2, \ldots \tag{4.89}$$

as $M \to \infty$.

6. As still another example, let the sequence $\{P_m : m = 1, 2, \ldots\}$ accumulate to a point $(\neq \infty)$, then (4.89) yields modal functions.
7. As a final example, let $\{\theta_m : m = 1, 2, \ldots\}$ be a sequence of complex valued phase angles accumulating to a finite value, then the plane waves

$$\varphi_m(P) = \exp[-jk\rho\cos(\theta - \theta_m)] \quad m = 1, 2, \ldots \tag{4.90}$$

form a set of modal functions.

APPENDIX 4B. DERIVATION OF EQUATION (4.55)

Differentiating I_N^p defined by (4.53) with respect to $\overline{\lambda_N^p}$ and putting the results to be zero, we have the second equation of (4.55). Before differentiating I_N^p with respect to $\overline{A_m^p(N)}$, we make some modifications to the $\Omega^p(N)$. Because

$$\boldsymbol{K}f(0) = \boldsymbol{K}f(1), \quad d\boldsymbol{K}^p f(s)/ds = \boldsymbol{K}^{p-1} f(s) \tag{4.91}$$

ve have

$$\Omega^p(N) = (-1)^p \int_0^1 \int_0^1 \overline{[E_N(s) + F(s)]} K_{2p}(s, t)[E_N(t) + F(t)]dsdt \quad (4.92)$$

hrough integration by parts. Hence, $\Omega^p(N)$ can be represented in terms of a quadratic orm of $\{A_m^p(N) : |m| \leq N\}$ with $((\varphi_m, \varphi_n))_p$ as the coefficients. Inserting the modified $\Omega^p(N)$ into I_N^p, making the differentiation, and setting the results to be zero, we have he first equation of (4.55).

PPENDIX 4C. ITERATED KERNEL $K_p(s, t)$

et us denote the kernel function of the $\boldsymbol{K}^p$ by $K_p(s, t)$. Seeing (4.50), we can write

$$K_p(s, t) = \int_0^1 K_1(s, v)K_{p-1}(v, t)dv \quad (4.93)$$

his is called the pth order iterated kernel of $K_1(s, t)$. The functional form of $K_p(s, t)$ is obtained by solving a differential equation satisfied by $K_p(s, t)$ [14]. For $= 2q$ the result is

$$K_{2q}(s, t) = -[B_{2q}(|s - t|) - (-1)^q B_q(0)]/(2q)! \quad (4.94)$$

here B_p stands for the Bernoulli polynomial [47]. And for $p = 2q + 1$, we have

$$K_{2q+1}(s, t) = \text{sign}(t - s)B_{2q+1}(|s - t|)/(2q + 1)! \quad (4.95)$$

vith $\text{sign}(s) = s/|s|$. Sometimes the representation

$$K_p(s, t) = \sum_{l \neq 0} \exp[2l\pi j(s - t)]/(2l\pi j)^p \quad (4.96)$$

s also available [17].

PPENDIX 4D. APPLICATION OF THE ORTHOGONAL DECOMPOSITION METHODS

xamining the algorithm of the MMM (take the CMMM as an example) from a umerical point of view, we find that this eventually is nothing but a collocational east squares method, because the normal equation (4.26) with coefficients obtained by (4.38) and (4.39) determines the A_m coefficients. It is well known in the field of numerical analysis that a normal equation is not a good means for finding the least squares solution [48]. Instead the orthogonal decomposition methods, for example, the QR algorithm or the SVD algorithm, should be employed because of numerical stability.

Let us show an example. Figure 4.12 illustrates the decrement of errors of the solutions obtained by two methods: the normal equation method (equations (4.26), (4.38), and (4.39)) and the QR algorithm. We observe that the normal equation method fails for $N > 18$ whereas the QR algorithm yields good results even for $N > 30$. We also find that the two methods give the same results in the range where the normal equation is safe. The CPU time of the normal equation for $N = 18$ is almost the same as that for the QR algorithm for $N = 30$.

These facts, which are expected theoretically, demonstrate the effectiveness of the QR algorithm. We can employ the QR algorithm in the context of the MMM with the SP or the SSP because it can be understood as a weighted collocational least squares method. Also, note that the validity of (4.38) and (4.39) can be verified numerically using the SVD algorithm.

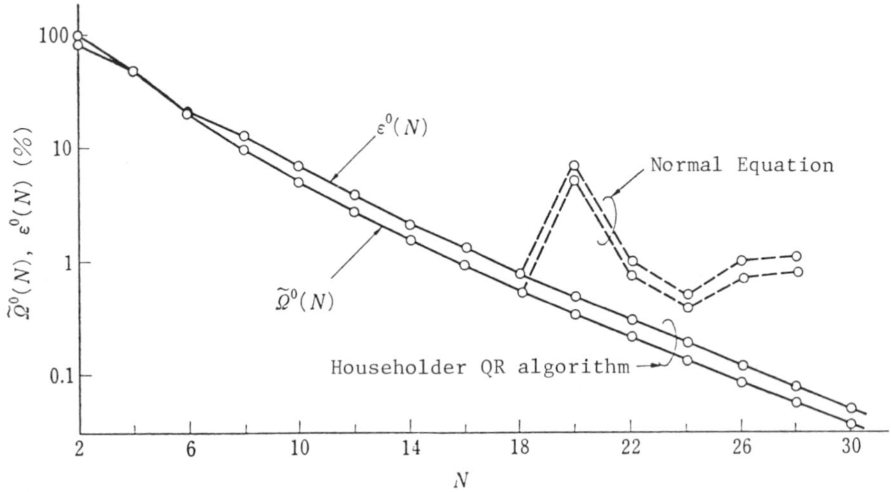

Figure 4.12 A comparison of two methods in solving the least squares problem: the Householder QR algorithm and the normal equation. Sinusoidal grating with $h/D = 0.15$, $\iota = 0$, and $kD = 15$. Similar results were obtained for $h/D = 0.25$ and 0.3.

REFERENCES

[1] Morse, P.M., and H. Feshbach, *Methods of Theoretical Physics,* McGraw-Hill, New York, 1953, pp. 495–522.

[2] Van Bladel, J., *Electromagnetic Fields,* McGraw-Hill, New York, 1964, p. 259.

[3] Strutt, J.W., (Lord Rayleigh), *The Theory of Sound,* Vol. 2, Macmillan, London, 1986, pp. 89, 297–311.

[4] Millar, R.F., "On the Rayleigh Assumption in Scattering by a Periodic Surface, Part II," *Proc. Cambridge Philos. Soc.,* Vol. 69, 1971, pp. 217–225.

[5] Petit, R., and M. Cadilhac: "Sur la diffraction d'une onde plane par un réseau infiniment conducteur," *Comptes Rendus Acad. Sci.* (Paris), 1966, Vol. B262, pp. 468–471.

[6] Van den Berg, P.M., and J.T. Fokkema, "The Rayleigh Hypothesis in the Theory of Diffraction by a Cylindrical Obstacle," *IEEE Trans. Antennas and Propagation,* Vol. AP-27, No. 5, Sept. 1979, pp. 577–583.

[7] Yasuura, K., and T. Itakura, "Approximation Method for Wave Functions (I), (II) and (III)," *Kyushu Univ. Tech. Rep.,* Vol. 38, No. 1, 1965, pp. 72–77; Vol. 38, No. 4, 1966, pp. 378–385; Vol. 39, No. 1, 1966, pp. 51–56 [in Japanese].

[8] Yasuura, K., "Numerical Methods for Boundary Value Problems," in *Denjikai no kindai kaisekiho* [Modern methods in electromagnetic fields], ed. T. Iijima, IECE Japan, 1979, pp. 57–117 [in Japanese].

[9] Yasuura, K., "New Method for Numerical Analysis of Diffraction Problems," *J. IECE Japan,* Vol. 57, No. 1, 1974, pp. 28–35 [in Japanese].

[10] Yasuura, K., and T. Itakura, "On Resonance in a Slightly Tilted Waveguide," *J. IECE Japan,* Vol. 48, No. 10, 1965, pp. 1629–1639 [in Japanese].

[11] Yasuura, K., "A View of Numerical Methods in Diffraction Problems," in *Progress in Radio Science 1966–1969,* ed. W.V. Tilson and M. Sauzade, URSI, Brussels, 1971, pp. 257–270.

[12] Ikuno, H., and K. Yasuura, "Improved Point-Matching Method with Application to Scattering from a Period Surface," *IEEE Trans. Antennas and Propagation,* Vol. AP-21, No. 5, September 1973, pp. 657–662.

[13] Meecham, W.C., "Variational Method for the Calculation of the Distribution of Energy Reflected from a Periodic Surface," *J. Appl. Phys.,* Vol. 27, 1956, pp. 361–367.

[14] Yasuura, K., and H. Ikuno, "Smoothing Process on the Mode-Matching Method for Solving Two-Dimensional Scattering Problems," *Mem. Fac. Eng. Kyushu Univ.,* Vol. 37, No. 4, Dec. 1977, pp. 175–192.

[15] Yasuura, K., Y. Okuno, and H. Ikuno, "Numerical Analysis of Scattering from Echelette Grating—Smoothing Process on Mode-Matching Method," *Trans. IECE Japan,* 1977, Vol. J60-B, No. 3, pp. 189–196 [in Japanese].

[16] Ikuno, H., and K. Yasuura, "Numerical Calculation of the Scattered Field from a Periodic Deformed Cylinder Using the Smoothing Process on the Mode-Matching Method," *Radio Sci.,* Nov.–Dec. 1978, Vol. 13, No. 6, pp. 937–946.

[17] Okuno, Y., and T. Matsuda, "An Efficient Technique for the Numerical Solution of Diffraction by a Fourier Grating," *J. Opt. Soc. Am. A,* March 1987, Vol. 4, No. 3, pp. 465–472.

[18] Yasuura, K., and Y. Okuno, "Singular-Smoothing Procedure on Fourier Analysis," *Mem. Fac. Eng. Kyushu Univ.,* June 1981, Vol. 41, No. 2, pp. 123–141.

[19] Yasuura, K., and Y. Okuno, "The Algorithm for the Mode-Matching Method Considering Edge Condition," *Trans. IECE Japan,* 1977, Vol. J60-B, No. 11, pp. 820–827 [in Japanese].

[20] Okuno, Y., and K. Yasuura, "Numerical Algorithm Based on the Mode-Matching Method with a Singular-Smoothing Procedure for Analysing Edge-Type Scattering Problems," *IEEE Trans. Antennas and Propagation,* July 1982, Vol. AP-30, No. 4, pp. 580–587.

[21] Okuno, Y., and T. Matsuda, "The Mode-Matching Method with a Generalized Singular-Smooth-

ing Procedure for Analysing the Edge-Type Scattering Problem with Several Edge Points," *Radio Sci.,* Sept.–Oct. 1984, Vol. 19, No. 5, pp. 1148–1154.

[22] Yasuura, K., Y. Okuno, and K. Ide, "On Numerical Technique for Calculating the Surface Current Density," *Trans. IECE Japan,* 1980, Vol. J63-B, No. 3, pp. 179–186 [in Japanese].

[23] Yasuura, K., and Y. Okuno, "Numerical Method for Calculating Surface Current Density on a Two-Dimensional Scatterer with Smooth Contour," *IEEE Trans. Antennas and Propagation,* Dec. 1985, Vol. AP-33, No. 12, pp. 1369–1378.

[24] Okuno, Y., "An Improved Algorithm for Calculating the Current Density on an Edge-Type Scatterer," *Proc. Int. U.R.S.I. Symp. 1980 on Electromagnetic Waves,* 1980, Vol. 231B. Note that the numerical example in this report contains some error.

[25] Stratton, J.A., *Electromagnetic Theory,* McGraw-Hill, New York, 1941, pp. 488–490.

[26] Sommerfeld, A., *Partielle Differentialgleichungen,* Dieterich'sche Verlag, Wiesbaden, 1947.

[27] Van Bladel [2], pp. 368–372.

[28] Van Bladel [2], pp. 372–375.

[29] Sommerfeld [26].

[30] Yoshida, K., *Functional Analysis,* Springer-Verlag, Berlin, 1965.

[31] If C is sufficiently smooth, then $E_N^0(s)$ converges to $-F(s)$ uniformly on C; otherwise (4.30) does not hold.

[32] Morse and Feshbach [1], p. 623.

[33] Considering the asymptotic form of the Hankel function for $|m| \gg k\rho$, we find that the series corresponds to a Taylor series with the center at ∞. We agree that the outside of the circle is in the convergence area of the series because there are no singularities. The asymptotic form can be found in Morse and Feshbach [1], p. 1564.

[34] Morse and Feshbach [1], p. 278.

[35] Application of the first order SP to an H-wave problem corresponds to the introduction of a stream function, which is well known in potential theory; see [8], pp. 97–105.

[36] Ikuno, H., and K. Nakata, "Transient Scattering from a Cylindrical Target with Inflection Points," *Trans. IECE Japan,* Vol. J70-B, No. 5, 1987, pp. 593–601 [in Japanese].

[37] R. Petit *et al.,* "Electromagnetic Theory of Gratings," ed. R. Petit. Springer-Verlag, Berlin, 1980.

[38] Ibid., p. 165.

[39] Okuno, Y., "A Numerical Method for Solving Edge-Type Scattering Problems," *Radio Sci.,* Vol. 22, No. 6, 1987, pp. 941–946.

[40] Tomita, M., and K. Yasuura, "Numerical Analysis of Plane Wave Scattering from Dielectric Cylinders," *Trans. IECE Japan,* 1979, Vol. J62-B, No. 2, pp. 132–139 [in Japanese].

[41] Yasuura, K., and M. Murayama, "Numerical Analysis of Diffraction from a Sinusoidal Metal Grating," *Trans. IECE Japan,* 1986, Vol. J69-B, No. 2, pp. 198–205 [in Japanese].

[42] Yasumoto, K., and M. Murayama, "A Rigorous Analysis of a Thin-Film Dielectric Waveguide with a Sinusoidal Surface Corrugation," *Radio Sci.,* to be published.

[43] Yasuura, K., K. Shimohara, and T. Miyamoto, "Numerical Analysis of a Thin-Film Waveguide by Mode-Matching Method," *J. Opt. Soc. Am.,* Feb. 1980, Vol. 70, No. 2, pp. 183–191.

[44] Miyamoto, T., and K. Yasuura, "Rigorous Analysis of an Embedded Optical Waveguide," *IEE Proc. H,* Feb. 1984, Vol. 131, No. 1, pp. 45–48.

[45] Calderon, A.P., "The Multipole Expansion of Radiaton Fields," *J. Ration. Mech. Anal. (J. Math. Mech.),* 1954, Vol. 3, pp. 523–537.

[46] If the contour C is strongly deformed from a circle, we should employ the functions given in example 2 or 3 in Appendix A to assure a rapid convergence.

[47] Erdelyi, A., W. Magnus, F. Oberhettinger, and F.G. Tricomi, *Higher Transcendental Functions,* Vol. 1, McGraw-Hill, New York, 1953, pp. 35–39.

[48] Lawson, C.L., and R.J. Hanson, *Solving Least Squares Problems,* Prentice-Hall, Englewood Cliffs, NJ, 1974.

Chapter 5
The Spatial Network Method

Norinobu Yoshida

5.1 INTRODUCTION

Recently, rapid developments in digital computing have made possible high-speed and large-scale computation. The digitalization of electronic equipment can be seen to use the computer as the index of its peak technology. Using these advances in the analyses of electromagnetic fields remarkably increases the effectiveness of numerical analysis methods in solving problems with complex boundary and medium conditions. Furthermore, this digitalization, accompanied by progress in ultra-high-speed pulse techniques, has increased the importance of the overall analyses of the system with respect to the time domain. These analyses are demanded specifically for obtaining such propagation characteristics as reflection and crosstalk.

Such analyses need to consider the direct relationship between field variations at each time and spatial point, rather than a formulation supposing only a sinusoidal incident wave. In addition, by utilizing a formulation that considers such relationships, Maxwell's equation can be used directly in the time domain, and simulations of electromagnetic phenomena can be performed.

Digital simulations using numerical methods necessarily demand the discretization of the space and time domains. In addition, the difference expression of the basic formula for the associated electromagnetic fields is given by the differential form of Maxwell's equation. This formulation describes the spatial structure of electromagnetic fields in which the field components are arranged at each discrete point to satisfy Maxwell's equation. In the 1940s, such a structure was theoretically proposed by Y. Okada, M. Iwashita, and T. Fuji; and it is known as an *electromagnetic lattice* [1]. At about the same time, G. Kron proposed an equivalent circuit using Maxwell's equation by corresponding field variables to circuit variables [2]. Three-dimensional numerical analyses based on the circuit could not be carried out at that time because of the undeveloped computer technology of those days. However, the proposition provided the theoretical foundation for a two-dimensional network analyzer with discrete *LC* elements and was later applied to solve waveguide problems by J.R. Whinnery and S. Ramo [3]. More recently, T.A. Tumolillo and J.P. Wondra formulated a method for the analysis of an electromagnetic pulse [4].

In 1966, F.H. Branin studied the topological properties of Maxwell's equation and clarified the relationship between differential operators and definitions in graph theory based on an electric and magnetic field duality [5]. At about this time, the appearance of the mainframe computer dramatically expanded the application of numerical analysis methods to realistic problems in various fields. Consequently, in 1966, K.S. Yee proposed a numerical analysis method [6] based mainly on the following two points: first, the expression of Maxwell's equation in the difference form was obtained by assigning each field variable to a discrete point in space; second, the iterative computation was realized by alternatively using the first and second formulas of Maxwell's equation in the time domain. This method, now known as the *finite difference time domain method* (FD-TD), has become one of the most practical and effective methods for time-dependent analysis of electromagnetic fields. A. Taflove and M.E. Browdin, R. Holland, K.S. Kunz and K.-M. Lee, and others have used this method in analyzing such problems as scattering in an electromagnetic pulse and absorbing in complex mediums [7, 8, 9].

Another effective method is the *transmission line matrix method* (TLM), which was initially proposed for analysis in a two-dimensional space and later expanded to analyze three-dimensional space by P.B. Johns and R.L. Beurle [10]. In this method, each component equation of Maxwell's equation for each spatial axis is assigned to a discrete point in space, and the branch between the points is considered a one-dimensional transmission line. The simulating process of generating secondary waves by the scattering matrix defined at each point is regarded as the modeling of Huygens's principle for wave propagation. This method has also been applied to many problems and further extended by S. Akhtarzad, W.J.R. Hoeffer, and others [11, 12].

For the analysis of electromagnetic fields in the time domain, N. Yoshida, *et al*. proposed the *spatial network method* (SNM) [13–14]. In this method, the one-dimensional line between points is expressed by the Bergeron method as an expansion of the treatment of the line in TLM. This formulation realizes a total analysis, specifically including medium conditions expressed as lumped elements. The Bergeron method was first proposed for the analysis of a water hammer by L.J.B. Bergeron. It was based on a finding by d'Alembert, where the quantity made up by the voltage and current components is seen to be conserved in a forward- and backward-traveling wave [15]. Later, in the 1960s, this method was applied to surge analysis in electric power systems by H.W. Dommel and programs for practical applications were developed [16, 17]. G. Metzger and J.-P. Pabre and others also expanded this method to distributed transmission lines with nonlinear active elements, by utilizing the method's iterative computation [18]. Such applications of the Bergeron method to the equivalent circuit of electromagnetic fields led to the definition of both voltage and current variables.

By using both variables, we can formulate a kinetic equation of polarization and the characteristic differential equation of anisotropy directly at each node [19, 20]. Such formulations are difficult in both the FD-TD and TLM method. In the following section, the spatial network method will be explained further.

5.2 THE RANGE OF APPLICATIONS

The spatial network method presents a model of a wave propagation mechanism based on the difference form of Maxwell's equation. This method is quite versatile, as it can be applied to various other problems. In applying the method to an individual problem, the medium and boundary conditions are expressed as lumped elements at the corresponding nodes in the equivalent circuit used in the spatial network method. An input correspondent to an incident condition is then applied. The electromagnetic wave form consequently propagates and generates a secondary wave due to the conditions at each node reached by the wave. The graph of the process shows various wave phenomena, such as reflection, penetration, diffraction, and radiation. As a result, the spatial and time-dependent variations of the electromagnetic fields in the total system are simulated. For example, the excitation of a continuous sinusoidal wave of a constant frequency brings about a steady state when a constant oscillation appears. This is apparent after the various transient wave phenomena have receded at each node. From the spatial characteristics of the electromagnetic fields in this steady state, the characteristic impedance and the propagation constant for the transmission line system can be evaluated. However, the excitation of an isolated wave such as a pulse waveform makes it possible to observe the wave propagation process; that is, the wave changing its form iteratively due to boundary and medium conditions. Thus, it becomes feasible to understand the influences of both conditions on wave propagation. Furthermore, by applying the Fourier transformation theorem to each time-dependent waveform at the input and output points, the ratio of those results gives the frequency response of the transmission characteristics.

As can be seen from this, analysis by the spatial network method is analogous to performing the experiment and obtaining time or frequency characteristics by computer analysis of a theoretical model. In addition, computation with different shapes and medium conditions corresponds to the estimation of characteristics of various test models in real world experiments. However, as an analyzed model constructed in a computer is discrete in the space and time domains, the spatial and temporal patterns of the computed field necessarily suffer the limitations of the sampling theorem. It also should be considered that the periodic structure of nodes in the spatial network has a low-pass property. Furthermore, the treatment of a medium condition as a lumped element at each node can be regarded as satisfying the distributed system in the same way as a loading coil. The relationship between the wavelength and the insertion distance of elements is important. These conditions are satisfied fundamentally by decreasing the discretized width in the space and time domains to bring about the necessary resolution. Factors that may cause errors or impose limitations are the same as those in TLM. The formulation on the characteristic line, based on the time increment being adjusted to the propagation time between nodes, does not cause a numerical instability due to the disagreement between the space and time differences. It is also known that the trapezoidal approximation of differential equations for medium conditions has numerical stability, because the expression of the trapezoidal equation includes both a forward- and backward-difference form. Still more, as the iterative computation at each node in the time domain uses only the four adjacent values obtained at previous time steps, a linear equation for the total system at each time step is not needed. Consequently, problems are not likely to occur in the computation of a large scale matrix. Limitation of either memory storage or computation time, as well as the generation of errors, can be avoided.

As mentioned previously, the spatial network method has characteristics based on both the discrete model of Maxwell's equation and formulations by the Bergeron method in the time domain. Nevertheless, the memory storage and the speed of the computer ultimately decide the effectiveness and precision of the method in each analysis of an individual problem. However, the recent improvement in memory capacity and high-speed computation, along with the development of the vector processor, especially in the supercomputer, makes it possible to perform ultra-high-speed computation of the iterative procedure in the spatial network method. This clearly reduces the limitation and increases the effectiveness of the method.

5.3 SPATIAL NETWORK FOR THREE-DIMENSIONAL MAXWELL'S EQUATION

In considering Maxwell's equation in a three-dimensional space, at each discrete point, each field variable is assigned to satisfy the mutual relationship between the variables derived from the equation. The resultant arrangement of the variables is the same as in the FD-TD and TLM methods, as is the correspondence of each component equation to each point. In Figure 5.1, a three-dimensional lattice network model of Maxwell's equation is shown. Table 5.1 presents the correspondence of each component in Maxwell's equation to each of the respective points, which are identified in the figure as A, B, C, D, E, and F.

In the spatial network method, an equivalent circuit is constructed on the following three principles: first, we assume that the interval between the discrete points is a one-dimensional line; second, every point is treated as a node where the con-

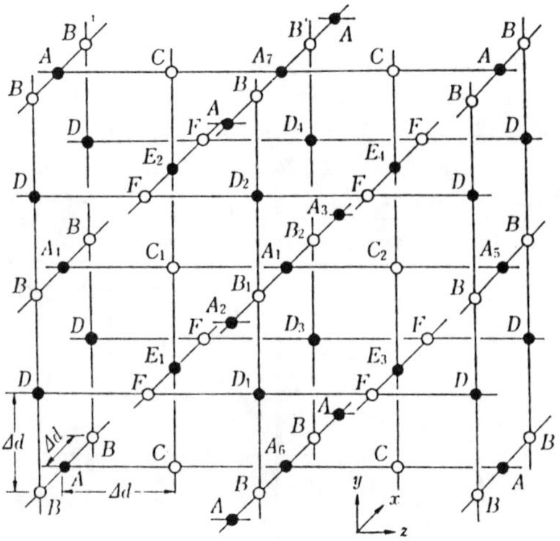

Figure 5.1 Three-dimensional cubic lattice network (Δd: the interval of spatial discretization).

tinuity law about electric or magnetic currents occurs; and finally, the medium conditions are expressed as lumped elements. The second formulation is made possible by defining the magnetic current by virtue of the duality between the electric and magnetic fields. These principles allow a wide application of the method.

To realize the network on these principles, all electromagnetic variables correspond to circuit variables at each node. As shown in Table 5.1, the equivalent voltage variables are given by V_u and V_u^*, and the equivalent current and equivalent magnetic current are given by I_u and I_u^* ($u = x, y, z$), respectively. By using these equivalent circuit variables, each component in Maxwell's equation assigned at each node is transformed into a two-dimensional transmission equation. This equation expresses the transmission of the plane wave perpendicular to the direction of the equivalent voltage defined at each node. This property coincides with the connecting structure of the one-dimensional lines at each node.

The usual node at which an electric field component is treated as a voltage variable is called an *electric node,* and the dual node at which a magnetic field component is treated as a voltage variable is called a *magnetic node*. In the magnetic nodes, the correspondence of the circuit variables to the field variables has a duality to those in the electric nodes. That is, the two variables exchange their correspondences. The symbol * represents the difference of correspondence at each kind of node. In these correspondences between the field variables and circuit variables, the polarity is decided by the direction of the electromagnetic field that constructs the Poynting vector. The Poynting vector has the same direction as the positive coordinate of each one-dimensional line. Therefore, the direction of the electric and mag-

Table 5.1
Correspondence between the Field Variables and Medium Constants and the Circuit Variables and Constants at Each Node in the Equivalent Circuit

	Electric Node			Magnetic Node	
	Maxwell's equs.	Variables		Maxwell's equs.	Variables
A_n	$\frac{\partial H_x}{\partial z}-\frac{\partial H_z}{\partial x}=\varepsilon_0\frac{\partial E_y}{\partial t}$	$V_y\equiv E_y$	F_n	$\frac{\partial E_x}{\partial z}-\frac{\partial E_z}{\partial x}=-\mu_0\frac{\partial H_y}{\partial t}$	$V_y^*\equiv H_y$
	$-\frac{\partial E_y}{\partial z}=-\mu_0\frac{\partial H_x}{\partial t}$	$I_z\equiv -H_x$		$-\frac{\partial H_y}{\partial z}=\varepsilon_0\frac{\partial E_x}{\partial t}$	$I_z^*\equiv E_x$
	$\frac{\partial E_y}{\partial x}=-\mu_0\frac{\partial H_z}{\partial t}$	$I_x\equiv H_z$		$\frac{\partial H_y}{\partial x}=\varepsilon_0\frac{\partial E_z}{\partial t}$	$I_x^*\equiv -E_z$
D_n	$\frac{\partial H_z}{\partial y}-\frac{\partial H_y}{\partial z}=\varepsilon_0\frac{\partial E_x}{\partial t}$	$V_x\equiv E_x$	B_n	$\frac{\partial E_y}{\partial x}-\frac{\partial E_x}{\partial y}=-\mu_0\frac{\partial H_z}{\partial t}$	$V_z^*\equiv H_z$
	$\frac{\partial E_x}{\partial z}=-\mu_0\frac{\partial H_y}{\partial t}$	$I_z\equiv H_y$		$\frac{\partial H_z}{\partial y}=\varepsilon_0\frac{\partial E_x}{\partial t}$	$I_y^*\equiv -E_x$
	$-\frac{\partial E_x}{\partial y}=-\mu_0\frac{\partial H_z}{\partial t}$	$I_y\equiv -H_z$		$-\frac{\partial H_z}{\partial x}=\varepsilon_0\frac{\partial E_y}{\partial t}$	$I_x^*\equiv E_y$
E_n	$\frac{\partial H_y}{\partial x}-\frac{\partial H_x}{\partial y}=\varepsilon_0\frac{\partial E_z}{\partial t}$	$V_z\equiv -E_z$	C_n	$\frac{\partial E_z}{\partial y}-\frac{\partial E_y}{\partial z}=-\mu_0\frac{\partial H_x}{\partial t}$	$V_x^*\equiv -H_x$
	$\frac{\partial E_z}{\partial y}=-\mu_0\frac{\partial H_x}{\partial t}$	$I_y\equiv -H_x$		$\frac{\partial H_x}{\partial z}=\varepsilon_0\frac{\partial E_y}{\partial t}$	$I_z^*\equiv E_y$
	$-\frac{\partial E_z}{\partial x}=-\mu_0\frac{\partial H_y}{\partial t}$	$I_x\equiv H_y$		$-\frac{\partial H_x}{\partial y}=\varepsilon_0\frac{\partial E_z}{\partial t}$	$I_y^*\equiv -E_z$
Permittivity		$C_0=\varepsilon_0/2$	Permittivity		$L_0^*=\varepsilon_0/2$
Permeability		$L_0=\mu_0/2$	Permeability		$C_0^*=\mu_0/2$
Polarization		$\Delta C=\varepsilon_0\chi_e/2\cdot\Delta d$	Magnetization		$\Delta C^*=\mu_0\chi_m/2\cdot\Delta d$
Conductivity		$G=\sigma/2\cdot\Delta d$	Magnetic Conductivity		$G^*=\sigma^*/2\cdot\Delta d$

netic currents coincides with that of the Poynting vector; that is, the direction of the energy flow. Thus, both currents have the same property as the real conduction current in the line.

In the equivalent circuit, to match the duality of the circuit variables at both the electric and magnetic nodes, a gyrator (as shown in Figure 5.2(a) and (b)) is inserted in series with each magnetic node. As a result, in the one-dimensional transmission line, the usual correspondences of both the electric field to the voltage and the magnetic field to the current are realized. As a special case exists between nodes B and D, a negative gyrator is inserted. The earlier-mentioned correspondence between the equivalent circuit variables and electromagnetic-field variables, the gyrators, and the Poynting vectors is expressed in the basic lattice network shown in Figure 5.2(c), which is composed of each of the nodes presented in Table 5.1.

Next, it should be confirmed that the lattice network shown in Figure 5.1 can

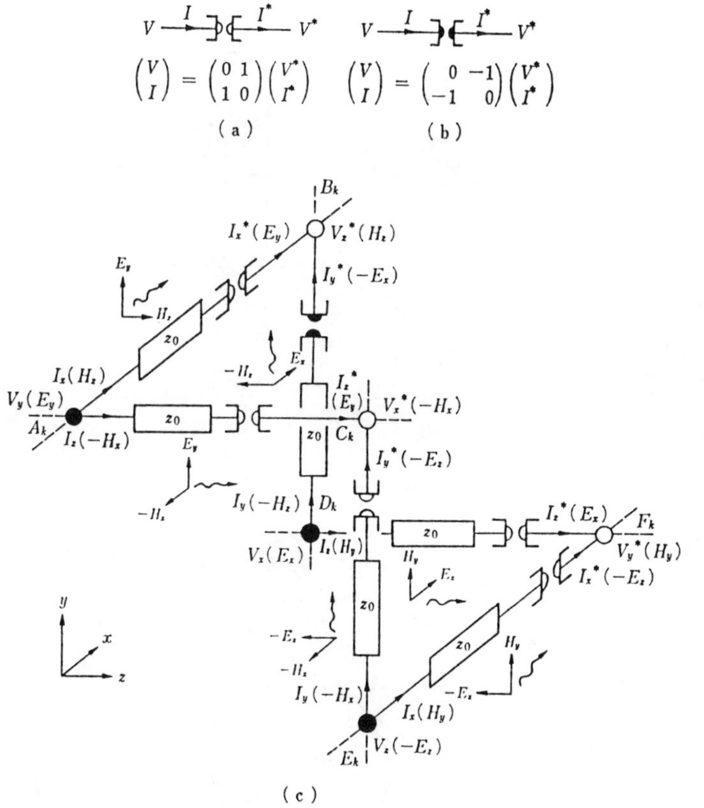

Figure 5.2 Gyrators: (a) positive gyrator; (b) negative gyrator; (c) the fundamental equivalent circuit with these gyrators.

equivalently express the wave field of electromagnetic fields. As the space is assumed to be linear and isotropic, the lattice network must satisfy the wave equation for each component of the electromagnetic fields. Here we prove that the wave equation about the y-directed component of the electric field E_y can be derived. In Figure 5.3, a detailed representation of the nodes arranged around node A_1 (as presented in Figure 5.1) is shown. Node A_1 is selected as representative of the nodes treating the electric field E_y as a voltage variable. Since each line between the nodes is assumed to be a one-dimensional transmission line, the following equations between voltage and current can be given as

$$\frac{\partial V_v^{(*)}}{\partial u} = -L_0^{(*)} \frac{\partial I_w^{(*)}}{\partial t} \tag{5.1a}$$

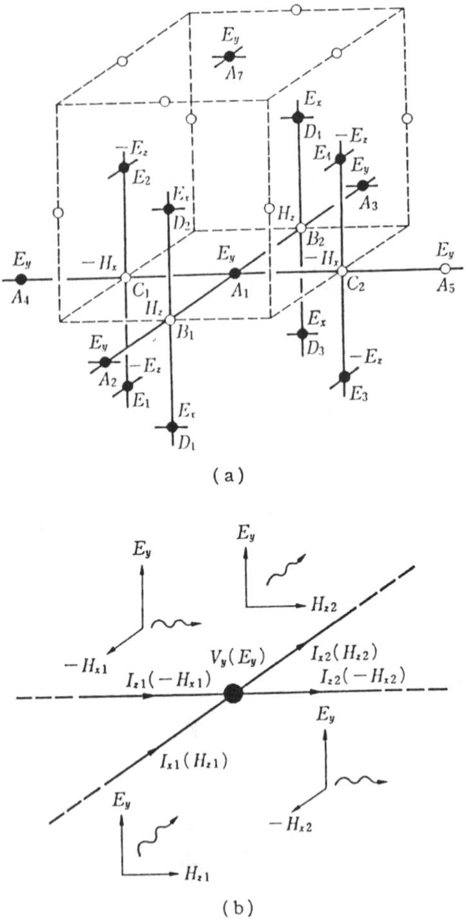

Figure 5.3 (a) Arrangement of nodes around electric node A_1 where the electric field corresponds t
the voltage variable and (b) definition of the circuit variables at node A_1.

$$\frac{\partial I_w^{(*)}}{\partial u} = -C_0^{(*)} \frac{\partial V_v^{(*)}}{\partial t} \tag{5.1b}$$

where, u, v, and w correspond to each component of the orthogonal coordinate sys tem and t is time. The variables and constants $V_v^{(*)}$, $I_w^{(*)}$, $L_0^{(*)}$, and $C_0^{(*)}$ have the sam correspondences as those of the electromagnetic field shown in Table 5.1. Initially according to equation (5.1b), the central difference equations at central points B an C are obtained. (The equations are expressed using the variables of the correspondin electromagnetic fields.) The difference equation in the x-direction at point B_1 is given as

$$E_y(A_2) - E_y(A_1) = 2\Delta x(\mu_0/2)\Delta_t H_z(B_1) \tag{5.2a}$$

where $\Delta x(= \Delta y = \Delta z = \Delta d)$ is the spatial interval and Δ_t denotes the difference form of the partial differential operator in the time domain $\partial/\partial t$ for the time discretization Δ_t. The variable in the parenthesis indicates the position of the node. At point B_2, the next equation is derived.

$$E_y(A_1) - E_y(A_3) = 2\Delta x(\mu_0/2)\Delta_t H_z(B_2) \tag{5.2b}$$

The difference equations in the y-direction at points B_1 and B_2 are derived as follows:

$$[-E_x(D_1)] - [-E_x(D_2)] = 2\Delta y(\mu_0/2)\Delta_t H_z(B_1) \tag{5.2c}$$

$$[-E_x(D_3)] - [-E_x(D_4)] = 2\Delta y(\mu_0/2)\Delta_t H_z(B_2) \tag{5.2d}$$

Similarly, in the z-direction, at points C_1 and C_2, the equations

$$E_y(A_4) - E_y(A_1) = 2\Delta z(\mu_0/2)\Delta_t[-H_x(C_1)] \tag{5.2e}$$

$$E_y(A_1) - E_y(A_5) = 2\Delta z(\mu_0/2)\Delta_t[-H_x(C_2)] \tag{5.2f}$$

are obtained; and in the y-direction, the following equations are given:

$$[-E_z(E_1)] - [-E_z(E_2)] = 2\Delta y(\mu_0/2)\Delta_t[-H_x(C_1)] \tag{5.2g}$$

$$[-E_z(E_3)] - [-E_z(E_4)] = 2\Delta y(\mu_0/2)\Delta_t[-H_x(C_2)] \tag{5.2h}$$

To express the variables H_z and $-H_x$ at points B_1, B_2 and C_1, C_2 in terms of the magnetic-field components H_{z1}, H_{z2} and $-H_{x1}$, $-H_{x2}$ at node A_1 (shown in Figure 5.3(b)), equation (5.1a) is used. For example, the variable H_z at node B_1 is given as follows:

$$H_z(B_1) = H_{z1} - \Delta x \cdot \Delta_x H_z(B_1) = H_{z1} + (\varepsilon_0/2)\Delta x \Delta_t E_y(B_1) \tag{5.3a}$$

where Δ_x, Δ_y, and Δ_z denote the difference forms of the partial differential operators in the space domain; namely, $\partial/\partial x$, $\partial/\partial y$ and $\partial/\partial z$. At the other nodes B_2, C_1 and C_2, the following equations are also formulated in the same manner:

$$H_z(B_2) = H_{z2}(A_1) - (\varepsilon_0/2)\Delta x \Delta_t E_y(B_2) \tag{5.3b}$$

$$[-H_x(C_1)] = [-H_{x1}(A_1)] + (\varepsilon_0/2)\Delta z \Delta_t E_y(C_1) \tag{5.3c}$$

$$[-H_x(C_2)] = [-H_{x2}(A_1)] - (\varepsilon_0/2)\Delta z\Delta_t E_y(C_2) \tag{5.3d}$$

By substituting equations (5.3a–d) with the respective corresponding magnetic-field variables in the right-hand side of (5.2a–h), the following formulas are given. For example, equations (5.2a) and (5.2c) are expressed as follows by (5.3a):

$$E_y(A_2) - E_y(A_1) = 2\Delta x(\mu_0/2)\Delta_t H_{z1}(A_1) + 2\Delta x^2(\varepsilon_0\mu_0/4)\Delta_t^2 E_y(B_1) \tag{5.4a}$$

$$[-E_x(D_1)] - [-E_x(D_2)] = 2\Delta y(\mu_0/2)\Delta_t H_{z1}(A_1) + 2\Delta x\Delta y(\varepsilon_0\mu_0/4)\Delta_t^2 E_y(B_1) \tag{5.4b}$$

Similarly, the following can be written

$$E_y(A_1) - E_y(A_3) = 2\Delta x(\mu_0/2)\Delta_t H_{z2}(A_1) - 2\Delta x^2(\varepsilon_0\mu_0/4)\Delta_t^2 E_y(B_2) \tag{5.4c}$$

$$[-E_x(D_3)] - [-E_x(D_4)] = 2\Delta y(\mu_0/2)\Delta_t H_{z2}(A_1) - 2\Delta x\Delta y(\varepsilon_0\mu_0/4)\Delta_t^2 E_y(B_2) \tag{5.4d}$$

$$E_y(A_4) - E_y(A_1) = 2\Delta z(\mu_0/2)\Delta_t[-H_{x1}(A_1)] + 2\Delta z^2(\varepsilon_0\mu_0/4)\Delta_t^2 E_y(C_1) \tag{5.4e}$$

$$E_y(A_1) - E_y(A_5) = 2\Delta z(\mu_0/2)\Delta_t[-H_{x2}(A_1)] - 2\Delta z^2(\varepsilon_0\mu_0/4)\Delta_t^2 E_y(C_2) \tag{5.4f}$$

$$[-E_z(E_1)] - [-E_z(E_2)] = 2\Delta y(\mu_0/2)\Delta_t[-H_{x1}(A_1)] + 2\Delta z\Delta y(\varepsilon_0\mu_0/4)\Delta_t^2 E_y(C_1) \tag{5.4g}$$

$$[-E_z(E_3)] - [-E_z(E_4)] = 2\Delta y(\mu_0/2)\Delta_t[-H_{x2}(A_1)] - 2\Delta z\Delta y(\varepsilon_0\mu_0/4)\Delta_t^2 E_y(C_2) \tag{5.4h}$$

where Δ_t^2 denotes the difference form of the second-order differential operator in the time domain $\partial^2/\partial t^2$. Because the components of the magnetic fields at node A_1 correspond to the currents shown in Figure 5.3(b), the direction of the currents (i.e., the inflow and the outflow) is defined according to the node. Thus, considering the current variable at node A_1 in the right-hand side of each equation in (5.4), previously obtained, the equations dealing with the inflowing current and outflowing current are respectively added and subtracted. This operation yields the following equation:

$$\begin{aligned}&\{E_y(A_2) + E_y(A_3) - 2E_y(A_1)\} + \{E_y(A_4) + E_y(A_5) - 2E_y(A_1)\}\\ &\quad - [\{E_x(D_4) - E_x(D_2)\} - \{E_x(D_3) - E_x(D_1)\}\\ &\quad + \{E_z(E_4) - E_z(E_2)\} - \{E_z(E_3) - E_z(E_1)\}]\\ &= 4\Delta d(\mu_0/2)\Delta_t\{H_{z1} - H_{z2} - H_{x1} + H_{x2}\}_{A1}\\ &\quad + 4\Delta d^2(\varepsilon_0\mu_0/4)\Delta_t^2\{E_y(B_1) + E_y(B_2) + E_y(C_1) + E_y(C_2)\} \qquad (5.5)\end{aligned}$$

In this equation, by supposing the existence of Kirchhoff's first law at node A_1, the first term on the right-hand side is eliminated from the continuity equation of currents:

$$H_{z1} - H_{z2} - H_{x1} + H_{x2} \equiv I_{x1} - I_{x2} + I_{z1} - I_{z2} = 0 \qquad (5.6)$$

Furthermore, after dividing both sides of equation (5.5) by $4\Delta d^2$, deriving the difference form of the left-hand side by considering the mutual relation between the components of the electromagnetic fields expressed in Figure 5.3(a) gives the following:

$$\begin{aligned}&\frac{E_y(A_2) + E_y(A_3) - 2E_y(A_1)}{4\Delta x^2} + \frac{E_y(A_4) + E_y(A_5) - 2E_y(A_1)}{4\Delta z^2}\\ &- \frac{\dfrac{E_x(D_4) - E_x(D_2)}{2\Delta x} - \dfrac{E_x(D_3) - E_x(D_1)}{2\Delta x}}{2\Delta y}\\ &- \frac{\dfrac{E_z(E_4) - E_z(E_2)}{2\Delta z} - \dfrac{E_z(E_3) - E_z(E_1)}{2\Delta z}}{2\Delta y} \qquad (5.7)\end{aligned}$$

Also, the second term on the right-hand side is given as follows:

$$\varepsilon_0\mu_0\Delta_t^2\,\frac{E_y(B_1) + E_y(B_2) + E_y(C_1) + E_y(C_2)}{4} \qquad (5.8)$$

When Δd and Δt approach zero, equation (5.7) is transformed as follows:

$$\frac{\partial^2 E_y}{\partial x^2} + \frac{\partial^2 E_y}{\partial z^2} - \frac{\partial}{\partial y}\left(\frac{\partial E_x}{\partial x} + \frac{\partial E_z}{\partial z}\right) \tag{5.9}$$

In this equation, the expression in parentheses corresponds to the divergence of electric fields shown by the dashed lines in Figure 5.3(a). Under the space charge free condition (div$\boldsymbol{E}$ = 0), the expression becomes $-\partial E_y/\partial y$. In such a case, the preceding equation becomes

$$\frac{\partial^2 E_y}{\partial x^2} + \frac{\partial^2 E_y}{\partial y^2} + \frac{\partial^2 E_y}{\partial z^2} \tag{5.10}$$

Furthermore, as equation (5.8) yields the average value according to the values at the adjacent four nodes, it gives the second-order derivative $\partial^2 Ey/\partial t^2$ in the time domain. As the difference interval becomes sufficiently small, equation (5.5) satisfies the wave equation of E_y at node A_1, thus converging to the following equation:

$$\frac{\partial^2 E_y}{\partial x^2} + \frac{\partial^2 E_y}{\partial y^2} + \frac{\partial^2 E_y}{\partial z^2} = \varepsilon_0 \mu_0 \frac{\partial^2 E_y}{\partial t^2} \tag{5.11}$$

At other A_n nodes, a similar wave equation may be derived. At nodes D_n and E_n, the wave equations about the electric fields E_x and E_z are obtained, respectively. Also, at the magnetic nodes B_n, C_n, and F_n (where the magnetic field corresponds to the voltage variable), using the continuity law of magnetic currents at each node, the wave equations according to the magnetic fields H_z, H_x, and H_y can be derived, respectively. As a result, the spatial lattice network shown in Figure 5.1 can realize the wave equation according to each component of the electromagnetic field. This is based on the treatment of each branch between nodes as a one-dimensional transmission line and each lattice point as the node where the continuity law of (magnetic) currents occurs. By combining these wave fields, the network satisfies the following vector wave equations for all the electromagnetic fields:

$$\nabla^2 \boldsymbol{E} = \varepsilon_0 \mu_0 \frac{\partial^2 \boldsymbol{E}}{\partial t^2} \tag{5.12}$$

$$\nabla^2 \boldsymbol{H} = \varepsilon_0 \mu_0 \frac{\partial^2 \boldsymbol{H}}{\partial t^2} \tag{5.13}$$

In the network satisfying these wave equations, both permittivity and permeability per unit length (at each one-dimensional transmission line) have half the worth of the real values designated by $C_0(*)$ and $L_0(*)$ in Table 5.1. The propagation time in the line is also equivalent to half the real value t_0. Thus, the propagation velocity becomes twice the real speed.

The medium condition can also be included in the wave equation in the cubic lattice network. This is done by expressing the medium conditions in terms of lumped elements at each node. For example, the lossy dielectric medium can be expressed by the parallel connection of a capacitance and a conductance at an electric node. Similarly, the magnetic medium can be expressed by the parallel connection of an equivalent magnetic capacitance and a magnetic conductance at a magnetic node. For a lossy dielectic with relative polarization χ_e and conductivity σ, the displacement current I_c and the conduction current I_d are given at node A_1 (using Table 5.1) as follows:

$$I_c = 4\Delta d \frac{\varepsilon_0 \chi_e}{2} \Delta_t E_y(A_1) \tag{5.14}$$

$$I_d = 4\Delta d \frac{\sigma}{2} E_y(A_1) \tag{5.15}$$

If we apply these equations to the bracketed expression of the first term on the right-hand side in equation (5.5), (5.6) can be rewritten as follows:

$$I_{x1} - I_{x2} + I_{z1} - I_{z2} = I_c + I_d \tag{5.16}$$

Then, the first term becomes

$$4\Delta d^2 \sigma \mu_0 \Delta_t E_y(A_1) + 4\Delta d^2 \mu_0 \varepsilon_0 \chi_e \Delta_t^2 E_y(A_1) \tag{5.17}$$

Using the relation $\varepsilon = \varepsilon_0(1 + \chi_e)$ at the limit of Δd and $\Delta t \to 0$, the wave equation according to the electric field E_y in the dielectic with conductive loss is given as follows:

$$\nabla^2 E_y = \mu_0 \sigma \frac{\partial E_y}{\partial t} + \varepsilon \mu_0 \frac{\partial^2 E_y}{\partial t^2} \tag{5.18}$$

At the electric nodes D_n and E_n, a similar expression by lumped elements gives the wave equations for the electric fields E_x and E_z in the dielectric medium. Combining these equations satisfies the equivalent circuit, giving the following wave equation:

$$\nabla^2 \mathbf{E} = \mu_0 \sigma \frac{\partial \mathbf{E}}{\partial t} + \varepsilon \mu_0 \frac{\partial^2 \mathbf{E}}{\partial t^2} \tag{5.19}$$

Consider the magnetic medium. By using the relative magnetic susceptibility χ_m and the conductivity of the magnetic current σ^*, the reverse electromotive force I_c^* and the magnetic conductive current I_d^* can be defined at magnetic nodes B_n, C_n, and F_n. For example, the following equations are derived at node F_1 (see Table 5.1).

$$I_c^* = 4\Delta d\left(\frac{\mu_0 \chi_m}{2}\right)\Delta_t H_y(F_1) \tag{5.20}$$

$$I_d^* = 4\Delta d\left(\frac{\sigma^*}{2}\right)H_y(F_1) \tag{5.21}$$

By referring to (5.6), the resultant equation of the continuity law of magnetic currents can be obtained as follows:

$$I_{x1}^* - I_{x2}^* + I_{z1}^* - I_{z2}^* = I_c^* + I_d^* \tag{5.22}$$

Then, the application of this equation and the relation $\mu = \mu_0(1 + \chi_m)$ derive the following equation:

$$\nabla^2 H_y = \varepsilon_0 \sigma^* \frac{\partial H_y}{\partial t} + \varepsilon_0 \mu \frac{\partial^2 H_y}{\partial t^2} \tag{5.23}$$

At other magnetic nodes, similar equations can be obtained. So, according to the magnetic fields, a general vector wave equation can be given in the lossy magnetic medium, thus,

$$\nabla^2 \boldsymbol{H} = \varepsilon_0 \sigma^* \frac{\partial \boldsymbol{H}}{\partial t} + \varepsilon_0 \mu \frac{\partial^2 \boldsymbol{H}}{\partial t^2} \tag{5.24}$$

Clearly, the cubic lattice network shown in Figure 5.1 can comprehensively model electromagnetic wave fields under various medium conditions.

5.4 THE BERGERON METHOD

In the Bergeron method, the propagation characteristics in a one-dimensional line of a finite length are formulated in the time domain using the voltages and currents at both ports of the line. The formulation is closely related to the d'Alembert solution, and the fundamental properties of the wave fields formulated by the wave equation are iteratively computed along a characteristic line in the space and time domain, determined by the propagation velocity. By these principles, the formulation and the computation of the circuit equation in terms of the connected lumped elements can be realized at each discrete point. This is done by utilizing the two variables of voltage and current and the independent relation of each discrete point at each given time by the existence of a propagation time. This method also has general characteristics, such that it can be applied extensively in other areas. For example, the treatment of reactance elements can be undertaken by formulating their characteristic differential equation by the trapezoidal rule in the time domain. Similarly, the treatment of nonlinear properties can be considered by piecewise approximation. In the following text, the derivation of Bergeron's formulation for a one-dimensional line and the formulation of reactance elements by the trapezoidal rule will be considered.

In a line of finite length d obtained by discretizing the original line, the Bergeron method evaluates the value of the voltage and current at each discrete point; namely, the node. This is conducted in the time domain by using an initial condition. The equation for the method is derived from the basic equation of a lossless line. The total value of the voltage and current in the line are given by the following differential equations:

$$\frac{d}{dt} v(x, t) = \frac{\partial v}{\partial x}\frac{dx}{dt} + \frac{\partial v}{\partial t} \tag{5.25a}$$

$$\frac{d}{dt} i(x, t) = \frac{\partial i}{\partial x}\frac{dx}{dt} + \frac{\partial i}{\partial t} \tag{5.25b}$$

By substituting the following transmission equations of a one-dimensional line to the preceding equations,

$$-\frac{\partial v}{\partial x} = L_0 \frac{\partial i}{\partial t} \tag{5.26a}$$

$$-\frac{\partial i}{\partial x} = C_0 \frac{\partial v}{\partial t} \tag{5.26b}$$

the following can be derived [21]:

$$\frac{d}{dt}\left\{v \pm \sqrt{\frac{L_0}{C_0}}\, i\right\} = \left\{\frac{\partial v}{\partial x} - \sqrt{\frac{L_0}{C_0}}\frac{\partial i}{\partial x}\right\}\left\{\frac{dx}{dt} \mp \frac{1}{\sqrt{L_0 C_0}}\right\} \tag{5.27}$$

The characteristic impedance z_0 and the propagation velocity c_0 are expressed by the following equations with the inductance L_0 and the capacitance C_0 per unit length of the line:

$$z_0 = \sqrt{\frac{L_0}{C_0}} \tag{5.28}$$

$$c_0 = \frac{1}{\sqrt{L_0 C_0}} \tag{5.29}$$

Using the following relation,

$$\frac{dx}{dt} = \pm \frac{1}{\sqrt{L_0 C_0}} = \pm c_0 \tag{5.30}$$

the expression in the second braces becomes zero for any moving coordinate having the same velocity as that of the traveling wave. Thus, equation (5.27) becomes

$$\frac{d}{dt}(v \pm z_0 i) = 0 \tag{5.31}$$

The positive and negative signs correspond respectively to the forward- and backward-traveling waves. By including this polarity in the current variable as the parameter indicating its direction, (5.31) shows that the quantity $v + z_0 i$ is conserved on the moving coordinate; that is, on the characteristic line. By integrating the preceding equation by the time interval of the propagation time Δt and by supposing $\Delta t = d\sqrt{L_0 C_0}$, the following equation is obtained:

$$[v \pm z_0 i]_{t-\Delta t}^{t} = 0 \tag{5.32}$$

For forward- and backward-traveling waves, the following two equations respectively are derived.

$$v(d, t) + z_0 i(d, t) = v(0, t - \Delta t) + z_0 i(0, t - \Delta t) \tag{5.33a}$$

$$v(0, t) - z_0 i(0, t) = v(d, t - \Delta t) - z_0 i(d, t - \Delta t) \tag{5.33b}$$

These algebraic equations show that the voltage and the current at both ends of a line with finite length are related to each other through the characteristic line. The d'Alembert solution is given as follows:

$$v + z_0 i = 2f(x - c_0 t) \tag{5.34a}$$

$$v - z_0 i = 2g(x + c_0 t) \tag{5.34b}$$

The characteristics of the waves expressed by equations (5.31) and (5.32) correspond to the fact that the functions of the forward-traveling wave f and the backward traveling wave g are conserved in a moving coordinate system with traveling velocity c_0. The earlier-mentioned characteristics of electromagnetic waves in a space and time domain are expressed by the group of characteristic lines on the $x = t$ plane, shown in Figure 5.4. In this figure, the axis of abscissa represents the spatial coordinates of a line of finite length where the ends are respectively indicated as $x = 0$ and $x = d$. The vertical axis indicates the time. The lines ascending to the right represent the characteristic lines of traveling waves that start from the end indicated as $x = 0$, at an arbitrary time. The gradient of this line equals the inverse of the velocity of wave c_0^{-1}. Along the line, the quantity $v + z_0 i$ is conserved. Conversely, the lines ascending to the left side show the characteristic lines of backward traveling waves. They begin at the end indicated as $x = d$. The gradient equals $-c_0^{-1}$ and the quantity $v - z_0 i$ is conserved. For both groups of lines, the relation to the correspondent lines at any given time can be formulated by a pair of difference equations ((5.33a) and (5.33b)), using the voltages and currents at each end. Furthermore, as the equivalent circuit of a one-dimensional transmission line, this relation is expressed by a series circuit composed of a constant voltage source and resistance, as shown in Figure 5.5(b). The constant voltage source E_r creates a backward traveling wave as observed at the input port indicated by $x = 0$ at time t in the line shown in Figure 5.5(a). This function and value are also given by the right-hand side of equation (5.33b); thus,

$$E_r = v(d, t - \Delta t) - z_0 i(d, t - \Delta t) \tag{5.35a}$$

Source E_i creates a forward traveling wave, as observed at the other port indicated by $x = d$ and given by the right-hand side of equation (5.33a):

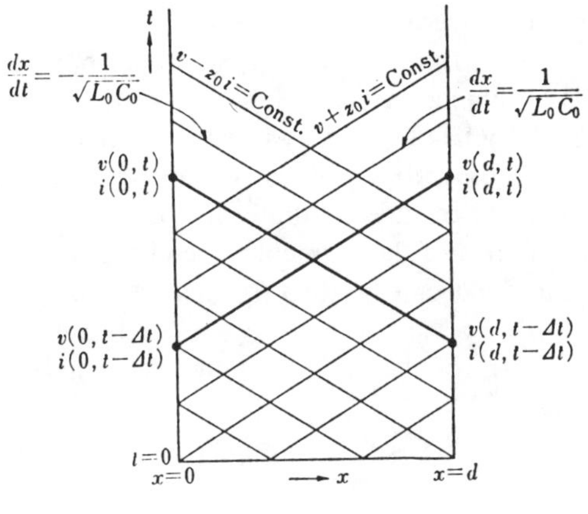

Figure 5.4 Characteristic lines of a one-dimensional line.

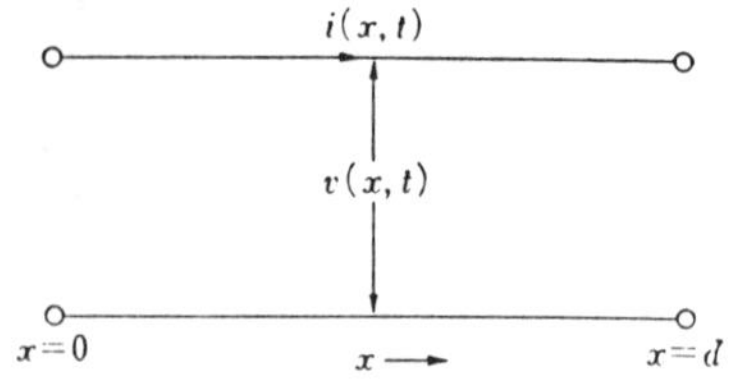

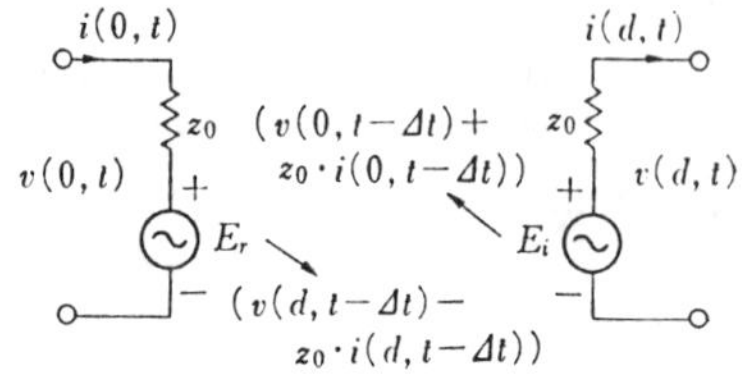

Figure 5.5 (a) A one-dimensional line, and (b) its equivalent circuit with the voltage source in Bergeron's expression.

$$E_i = v(0, t - \Delta t) + z_0 i(0, t - \Delta t) \tag{5.35b}$$

Each respective source is determined by the values of the voltage and current obtained at propagation time $t - \Delta t$ in the other port. Thus, using such controlled sources, both nodes of the line are separated and the physical states at both ends can be expressed respectively by the independent circuit.

By considering the forward and backward traveling waves, the characteristics of the voltage and current in the analyzed system can be formulated by linear algebraic equations. The number of such equations is twice the number of discrete points in the system and can be obtained by iterative computation in the time domain. In the Bergeron method, the two variables of voltage and current are used as the circuit variables at each end of the one-dimensional transmission line. Then, using the voltage-current properties of the elements, the characteristics of the lumped elements connected at each end can be directly included in the iteratively computed equations of the line. Giving the voltage drops at each time as $v_R(t)$ and $v_G(t)$, and the current as $i_R(t)$ and $i_G(t)$, resistance and conductance can be expressed as follows:

$$v_R(t) = R \cdot i_R(t) \tag{5.36}$$

$$i_G(t) = G \cdot v_G(t) \tag{5.37}$$

Here, R and G are the values of the resistance and conductance, respectively. In addition, the reactance elements such as inductance and capacitance are formulated by approximating their voltage-current characteristics by the trapezoidal rule in the time domain. The resultant equations obtained by the previous formulations are combined with Bergeron's expression of the line to make a total equation. The characteristic differential equation of inductance is given by the following, defining the voltage between terminals v_L with a current of i_L as shown in Figure 5.6(a):

$$v_L(t) = L\frac{di_L(t)}{dt} \tag{5.38}$$

By integrating this equation from the time point $t - \Delta t$ to the present time t, the following equation is obtained:

$$i_L(t) = \frac{1}{L}\int_{t-\Delta t}^{t} v_L(\tau)d\tau + i_L(t - \Delta t) \tag{5.39}$$

This integral can be expressed by the trapezoidal approximation as follows:

$$i_L(t) = \frac{v_L(t) + v_L(t - \Delta t)}{2L}\Delta t + i_L(t - \Delta t) \tag{5.40}$$

By dividing this equation into terms involving the present t and the past $t - \Delta t$, the next difference equation for the inductance can be derived:

$$G_L v_L(t) - i_L(t) = -\{G_L v_L(t - \Delta t) + i_L(t - \Delta t)\} \tag{5.41a}$$

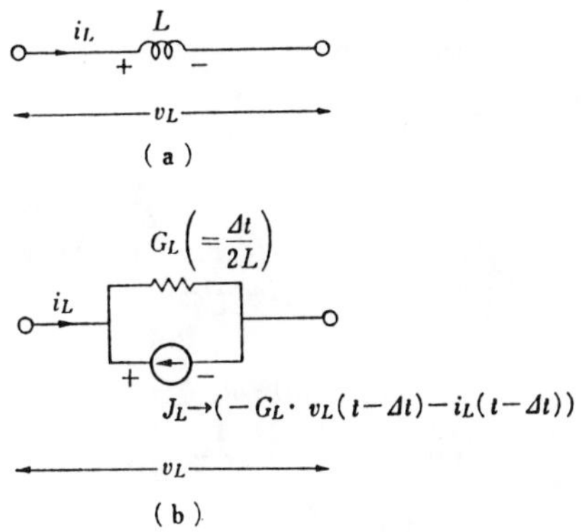

Figure 5.6 (a) An inductance and (b) its equivalent circuit with the current source given by the trapezoidal approximation in the time domain.

$$G_L = \frac{\Delta t}{2L} \tag{5.41b}$$

As G_L represents the unit of admittance, it is considered the characteristic conductance of the inductance in the time domain. From equation (5.41a), the equivalent circuit of the inductance is expressed using the current source J_L and the characteristic conductance G_L (see Figure 5.6(b)). J_L is given using the right-hand side of (5.41a):

$$J_L = -\{G_L v_L(t - \Delta t) + i_L(t - \Delta t)\} \tag{5.42}$$

This current source is also the controlled source determined by the values of the voltage and current at time $t - \Delta t$.

Also, for the capacitance, a differential equation of the voltage-current characteristics is given by the following equation. The voltage is defined between terminals as v_C, with the current defined as i_C, shown in Figure 5.7(a).

$$i_C(t) = C\frac{dv_C(t)}{dt} \tag{5.43}$$

Integrating this equation yields

$$v_C(t) = \frac{1}{C}\int_{t-\Delta t}^{t} i_C(\tau)d\tau + v_C(t - \Delta t) \tag{5.44}$$

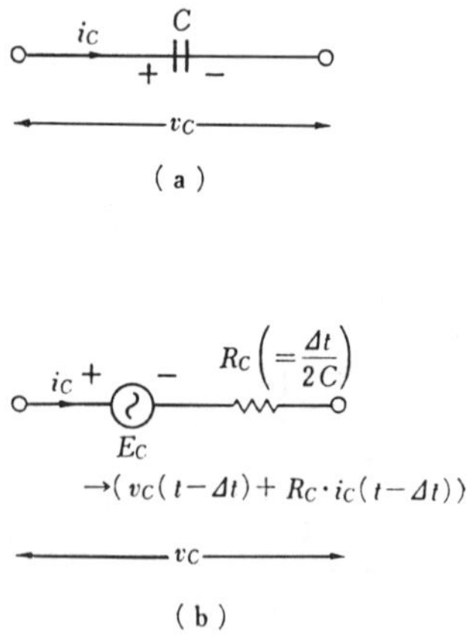

Figure 5.7 (a) A capacitance and (b) its equivalent circuit with the voltage source given by the trapezoidal approximation in the time domain.

The approximation equation by the trapezoidal rule is derived as

$$v_C(t) = \frac{i_C(t) + i_C(t - \Delta t)}{2C} \Delta t + v_c(t - \Delta t) \tag{5.45}$$

By dividing this equation into terms related to either present time t or the past $t - \Delta t$, the following equations result.

$$v_C(t) - R_C i_C(t) = v_C(t - \Delta t) + R_C i_C(t - \Delta t) \tag{5.46a}$$

$$R_C = \frac{\Delta t}{2C} \tag{5.46b}$$

R_C represents the unit of resistance and is considered the characteristic resistance in the difference formulation of the capacitance in the time domain. From (5.46a), the equivalent circuit of the capacitance is expressed using the voltage source E_C and the characteristic resistance R_C, as shown in Figure 5.7(b). E_C is given as follows, using the right-hand side of equation (5.46a):

$$E_c = v_c(t - \Delta t) + R_C i_C(t - \Delta t) \tag{5.47}$$

This voltage source is also the controlled source determined by the values of the voltage and current, obtained at time interval Δt, which is similar to equations (5.35) and (5.42).

Thus, the equations for inductance and capacitance approximated to their difference forms by the trapezoidal rule, have an iteratively computed form similar to the Bergeron expression of a line, given by equation (5.33). In the difference form, the values of the voltage and current at the present time t are determined by the characteristic conductance or resistance and using those values at $t - \Delta t$. Moreover, using the common interval of the time difference, the unified circuit equation at each node (including the characteristics of the lines and the lumped elements) can be derived in the time domain, and the time response can consequently be calculated.

Because the Bergeron expression of a line describes the precise characteristics of a wave propagation on a moving coordinate system, that is, on the characteristic line, the interval of the spatial and time difference must satisfy the rigid relation $\Delta t = d/c_0$ by considering the propagation velocity. Furthermore, the trapezoidal approximation, which gives the relation between the voltage and current by difference forms in the time domain, is considered the hybrid form of the forward and backward difference. This approximation is known to exhibit stability for numerical analysis, and its precision is fundamentally decided by the size of the sampling interval of the wave phenomena in the space and time domain.

5.5 BERGERON'S EXPRESSION IN THE THREE-DIMENSIONAL SPATIAL NETWORK

In this section, a time-dependent analysis method of three-dimensional electromagnetic fields is derived by applying the Bergeron method (described in Section 5.4) to the three-dimensional spatial network (shown in Section 5.3). In Figure 5.8, the model of a one-dimensional transmission line is shown with a characteristic line impedance of z_0 and a propagation time Δt; thus,

$$z_0 = \sqrt{\frac{L_0}{C_0}} = \sqrt{\frac{\mu_0}{\varepsilon_0}} = Z_0 \tag{5.48a}$$

$$\Delta t = \Delta d\sqrt{L_0 C_0} = \Delta d\sqrt{\varepsilon_0 \mu_0}/2 = \Delta t_0/2 \tag{5.48b}$$

where L_0 and C_0 represent the inductance and the capacitance per unit length of the one-dimensional line in the three-dimensional spatial network given in Table 5.1. The terms z_0 and Δt_0 are the characteristic impedance and the propagation time of the plane wave in the free space, respectively. Especially note the relation given in equation (5.48b), as it corresponds to the real physical values. The term Δd is the length of the line and gives the interval of the discretization in the three-dimensional spatial network. By supposing the parameter k as the number of discrete nodes, the following equations can be given for the forward- and backward-traveling waves expressed in equations (5.33a) and (5.33b), respectively.

$$V(k, t) + z_0 I(k, t) = V(k - 1, t - \Delta t) + z_0 I(k - 1, t - \Delta t) \tag{5.49a}$$

$$V(k - 1, t) - z_0 I(k - 1, t) = V(k, t - \Delta t) - z_0 I(k, t - \Delta t) \tag{5.49b}$$

Here, V and I are the voltage and current at each discrete node. In Figure 5.9, the equivalent circuit is shown at the electric node A where the electric field E_y is supposed to be a voltage variable. The spatial position in the directions x, y, and z

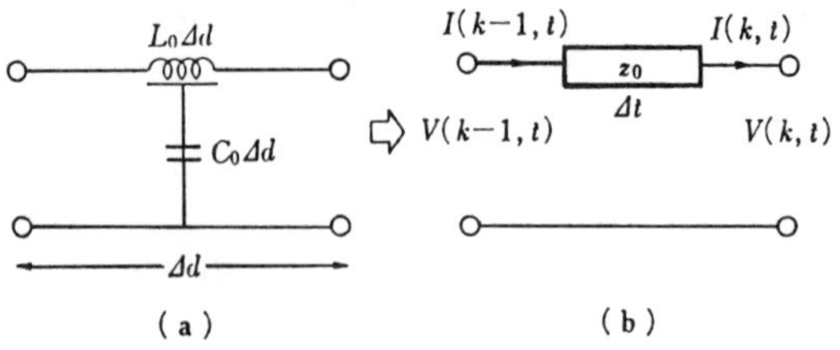

Figure 5.8 (a) A one-dimensional line and (b) its Bergeron expression.

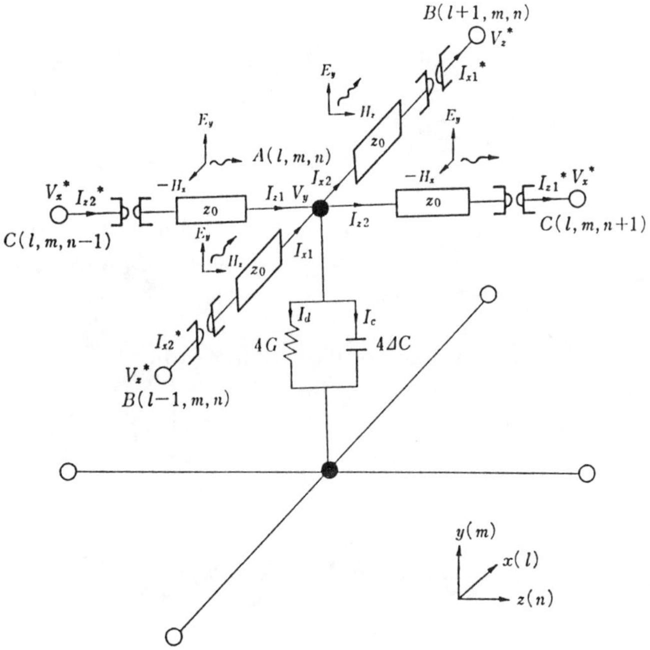

Figure 5.9 The equivalent circuit at node $A(l, m, n)$ including the medium condition.

is denoted by the discrete numbers l, m, and n, respectively. Time is expressed by t. In addition, in this figure, the one-dimensional lines between nodes are expressed temporarily as unbalanced lines. These are shown as solid lines in Figure 5.1 for the sake of simplicity. In this circuit, the voltage between the lines corresponds to the electric field E_y; and the values of the medium conditions, ΔC and G, correspond to the polarization and conductivity listed in Table 5.1. These values are quadruple those at each line because of the concentration of values at the node, as can be seen in equations (5.14) and (5.15). Application of the Bergeron equation, represented in equation (5.49), to each line connected to node A yields the following equations. These equations indicate the direction of the wave in relation to the node by considering the series gyrators at the magnetic nodes.

$$V_y(l, m, n, t) + z_0 I_{z1}(l, m, n, t) = I_{z2}^*(l, m, n-1, t-\Delta t) + z_0 V_x^*(l, m, n-1, t-\Delta t) \quad (5.50a)$$

$$V_y(l, m, n, t) - z_0 I_{z2}(l, m, n, t) = I_{z1}^*(l, m, n+1, t-\Delta t) - z_0 V_x^*(l, m, n+1, t-\Delta t) \quad (5.50b)$$

$$V_y(l, m, n, t) + z_0 I_{x1}(l, m, n, t) = I^*_{x2}(l-1, m, n, t-\Delta t) + z_0 V^*_z(l-1, m, n, t-\Delta t) \quad (5.50c)$$

$$V_y(l, m, n, t) - z_0 I_{x2}(l, m, n, t) = I^*_{x1}(l+1, m, n, t-\Delta t) - z_0 V^*_z(l+1, m, n, t-\Delta t) \quad (5.50d)$$

Next, the lumped elements making up the medium condition can also be expressed by the Bergeron method. By defining the conductance current as I_d, the characteristics of the conductance G can be written as follows:

$$I_d(l, m, n, t) = 4G(l, m, n)V_y(l, m, n, t) \quad (5.51)$$

Also, the characteristics of the capacitance can be defined by using the trapezoidal expression of equation (5.46a), as follows:

$$V_y(l, m, n, t) - R_c(l, m, n)I_c(l, m, n, t) = V_y(l, m, n, t-\Delta t) + R_c(l, m, n)I_c(l, m, n, t-\Delta t) \quad (5.52a)$$

where

$$R_c(l, m, n) = \frac{\Delta t}{2[4\Delta C(l, m, n)]} \quad (5.52b)$$

Here, I_c and R_c respectively stand for the current through the capacitance and the characteristic resistance of the capacitance in the time domain. In the preceding equations, the expression of each parameter of the medium conditions as a function of position shows the eligibility of applying the method to nonuniform fields by changing the values of the parameters corresponding to the distribution of the medium. The right-hand sides of equations (5.50) and (5.52a) are the already known values obtained from the voltage and current estimations, taken from the previous time step. By denoting each respective right-hand side of these equations as ψ_1, ψ_2, ψ_3, ψ_4, and ψ_c and by using the following equation of the continuity of current at node A

$$I_{z1} - I_{z2} + I_{x1} - I_{x2} - I_c - I_d = 0 \quad (5.53)$$

we can solve equations (5.50), (5.51), and (5.52a) according to the voltage V_y, yielding the next nodal equation at time t:

$$V_y(l, m, n, t) = \frac{R_c(l, m, n)(\psi^*_1 + \psi^*_2 + \psi^*_3 + \psi^*_4) + z_0\psi_c}{z_0 + R_c(l, m, n)\{4 + 4G(l, m, n)z_0\}} \quad (5.54)$$

Each component of the current at time t is evaluated by substituting this V_y for that defined in each of equations (5.50), (5.51), and (5.52a), respectively. At other A nodes, similar formulations can be performed. Also, at the magnetic nodes, using the condition of the continuity of magnetic currents, a similar nodal equation can be derived thus:

$$V_u^*(l, m, n, t) = \frac{R_c^*(l, m, n)(\psi_1 + \psi_2 + \psi_3 + \psi_4) + z_0^* \psi_c^*}{z_0^* + R_c^*(l, m, n)\{4 + 4G^*(l, m, n)z_0^*\}} \tag{5.55}$$

Here, $z_0^* = z_0^{-1}$, and ΔC^* and G^* are the magnetic polarization and conductance listed in Table 5.1. Also, u corresponds to one of each (x, y, z) coordinate axis. In equations (5.54) and (5.55), every variable and parameter at a magnetic node is identified by the symbol *. These nodal equations formulated at every node are evaluated at time t from the previously obtained values. The values at the next time step are computed using those obtained from the present. Thus, the time response of the field is evaluated iteratively. For all voltages and currents at each node, this procedure demands only two memory storages: one for the values for the previous time step, and one for the present values. The advantage of the Bergeron method is that the independence of each node at each time does not require solving the simultaneous equation for the total system. Thus, the construction of a matrix in memory storage of magnitude N^2 is unnecessary for the number of nodes N. Computation time clearly also is saved. However, as the computation of the overall region is performed synchronously in the time domain, the interval of the discrete time appears in both the line and medium equations. Consequently, this interval must be both equal in both equations and small in order to approximate the smallest time variations in the analyzed system.

Consider the computation of the dispersive characteristics. By applying the trapezoidal approximation, a similar iterative calculation to the nodal equation can be performed. As an example, the orientation polarization can be thus treated, as it has an important effect on the property of the dielectric medium in the microwave band. The fundamental equation for the orientation polarization characteristics can be expressed using the electric field E_u and polarization P_u in the direction u:

$$\frac{dP_u(t)}{dt} + \frac{P_u(t)}{\tau} = \frac{\varepsilon_0 \chi_e}{\tau} E_u(t) \tag{5.56}$$

The term τ is the relaxation time and χ_e is the relative polarization. Using the trapezoidal rule, the following difference form of the preceding equation is obtained:

$$P_u(t) - P_u(t - \Delta t) + \frac{P_u(t) + P_u(t - \Delta t)}{2\tau} \Delta t = \frac{\varepsilon_0 \chi_e \{E_u(t) + E_u(t - \Delta t)\}}{2\tau} \Delta t \tag{5.57}$$

Arranging this equation according to time t and $t - \Delta t$ yields

$$\left(1 + \frac{\Delta t}{2\tau}\right)P_u(t) - \frac{\varepsilon_0\chi_e\Delta t}{2\tau}E_u(t) = \left(1 - \frac{\Delta t}{2\tau}\right)P_u(t - \Delta t) + \frac{\varepsilon_0\chi_e\Delta t}{2\tau}E_u(t - \Delta t) \quad (5.58)$$

The relation between the time variation of the polarization and the displacement current I_c is given as

$$I_c(t) = \frac{dP_u(t)}{dt} \quad (5.59)$$

Applying the trapezoidal approximation to the preceding equation yields

$$P_u(t) = \frac{I_e(t) + I_e(t - \Delta t)}{2}\Delta t + P_u(t - \Delta t) \quad (5.60)$$

$P_u(t)$ can be eliminated by substituting equation (5.60) for (5.58) to derive an equation with respect to $E_u(t)$ and $I_c(t)$. Thus, eliminating $P_u(t)$ and rewriting E_u as V_u yields the following equation of the orientation polarization approximated by the trapezoidal rule:

$$V_u(t) - R_pI_c(t) = -V_u(t - \Delta t) + R_pI_c(t - \Delta t) + V_p(t - \Delta t) \quad (5.61a)$$

$$R_p = \left(\frac{\tau}{\varepsilon_0\chi_e} + \frac{\Delta t}{2\varepsilon_0\chi_e}\right) \quad (5.61b)$$

$$V_p(t - \Delta t) = \frac{2P_u(t - \Delta t)}{\varepsilon_0\chi_e} \quad (5.61c)$$

R_p is the characteristic resistance of the polarization in the time domain, and $V_p(t - \Delta t)$ corresponds to the polarization of the previous time step. All the terms on the right-hand side of equation (5.61a) are obtained from the previous time step. Thus, by supposing this right-hand side to be ψ_c and rewriting R_p as R_c, the characteristics of the orientation polarization can be expressed in a form quite similar to that of the usual dielectric medium. It can also be computed iteratively using equation (5.54) in the time domain. At each time (t), the displacement current $I_c(t)$ and the polarization $P_u(t)$ are calculated by substituting the voltage $V_u(t)$, obtained from equation (5.54), for $E_u(t)$ in (5.61a) and (5.58).

This treatment of the medium characteristics by the trapezoidal rule in the time domain can be extended further to the case of a nonmagnetized plasma and resonance absorption. Also for a ferrite and a magnetized plasma, a similar formulation by the trapezoidal rule can be applied by expressing their characteristic equations (including the bidirectional connections between variables) as a mutually connected circuit.

Such treatments clearly show the efficacy of the Bergeron method in using a circuit and the two variables of voltage and current, along with the formulation by trapezoidal approximation in the time domain.

5.6 ANALYZED RESULTS AND DISCUSSION

Using the three-dimensional analysis method formulated from the Bergeron method in the time domain, an L-shaped stripline is analyzed [22]. The analyzed model is shown in Figure 5.10. The term Δd denotes the interval of the spatial discretization. The analyzed region has a rectangular shape with dimensions $40\Delta d$, $10\Delta d$, and $40\Delta d$ in the x, y, and z directions, respectively. The dielectric substrate thickness on the ground plate is $4\Delta d$; and in the air region, a thickness of $6\Delta d$ is supposed. The upper plane and the side planes of the region are assumed to be a free boundary. The treatment of the boundary conditions of the conductor, along with the medium condition of the dielectric and the free boundary condition will now be considered. Finally, time responses within the electromagnetic fields will be discussed.

5.6.1 The Boundary Condition of the Conductor System

In this analysis, the conductors are supposed to have perfect conductivity as a fundamental condition. As a result, on the surface of the conductor, two boundary con-

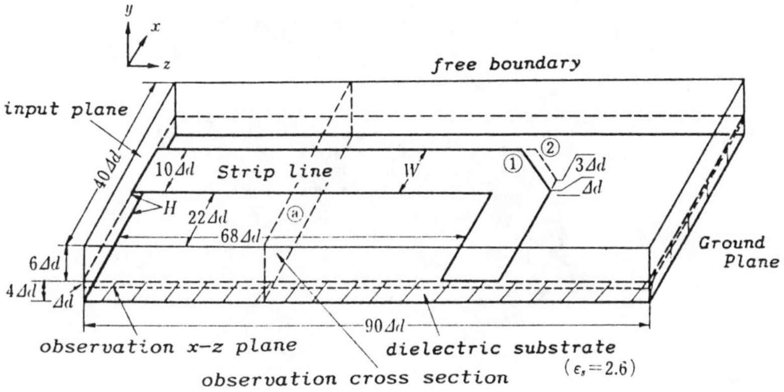

Figure 5.10 Analyzed model of an L-shaped stripline (copyright © 1984 IEEE).

ditions must be satisfied; that is, the tangential component of the electric fields is zero, and the normal component of the magnetic fields is zero. The general equivalent circuit of the boundary conditions for the conductor's surface is shown in Figure 5.11. The figure shows a very thin strip conductor attached to the surface of the dielectric. The dashed line designates the shape of the strip conductor. The surface is positioned on the *xz* plane and consists only of nodes *D*, *E*, and *F* (called the *DEF* plane) from the possible six node types: *A*, *B*, *C*, *D*, *E*, and *F*. The black dots in the figure denote nodes in which the voltage corresponds to the electric fields (electric nodes), whereas the open dots signify nodes in which the voltage corresponds to the magnetic fields (magnetic nodes). The boundary condition at each node is specified as follows:

1. It is given as either a short circuited or open terminal at the node where the tangential electric field is respectively either a voltage or current variable.
2. It is given as either a short circuited or open terminal at the node where the normal magnetic field is respectively either a voltage or current variable.

For instance, at nodes *D* and *E* on the conductive surface in the figure, both voltages respectively correspond to the tangential components of electric fields E_x and E_z on the conductive surface. They are short-circuited points from the preceding condition (1) and are equivalently expressed as the grounded nodes. As a result, the upper and lower surfaces at nodes *D* and *E* are separated because of no current passing through them in the *y* direction. This situation models the very thin conductive strip. However, at node *F*, the voltage corresponds to the normal magnetic field H_y and the node is short circuited due to condition (2). Furthermore, the currents in the *x* and *z* directions correspond to the tangential electric-field components E_z and E_x. Thus, the node is opened due to condition (1). The node is then identified as the isolated point where all circuit variables are zero, and no calculation is required as compu-

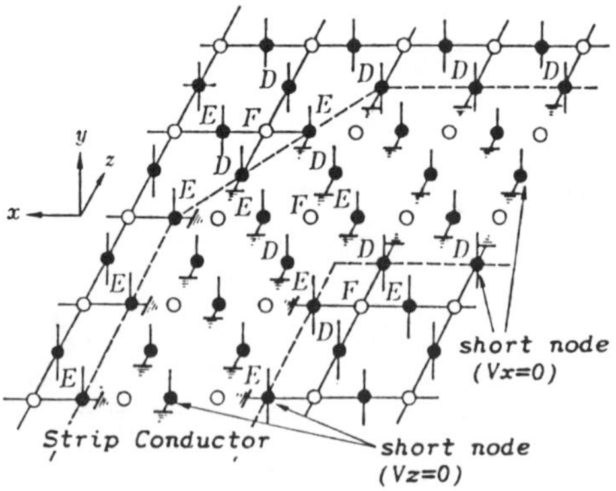

Figure 5.11 Equivalent circuit of the strip conductor supposed to have perfect conductivity (copyrigh
© 1984 IEEE).

tation time is eliminated. As a result, efficiency is increased. Moreover, the boundary condition at the edges of the strip conductor can be given in a similar manner. For example, consider the *E* and *D* nodes aligned along the edge of the strip and respectively pointing in the *x* and *z* directions. As the electric fields E_z and E_x correspond respectively to the voltage of each node, they are short circuited not only in the *y* direction but also in the *x* and *z* directions, respectively. In this analysis, as the ground conductor is also in the *DEF* plane, the boundary condition can be expressed similarly, as only the upper surface is involved in the analyzed region.

5.6.2 Treatment of Dielectric Materials

The dielectric material can be expressed by the equivalent circuit of the electric node in conjunction with the capacitance and conductance. As shown in Figure 5.9 for the equivalent circuit where the electric field E_y is represented as a voltage variable, the capacitance expresses the polarization and the conductance expresses the conductivity. Because the loss in the dielectric material is neglected in this analysis, the conductance *G* is zero. The condition on the boundary surface between the dielectric material and the air is expressed equivalently as follows. By positioning the boundary plane on the *DEF* surface, the continuity condition of the tangential component in both mediums is satisfied. This is because the voltage variables (E_x in node *D* and E_z in node *E*) are uniquely determined due to their single values at the node. Also, the medium value of the node on the surface has half that of the node in the medium, because the dielectric occupies only half the volume of that which the node represents.

5.6.3 Treatment of the Free Boundary

Because the analysis is performed at each node obtained by discretizing a finite spatial region, the size of the region is decided fundamentally by the relation between the number of the variables correspondent to that of the nodes, and the size of the memory storage of the computer used. As a result, for a finite region having a structure that is not closed by conductors, it is necessary to assume an arbitrary closed space. The surface of such a closed region must have no influence on the internal fields due to reflection. This is known as the *free boundary*. As the equivalent circuit of the surface, the nonreflective condition is realized by introducing a matching load impedance. In Figure 5.12, the equivalent circuit representations for the upper plane and the side plane in this analysis are given as (a) and (b), respectively. For the upper *xz* plane, assigned the *DEF* surface, a matched load resistance R_L is connected at nodes *D* and *E* in the *y* direction. Similarly, for the side *yz* plane assigned the *ACE* surface, the matched load impedance R_L is connected at nodes *A* and *E* in the *x* direction. The free space characteristic impedance is used for the matching resistance R_L by supposing the outward propagation of the plane wave. Generally, in the

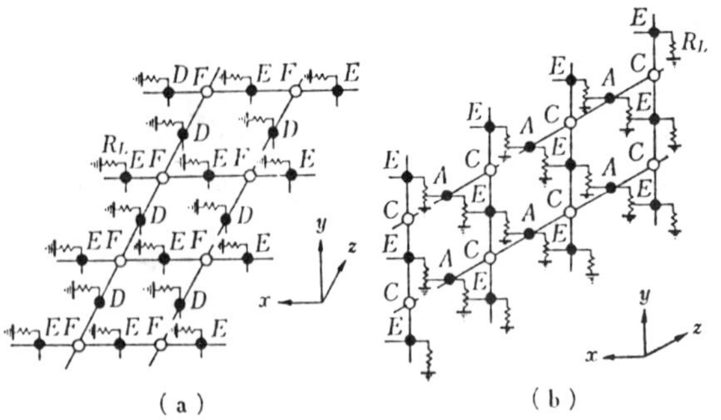

Figure 5.12 Equivalent circuit of the free boundary: (a) the top plane (xz plane); (b) the side plane (yz plane) (copyright © 1984 IEEE).

spatial region very close to the stripline, higher-order wave modes constructing complicated electromagnetic fields are generated. Consequently, the matching condition cannot be simply satisfied by only the free space impedance. Therefore, it becomes necessary to either place the free boundaries at a distance sufficiently far to neglect such complicated electromagnetic fields or to use a simulated wave absorber.

However, for the analysis of the transmission line in this case, the equivalent circuits for the input and output ports are important. In Figure 5.13, the equivalent circuit of the input port is shown. For the stripline structure, the field E_y is the dominant propagation electric field. The following input model is assumed: Sinusoidal voltage sources are applied through the source impedance R_s at the A nodes. At these points the electric field E_y corresponds to the voltage variable, positioned between the stripline and ground conductive plane. Each voltage source has a constant amplitude for simplicity and generates a sinusoidal wave with a constant period T in the time domain. The period T is normalized by the discrete time step Δt. The source impedance has different values due to the manner of modeling the real ex-

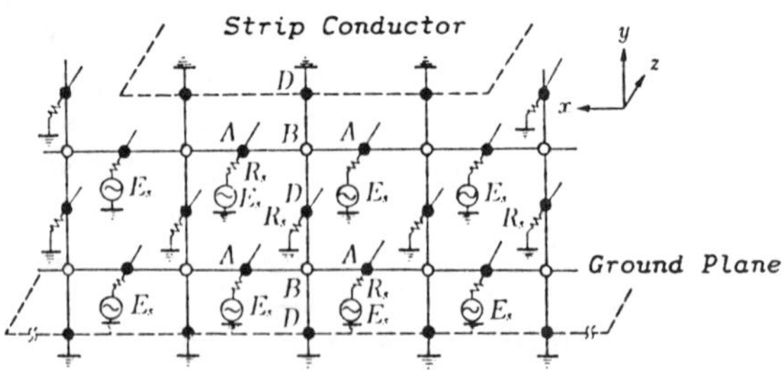

Figure 5.13 Equivalent circuit of the input condition, where E_s is the sinusoidal voltage source and R_s is the source impedance (copyright © 1984 IEEE).

citation system. In this analysis, by considering the matching conditions of the line, a characteristic impedance for the propagation of the fundamental mode is used for simplicity. Also, the output plane is modeled in a manner similar to the input plane, and its equivalent circuit is given by that of the input plane without the voltage sources.

5.6.4 Analyzed Results

In this analysis, an L-shaped stripline is analyzed. The time variation of the basic electromagnetic field distribution for a sinusoidal wave is computed and discussed using each field component obtained by vector analysis of the method. The sinusoidal wave of the input is expressed by a pulse train with Δt spacing in the time domain. In this numerical evaluation, the spatial interval Δd between adjacent nodes in the equivalent circuit is arbitrarily chosen to be 0.005 cm. Thus, the time interval Δt, from (5.48b), becomes 8.3×10^{-5} ns. The period of the applied sinusoidal wave is $213\Delta t$, so its frequency is about 56 GHz. The values of Δd and Δt have a direct effect on the precision of approximating the spatial structure of the subject and the applied sinusoidal wave. In this analysis, these values are sufficiently small that the resolution of the spatial and time function is satisfactory.

In Figure 5.14, the numerical results for the characteristic impedance and wavelength are plotted as a function of the line width. This is given as the ratio of line width W to the height of the line from ground plane H. In the figure, the dots and the lines represent the computed and the analytical values, respectively. Both results closely agree with each other. For the time response, the impedance is given by the

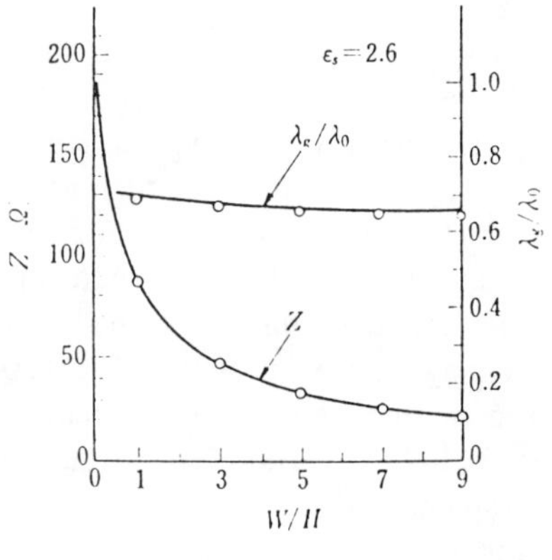

igure 5.14 Variations of characteristic impedance and wavelength of the straight stripline as a function of W/H, where dots represent the result computed by the present method and lines are the analytical results by E. Yamashita and R. Mittra (copyright © 1984 IEEE).

ratio of the electric potential of the strip conductor to the current in the z direction. The former is calculated by integrating the electric fields from the ground conductor to that of the strip. The latter is evaluated by the line integral of the magnetic fields around the strip conductor in the xy plane. The wave length is obtained from the instantaneous electric field distribution along the z axis. This computing method is based on the supposition of a fundamental propagation mode; namely, the quasi TEM. In the Figures 5.15, 5.16, 5.17, and 5.18, the spatial distribution of the electromagnetic field is obtained as the envelope by using the maximum values at every node in the half period of the applied wave. In each figure, the observation time is shown in the top right corner. T is the period of the applied wave, and t_0 is the initial time at which the incident wave is applied to the input plane. The observation plane is the xz plane Δd beneath the strip conductor. This is chosen for easy viewing of the propagation characteristics of the field in the space and time domains. This plane is indicated as the observation plane (xz plane) in Figure 5.10. Figures 5.15, 5.17, and 5.18 show the field distributions in the case of stripline 1. Stripline 1 resembles an $\mathcal{L}$ where the right angle is formed by the horizontal meeting, and the vertex is diagonally cut off. Stripline 2 resembles stripline 1, but with a smaller diagonal cut (see Figure 5.10).

Figure 5.15 shows the time variation of the electric field E_y. The following phenomena are observed. Before the wave arrives at the corner, a uniform amplitude distribution is observed along the direction of the line because of the existence of only a forward-traveling wave. After reaching the corner, the wave changes its propagation to the x direction. In addition, a standing wave is observed on the line due to the generation of a reflective wave in the line on the input side. The electric field rapidly drops away from the strip conductor. This clearly shows the propagation of the wave along the conductor.

Figure 5.16 shows the electric field distribution in the steady state in the case of stripline 2. The standing wave is comparatively larger than that for the case shown in Figure 5.15. This result clearly shows that shape and size of the diagonally cut corner influences the propagation characteristics noticeably. Thus, by such time-dependent analysis, the influences of a structure, such as a corner, can be iteratively observed. The steady-state characteristic is obtained after the transient time variations have finished. In this analysis, after about two periods, the field over the total analyzed region attains the steady state. Figure 5.17 and 5.18 show the time variation of the spatial distribution of magnetic fields H_x and H_z, respectively (for stripline 1). Each magnetic component constructs the Poynting vector, in the z and x directions respectively, with the electric field E_y. Therefore, Figure 5.17 shows that the magnetic field component H_x propagates in the z direction but the amplitude on the stripline rapidly decreases in the x direction at the back corner, where the component does not construct the propagation wave. Conversely, however, Figure 5.18 shows that the magnetic field component H_z appears abruptly at the back corner, because the component constructs the wave in the x direction, but scarcely appears on th

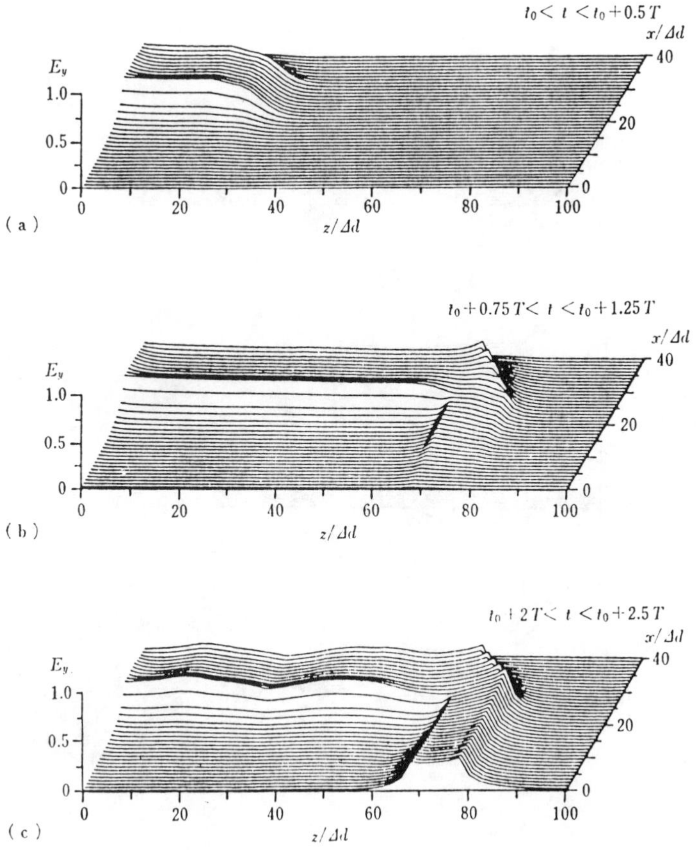

Figure 5.15 Time variation in the spatial distribution of the electric field E_y in the case of stripline 1 of Figure 5.10, where T is the period of an incident wave and t_0 is the beginning time of applying input (copyright © 1984 IEEE).

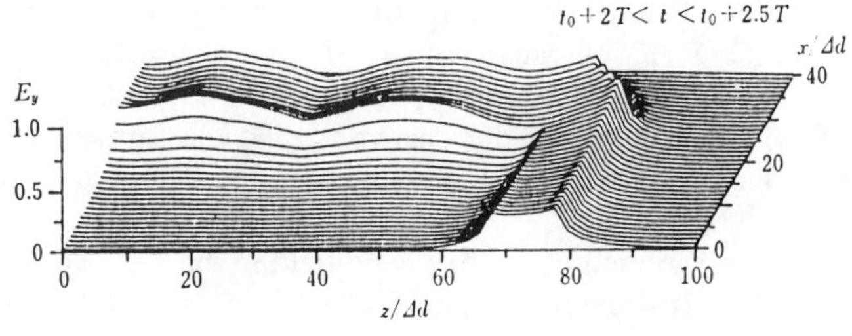

igure 5.16 Spatial distribution in the electric field E_y in the case of stripline 2 (copyright © 1984 IEEE).

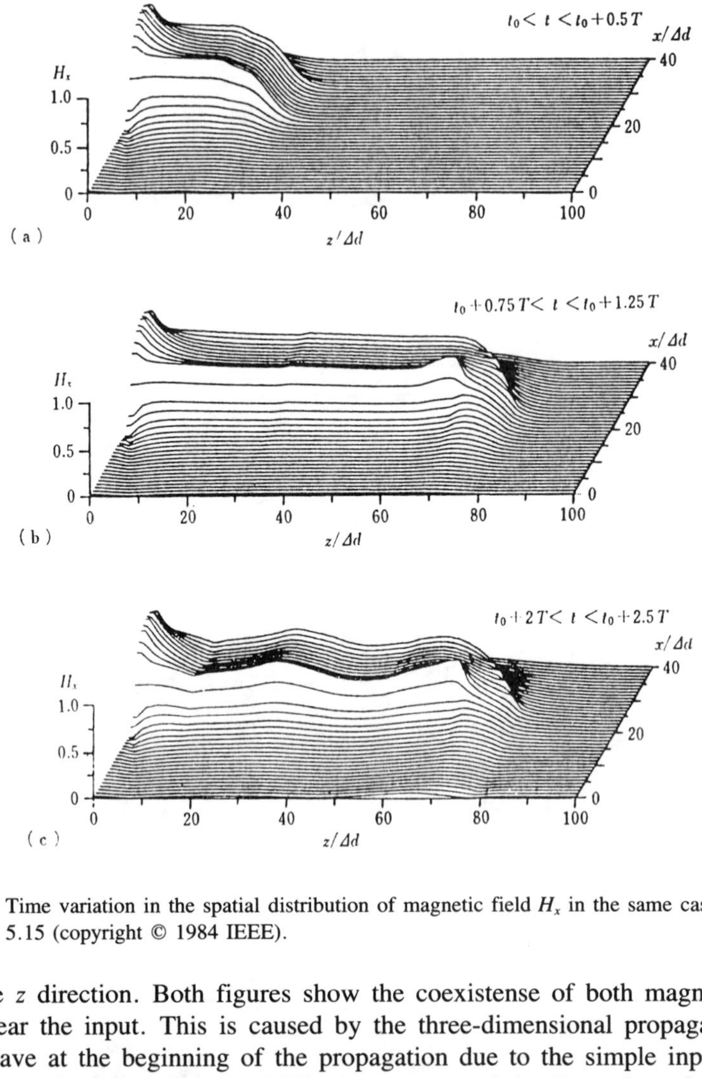

Figure 5.17 Time variation in the spatial distribution of magnetic field H_x in the same case as Figure 5.15 (copyright © 1984 IEEE).

line in the z direction. Both figures show the coexistense of both magnetic components near the input. This is caused by the three-dimensional propagation of a circular wave at the beginning of the propagation due to the simple input model. However, the field's amplitude rapidly decreases, and the formation of the quasi-TEM wave as the basic propagation mode is observed in the variations of the magnetic fields.

Finally, the time variation of the magnetic field in the x direction on the lower surface of the stripline is shown in Figure 5.19. The observation position in the z direction is the same as the observed cross section (x-y plane) in Figure 5.10. The magnitude corresponds to the current in the z direction on the surface and shows an edge effect. In this effect, the spatial distribution shows the magnitude increase a

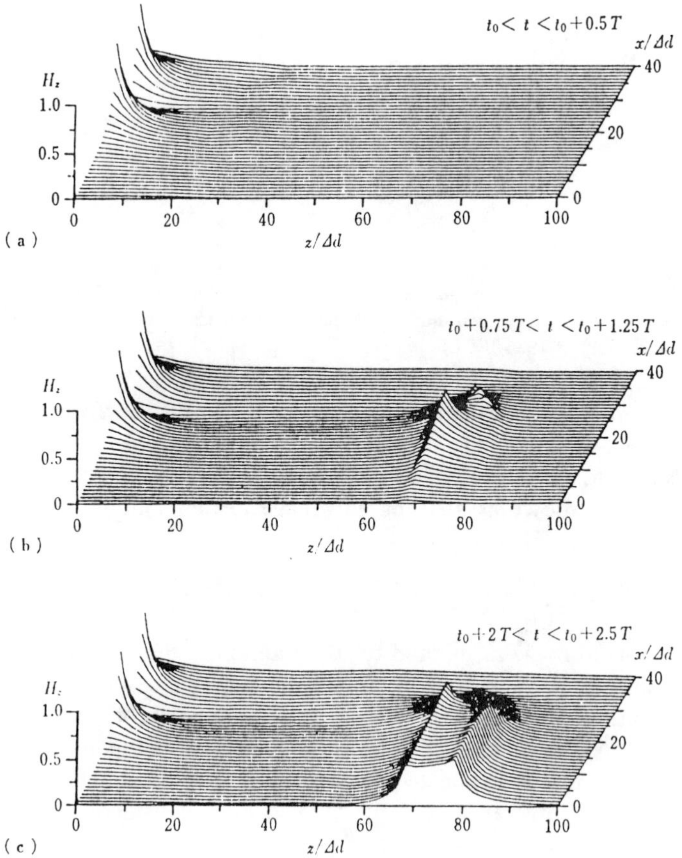

Figure 5.18 Time variation in the spatial distribution of magnetic field H_z in the same case as Figure 5.15 (copyright © 1984 IEEE).

both edges. The time variation shows that the magnitude of the distribution increases while conserving a similar shape.

Thus, the $\mathcal{L}$ shape, which is a fundamental structure in the stripline, is analyzed by the spatial network method. Not only the characteristics of the line in the steady state but also the accessibility of iterative observation of the three-dimensional transient variations in the fields are shown. As a result, the simulation in the time domain clearly shows that the *L*-shaped stripline operates as the converter that transforms the magnetic field component from the *x* direction to the *z* direction. Thus, the vector analysis in the time domain offers a new viewpoint to the estimation of the characteristics of microwave circuitry. Recently, S. Koike, N. Yoshida, and I. Fukai applied this spatial network method to the analysis of some kinds of elements and

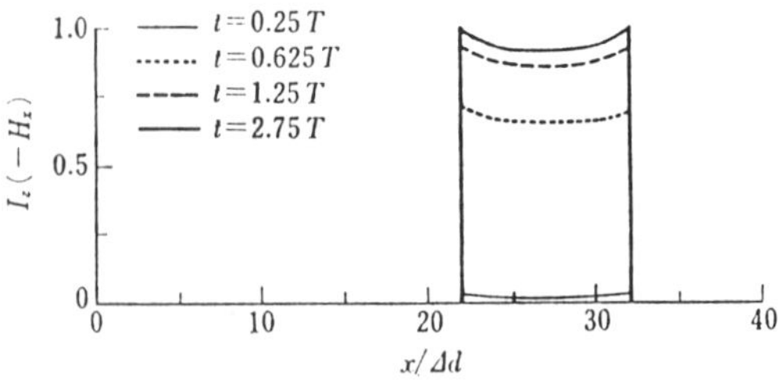

Figure 5.19 Time variation of the change in the x-direction of magnetic field H_x (corresponding to current I_z) on the lower surface of the strip conductor in the same case as Figure 5.15 (copyright © 1984 IEEE).

devices in the microstrip structure [23, 24]. T. Kashiwa, N. Yoshida, and I. Fukai have also applied the method to the analysis of complicated near fields in microwave exciting systems for heating [25]. In both applications, useful results were obtained, and a generality is shown as a strong advantage of this method.

5.7 SOME PRECAUTIONS FOR PROGRAMMING

In using the spatial network method for programming, the following considerations and recommendations are useful to extract the advantages of the method. For numerical methods based on discretization, the determination of the difference interval is of primary importance. In formulations by the Bergeron method, because the spatial and time intervals are connected by the relation given in equation (5.48b), either the spatial or time interval should be decided after considering some conditions. Usually, the spatial interval Δd is decided. If sufficient memory storage is available, an interval able to divide the free-space wavelength to more than 20 parts is desirable. However, in regions including dielectric or magnetic materials, the shortened wavelength demands a smaller interval. Corresponding to the expression of the free-space wavelength $N\Delta d$ (N being the number of divisions), the period of the applied wave T is given as $2N\Delta t$, from the relation between the space and time intervals.

Next, at each node without medium conditions, sufficient memory for ten vari-ables in all is necessary. This covers the equivalent voltage and the equivalent cur-rents in four lines in both a present and previous situation. Therefore, the saving o the memory region is indispensable for realizing smaller divisions. By consecutively numbering in each of the x, y, and z directions, the lattice points where the node i not defined in the lattice network (in Figure 5.1) can also be stored in the memor region. That is, two vertices of the eight vertices of the fundamental lattice show in Figure 5.2 are not used. To avoid this waste in the memory region, two numberin systems are provided in each direction. By combining these, each of six kinds c nodes can be indicated. For instance, each coordinate passing through node A use the numbering system (*IE*, *JE*, *KE*) for each direction. For node F, which is diagonally opposite node A, the other numbering system (*IM*, *JM*, *KM*) is used. The positions of all other respective nodes are individually decided from nodes A and F.

To analyze the three-dimensional electromagnetic fields in the time domain, it is also indispensable to perform high-speed computation of the field at each time step. To this end, the normalization of each constant used is an effective means. In this spatial network, by normalizing the medium constants of each line $\varepsilon_0/2$ and $\mu_0/2$ as 1, the values of the characteristic impedance and the propagation time become equal to 1. The resultant Bergeron expresson can then be simplified to the equation ($V \pm I$), needing only addition or subtraction. This expression increases the efficiency of the computation. The iterative computation of the Bergeron method adapts well to either the mechanism of the *internal array processor* (IAP) or the vector computation mechanism of the supercomputer, which have both become quite useful in modern scientific mainframe computers. Still more, to utilize the performance of these machines sufficiently, the arrangement of the order of the subscripts in the array, the maximization of the most inner loop in the DO loops, and the avoidance of the use of the statement function are required to optimize the program.

Furthermore, for increasing the generality of the program, the development of back- and front-processing programs is necessary. In the latter, for instance, automatic arrangement of nodes for boundary and medium conditions and the selection of the appropriate equation are performed. In the former, for example, the required characteristic parameters are calculated from the output results.

REFERENCES

[1] Okada, Y., M. Iwashita, and T. Fuji, "Electromagnetic Theory (II), Theory of Electromagnetic Lattice, Summary, Electromagnetic Theory (IV), and Theory of Electromagnetic Lattice (II)," *Proc. ICE Japan,* Vol. 28, No. 9, Sept. 1944, pp. 15–18 [in Japanese].

[2] Kron, G., "Equivalent Circuit of the Field Equations of Maxwell-I," *Proc. IRE,* Vol. 32, No. 5, May 1944, pp. 289–299.

[3] Whinnery, J.R., and S. Ramo, "A New Approach to the Solution of High-Frequency Field Problems," *Proc. IRE,* Vol. 32, No. 5, May 1944, pp. 284–288.

[4] Tumolillo, T.A., and J.P. Wondra, "MEEC-3D: A Computer Code for Self-Consistent Solution of the Maxwell-Lorentz Equations in Three Dimensions," *IEEE Trans. Nuclear Science,* Vol. NS-24, No. 6, Dec. 1977, pp. 2449–2455.

[5] Branin, Jr., F.H., "The Algebraic-Topological Basis for Network Analogies and the Vector Calculas," *Proc. the Symp. on Generalized Networks,* Polytechnique Press of the Polytechnique Inst. of Brooklyn, New York, 1966, pp. 453–491.

[6] Yee, K.S., "Numerical Solution of Initial Boundary Value Problems Involving Maxwell's Equations in Isotropic Media," *IEEE Trans. Antennas & Propagation,* Vol. AP-14, No. 3, May 1966, pp. 302–307.

[7] Taflove, A., and M.E. Brodwin, "Numerical Solution of Steady-State Electromagnetic Scattering Problems Using the Time-Dependent Maxwell's Equations," *IEEE Trans. Microwave Theory & Tech.,* Vol. MTT-23, No. 8, Aug. 1975, pp. 623–630.

[8] Holland, R., "THREDE: A Free-Field EMP Coupling and Scattering Code," *IEEE Trans. Nuclear Science,* Vol. NS-24, No. 6, Dec. 1977, pp. 2416–2421.

[9] Kunz, K.S., and K.-M. Lee, "A Three-Dimensional Finite-Difference Solution of the External Response of an Aircraft to a Complex Transient EM Environment: Part I—the Method and Its Implementation," *IEEE Trans. Electromagnetic Compatibility,* Vol. EMC-20, No. 2, May 1978, pp. 328–341.

[10] Johns, P.B., and R.L. Beurle, "Numerical Solution of 2-Dimensional Scattering Problems Using a Transmission-Line Matrix," *Proc. IEE,* Vol. 118, No. 9, Sept. 1971, pp. 1203–1208.

[11] Akhtarzad, S., and P.B. Johns, "Solution of Maxwell's Equations in Three Space Dimensions and Time by the t.l.m Method of Numerical Analysis," *Proc. IEE,* Vol. 122, No. 12, Dec. 1975, pp. 1344–1348.

[12] Hoefer, W.J.R., "The Transmission-Line Matrix Method—Theory and Applications," *IEEE Trans. Microwave Theory & Tech.,* Vol. MTT-33, No. 10, Oct. 1985, pp. 882–893.

[13] Yoshida, N., I. Fukai, and J. Fukuoka, "Transient Analysis of Two-Dimensional Maxwell's Equations by Bergeron's Method," *Electronics & Communication in Japan,* Vol. 62-B, No. 6, June 1979, pp. 34–42.

[14] Yoshida, N., I. Fukai, and J. Fukuoka, "Transient Analysis of Three-Dimensional Electromagnetic Fields by Nodal Equations," *Trans. IECE Japan(B),* Vol. J63-B, No. 9, Sept. 1980, pp. 876–883 [in Japanese].

[15] Bergeron, L.J.B., "Du Coup de Bélier en Hydraulique au Coup de Foudre en Électricitè," Dunod, Paris, 1949.

[16] Dommel, H.W., "Digital Computer Solution of Electromagnetic Transients in Single- and Multiphase Network," *IEEE Trans. Power Apparatus & Systems,* Vol. PAS-88, No. 4, April 1969, pp. 388–396.

[17] Dommel, H.W., and W.S. Mayer, "Computation of Electromagnetic Transient," *Proc. IEEE,* Vol. 62, No. 7, July 1974, pp. 983–993.

[18] Metzger, G., and J-P. Pabre, *Transmission Lines with Pulse Excitation,* Academic Press, New York, 1969.

[19] Yoshida, N., I. Fukai, and J. Fukuoka, "Adaptation of Bergeron's Method to Complicated Boundary Problems," *Trans. IECE Japan(E),* Vol. E64, No. 7, July 1981, pp. 455–462.

[20] Yoshida, N., I. Fukai, and J. Fukuoka, "Application of Bergeron's Method to Anisotropic Media," *Trans. IECE Japan (B),* Vol. J64-B, No. 11, Nov. 1981, pp. 1242–1249 [in Japanese].

[21] Dvorak, V., "Computer Simulation of Signal Propagation through a Nonuniform Transmission Line," *IEEE Trans. Circuit Theory,* Vol. CT-20, No. 9, Sept. 1973, pp. 580–583.

[22] Yoshida, N., and I. Fukai, "Transient Analysis of a Stripline Having a Corner in Three-Dimensional Space," *IEEE Trans. Microwave Theory & Tech.,* Vol. MTT-32, No. 5, May 1984, pp. 491–498.

[23] Koike, S., N. Yoshida, and I. Fukai, "Transient Analysis of Microstrip Gap," *Electronics & Communications in Japan,* Vol. 67-B, No. 11, Nov. 1984, pp. 76–83.

[24] Koike, S., N. Yoshida, and I. Fukai, "Transient Analysis of Microstrip Side-Coupled Filter in Three-Dimensional Space," *Trans. IECE Japan* (E), Vol. E69, No. 11, Nov. 1986, pp. 1199–1205.

[25] Kashiwa, T., N. Yoshida, and I. Fukai, "Transient Analysis of Cutoff Waveguide Antenna in Three-Dimensional Space," *Trans. IECE Japan (B),* Vol. J69-B, No. 7, July 1986, pp. 732–740 [in Japanese].

Chapter 6
The Equivalent Source Method

Hiroshi Shigesawa

6.1 HISTORICAL BACKGROUND AND APPLICATIONS

The solutions to the steady-state electromagnetic-field problems are obtained by solving the basic Maxwell's equations, the simultaneous partial differential equations of the first order with respect to the space variables (x, y, z). Those problems are usually accompanied by boundary conditions in that the electromagnetic field takes the known values on the specified boundaries. The boundary-value problems including real sources are solved by considering effects generated by the sources of the electromagnetic field on the specified boundaries. This field is, of course, the same as the one generated by the same sources in the homogeneous space without boundaries. Our problem is to solve the Helmholtz equation subject to specified boundary values.

This chapter discusses the equivalent-source method, an efficient numerical method to solve boundary-value problems including real sources or not. In this method, the unknown field to be solved is approximately expressed in terms of the fields generated by the hypothetical sources (or equivalent sources).

Similar approaches have been developed independently in the fields of the electrostatics [1], seismology [2], fluid mechanics [3], and electromagnetics [4]. This approach is widely known as the charge-simulation method [5] in electrostatics. In the field of the electromagnetic waves, however, no generic name encompasses all the methods, and it is generally understood by several names; for example, the discrete-singularity method [6], the charge-simulation method [7], the current-simulation method [8], the multiple-multipole method [9], and so on. Nevertheless, these methods have several common properties in their methodology, and there are many ideas common to all in their numerical approach. Therefore, the term *equivalent-source method* is applied throughout this chapter to any method based on "the use of hypothetical or equivalent sources." We no longer constrain this term by other auxiliary conditions [1].

The equivalent-source method has been developed mainly in connection with the analysis of electromagnetic scattering problems. A typical example is the scattering of a plane wave from scatterers of the perfect conductor, where the main problem is to calculate the secondary field (the scattered field) generated by the scatterers. This scattered field should be solved so that the total field (the incident plane wave plus the scattered field) fits the prescribed condition on the boundary C separating the scatterers from the exterior space region. Then we may understand that there are unknown electric or magnetic currents (secondary sources) on C, generating the scattered field. As for the analysis of such a problem, the integral-equation method has been widely used [10]. In this familiar method, the boundary-value problem comes to the problem of solving the inhomogeneous Fredholm equation of the second kind for an unknown source distribution on C. Evidently, this method represents the unknown scattered field outside C in terms of the real-current sources on C. Such an integral equation does not always guarantee the sufficient conditions of the problem. Therefore, when it is used, we should guard against the undesirable effects of spurious solutions. In addition, we must always consider a numerical technique to avoid the singular behavior of the integral kernel.

Such weak points of the integral-equation method, however, become a trigger to developing an alternative method, the equivalent-source method. In this method, equivalent sources are assumed on a closed contour Γ taken in the interior region of scatterers so that the field generated by them in the region outside C is identical with the scattered field produced by the original sources on C. This is rightly the essential of the equivalent-source method. In numerical calculations, equivalent sources distributed on Γ are usually quantized by the method of moment based simply on, for example, a delta-function array, whereas it is typical to use the least-squares [6] or the point-matching method [7] for fulfilling the boundary condition on C. An important point here is whether an approximate solution obtained by the equivalent-source method uniformly converges to the true solution, as the number of the quantized sources increases. In practical calculations, it is inevitable to approximate scattered fields, and we must clarify the mathematical basis of the numerical analysis. If we neglect it, the error estimation always remains uncertainly in approximate calculations. Therefore, the outline of the equivalent-source method and its numerical basis are described in Section 6.2.

We have discussed a number of scattering problems by the equivalent-source method and found that this method easily applies to two-dimensional scattering problems of perfect conductors, loss-less dielectrics, lossy dielectrics, and their combinations. These examples will provide good clues to solving a large number of problems of wave scattering from, for example, missiles with complicated structures, multiple high rises, and so on. In Sections 6.3 and 6.4, therefore, we describe the application of the equivalent-source method to several scattering problems, investigated mainly by our group. Other applications (for example, as seen in [8]) will be omitted here due to the limited space.

For the equivalent-source method, there are a large number—theoretically an infinite number—of freedoms to arrange the discretized equivalent sources. Therefore, the equivalent-source arrangement can be defined flexibly by considering the boundary shape of arbitrary scatterers. On the other hand, such a flexibility often becomes a weak point as the precision of solutions is strongly affected by the source arrangement if the number of numerical steps in calculations is kept constant. Contrary to the case in electrostatics, electromagnetic problems are always accompanied by energy flow, so that it may be desirable to define the source arrangement by considering such an unknown energy flow, too. However, no published paper has discussed a reasonable way of arranging equivalent sources. Most studies, therefore, had to fall back on experience stocked by a large number of practical calculations and often found it difficult to maintain a necessary precision in the calculations. Recently, a powerful solution to this difficulty has been developed by the author's group [11]. Its outline is also described here.

Another application of the equivalent-source method can be seen in the eigenvalue problem of waveguides [12]. However, there is no distinct difference from other methods (for example, the finite-element method) in either the number of numerical steps or the achieved precision. In the scattering problem, the scattered field at any observation point may be roughly considered a sum of fields diffracted or scattered at local areas on scatterers. Such a locally scattered field, of course, is directly affected by the structural nature of the corresponding local area. Therefore, the characteristics of a scatterer are roughly defined by the field scattered from an effective extent of such a local area, which has the predominant contribution on the scattered field at a given observation point. Such thinking is justified when the effective extent of the predominant local area (or "the characteristic dimension") of a scatter becomes large relative to the wavelength.

An effective approximate method for such a case will be the *geometric theory of diffraction* (GTD) discussed in Chapter 7. This method consists of both the geometric optics and the diffraction theory, whose individual contributions depend on the local nature of scatterers. However, we must neatly consider boundary conditions in the electromagnetic-wave theory, if scatterers have a complicated structure and locally diffracted fields intricately interfere with one another. The equivalent-source method will show its real ability in such a problem. Of course, there are other effective candidates for numerical analysis; for example, the integral-equation method (discretizing an integral by the boundary element method), discussed in Chapter 2, or the mode-matching method, discussed in Chapter 4. However, these methods are not always efficient if we consider the relation between the number of numerical steps in practical calculations and the achieved precision of the solution. The tractability of the equivalent-source method can be summarized in the following two points: one is that the equivalent-source method incorporates the merits of both the integral-equation method and the mode-matching method, and the other is that it disperses sources in a limited space, thereby making it easy to apply the method to scatterers associated with peculiar local phenomena as considered in GTD approach.

However, the equivalent-source method is not always better than other methods for scatterers that are not as large relative to wavelength and also with a quite smooth boundary. For such special scatterers, the use of the integral-equation method or the mode-matching method may be more effective than the use of the equivalent-source method; but if a small-sized scatterer has a complicated boundary shape, the equivalent-source method still will be powerful.

The equivalent-source method can be applied to three-dimensional vector boundary-value problems. In practical calculations, however, we have to relax some of the difficulties caused by their prohibitively large storage needs on the computer. At present, this method is applicable only to three-dimensional scatterers with simple boundary shapes. Advanced investigations should be expected in future.

6.2 BASIC THEORY OF THE EQUIVALENT SOURCE METHOD

For an easy understanding of the essentials of this method, we assume here that the electromagnetic field in the two-dimensional space is uniformly extended along the z direction. The time dependence $\exp(j\omega t)$ is considered throughout this chapter. As a simple example, let us consider a cylindrical scatterer of the perfect conductor, as shown in Figure 6.1. This cylinder is uniform along the z axis and has an arbitrary cross section described by a closed smooth contour C on the xy plane. The exterior space region of the conducting scatterer is denoted by S, and $\tilde{S}$ represents its interior region. Then, the problem is to calculate the scattering field generated by a plane wave impinging on the scatterer at an arbitrary angle (θ_i is measured from the x axis)

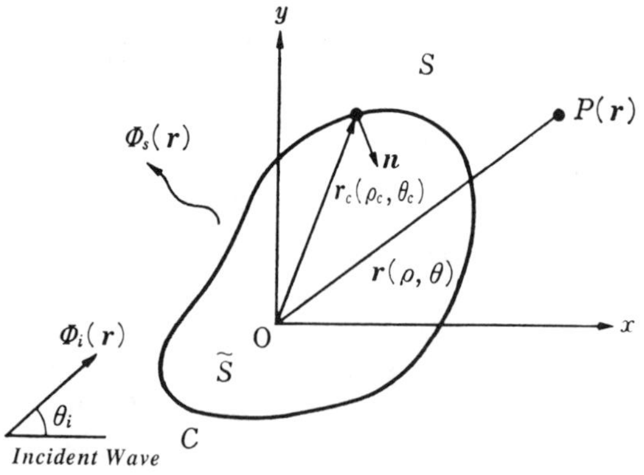

Figure 6.1 Infinitely extended external region S, the interior region $\tilde{S}$ of the perfect conductor, and the boundary contour C between S and $\tilde{S}$.

from the region S. We further assume that such an incident plane wave can be expressed by the scalar wave function $\Phi_i(\boldsymbol{r})$ ($\boldsymbol{r} \in S$), where $\boldsymbol{r}(\rho, \theta)$ denotes the position vector from the origin O to an arbitrary observation point P in S. The main problems of the equivalent-source method are choosing which kinds of secondary fields (that is, approximated wave functions) associated with equivalent sources best express the unknown scattered field and how to fit such a scattered field to the given boundary condition.

6.2.1 Approximated Wave Function for the Equivalent Source Method

Let us indicate the scattered field in S by $\Phi_s(\boldsymbol{r})$, which is generated by $\Phi_i(\boldsymbol{r})$ illuminating the perfect conductor. Then the total field $\Phi(\boldsymbol{r})$ in S can be expressed as follows:

$$\Phi(\boldsymbol{r}) = \Phi_i(\boldsymbol{r}) + \Phi_s(\boldsymbol{r}) \tag{6.1}$$

Each of these fields satisfies the Helmholtz equation ($n = 1$) in S:

$$\left(\frac{\partial^2}{\partial x^2} + \frac{\partial^2}{\partial y^2} + n^2k^2\right)\Psi(\boldsymbol{r}) = 0 \tag{6.2}$$

and the scattered field $\Phi_s(\boldsymbol{r})$ must satisfy the following type of the radiation condition at $\rho \to \infty$:

$$\lim_{\rho\to\infty} \sqrt{\rho}\left[\frac{\partial \Phi_s(\boldsymbol{r})}{\partial \rho} + jk\Phi_s(\boldsymbol{r})\right] = 0 \tag{6.3}$$

In the present example, the boundary condition on the given contour C is expressed as

$$\Phi(\boldsymbol{r}_c) = \Phi_i(\boldsymbol{r}_c) + \Phi_s(\boldsymbol{r}_c) = 0, \quad \boldsymbol{r}_c \in C \tag{6.4}$$

This is a Dirichlet problem, in which the scattered field $\Phi_s(\boldsymbol{r}_c)$ on C must take the specified boundary value $f(\boldsymbol{r}_c) = -\Phi_i(\boldsymbol{r}_c)$. The following discussions can be basically applied to Neumann problems and also to dielectric scatterers or dielectric-coated conductors. In case of dielectrics, however, there is a complexity to consider simultaneously in both the exterior (S) and interior ($\tilde{S}$) regions. Because such a problem is quite important in practice, we discuss it in detail in Section 6.4 along with applications.

Now, there are many approaches to solving the boundary-value problem mentioned earlier, as seen in the other chapters of this book. Actually, we may understand the equivalent-source method to be an approach of the mode-matching method discussed in Chapter 4 on its basis. Also, it will be understood later that the equivalent-source method is in the category of the generalized multipole expansion method [13]. (Recently, this method often has been called Yasuura's *mode-matching method*.) The latter method basically uses the mode-matching technique on C, after expanding $\Phi_s(\boldsymbol{r})$ into a complete set of modal solutions of the Helmholtz equation in S. The reader can find discussions on both the uniqueness and the convergence nature of the solutions to this method in Chapter 4. However, it is useful to describe here one important result.

Let us consider a complete set of wave functions (or modal functions) $\{\phi_n(\boldsymbol{r})\}$ in S. It always is possible for an arbitrary regular wave function $\Phi_s(\boldsymbol{r})$ to be uniformly approximated in wider sense in S by a linear combination of the form

$$\Phi_{sN}(\boldsymbol{r}) = \sum_{n=1}^{N} a_n(N)\phi_n(\boldsymbol{r}) \tag{6.5}$$

The series in equation (6.5) is truncated, and $\Phi_s(\boldsymbol{r})$ is precisely represented as the limiting wave function of the sequence $\{\Phi_{sN}(\boldsymbol{r})\}$:

$$\lim_{N\to\infty} |\Phi_{sN}(\boldsymbol{r}) - \Phi_s(\boldsymbol{r})| = 0 \tag{6.6}$$

Note here that the coefficients $\{a_n(N)\}$ generally depend on the number N of truncations. The complete set of wave functions that appears here must satisfy the Helmholtz equation with a wavenumber k identical to (6.2) and also must satisfy the radiation condition given by (6.3). Such a set has the following important nature.

Let $\{\phi_n(\boldsymbol{r}_c),\ \boldsymbol{r}_c \in C\}$ denote a complete set of modal functions $\{\phi_n(\boldsymbol{r})\}$ on C. There exist the coefficients $\{a_n(N)\}$ such that, for an arbitrary positive quantity ε, the boundary value $f(\boldsymbol{r}_c) = -\Phi_i(\boldsymbol{r}_c)$ for a regular wave function $\Phi_s(\boldsymbol{r})$ on C is represented as

$$\int_C \left| f(\boldsymbol{r}_c) - \sum_{n=1}^{N} a_n(N)\phi_n(\boldsymbol{r}_c)\right|^2 dl < \varepsilon \tag{6.7}$$

According to these results, we should next discuss the complete set of modal functions in S [14], available for the scattering problem of Figure 6.1. The most primitive but widely used example of the complete set is the set of multipole functions with respect to the origin in $\tilde{S}$:

$$\phi_n(\boldsymbol{r}) = \frac{(-1)^n}{4j} H_n^{(2)}(k\rho)e^{jn\theta}, \quad n = 0, \pm 1, \pm 2, \ldots \tag{6.8}$$

where $H_n^{(2)}(k\rho)$ is the nth-order Hankel function of the second kind. According to the generalized multipole expansion method, the set of (6.8) is not always a unique set for our use. The following examples of the complete set are rather useful for the equivalent-source method.

Example 1. Let us take the range of points $\{Q_m\}$ in $\tilde{S}$. If it has a cluster point in $\tilde{S}$, a set of following functions allocated to each point Q_m is a complete set in S:

$$H_0^{(2)}(k\rho_m),\ H_1^{(2)}(k\rho_m)\mathrm{e}^{j\theta},\ H_1^{(2)}(k\rho_m)\mathrm{e}^{-j\theta}, \quad m = 1, 2, \ldots \tag{6.9}$$

where ρ_m is the distance between Q_m and an observation point P in S.

Example 2. Let Q_m ($m = 1, 2, \ldots, M$) denote the lattice points obtained by dividing the interior region $\tilde{S}$ into a grid. Then the following is a complete set in S for $M \to \infty$:

$$H_0^{(2)}(k\rho_m), \quad m = 1, 2, \ldots \tag{6.10}$$

According to the complete set mentioned earlier, we can obtain several kinds of approximated wave functions in S, which are applicable to the boundary-value problem of Figure 6.1. For this purpose, the complete set of functions given by (6.8) is widely used and has already been discussed in Chapter 4. This type of modal function, however, has some weak points. For example, it is not always effective for scatterers with complicated boundary shapes, or its use often makes it difficult to ensure a successful convergence of solutions even when the number of expansion terms is increased, because we meet with the degeneracy of Hankel functions as the characteristic dimension of scatterers becomes large relative to the wavelength. This degeneracy is due to the asymptotic nature of Hankel function of the form $H_m^{(2)}(z) \approx \sqrt{2/(\pi z)}\,\exp\{-j[z - (m\pi)/2 - \pi/2]\}$ as $|z| \to \infty$.

On the other hand, the complete set of modal functions given by (6.10) supplies another type of approximated wave functions based only on the 0th-order Hankel function of the second kind with the argument $k\rho_m$. However, when these wave functions are used for expressing the scattered field at an observation point in S, the necessary space domain for calculations extends over the two-dimensional space of the interior region $\tilde{S}$, instead of the one-dimensional one on C of (6.9), and (6.10) is not useful to reduce the number of calculation steps, ensuring the uniform convergence of solutions. Therefore, here we construct the approximate wave function $\Phi_{sM}(\boldsymbol{r})$ using the modal functions of (6.9). For this purpose, we consider a closed contour Γ with no holding point (i.e., a Jordan contour) in the interior region $\tilde{S}$, as shown in Figure 6.2, and express $\Phi_{sM}(\boldsymbol{r})$ as follows by taking the fixed points Q_m($m = 1, 2, \ldots, M$) on Γ:

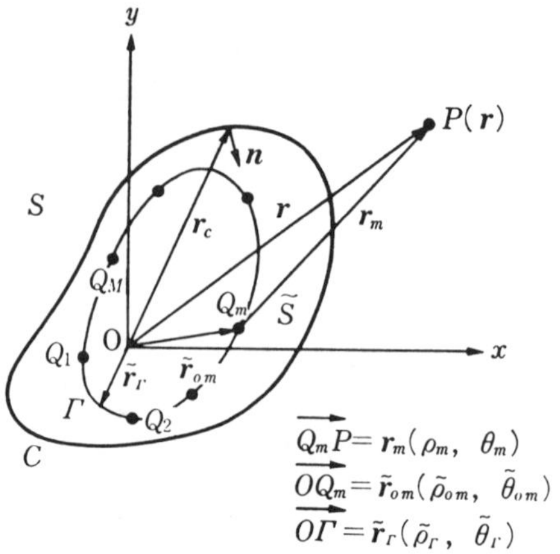

Figure 6.2 Equivalent sources arranged on a closed contour $\Gamma(\tilde{S})$, which are used for constructing an approximated wave function in the exterior region S.

$$\Phi_{sM}(\boldsymbol{r}) = \sum_{m=1}^{M} \{a_m(M)H_0^{(2)}(k\rho_m) + b_m^{+}(M)H_1^{(2)}(k\rho_m)e^{j\theta} + b_m^{-}(M)H_1^{(2)}(k\rho_m)e^{-j\theta}\}, \quad \boldsymbol{r} \in S \tag{6.11}$$

where the range of points $\{Q_m\}$ has a cluster point on the contour Γ.

According to the wave functions $\Phi_{sM}(\boldsymbol{r})$ given by (6.11), we have only to consider the one-dimensional space domain in calculations and also may take the arbitrary contour Γ inside C. It is then expected in some cases that the choice of (6.11) will reduce the number of calculation steps more than other choices for $\Phi_{sM}(\boldsymbol{r})$. Furthermore, the simplified equivalent-source method often approximates the $\Phi_{sM}(\boldsymbol{r})$ of (6.11) only by the first term of its right-hand side as follows:

$$\Psi_{sM}(\boldsymbol{r}) = \sum_{m=1}^{M} a_m(M)H_0^{(2)}(k\rho_m), \quad \boldsymbol{r} \in S \tag{6.12}$$

This approximation indeed effectively reduces the number of calculation steps, but it misses the completeness of (6.11), because (6.12) consists of only $H_0^{(2)}(k\rho_m)$ allocated on each point Q_m on a Jordan contour.

On the other hand, $\Psi_{sM}(\boldsymbol{r})$ of (6.12) is often understood to be an approximated wave function based on the complete set of modal functions given by (6.10). However, $\Psi_{sM}(\boldsymbol{r})$ of (6.12) does not have the completeness in S of the set of (6.10) because we can not fill the area $\tilde{S}$ with a single Jordan contour [15]. Then, $\Psi_{sM}(\boldsymbol{r}_c$

given by (6.12) does not rightly approximate the boundary value $f(\boldsymbol{r}_c)$ in the sense of (6.7). Therefore, we should use the approximated wave function (6.12) in practice, knowing that it misses the nature of (6.6) altogether.

The use of the approximated wave function (6.12), on the other hand, can be understood as follows by referring to the formulation based on the integral-equation method. The Green's function in the two-dimensional free space is given by

$$G(\boldsymbol{r}, \boldsymbol{r}_m) = \frac{1}{4j} H_0^{(2)}(k\rho_m) \tag{6.13}$$

Letting the closed contour Γ coincide with the boundary $C(\boldsymbol{r}_\Gamma = \boldsymbol{r}_c)$, $\Phi_s(\boldsymbol{r})$ in S can be expressed by the generalized secondary source $I(\boldsymbol{r}_c)$ on C as follows [16]:

$$\Phi_s(\boldsymbol{r}) = \int_C I(\boldsymbol{r}_c) G(\boldsymbol{r}, \boldsymbol{r}_c) dl, \quad \boldsymbol{r} \in S, \quad \boldsymbol{r}_c \in C \tag{6.14}$$

where $I(\boldsymbol{r}_c) = \partial_n \Phi(\boldsymbol{r}_c)$, and ∂_n means the partial derivative with respect to the outward unit normal on C. If we discretize $I(\boldsymbol{r}_c)$, which is often called the single-layer intensity distribution, by the Dirac delta-function array $\{\delta(\boldsymbol{r}_c - \boldsymbol{r}_m)\}$ on C, we can obtain an approximated $\Phi_s(\boldsymbol{r})$ that has the same expression as the $\Psi_{sM}(\boldsymbol{r})$ of (6.12). Therefore, we may understand that the wave functions used in the equivalent-source method are identical with the approximated field in the integral-equation method, as long as a closed contour Γ is coincided with the boundary C of scatterers.

As a simple extension of this understanding, let us assume a secondary source distribution $\tilde{I}(\boldsymbol{r}_\Gamma)$ on Γ, instead of $I(\boldsymbol{r}_c)$ on C and express the approximated field in S using the Green's function of (6.13) as follows:

$$\Phi_s'(\boldsymbol{r}) = \int_\Gamma \tilde{I}(\boldsymbol{r}_\Gamma) G(\boldsymbol{r}, \boldsymbol{r}_\Gamma) dl, \quad \boldsymbol{r} \in S, \quad \boldsymbol{r}_\Gamma \in \Gamma \tag{6.15}$$

If $\Phi_s'(\boldsymbol{r})$ of (6.15) is identical with $\Phi_s(\boldsymbol{r})$ of (6.14) in S, we may recognize that (6.12) is derived by discretizing $\tilde{I}(\boldsymbol{r}_\Gamma)$ by means of Dirac delta-function array $\{\delta(\boldsymbol{r}_\Gamma - \boldsymbol{r}_m)\}$ on Γ. Then we may understand that $\Phi_s(\boldsymbol{r})$ in S, generated by $I(\boldsymbol{r}_c)$ on C, can be described using the source $\tilde{I}(\boldsymbol{r}_\Gamma)$ on Γ. In this sense, the method based on (6.12) may be recognized as the equivalent-source method. However, equation (6.14) is derived from Green's second identity and the integral included there is performed over the boundary contour C of a perfect-conducting scatterer. Therefore, if the integral of (6.14) is simply performed over an arbitrary closed contour Γ, not identical with C, the value $\Phi_s'(\boldsymbol{r}_c)$ on C will not always fulfill $f(\boldsymbol{r}_c)$, the given boundary value of $\Phi_s(\boldsymbol{r})$. In other words, an arbitrary $\Phi_s'(\boldsymbol{r})$ cannot be sufficiently expressed only by the equivalent source $\tilde{I}(\boldsymbol{r}_\Gamma)$ corresponding to the single-layer intensity distribution on Γ; rather, it is necessary to use one more type of equivalent source $\tilde{K}(\boldsymbol{r}_\Gamma)$, which is often called the *double-layer intensity distribution*. As a result, it is obvious that (6.12) applied to an array of points on Γ is insufficient for expressing any field in S from the standpoint of the completeness, although we can prove the completeness of (6.11), though its detail is omitted here.

6.2.2 Boundary Conditions and Scattered Field

For the sake of a simple explanation of the method, we hereafter use the approximated wave functions given by (6.12). Let us here define the norm of an arbitrary function $A(\boldsymbol{r}_c)$ on the boundary C as follows:

$$\|A(\boldsymbol{r}_c)\| = \left[\int_C |A(\boldsymbol{r}_c)|^2 dl\right]^{1/2} \tag{6.16}$$

and the inner product of arbitrary two vector functions $A(\boldsymbol{r})$ and $B(\boldsymbol{r})$ on C as follows:

$$\langle A(\boldsymbol{r}_c), B(\boldsymbol{r}_c)\rangle = \int_C A^*(\boldsymbol{r}_c)B(\boldsymbol{r}_c)dl \tag{6.17}$$

where the mark * denotes a complex conjugate.

Usually, the approximated wave function $\Psi_{sM}(\boldsymbol{r})$ does not always coincide perfectly with the boundary value $f(\boldsymbol{r}_c)$ for all values of $\boldsymbol{r}_c$ on C. Therefore, let us define the relative mean square error for the boundary condition on C as follows:

$$\varepsilon(M) = \|f(\boldsymbol{r}_c) - \Psi_{sM}(\boldsymbol{r}_c)\|^2/\|f(\boldsymbol{r}_c)\|^2 \tag{6.18}$$

Then we minimize (6.18) with respect to the unknown coefficients $a_m(M)$ to obtain

$$\partial\varepsilon(M)/\partial a_m^*(M) = 0, \quad m = 1, 2, \ldots, M \tag{6.19}$$

As a result, an optimum solution for $\{a_m(M)\}$ can be obtained by solving the following set of linear equations:

$$\sum_{m=1}^{M} < \phi_n(\boldsymbol{r}_c), \quad \phi_m(\boldsymbol{r}_c) > a_m(M) = \langle\phi_n(\boldsymbol{r}_c), f(\boldsymbol{r}_c)\rangle, \quad n = 1, 2, \ldots, M \tag{6.20}$$

where $\phi_n(\boldsymbol{r}_c) = H_0^{(2)}(k|\boldsymbol{r}_c - \boldsymbol{r}_n|)$.

Putting H_M as the positive Hermitian matrix of the order M of which the (n, m) element is given by $\langle \phi_n(\boldsymbol{r}_c), \phi_m(\boldsymbol{r}_c) \rangle$, we can rewrite (6.20) into the following matrix form:

$$H_M X(M) = F_M \tag{6.21}$$

where $X(M)$ and F_M are the Mth-order column vectors with $a_n(M)$ and $\langle \phi_n(\boldsymbol{r}_c), f(\boldsymbol{r}_c) \rangle$, respectively, as the nth elements.

By the way, let us approximately express the inner products in (6.20) as follows:

$$\langle \phi_n(\boldsymbol{r}_c), \phi_m(\boldsymbol{r}_c) \rangle \doteqdot \sum_{k=1}^{M} \phi_n^*(\boldsymbol{r}_{ck}) \phi_m(\boldsymbol{r}_{ck}) \Delta l \tag{6.22}$$

$$\langle \phi_n(\boldsymbol{r}_c), f(\boldsymbol{r}_c) \rangle \doteqdot \sum_{k=1}^{M} \phi_n^*(\boldsymbol{r}_{ck}) f(\boldsymbol{r}_{ck}) \Delta l \tag{6.23}$$

where Δl means an arc length between two neighboring points that are equally spaced on C. Each point is expressed by a vector $\boldsymbol{r}_{cn}$ ($n = 1, 2, \ldots, M$). Substituting (6.22) and (6.23) into (6.20), we have

$$\sum_{m=1}^{M} \phi_m(\boldsymbol{r}_{ck}) a_m(M) = f(\boldsymbol{r}_{ck}), \quad k = 1, 2, \ldots, M \tag{6.24}$$

This is rightly equivalent to the formulation of the collocation method in which $\Psi_{sM}(\boldsymbol{r})$ of (6.12) coincides with the specified boundary value $f(\boldsymbol{r}_c)$ at a finite number M of discrete points on C. Although often in the equivalent-source method we get solutions by solving (6.24) rather than (6.20), we should be careful in using (6.24) to avoid the unexpected error caused mainly by fitting the field solely on a finite number M of discrete points on C. When (6.24) has to be used, it is desirable to take into account special precautions in calculation; for example, the one described in [17].

Now, let us hereafter consider a problem in which the incident field of Figure 6.1 is a plane wave with the unit amplitude as given by the following:

$$\Phi_i(\boldsymbol{r}) = \mathrm{e}^{-jk(x\cos\theta_i + y\sin\theta_i)} \tag{6.25}$$

Then, solving (6.20) for $\{a_m(M)\}$ ($m = 1, 2, \ldots, M$), we can apply (6.12) to obtain the scattered field at either the near range or the far range. Especially, when an observation point P is sufficiently far from scatterers, we may apply the asymptotic expansion of Hankel function, and the scattered field is approximately expressed as follows:

$$\Psi_{sM}(\boldsymbol{r}) \doteq \sqrt{\frac{2}{\pi k\rho}}\, e^{-jk\rho-(\pi/4)} f(\theta) \tag{6.26}$$

where

$$f(\theta) = \sum_{m=1}^{M} a_m(M) e^{jk\bar{\rho}_{0m}\cos(\theta-\theta_{0m})} \tag{6.27}$$

is the directivity function of the scattered field. Using (6.25) and (6.26), the scattering cross section $\sigma(\theta;\,\theta_i)$ and the total-scattering cross section $\sigma_s(\theta_i)$ are given as follows [18]:

$$\sigma(\theta;\,\theta_i) = \lim_{\rho\to\infty} 2\pi\rho \frac{|\Psi_{sM}(\boldsymbol{r})|^2}{|\Phi_i(\boldsymbol{r})|^2} = \frac{4}{k}|f(\theta)|^2 \tag{6.28}$$

$$\sigma_s(\theta_i) = \frac{1}{|\Phi_i(\boldsymbol{r})|^2}\int_0^{2\pi} |\Psi_{sM}(\boldsymbol{r})|^2 \rho d\theta = \frac{1}{2\pi}\int_0^{2\pi} \sigma(\theta;\,\theta_i) d\theta \tag{6.29}$$

Additionally, $\sigma_b(\theta_i) = \sigma(\theta_i + \pi;\,\theta_i)$ and $\sigma_f(\theta_i) = \sigma(\theta_i;\,\theta_i)$ are called the *backward-scattering* and the *forward-scattering cross sections,* respectively; and they specify the characteristics peculiar to a scatterer.

6.2.3 Error Estimation of the Scattered Field

According to the numerical approach mentioned earlier, the relative mean square error $\varepsilon(M)$ given by (6.18) becomes a good measure for the accuracy that the calculated field fits the boundary condition on C. In addition to this error, it is useful to introduce the error $\varepsilon_{\text{opt}}(M)$ relating to the energy conservation of the total field in order to estimate the accuracy of solutions from the physical point of view. For this purpose, we can use the optical theory [19], which generally denotes that, for a scatterer either with or without loss, the sums of the scattered energy and the absorbed energy are closely related with only both the amplitude and phase of scattered field $\Psi_{sM}(\boldsymbol{r})$ at a point $\boldsymbol{r}(\rho \to \infty,\ \theta = \theta_i)$ in the shadow zone for a plane-wave incidence. This interesting relation is derived from the fact that the absorbed energy in a lossy scatterer must be equal to the difference between the energy of the incident field and that of the scattered field that flows across a closed surface enclosing the scatterer. For a perfect-conducting scatterer, we can obtain both the scattered energy and the mean value of the stored energy of the scattered field from the amplitude and phase of the far field observed at a point in the shadow zone. Then it is easy to prove the following relationship between the total-scattering cross section given by (6.29) and the directivity function of the scattered field given by (6.27):

$$\sigma_s(\theta_i) = \frac{4}{k} \operatorname{Re}[f(\theta_i)] \tag{6.30}$$

Therefore, we can introduce the relative error $\varepsilon_{\text{opt}}(M)$ for the optical theory as for the approximated wave function as follows:

$$\varepsilon_{\text{opt}}(M) = 2|\sigma_s(\theta_i) - 4 \operatorname{Re}[f(\theta_i)]/k| / |\sigma_s(\theta_i) + 4 \operatorname{Re}[f(\theta_i)]/k| \tag{6.31}$$

We can usually estimate the accuracy of solutions by using the error functions given by both (6.18) and (6.31).

However, these errors are defined in the mean sense on the boundary C or from the macroscopic point of view for far-field behavior. Therefore, they often are not effective estimates of the boundary error, which becomes quite large at several local points on C where the boundary shape sharply changes. This type of error does not always decrease, even if the number M of equivalent sources is simply increased on Γ. In such a case, it is effective to arrange the equivalent sources by considering the boundary shape to reduce the error just mentioned. Thus, the successful use of the equivalent-source method often calls for experience in efficiently arranging sources, which can be a drawback of the equivalent-source method. (A method to overcome this drawback will be discussed later.)

Now, we return our attention to the error at a local point on the boundary C and define the relative error function on C as follows:

$$\Omega_M(\boldsymbol{r}_c) = |f(\boldsymbol{r}_c) - \Psi_{sM}(\boldsymbol{r}_c)| / |f(\boldsymbol{r}_c)| \tag{6.32}$$

Then, we can generally estimate the accuracy of solutions by using the error function of (6.32) along with those of (6.18) and (6.31). Especially, the use of $\Omega_M(\boldsymbol{r}_c)$ of (6.32) will be inevitable, for example, when the current-density distribution on perfect-conducting scatterers is discussed with sufficient accuracy [20].

6.2.4 Optimum Arrangement of Equivalent Sources

As mentioned earlier, the accuracy of solutions obtained from the equivalent-source method is severely affected by both the location of a closed contour Γ and the arrangement of M sources on Γ. Although it is desirable to define beforehand an arrangement of sources by assuming an energy flow in the electromagnetic field, it will be impossible in most practical cases. A substitutive method will be the iterative method in which an optimum arrangement is iteratively found from a good initial guess for the arrangement. A necessary condition in this approach is that each of the source points $\{Q_m\}$ must be on different positions in the interior region of scatterers at each iteration. In addition, the amplitude $\{a_m(M)\}$ of sources necessarily vary in each iteration step. These unknown amplitudes behave as the linear variables in (6.20), whereas the location of each source behaves as the nonlinear variables. Therefore, the approach discussed in Section 6.2.2 should be modified into a nonlinear optimization approach.

As a simplified method, we often use a linear search. In this method, we define beforehand a closed contour Γ_α similar to C with a similarity ratio $\alpha(<1)$ or a closed contour Γ_d set apart from C by a distance d. Then the linear search is applied to solve α or d by minimizing the error $\varepsilon(M)$. In most cases, M sources will be equally spaced on Γ_α or Γ_d for simplicity. We will show several numerical examples obtained by this method later. However, the method usually needs a large number of iterations to find an optimum solution and does not show the real ability of the equivalent-source method, which has been developed for efficiently solving problems by optimizing the location of each source. For this purpose, the nonlinear optimizaton will become a powerful approach.

In applications of a nonlinear optimization method (see [11] for details), the unknown amplitude $a_m(M)$ of each equivalent source varies together with the change of the source location. Therefore, it is necessary to develop a numerical approach to solve simultaneously a set of unknowns $\{a_m(M), \tilde{\boldsymbol{r}}_{0m}\}$ $(m = 1, 2, \ldots, M)$. For this purpose, we can apply the nonlinear least-mean-square method based on the standard iteration method. For example, consider that the solution for a set of unknowns and the corresponding wave function are solved at the lth iteration as $\{a_m^{(l)}(M), \tilde{\boldsymbol{r}}_{0m}^{(l)}\}$ and $\Psi_{sM}^{(l)}(\boldsymbol{r})$, respectively. By using these results, we write the unknown wave function $\Psi_{sM}^{(l+1)}(\boldsymbol{r})$ at the $(l + 1)$th iteration as follows:

$$\Psi_{sM}^{(l+1)}(\boldsymbol{r}) = \Psi_{sM}^{(l)}(\boldsymbol{r}) + \psi_{sM}^{(l)}(\boldsymbol{r}) \tag{6.33}$$

where $\psi_{sM}^{(l)}(\boldsymbol{r})$ denotes an unknown small compensator to be introduced in the $(l + 1)$th numerical process.

We assume here $\{a_m^{(l+1)}(M), \tilde{\boldsymbol{r}}_{0m}^{(l+1)}\}$ to be a set of unknowns associated with $\Psi_{sM}^{(l+1)}(\boldsymbol{r})$, and let us here define a set of new unknowns $\{\Delta a_m^{(l)}(M), \Delta\tilde{\boldsymbol{r}}_{0m}^{(l)}\}$ that denotes the difference between $\{a_m^{(l+1)}(M), \tilde{\boldsymbol{r}}_{0m}^{(l+1)}\}$ and $\{a_m^{(l)}(M), \tilde{\boldsymbol{r}}_{0m}^{(l)}\}$. After expanding $\Psi_{sM}^{(l+1)}(\boldsymbol{r})$ into Talor series around $\Psi_{sM}^{(l)}(\boldsymbol{r})$ and neglecting the higher-order terms beginning at the second one, we can rewrite an unknown compensator $\psi_{sM}^{(l)}(\boldsymbol{r})$ in (6.33) in terms of a set of unknowns $\{\Delta a_m^{(l)}(M), \Delta\tilde{\boldsymbol{r}}_{0m}^{(l)}\}$. Referring to (6.18), we can derive the relative mean-square error $\varepsilon^{(l+1)}(M)$ on C at the $(l + 1)$th iteration as follows:

$$\varepsilon^{(l+1)}(M) = \|f(\boldsymbol{r}_c) - \Psi_{sM}^{(l)}(\boldsymbol{r}_c) - \psi_{sM}^{(l)}(\boldsymbol{r}_c)\|^2 / \|f(\boldsymbol{r}_c)\|^2 \tag{6.34}$$

Therefore, we have only to solve a set of unknowns $\{\Delta a_m^{(l)}(M), \Delta\tilde{\boldsymbol{r}}_{0m}^{(l)}\}$ or $\psi_{sM}^{(l)}(\boldsymbol{r})$ by minimizing $\varepsilon^{(l+1)}(M)$. This process is equivalent to solving a set of linear equations given by

$$\partial \varepsilon^{(l+1)}(M)/\partial \Delta a_m^{(l)}(M)^* = \partial \varepsilon^{(l+1)}(M)/\partial \Delta \tilde{\boldsymbol{r}}_{0m}^{(l)*} = 0, \quad m = 1, 2, \ldots, M \qquad (6.35)$$

where the differentiation with respect to $\Delta \tilde{\boldsymbol{r}}_{0m}^{(l)*}$ is equivalent to it with respect to the complex notation $(\Delta \tilde{x}_{0m}^{(l)} + j\Delta \tilde{y}_{0m}^{(l)})^*$ of a vector $\Delta \tilde{\boldsymbol{r}}_{0m}^{(l)*}$. Such a set of linear equations can be rewritten by using the following matrix form, as seen in (6.21):

$$H_M^{(l)} \tilde{X}^{(l)}(M) = F_M^{(l)} \qquad (6.36)$$

where $H_M^{(l)}$ is also a positive Hermitian matrix and $\tilde{X}^{(l)}(M)$ is a column vector of which the mth and $(M + m)$th elements are $\Delta a_m^{(l)}(M)$ and $\Delta \tilde{\boldsymbol{r}}_{0m}^{(l)}$, respectively.

For obtaining an optimum solution for $\tilde{X}^{(l)}(M)$ of (6.36), the modified Gauss method is widely used. However, it is also true that this iteration is not always effective for every case of scatterers, and a method [21] developed by Marquardt, Morrison, and Levenberg (called the *Marquardt method* hereafter) is effective for many cases. In this method, the solution $\tilde{X}^{(l)}(M)$ for the compensator in the modified Gauss iteration is replaced with the solution $\tilde{Z}^{(l)}(M)$ of the following equation:

$$(H_M^{(l)} + \mu \boldsymbol{I})\tilde{Z}^{(l)}(M) = F_M^{(l)} \qquad (6.37)$$

where μ is Marquardt's positive constant and $(H_M^{(l)} + \mu \boldsymbol{I})$ is also the positive Hermitian matrix because $\boldsymbol{I}$ denotes the unit matrix. It is proved that (6.37) shows a stable convergence when a proper number is used for μ. We will discuss an optimum value for μ in the numerical examples described in the next section. Additionally, Section 6.5 will touch on the practical use of Marquardt iteration.

6.3 APPLICATION TO ANALYSES OF AN ELECTROMAGNETIC FIELD SCATTERED BY PERFECT-CONDUCTING CYLINDERS

6.3.1 Numerical Examples by the Linear-Search Method

Let us consider the rectangular cylinder of perfect conductor as a scatterer (uniform in the z direction) as shown in Figure 6.3. We now assume that the wave incident at an angle θ_i is the plane wave of the TE-mode or the TM-mode, which has only the z component of the magnetic or the electric field, respectively. Hereafter, the wave functions of both types of incident wave are commonly expressed by (6.25). On the other hand, by arranging the M equivalent sources on a closed contour Γ_d equally spaced as mentioned earlier, we use $\Psi_{sM}(\boldsymbol{r})$ of (6.12) as an approximated wave function for the scattered field. Then an optimum set of coefficients $\{a_m(M)\}$ can be obtained by solving (6.21) under the condition of keeping M constant and varying d.

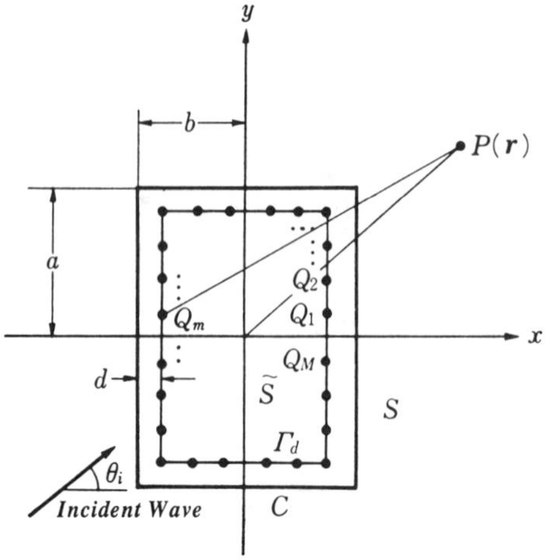

Figure 6.3 Plane-wave incidence on a rectangular cylinder of the perfect conductor and the arrangement of equivalent sources for expressing the scattered field.

Figure 6.4 shows the dependence of $\varepsilon(M)$ on d in the case of the TM-wave incidence, with M and the incident angle θ_i as parameters. We find from these results that an optimum contour Γ_d is determined by an optimum value of d, where the error $\varepsilon(M)$ is minimized and the dimension and shape are kept constant ($ak = 10$, $a = b$). Such a location of Γ_d approaches the boundary C by increasing the number M of sources, and this tendency depends little on the incident angle θ_i.

Results similar to Figure 6.4 are obtained for the TE-wave incidence or Neumann boundary-value problem, shown in Figure 6.5. These results clearly show that

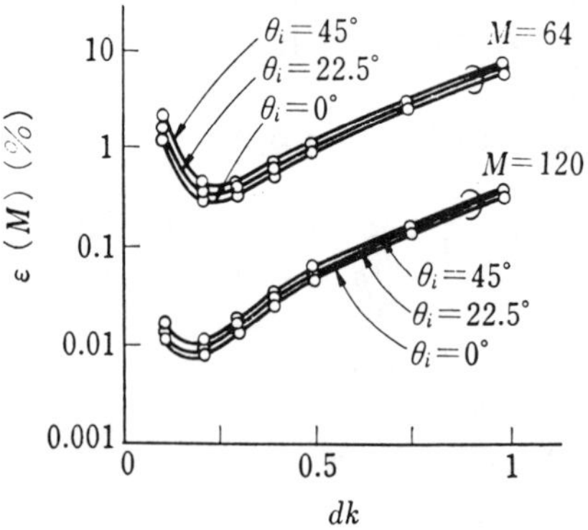

Figure 6.4 Mean square error *versus* the location of equivalent sources in the case of TM-wave incidence: $ak = 10$; $a = b$ [22].

this type of problem is more difficult in fitting the field to the boundary condition than the Dirichlet one. Indeed, it is obvious from Figure 6.5 that the error in the TE-mode changes more slightly with d than in the TM-mode and an optimum value for d does not appear clearly. What is worse, the magnitude of $\varepsilon(M)$ becomes the value of a figure or higher than in the TM-mode. However, the value d minimizing $\varepsilon(M)$ still exists even for the TE-mode, and it depends little on θ_i like the case of the TM-wave incidence.

Figure 6.6 shows the dependence of both the minimum value of $\varepsilon(M)$ and the corresponding $\varepsilon_{opt}(M)$ on the dimension a of the scatterer (when $\theta_i = 0$ and $a = b$, see Figure 6.3). We can find that the magnitude of $\varepsilon_{opt}(M)$ is almost the same value as that of $\varepsilon(M)$. Figure 6.7 shows numerical examples of the directivity pattern of the scattering cross section $\sigma(\theta;\ \theta_i)$ for ($ak = 30$, $b = 4a/3$, and $\theta_i = 60°$). We confirm that these results agree very well with those obtained by both the point-matching method [23] and the Wiener-Hopf method [24]. By the way, the error $\varepsilon(M)$ in the calculations of Figure 6.7 with $M = 160$ are 0.82 percent and 1.08 percent for the TM-wave and the TE-wave incidences, respectively.

Figure 6.8 shows the dependence of the normalized total-scattering cross section $\sigma_s(\theta_i)/4a$ at $\theta_i = 0°$ on the dimension a of a scatterer, and the result is compared with those obtained by both the point-matching method [24] and the integral-equation method [16]. Those results agree very well with one another except when a scatterer has a flat shape (like $b/a = 5$) and a large size (like $ak > 4$). In such an exceptional range of variables, the integral-equation method has a tendency to show a result different from other results. By the way, the error $\varepsilon(M)$ of 0.65 percent (when $M = 120$) is attained by the equivalent-source method in the calculation for the case of ($b/a = 5$, $ak = 8$).

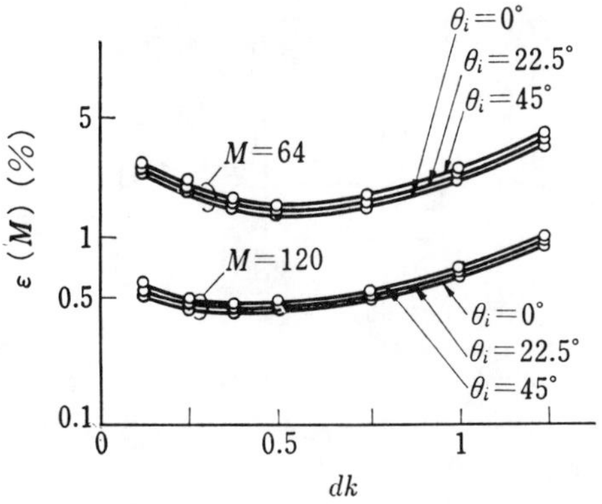

Figure 6.5 Mean square error *versus* the location of equivalent sources in the case of TE-wave incidence: $ak = 10$; $a = b$ [22].

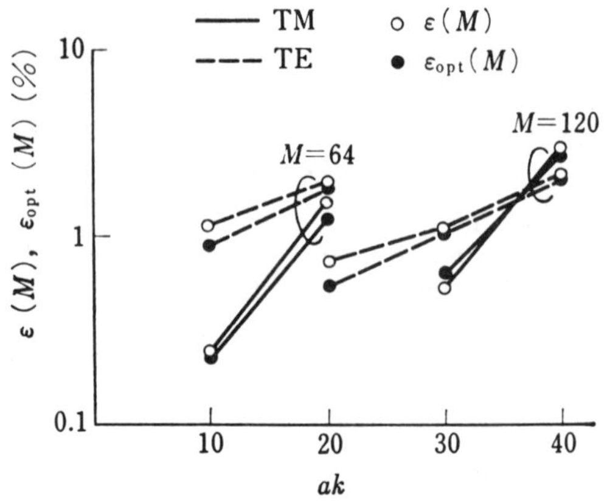

Figure 6.6 Mean square error *versus* the size a (see Figure 6.3) of a rectangular cylinder of the perfect conductor: $a = b$; $\theta_i = 0°$ [22].

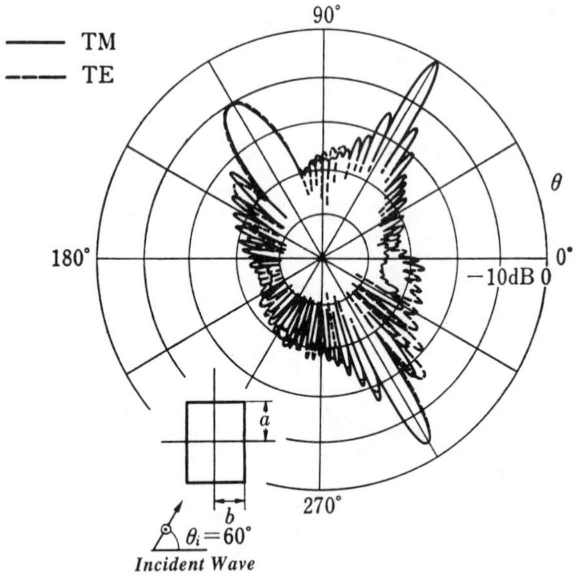

Figure 6.7 Directivity patterns of the scattering cross section $\sigma(\theta;\ \theta_i)$ for a rectangular cylinder of the perfect conductor: $ak = 30$; $b = 4a/3$; $\theta_i = 60°$; $M = 160$.

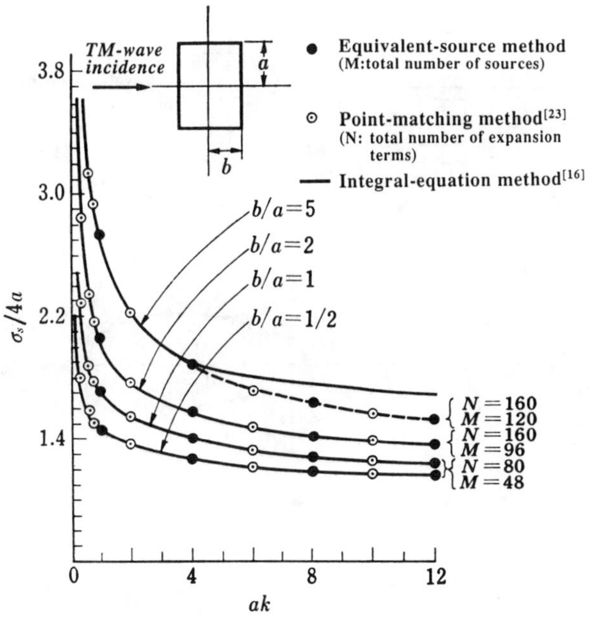

Figure 6.8 Total scattering cross section $\sigma_s(\theta_i)/4a$ *versus* the size a of a rectangular cylinder of the perfect conductor: $\theta_i = 0°$.

So far we have discussed a single scatterer, but the equivalent-source method is easy to apply to scatterers consisting of a number of perfect-conducting cylinders with arbitrary shapes. Figure 6.9 shows a numerical example of the directivity pattern for the scattering cross section when scatterers consist of two rectangular cylinders of the perfect conductor. In this calculation, we take a closed contour Γ_1 or Γ_2 in the interior region of each cylinder. The number M of the equivalent sources are divided into M_1 and M_2 $(= M - M_1)$ in proportion to the circumferential length of each cylinder; and they are arranged on Γ_1 and Γ_2, respectively. Excluding such a source arrangement, the numerical approach used in the case of a single scatterer applies to the present problem, too. For an example (Figure 6.9(a)) of scatterers with a small size, the error $\varepsilon(M)$ is 0.5 percent (when $M = 32$), whereas, for the case (Figure 6.9(b)) with a large size, $\varepsilon(M)$ becomes 1.5 percent, even if the number M is increased up to 160. Applications to other scatterers are described in [22] and so on.

6.3.2 Numerical Examples Obtained by Nonlinear Optimization

Although some types of the efficient Marquardt method have been conveniently installed in the scientific subroutine library, we follow its fundamental procedure

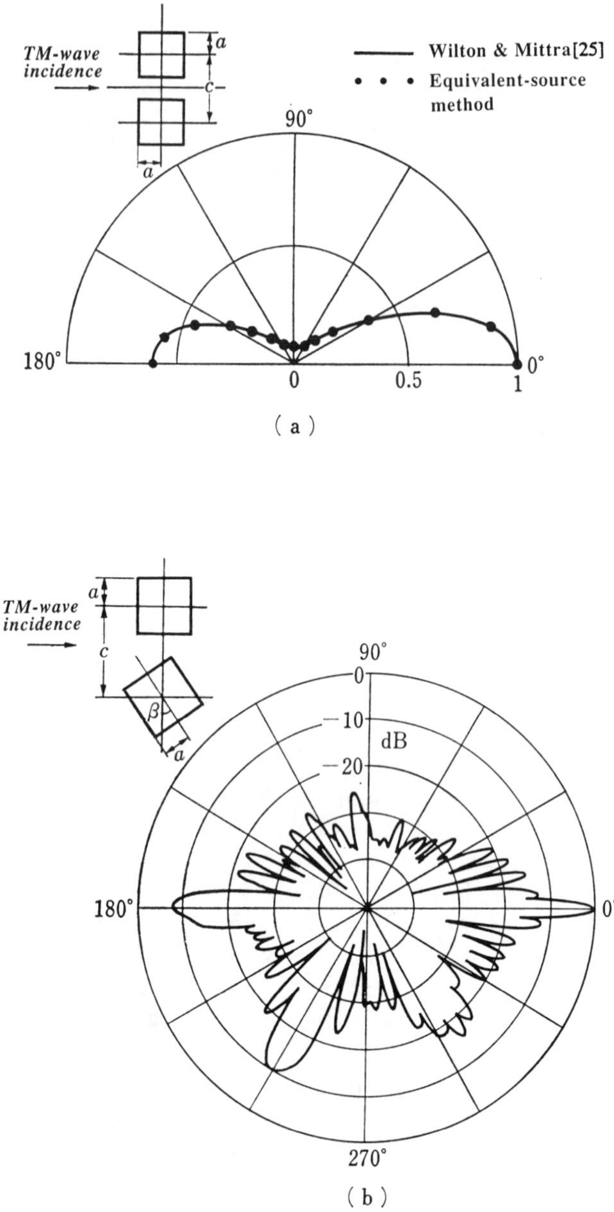

Figure 6.9 Directivity patterns of the scattering cross section $\sigma(\theta;\ \theta_i)$ for two parallel rectangular cylinders of the pertect conductor: (a) $ak = 1$, $c = 3a$, $\theta_i = 0°$, $M = 32$; (b) $ak = 20$, $c = 3a$; $\beta = 60°$, $M = 160$.

explained in Section 6.2.4 for calculations in this section. Here, we consider the rectangular cylinder of a perfect conductor with ($ak = 3$, $a = 2b$) and discuss the scattering problem when a plane wave of the TM-mode is incident with an angle $\theta_i = 0°$.

As described in Section 6.2.4, the convergence of solutions in this method depends significantly on the Marquardt constant μ. For example, Figure 6.10 shows the error $\varepsilon(M)$ as a function of the number of iteration for several μ values. As an example of the initial guess for the arrangement of sources, we locate them (now $M = 12$) on a closed contour $d = 0.1a$ away from the boundary surface. The case of $\mu = 0$ corresponds to the conventional Gauss iteration given by (6.36); and we find that the boundary error in this case rapidly increases with the number of iterations. The cases of $\mu \neq 0$ do not exhibit such an instability, and the error converges fairly for any μ, especially the error for $\mu = 0.1$ or so. When a large μ value is employed, the error indeed decreases uniformly with the increasing number of iterations, but the speed of the convergence becomes quite slow and the numerical efficiency worsens.

Figure 6.11 shows the loci of the positions of the typical equivalent sources when the nonlinear optimization is tried with $\mu = 0.1$. The black dots in each figure indicate the initial positions and the open dots indicate the positions obtained after the tenth iteration. The figure shows that a successful direction in the nonlinear search is obtained in any case at the beginning of the iterations. We also confirm that there is similar tendency in many kinds of cylindrical conductors with different shapes and sizes of cross section.

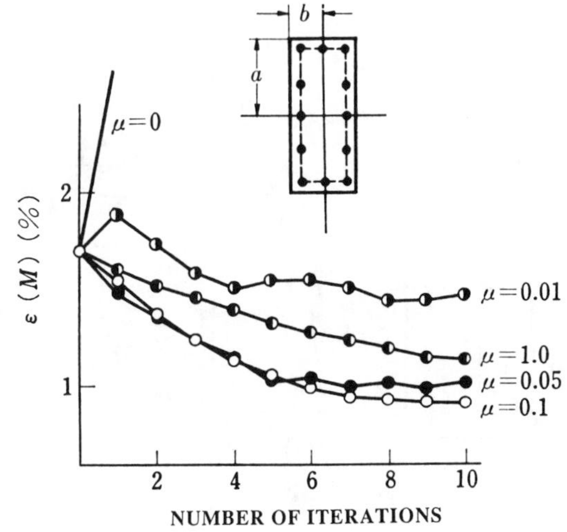

Figure 6.10 Mean square error *versus* the number of iterations in Marquardt method (Marquardt constant μ): $ak = 3$; $a = 2b$; $\theta_i = 0°$; $M = 12$.

It is interesting to note that the sources located near a rectangular corner (the points A and B in Figure 6.11) move toward the nearest corner as the number of iteration increases. Such behavior is closely related to the singular behavior of the electromagnetic field at a conducting corner. On the other hand, the source at the point C moves toward the inside of the scatterer. Such behavior is commonly observed for the sources near the specular points and the points in the shadow zone. This movement of sources, of course, causes some change in the scattered field. Figure 6.12 shows such a change in the directivity patterns of the scattering cross section. We find that the pattern at the initial guess indicated by the dashed curve changes slightly into the pattern indicated by the solid curve obtained after the tenth iteration.

The nonlinear optimization described here is effective for analyzing scatterers, even those with more complicated shapes [22]. As an example, it has been applied to a dish-aperture antenna including a primary radiator, and its receiving characteristics have been precisely solved from a boundary-value point of view [20].

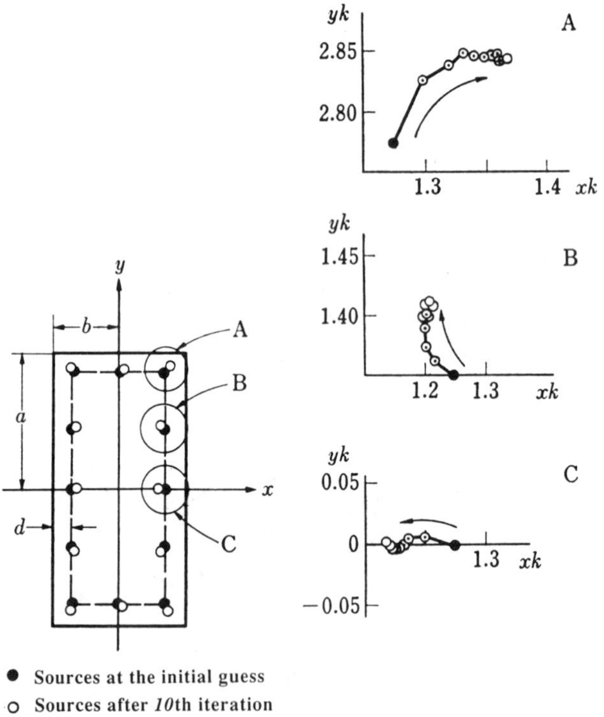

Figure 6.11 An example for the contours of equivalent sources obtained by nonlinear optimization: $ak = 3$; $a = 2b$; $\theta_i = 0°$, $M = 12$; $\mu = 0.1$.

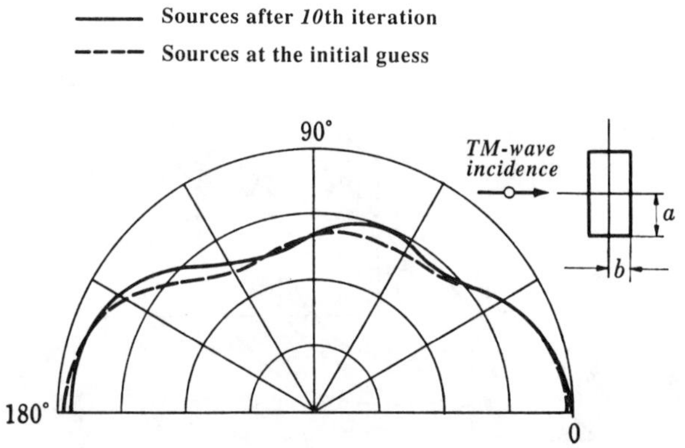

Figure 6.12 Change in the directivity patterns of the scattering cross sections observed in the calculations of Figure 6.11. The dashed curve shows the pattern at the initial guess, and the solid curve shows the pattern after tenth iteration.

6.4 PRACTICAL ANALYSES OF THE SCATTERED FIELD FROM DIELECTRIC CYLINDERS

The equivalent-source method applies easily not only to scatterers of a perfect conductor, but also dielectric scatterers of hybrid structures. Among these applicatons, a typical example is a lossless dielectric cylinder illuminated by a plane wave. To solve such a problem, it is necessary to find a complete set of the wave functions $\{\tilde{\phi}_n(\boldsymbol{r})\}$ $(\boldsymbol{r} \in \tilde{S})$ that constructs an approximated field in the interior region $\tilde{S}$ of a dielectric cylinder. If such work is successful, we have to solve only the simultaneous equations that express the continuity conditions of the electric and magnetic fields on the boundary C. For this purpose, the interior electric and magnetic fields are rightly derived directly using $\{\tilde{\psi}_n(\boldsymbol{r})\}$, mentioned earlier, or their approximation; and the exterior ones are obtained using the approximated wave functions discussed in Section 6.2.1. The discussions in this section will include the scattering from lossy dielectrics.

6.4.1 Approximate Wave Functions for Dielectric Region and Boundary Conditions

As shown in Figure 6.13, we assume a dielectric cylinder uniform along the z direction. Its cross section is divided into two regions by a smooth closed contour C on the xy plane: one region is the interior dielectric region (the refractive index is $\tilde{n}$), and the other is the exterior space region extending to the infinity (the refractive index is $n = 1$). Here, we consider that the plane wave of the TM-mode (its wave

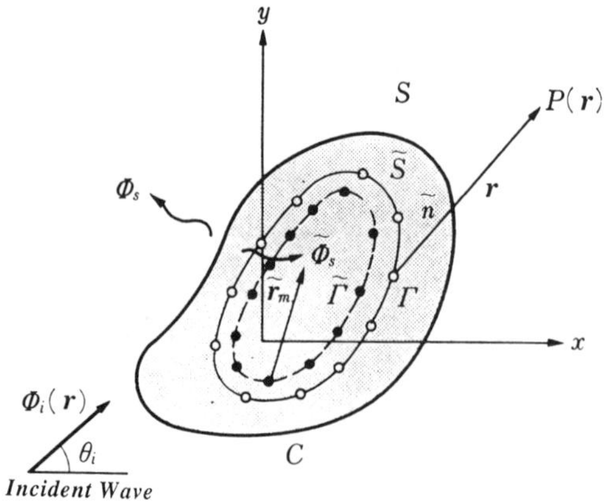

Figure 6.13 Two closed contours Γ and $\tilde{\Gamma}$ ($\in \tilde{S}$) on which the equivalent sources are arranged for expressing the wave function of the scattered field in exterior region S and interior region $\tilde{S}$ of a lossless dielectric, respectively.

function $\Phi_i(\boldsymbol{r})$ is given by (6.25)) is incident at an angle θ_i from the region S. We assume that the function $\Phi_i(\boldsymbol{r})$ denotes the z component of the electric field, whereas the scattered fields in the regions S and $\tilde{S}$ are denoted by $\Phi_s(\boldsymbol{r})$ ($\boldsymbol{r} \in S$) and $\tilde{\Phi}_s(\boldsymbol{r})$ ($\boldsymbol{r} \in \tilde{S}$), respectively. Then, $\Phi_s(\boldsymbol{r})$ satisfies (6.2) and (6.3) with $n = 1$ and $\tilde{\Phi}_s(\boldsymbol{r})$ does (6.2) with $n = \tilde{n}$; and these scattered fields must satisfy the boundary conditions on C in the following forms:

$$\Phi_i(\boldsymbol{r}_c) + \Phi_s(\boldsymbol{r}_c) + \tilde{\Phi}_s(\boldsymbol{r}_c) = 0, \quad \boldsymbol{r}_c \in C \tag{6.38}$$

$$\partial_n\Phi_i(\boldsymbol{r}_c) + \partial_n\Phi_s(\boldsymbol{r}_c) + \partial_n\tilde{\Phi}_s(\boldsymbol{r}) = 0, \quad \boldsymbol{r}_c \in C \tag{6.39}$$

where $\tilde{\Phi}_s(\boldsymbol{r})$ ($\boldsymbol{r} \in \tilde{S}$) is not necessary to satisfy the radiation condition, but enough to be finite in $\tilde{S}$. Because the conditions imposed on the exterior field $\Phi_s(\boldsymbol{r})$, of course, are identical with those for the scattered field discussed in Section 6.2.1, it is still effective to use the $\Phi_{sM}(\boldsymbol{r})$ of (6.11) or its simplified form, the $\Psi_{sM}(\boldsymbol{r})$ of (6.12), as an approximated wave function in this case. For such an approximated wave function, we arrange a set of source points $\{Q_m\}$ on the closed contour Γ as shown in Figure 6.13.

On the other hand, the idea of the generalized multipole expansion suggests a useful complete set of the modal functions in S in connection with the complete set of (6.9) as follows [14].

When S is bounded, let us replace Hankel functions in (6.9) with the 0th and first-order Bessel functions with the argument $nk\rho_m$. Then a set of the following functions allocated to each point Q_m is a complete set in S:

$$J_0(\tilde{n}k\tilde{\rho}_m),\ J_1(\tilde{n}k\tilde{\rho}_m)e^{j\theta},\ J_1(\tilde{n}k\tilde{\rho}_m)e^{-j\theta}, \quad m = 1, 2, \ldots \tag{6.40}$$

where ρ_m is the distance between Q_m and an observation point in S.

According to this result, a desirable wave function in $\tilde{S}$ can be constructed, as in (6.11). For this purpose, let us take one more Jordan contour, $\tilde{\Gamma}$ in $\tilde{S}$, and express $\tilde{\Phi}_{s\tilde{M}}(\boldsymbol{r})$ as follows by arranging the source points $\tilde{Q}_m$ ($m = 1, 2, \ldots, \tilde{M}$) on $\tilde{\Gamma}$:

$$\tilde{\Phi}_{S\tilde{M}}(\boldsymbol{r}) = \sum_{m=1}^{\tilde{M}} \{\tilde{a}_m(\tilde{M})J_0(\tilde{n}k\tilde{\rho}_m) + \tilde{b}_m^+(\tilde{M})J_1(\tilde{n}k\tilde{\rho}_m)e^{j\theta}$$
$$+ \tilde{b}_m^-(\tilde{M})J_1(\tilde{n}k\tilde{\rho}_m)e^{-j\theta}\}, \quad \boldsymbol{r} \in \tilde{S} \tag{6.41}$$

However, the simplified equivalent-source method often approximates (6.41) only in the first term of its right-hand side as follows:

$$\tilde{\Psi}_{S\tilde{M}}(\boldsymbol{r}) = \sum_{m=1}^{\tilde{M}} \tilde{a}_m(\tilde{M})J_0(\tilde{n}k\tilde{\rho}_m), \quad \boldsymbol{r} \in \tilde{S} \tag{6.42}$$

We carefully monitor the nature of the convergence, however, when $\tilde{\Psi}_{s\tilde{M}}(\boldsymbol{r})$ of (6.42) is used, because it has a drawback similar to that in $\Psi_{sM}(\boldsymbol{r})$ of (6.12).

Let us consider here the boundary conditions given by (6.38) and (6.39). For this purpose, we define the following functions on C, which are obtained using $\Psi_{sM}(\boldsymbol{r})$ and $\tilde{\Psi}_{s\tilde{M}}(\boldsymbol{r})$ in place of $\Phi_s(\boldsymbol{r})$ and $\tilde{\Phi}_s(\boldsymbol{r})$, respectively:

$$E_L(\boldsymbol{r}_c) = \Phi_i(\boldsymbol{r}_c) + \Psi_{sM}(\boldsymbol{r}_c) + \tilde{\Psi}_{s\tilde{M}}(\boldsymbol{r}_c) \tag{6.43}$$

$$\partial_n E_L(\boldsymbol{r}_c) = \partial_n \Phi_i(\boldsymbol{r}_c) + \partial_n \Psi_{sM}(\boldsymbol{r}_c) + \partial_n \tilde{\Psi}_{s\tilde{M}}(\boldsymbol{r}_c) \tag{6.44}$$

where $L = M + \tilde{M}$ is the total number of equivalent sources. Then, after defining the relative mean square error on C as follows:

$$\varepsilon(L) = \|E_L(\boldsymbol{r}_c)\|^2/\|\Phi_i(\boldsymbol{r}_c)\|^2 + \|\partial_n E_L(\boldsymbol{r}_c)\|^2/\|\partial_n \Phi_i(\boldsymbol{r}_c)\|^2 \tag{6.45}$$

we have only to minimize this error to obtain the unknowns. In practice, we may follow either the linear optimization to obtain both $a_m(M)$ and $\tilde{a}_m(\tilde{M})$, as discussed in Section 6.2.2, or the nonlinear optimization to obtain a set of $\{a_m(M), \boldsymbol{r}_{0m}, \tilde{a}_m(\tilde{M}), \tilde{\boldsymbol{r}}_{0m}\}$ as discussed in Section 6.2.4.

6.4.2 Numerical Examples of the Scattered Field From Lossless Dielectric Cylinders

We consider here the rectangular cylinder similar to that in Figure 6.3, but consisting of lossless dielectrics. This cylinder is illuminated by a plane wave of the TM-mode incident at an angle θ_i from the exterior region S. We settle two closed contours, Γ and $\tilde{\Gamma}$, constant distances d and $\tilde{d}$, respectively, away from C in the interior region. Some of the M and $\tilde{M}$ sources then are equally spaced on one side of the contours Γ and $\tilde{\Gamma}$ in proportion to the length of the corresponding side of the boundary C, respectively. An easy way to obtain solutions is to apply the linear search with respect to d and $\tilde{d}$, alternatively, and to find out the optimum d and $\tilde{d}$ that minimize the boundary error. It is shown in [22] that solutions converge sufficiently within a couple of iterations in some cases.

Figure 6.14 compares the convergence of the total scattering cross section σ_s for ($ak = 0.3$, $a = b$, $n = 1.1$, $\theta_i = 0°$). These results are obtained by both the present method and the integral-equation method [26], as a function of M ($= \tilde{M}$) or N (the number of the boundary division in practical calculations of the integral-equation method). The solution obtained by the equivalent-source method converges to $\sigma_s = 1.39 \times 10^{-3}/k$ for more than $M = 16$ and such a result is obtained at $N \approx 40$ when the integral-equation method is used. The errors of the solution obtained by the equivalent-source method with $M = 18$ are $\varepsilon(L) = 0.18$ percent and $\varepsilon_{\text{opt}}(L) = 0.11$ percent.

Figure 6.15 shows the dependence of the error on the number M of sources with the scatterer's size a as a parameter. Figure 6.16 shows the total scattering cross section of the dielectric cylinders ($\tilde{n} = 1.7$, $a = b$) with sizes larger than the wavelength. The number of sources and the total scattering cross sections are indicated on each figure. It is clearly shown that the directivity pattern of the scattering cross section becomes more complicated as the relative size of a scatterer to the wavelength increases. By the way, the errors in the case of Figure 6.16(c) ($ak = 20$) are $\varepsilon(L) = 2$ percent and $\varepsilon_{\text{opt}}(L) = 1.2$ percent. The magnitude of these errors is acceptably small, and we may conclude that the equivalent-source method is still

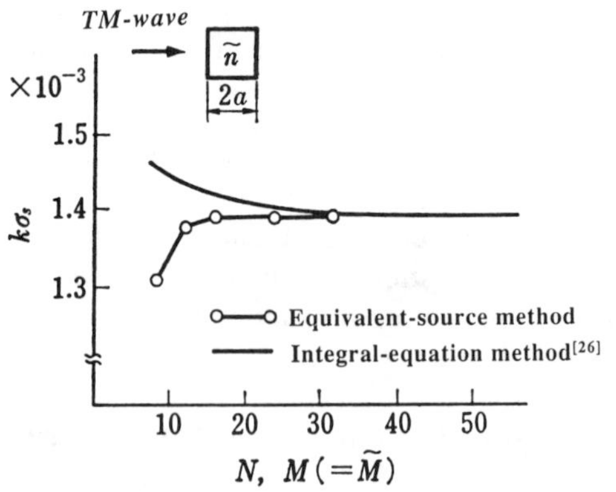

Figure 6.14 Total scattering cross section $\sigma_s(\theta_i)$ *versus* the number of equivalent sources M ($= \tilde{M}$) an the partition number in the integral-equation method: $ak = 0.3$; $a = b$; $\tilde{n} = 1.1$; $\theta_i = 0$

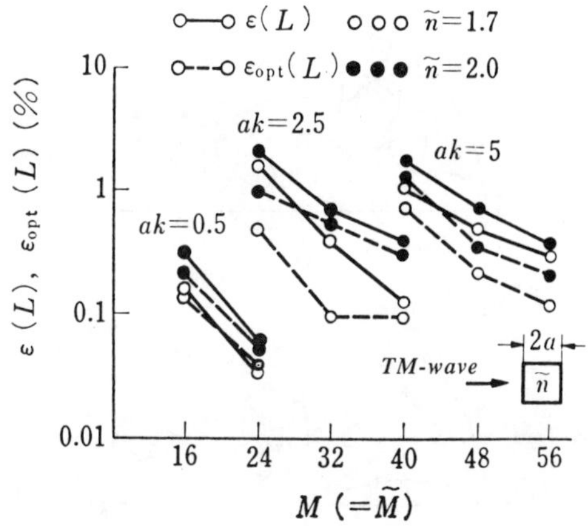

Figure 6.15 Mean square error *versus* the number of equivalent sources M with the size of a scatterer ak as a parameter: $a = b$; $\theta_i = 0°$ [22].

effective for analyzing lossless dielectric cylinders, even if their size becomes much larger than the wavelength.

6.4.3 Numerical Examples of the Scattered Field from Lossy Dielectric Cylinders

In the previous section, we showed that the equivalent-source method becomes an efficient one for discussing scattering problems from lossless dielectric cylinders. Our question here is whether this method still efficiently applies to lossy dielectric cylinders only replacing the real value of n with a complex value. To answer this question, let us first consider simply a lossy dielectric cylinder uniform along the z direction, which is illuminated by a plane wave of the TM-mode incident at an angle θ_i from the exterior region S, as shown in Figure 6.17. Figure 6.17(a) indicates the interior region $\tilde{S}$ filled with a homogeneous lossy dielectric with the relative permittivity $\varepsilon_r = \varepsilon' - j\varepsilon''$, whereas Figure 6.17(b) indicates a scatterer made of a center cylinder of the perfect conductor coated with a homogeneous lossy dielectric. Then it is necessary that $\tilde{\Phi}_s(\boldsymbol{r})$ ($\boldsymbol{r} \in \tilde{S}$) in the lossy dielectric region $\tilde{S}$ must satisfy (6.2) with $\tilde{n} = \sqrt{\varepsilon' - j\varepsilon''}$, instead of n. Let us discuss this problem in more detail in the case of Figure 6.17(a).

It is obvious that the scattered field in $\tilde{S}$ suddenly decays toward the center conductor as the dielectric loss becomes large. Nevertheless, if the field in $\tilde{S}$ is still expanded in terms of $J_0(\tilde{n}k\rho_m)$ like (6.42), it is quite difficult to efficiently approximate such a highly decaying field, and it is not always effective to apply the equivalent-source method discussed in Section 6.4.1 directly to this problem. Meanwhile,

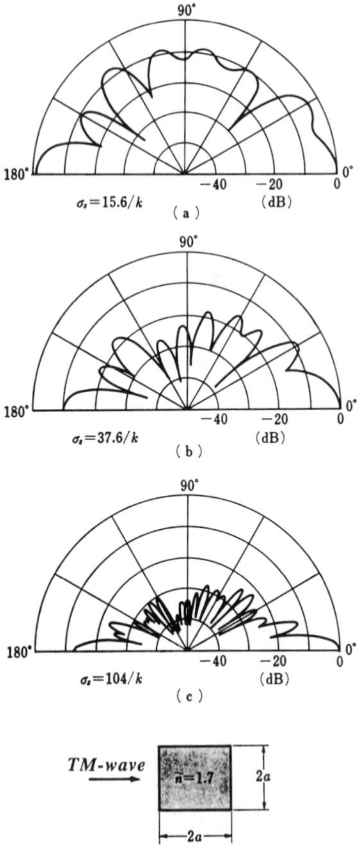

Figure 6.16 Change in directivity patterns of the scattering cross section due to the size of a lossless dielectric cylinder: $a = b$; $\tilde{n} = 1.7$; $\theta_i = 0°$; and in (a) $ak = 5$, in (b) $ak = 10$; and (c) $ak = 20$ [22].

the Hankel function $H_0^{(2)}(\tilde{n}k\rho_m)$ exhibits a rapid exponential decay in the ρ direction for a complex argument, and we can efficiently construct an approximated wave function. However, the field must be finite in interior region $\tilde{S}$, so that the source points should be arranged on a closed contour $\tilde{\Gamma}$ taken in exterior region S as shown by the black dots of Figure 6.18(a). As a result, the interior field is represented by

$$\tilde{\Psi}_{sM}(\boldsymbol{r}) = \sum_{m=1}^{\tilde{M}} \tilde{a}_m(\tilde{M}) H_0^{(2)}(\tilde{n}k\tilde{\rho}_m), \quad \boldsymbol{r} \in \tilde{S} \tag{6.46}$$

This equation is identical to (6.12) if the role of S is exchanged by $\tilde{S}$. Therefore we may follow Section 6.4.1 for further discussions on this problem.

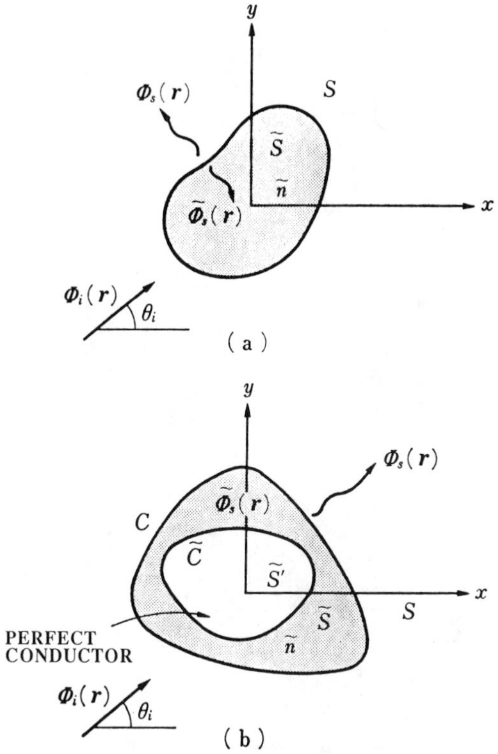

Figure 6.17 Infinitely extended external region S, the interior region $\tilde{S}$ of a lossy dielectric, the central region $\tilde{S}'$ of a perfect conductor, and their boundaries C and $\tilde{C}$: (a) a lossy dielectric cylinder; (b) a perfect-conducting cylinder coated with a lossy dielectric material.

Next, let us consider the problem of Figure 6.17(b). This problem can be solved by following the approach mentioned earlier. However, it is obvious in this case that the scattered field in $\tilde{S}$ caused by the conducting cylinder should be considered simultaneously. According to the discussions in Section 6.2.1, this scattered field is represented by the $\tilde{M}'$ equivalent sources arranged on a more closed contour $\tilde{\Gamma}'$ taken in the interior region of the conductor. As a result, the scattered field in $\tilde{S}$ can be approximated by the equivalent sources (shown by the black dots in Figure 6.18(b)) on the two closed contours $\tilde{\Gamma}$ and $\tilde{\Gamma}'$ as follows:

$$\tilde{\psi}_{s\tilde{L}}(\boldsymbol{r}) = \sum_{m=1}^{\tilde{M}} \tilde{a}_m(\tilde{M}) H_0^{(2)}(\tilde{n}k\tilde{\rho}_m) + \sum_{m=1}^{\tilde{M}'} \tilde{a}'_m(\tilde{M}') H_0^{(2)}(\tilde{n}k\tilde{\rho}'_m), \quad \boldsymbol{r} \in \tilde{S} \tag{6.47}$$

where $\tilde{L} = \tilde{M} + \tilde{M}'$ and $\tilde{\rho}'_m$ is the distance between a source on $\tilde{\Gamma}'$ and an observation point in S.

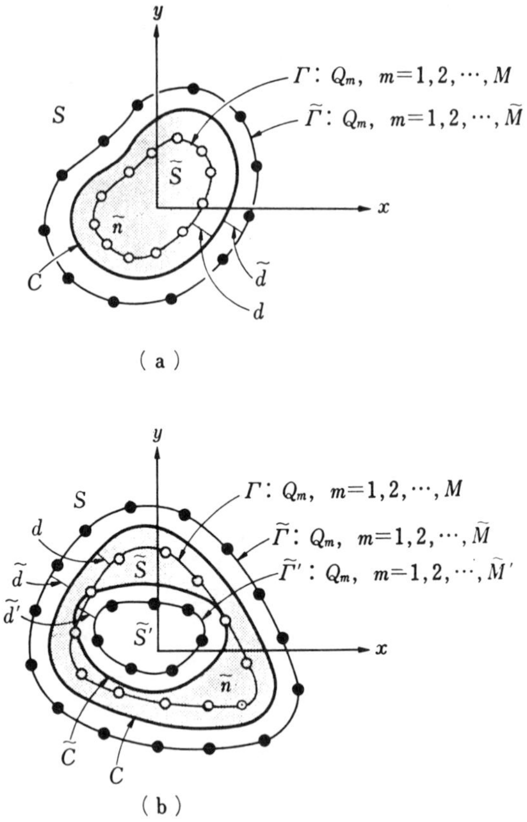

Figure 6.18 Closed contours on which equivalent sources are arranged for the problems of Figure 6.17: (a) the contours Γ and $\tilde{\Gamma}$ for expressing $\Phi_s(\boldsymbol{r})$ $(\boldsymbol{r} \in \tilde{S})$ and $\tilde{\Phi}_s(\boldsymbol{r})$ $(\boldsymbol{r} \in \tilde{S})$, respectively; (b) the contours Γ and $(\tilde{\Gamma}, \tilde{\Gamma}')$ for expressing $\Phi_s(\boldsymbol{r})$ and $\tilde{\Phi}_s(\boldsymbol{r})$, respectively.

For the present case, the error function on boundary C is obtained by replacing $\tilde{\Psi}_{s\tilde{M}}(\boldsymbol{r}_c)$ and $\partial_n \tilde{\Psi}_{s\tilde{M}}(\boldsymbol{r}_c)$ in (6.43) and (6.44) with $\tilde{\Psi}_{s\tilde{L}}(\boldsymbol{r}_c)$ of (6.47) and its corresponding normal derivative $\partial_n \tilde{\Psi}_{s\tilde{L}}(\boldsymbol{r}_c)$, respectively. In addition, the wave function $\tilde{\Psi}_{s\tilde{L}}(\boldsymbol{r})$ $(\boldsymbol{r} \in \tilde{S})$ must satisfy the boundary condition on the conductor surface $\tilde{C}$ $(\tilde{\Psi}_{s\tilde{L}}(\boldsymbol{r}_c))$ ideally must be zero), and we introduce the error function on $\tilde{C}$ given by

$$\tilde{E}_{\tilde{L}}(\tilde{\boldsymbol{r}}_c) = \tilde{\Psi}_{s\tilde{L}}(\tilde{\boldsymbol{r}}_c), \quad \tilde{\boldsymbol{r}}_c \in \tilde{C} \tag{6.48}$$

Then, after defining the following relative mean square error on C and $\tilde{C}$,

$$\begin{aligned} \varepsilon(L) = {} & \|E_L(\boldsymbol{r}_c)\|^2/\|\Phi_i(\boldsymbol{r}_c)\|^2 + \|\partial_n E_L(\boldsymbol{r}_c)\|^2/\|\partial_n \Phi_i(\boldsymbol{r}_c)\|^2 \\ & + \|\tilde{E}_{\tilde{L}}(\tilde{\boldsymbol{r}}_c)\|^2/\|\Phi_i(\tilde{\boldsymbol{r}}_c)\|^2, \quad \boldsymbol{r}_c \in C, \quad \tilde{\boldsymbol{r}}_c \in \tilde{C} \end{aligned} \tag{6.49}$$

we minimize it to obtain the solution. The detailed procedures have already been described in the preceding sections.

Now, it is obvious that the power absorbed in a dielectric scatterer is given by the integral value of Poynting power of the incident and scattered fields ($\Phi_i(\boldsymbol{r})$ + $\Psi_{sM}(\boldsymbol{r})$) ($\boldsymbol{r} \in S$) calculated over a cylindrical surface surrounding the scatterer. We usually define the cross section corresponding to this absorbed power in a dielectric scatterer and call it the *absorption cross section* $\sigma_a(\theta_i)$. Then, the total cross section $\sigma_t(\theta_i)$ is defined by the sum of $\sigma_a(\theta_i)$ and $\sigma_s(\theta_i)$ ($\sigma_t(\theta_i) = \sigma_a(\theta_i) + \sigma_s(\theta_i)$) [18].

Let us show here some numerical examples. Figure 6.19 shows the backward-scattering cross section $\sigma_b(\theta_i)$ of the lossy dielectric circular cylinder as a function of its radius a. The solid curves indicate the rigorous solutions obtained using the field based on the separation of variables, whereas the white dots show the results of the equivalent-source method. Both results show an excellent agreement in the wide range of variables. In calculations, the linear search is used as in lossless dielectric cylinders.

This example has also been calculated by the integral-equation method, but its solutions have included a spurious solution (or a resonant solution) at around ak = 2.4. Therefore, the integral-equation method needs to avoid such a solution. This usually becomes laborious in both the formulation of an improved method and numerical calculations.

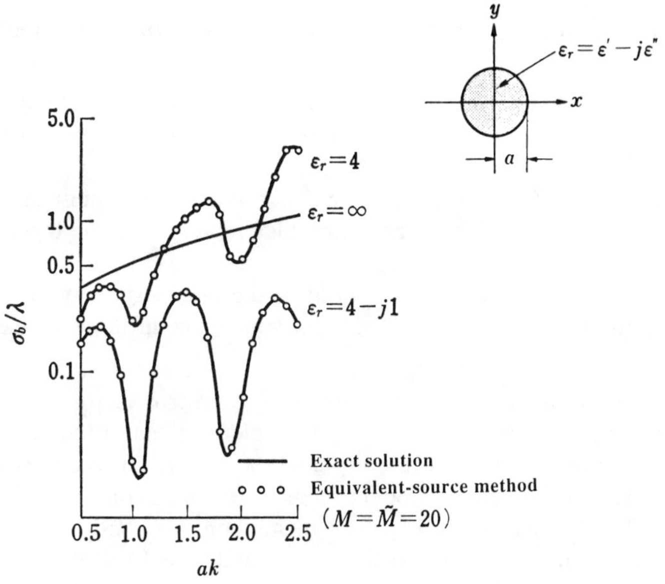

Figure 6.19 Backward scattering cross section $\sigma_b(\theta_i)$ *versus* the radius a of a lossy dielectric circular cylinder: $M = \tilde{M} = 20$ [22].

Figure 6.20(a) shows $\sigma_t(\theta_i)$, $\sigma_s(\theta_i)$, and $\sigma_a(\theta_i)$ for the rectangular cylinder ($ak = 2.5$, $\varepsilon_r = 4 - j\varepsilon''$) as a function of the loss term ε''. We may assume that the case of $\varepsilon'' = 0.01$ is almost lossless, whereas the case of $\varepsilon'' = 100$ is almost a perfect conductor. We then find that σ_s takes the minimum value at a certain value ε'' ($\varepsilon'' \approx 1$ for this case). Figure 6.20(b) shows the directivity pattern of the scattering cross section $\sigma(\theta;\ \theta_i)$ with ε'' as a parameter. By the way, all these results have been obtained with $\varepsilon(L) < 1.7$ percent.

Figure 6.21 shows $\sigma_s(\theta_i)$ and $\sigma_b(\theta_i)$ for the rectangular cylinder ($a = \lambda$) of the perfect conductor coated with a lossy dielectric ($\varepsilon_r = 2 - j2$) of thickness t. Figure 6.21(a) shows that the backward-scattering cross section σ_b greatly decreases for the coating thickness of about $t = 0.2\lambda$. Figure 6.21(b) shows the directivity pattern of the scattering cross section $\sigma(\theta;\ \theta_i)$, and it is obvious that $\sigma_b(\theta_i)$ is certainly about 15 dB lower in magnitude than the case without coating ($t = 0$). By the way, all these results have been obtained with $\varepsilon(L) < 0.8$ percent when $M = \tilde{M} = \tilde{M}' = 40$.

Some other types of scatterer consisting of lossy dielectrics have been investigated [22]. In practical calculations for scatterers with complicated shape, a high precision in numerical results is achieved even for a small number of sources when the nonlinear optimization is performed using a computer program installed in the scientific subroutine library.

6.5 SOME PRECAUTIONS FOR PROGRAMMING

The key points of the arithmetic based on the equivalent-source method are the following:

1. To solve the homogeneous linear equations, it is necessary to solve (6.21) by means of the linear optimization method, and (6.37) by means of the nonlinear optimization method.
2. For linear optimization, we apply the linear search for minimizing the boundary error with respect to some simple variables representing the location of equivalent sources as a whole.
3. For nonlinear optimization, we optimize the boundary error by varying both the amplitude and the position of each source by an iterative method with a good initial guess for them.

As is clear from (6.20), an element of the positive Hermitian matrix H_M in (6.21) or $(H_M^{(l)} + \mu \boldsymbol{I})$ in (6.37) is given by the inner product of the 0th-order Hankel function of the second kind with the argument $k|\boldsymbol{r}_c - \boldsymbol{r}_k|$. This poses no difficulty for the calculations of the inner product or the one-dimensional integral of the type given by (6.17). The matrix with such elements behaves gently even if the size of scatterers becomes large relative to the wavelength, because the elements do not exhibit the degeneracy due to Hankel functions with a large argument as seen in (6.8). Therefore, there is no problem for solving the homogeneous linear equations

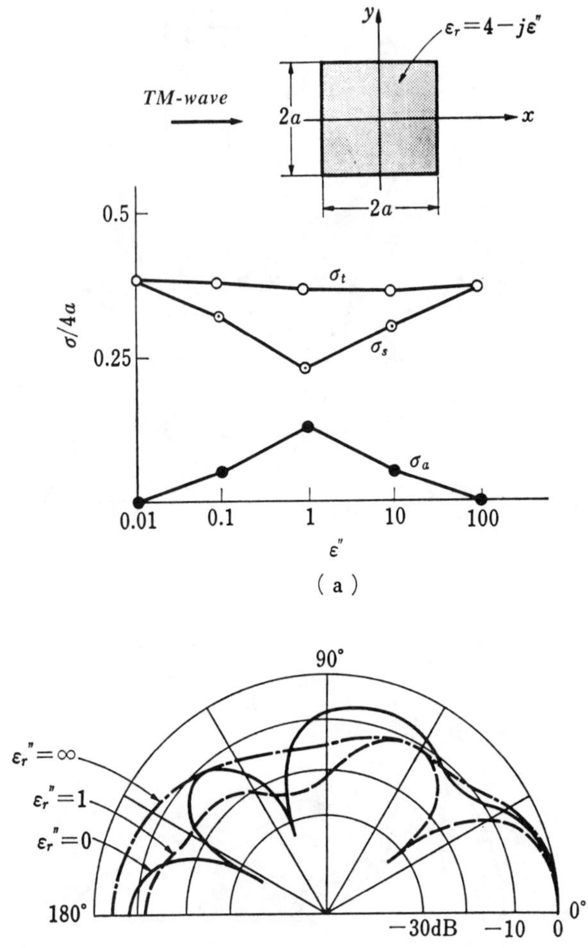

Figure 6.20 Dependence of the scattering characteristics of a rectangular cylinder of the lossy dielectric on the dielectric-loss term ε'': (a) cross sections σ_t, σ_s, and σ_a ($\varepsilon_r = 4 - j\varepsilon''$); (b) changes in the directivity pattern of the scattering cross section ($ak = 2.5$; $a = b$; $\theta_i = 0°$) [22].

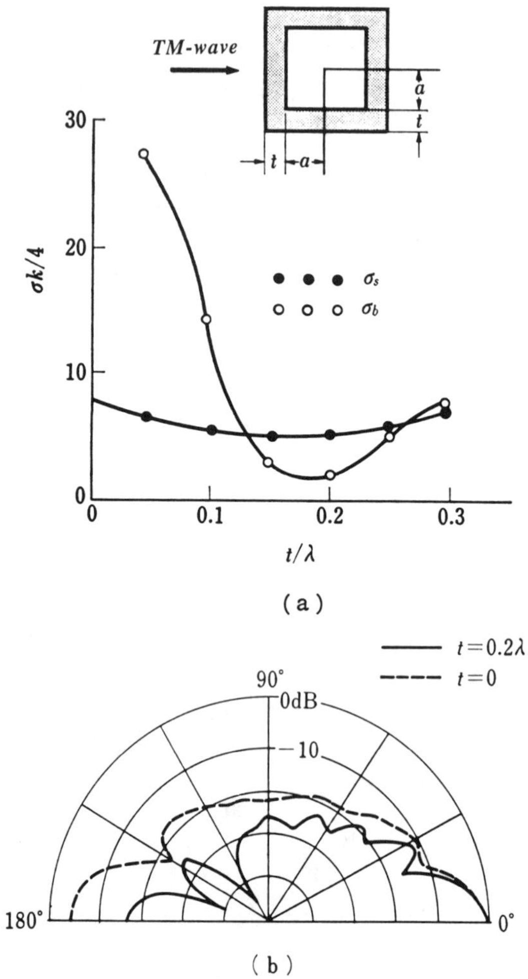

Figure 6.21 Scattering characteristics of the rectangular cylinder of a perfect conductor coated with lossy dielectric material: $ak = 6.28$; $a = b$; $\varepsilon_r = 2 - j2$; (a) dependence of cross section σ_s and σ_b on the dielectric thickness t; (b) change in directivity patterns of the scattering cross section [22].

The real problem in the linear search is a good choice of the variable to represent a group of a number of equivalent sources, by which (6.21) is optimized. A method for such a choice has been discussed in this chapter, but it is almost impossible to judge whether it is really effective. Therefore, we inevitably feel like utilizing the nonlinear optimization.

This approach seems to be complicated in its basis, as seen in Section 6.2.4, but recent computer programs installed in almost all scientific subroutine libraries (for example, the modified Marquardt method) have made such calculation easy. Using this, it is not necessary to solve (6.37) but only to provide the computer program that handles the vectors $\tilde{\boldsymbol{r}}_k = \tilde{x}_k + j\tilde{y}_k$ ($\in \tilde{S}$) as variables for calculating the inner products of Hankel functions $H_0^{(2)}(k|\boldsymbol{r}_c - \tilde{\boldsymbol{r}}_k|)$ in (6.20).

The only difficulty is an initial guess of $\tilde{\boldsymbol{r}}_k$. Usually, equivalent sources are arranged on a contour taken closely to a boundary and equally spaced. However, it is inevitable to change an initial guess for efficient calculations. Such a trial will provide the solutions with an expected precision and also effectively prevent improper behavior, where some sources run into the impermissible region.

REFERENCES

[1] Steinbigler, H., diss. Univ. Munchen, April 1969.

[2] Sanchez-Sesma, F.J., and E. Rosenblueth, "Ground Motion at Canyons of Arbitrary Shape under Incident SH Waves," *Earthquake Eng. Struct. Dyn.*, Vol. 7, 1981, p. 441.

[3] Yano, H. and A. Kieda, "An Approximate Method for Solving Two-Dimensional Low Reynolds-Number Flow Past Arbitrary Cylindrical Bodies," *J. Fluid Mech.*, Vol. 97, Part 1, 1980, pp. 157–179.

[4] Hafner, C., "Beitrage zur Berechnung der Ausbreitung electromagnetischer Wellen in zylindrischen Strukturen mit Hilfe des 'Point-Matching'-Verfahrens," diss. 6683, ETH, Zurich, 1980.

[5] Kohno, T., and T. Takuma, *Numerical Calculation Method of Electric Field,* Corona Publishing, Tokyo, 1980 [in Japanese].

[6] Nishimura, M., and H. Shigesawa, "A Numerical Analysis of Electromagnetic Scattering of Perfect Conducting Cylinders by Means of Discrete-Singularity Method," *Trans. on IECE of Japan,* Vol. J66-B, No. 2, Feb. 1983, pp. 245–252 [in Japanese].

[7] Komiyama, A., "Analysis of Scattering Problems by Means of the Charge-Simulation Method," *Trans. on IECE of Japan,* Vol. J66-B, No. 6, June 1983, pp. 790–797 [in Japanese].

8] Kodama, M., and K. Taira, "Calculation of Impedance of Posts in Waveguides by a Current-Simulation Method," *Trans. on IECE of Japan,* Vol. J66-B, No. 5, May 1983, pp. 689–690 [in Japanese].

9] Klaus, G., "The MMP-Method Applied to 3-D Scattering Problems," *Proc. Int. Symp. on Antennas and Propagation,* (Kyoto), Vol. 2, Aug. 1985, pp. 599–602.

0] Morita, N. "Numerical Methods Based on the Integral Equation for Electromagnetic Wave Problems," *Proc. of IECE of Japan,* Vol. 68, No. 3, March 1985, pp. 266–270 [in Japanese].

1] Nishimura, M., S. Takamatsu, and H. Shigesawa, "A Numerical Analysis of Electromagnetic Scattering of Perfect Conducting Cylinders by Means of Discrete Singularity Method Improved by Optimization Process," *Trans. on IECE of Japan,* Vol. J67-B, No. 5, May 1983, pp. 552–558 [in Japanese].

[12] Cory, H., Z. Altman, and Y. Leviatan, "Determination of a Dielectric Waveguide Propagation Constant Using a Multifilament-Current Model," *Opt. Lett.*, Vol. 14, Sept. 1989, pp. 1026–1028.

[13] Iijima, T., Ed., *Modern Analytical Methods for Electromagnetic Fields, IECE of Japan,* Tokyo, Chapter 2 [in Japanese].

[14] Yasuura, K., and T. Itakura, "Approximation Method for Wave Functions (I)," *Memo. Faculty of Eng., Kyushu Univ.*, Vol. 38, No. 1, March 1965, pp. 72–77 [in Japanese]; "Complete Set of Wave Functions—Approximate Method for Wave Functions (II)," Vol. 38, No. 4, Feb. 1966, pp. 378–385; "Approximate Algorithm by Complete Set of Wave Functions—Approximate Method for Wave Functions (III)," Vol. 39, No. 1, May 1966, pp. 51–56.

[15] Takagi, S., *The Fundamentals of Analysis,* third ed. Iwanami, Tokyo, 1961 [in Japanese].

[16] Mei, K.K., and J.G. Van Bladel, "Scattering by Perfect-Conducting Cylinders," *IEEE Trans. Antennas & Propagation,* Vol. AP-11, No. 2, March 1963, pp. 185–192.

[17] Hinata, T., and T. Hosono," "On the Scattering of Electromagnetic Wave by Plane Grating Placed in Homogeneous Medium—Mathematical Foundation of Point Matching Method and Numerical Analysis," *Trans. on IECE of Japan,* Vol. J59-B, No. 12, Dec. 1976, pp. 571–578 [in Japanese].

[18] Mushiake, Y., and S. Adachi, *The Fundamentals of Radio-Wave Engineering,* Kyoritsu, Tokyo, 1970 [in Japanese].

[19] Bolljahn, L.J., and W.S. Lucke, "Some Relationships between Total Scattered Power and the Scattered Field in the Shadow Zone," *IRE Trans. on Antennas & Propagation,* Vol. AP-4, No. 1, pp. 69–71.

[20] Shigesawa, H., and M. Nishimura, "Surface Current and Receiving Characteristics of Reflector Antennas," *1986 IEEE/AP-S Int. Symp. Digest,* Vol. 2, 1986, pp. 535–538.

[21] Kowalik, J., and M.R. Osborne, *Methods of Unconstrained Optimization Problems,* American Elsevier, New York, 1968.

[22] Nishimura, M., diss. Doshisha University, March 1985 [in Japanese].

[23] Hosono, T., T. Hinata, T. Yamasaki, and K. Esaki, "Scattering of Plane Electromagnetic Waves by Conducting Rectangular Cylinders—A Method to Predict the Areas with Poor Reception Owing to a Rectangular Building," *Trans. on IECE of Japan,* Vol. J62-B, No. 7, July 1980, pp. 690–697 [in Japanese].

[24] Aoki, K., and K. Uchida, "Scattering of a Plane Electromagnetic Wave by a Conducting Rectangular Cylinder," *Trans. on IECE of Japan,* Vol. J63-B, No. 6, June 1980, pp. 596–603 [in Japanese].

[25] Wilton, D.R., and R. Mittra, "A New Numerical Approach to the Calculation of Electromagnetic Scattering Properties of Two-Dimensional Bodies of Arbitrary Cross Section," *IEEE Trans. on Antennas & Propagation,* Vol. AP-20, No. 3, March 1972, pp. 310–317.

[26] Morita, N., "Resonant Solutions Involved in the Integral Equation Approach to Scattering from Conducting and Dielectric Cylinders," *IEEE Trans. on Antennas and Propagation,* Vol. AP-27, No. 11, Nov. 1979, pp. 869–871.

[27] Nishimura, M., and Shigesawa, H., "An Analysis of Electromagnetic Fields Scattered from Loss Dielectric Cylinders," *Rep. of Tech. Meeting on EM Theory* (IEE of Japan), Vol. EMT-82-76, October 1982, pp. 183–192 [in Japanese].

Chapter 7
The Geometrical Theory of Diffraction

Makoto Ando

7.1 HIGH-FREQUENCY APPROXIMATION OF ELECTROMAGNETIC FIELD AND GEOMETRICAL THEORY OF DIFFRACTION

7.1.1 Historical Background

In analyses of scattering and diffraction of electromagnetic waves, exact analytical solutions such as eigenfunction series are limited to a small class of simple shapes. Various numerical (approximation) methods have been proposed to cover a wider class of problems (Chapters 1 to 6 and Chapter 8). These exact or numerical solutions are effective only at lower frequencies. At higher frequencies, where scatterers are greater than a few wavelengths, they are not of practical use due to slow convergence, difficulties in numerical integration, the excessive size of matrices, and so on. Conversely, *geometrical optics* (GO) has been a popular, classic high-frequency approximation. GO is advantageous because of its applicability to a wider class of scatterers with arbitrary shapes. It provides easy and intuitive understanding of the phenomena, such as Fermat's principle. However, GO treats only a principal contribution (GO terms) in high-frequency asymptotics, and it cannot express diffraction in higher-order terms.

Geometrical theory of diffraction (GTD) was proposed by Keller [1], who extends and improves GO by introducing additional rays for diffraction. In GTD, the following are assumed:

1. Diffraction is a phenomenon that depends on the local environment, such as the shape of scatterer and type of incident waves in the vicinity of the ray path.
2. Diffracted ray paths satisfy the extended Fermat's principle. The amplitude and phase of the field in a diffracted ray varies in a manner similar to the geometrical optics field.

Detailed historical background and extensive references are in other works [2–5]; some essentials are surveyed here. Fock suggested (1) that diffraction is local phenomenon by showing the current distribution in the immediate vicinity of the shadow boundary, depending only on the local curvature of the surface. Sommerfeld analyzed the plane wave diffraction from a perfectly conducting half sheet and gave the solution in the form of Fresnel integral; this closed-form solution is unique in that it is exact and yet applicable for arbitrary frequencies. Its high frequency asymptotic indicates the diffracted cylindrical rays emanating from the edge. Afterwards, Pauli extended this to the wedge with an arbitrary angle. For diffraction by perfectly conducting smooth convex surfaces, such as a cylinder and a sphere, high-frequency solutions are obtained from the poorly convergent eigenfunction series using a Watson transform. The diffracted fields in the shadow region, derived by Franz and Depperman and Levy and Keller, decays exponentially away from the shadow boundary and is called a *creeping wave*.

These high-frequency solutions for wedges and smooth surfaces explicitly suggest the ray behavior of the diffracted waves (2); Keller utilized these solutions as well as assumptions (1) and (2) and constructed GTD as a generalized approximation method.

7.1.2 The Range of Applications

GTD is a high-frequency technique that expresses diffraction by perfectly conducting scatterers in ray forms. Hence, accuracy of GTD is better for larger scatterers relative to wavelength. Nonetheless, GTD often gives excellent results for the small scatterer. The local property of the phenomenon assures wide applicability of GTD for scatterers with arbitrary macroscopic shapes, provided the local shape of the scatterer can be regarded as a part of the canonical ones (wedge and cylinder, *et cetera*). An additional advantage of GTD as well as GO includes its simplicity: a physical meaning of the phenomenon is clear. This makes GTD a powerful tool not only for analysis but also for experiments and designs concerning waves. Irrespective of these merits, GTD possesses difficulties similar to GO, such as singularities of fields at shadow and reflection boundaries and caustics, infinities of geometrical optics fields in focusing systems, and errors occurring when the leading terms vanish. Various approaches are proposed for these defects, though they always degrade the simplicity of GTD. Studies toward novel canonical problems are also undertaken to extend and improve GTD. Diffraction by dielectric (impedance) bodies is also discussed recently in GTD, though only a few are included here.

7.1.3 Geometrical Optics

Some characteristics of geometrical optics rays are derived in this section, because the behavior of diffracted rays is similar to GO. The concepts of phase and vector do not appear in classic geometrical optics. Luneberg and Kline showed that the high frequency asymptotics for the Maxwell's equations give the vectorial geometrical optics term. Maxwell's equations in a homogeneous medium are written

$$\nabla^2 \boldsymbol{E} + k^2 \boldsymbol{E} = 0 \tag{7.1}$$

$$\nabla \cdot \boldsymbol{E} = 0 \tag{7.2}$$

$$\nabla \times \boldsymbol{E} = j\omega\mu \boldsymbol{H} \tag{7.3}$$

where $k(= \omega\sqrt{\varepsilon\mu})$ is the wavenumber that becomes larger at higher frequencies. For a large k, the electric field is asymptotically expanded in a polynomial of k^{-1} (Luneberg-Kline expansion):

$$\boldsymbol{E}(\boldsymbol{r}) \simeq \mathrm{e}^{-jk\Phi(\boldsymbol{r})} \sum_{m=0}^{\infty} (-jk)^{-m} \boldsymbol{E}_m(\boldsymbol{r}) \tag{7.4}$$

$\Phi(r)$ indicates the phase function and Φ = constant defines the wavefronts whereas $\nabla\Phi$ points the direction in which ray travels. Geometrical optics discusses only the first term of $m = 0$. Substituting equation (7.4) into (7.1) and equating like powers of k, we get

$$|\nabla\Phi|^2 = 1 \tag{7.5}$$

$$(\nabla\Phi \cdot \nabla)\boldsymbol{E}_0 + \frac{1}{2}(\nabla^2\Phi)\boldsymbol{E}_0 = 0 \tag{7.6}$$

Equation (7.5) indicates the phase velocity of geometrical optics term is k/ω; in other words, the phase variation along the ray equals the path length times k. It also indicates that rays in a homogeneous medium are straight lines. The relation (7.6), called the *transport equation,* expresses the amplitude variation of the field along the ray. By using (7.2), (7.3), and (7.4), we have

$$\nabla\Phi \cdot \boldsymbol{E}_0 = 0 \tag{7.7}$$

$$\boldsymbol{H}_0 = Y_0 \nabla\Phi \times \boldsymbol{E}_0 \tag{7.8}$$

These results reveal that the geometrical optics fields are TEM waves. Field vectors are in the plane perpendicular to the direction of ray propagation, where electric and magnetic fields are perpendicular to each other and the ratio between them equals the wave admittance $Y_0(\sqrt{\varepsilon/\mu})$. We next solve (7.6) for the wavefront with a Gaussian curvature with the principal radii R_1 and R_2, as is shown in Figure 7.1. The general solution for (7.6) is given as

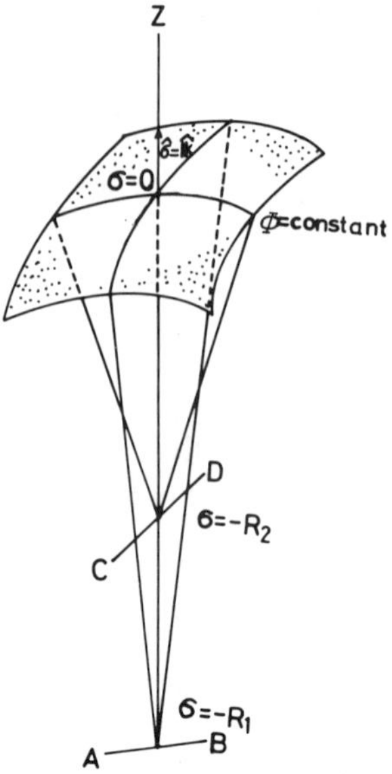

Figure 7.1 Wavefront with the principal radii of curvature, R_1 and R_2.

$$\boldsymbol{E}_0(\sigma) = \boldsymbol{E}_0(\sigma_0)\exp\left[-\frac{1}{2}\int_{\sigma_0}^{\sigma}\nabla^2\Phi d\sigma\right] \tag{7.9}$$

where σ corresponds to the arclength of a ray and $\nabla^2\Phi$ is expressed as follows using the parameters in Figure 7.1:

$$\nabla^2\Phi = \frac{1}{R_1+\sigma} + \frac{1}{R_2+\sigma} \tag{7.10}$$

By substituting (7.10) into (7.9), we arrive at a relation between the amplitude of the electric fields at different points σ_0 and σ:

$$\boldsymbol{E}_0(\sigma) = \boldsymbol{E}_0(\sigma_0)\left[\frac{S_0}{S}\right]^{1/2} = \boldsymbol{E}_0(\sigma_0)\left[\frac{(R_1+\sigma_0)(R_2+\sigma_0)}{(R_1+\sigma)(R_2+\sigma)}\right]^{1/2} \tag{7.11}$$

Figure 7.2 shows a ray tube in which the field strength varies in inverse proportion to the square root of the sectional area according to (7.11), conservation of energy. The points at $\sigma = -R_1$ and $-R_2$ are caustics where the sectional area vanishes and the field strength seems to be infinite. A complete relation of the

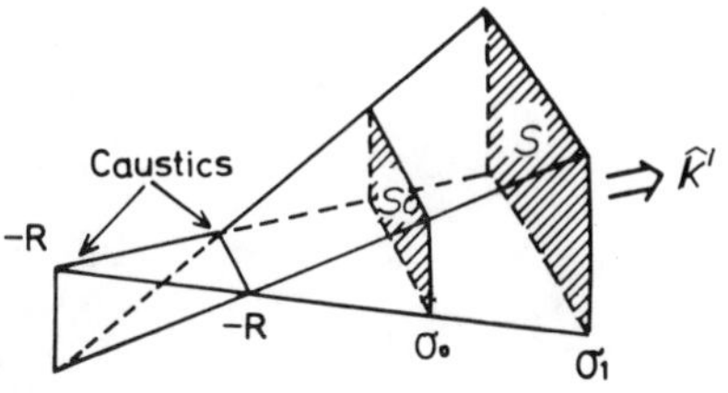

Figure 7.2 Geometrical optics ray.

geometrical optics fields at two points on a ray is obtained by adding the phase variation as

$$\boldsymbol{E}^{\mathrm{GO}}(\sigma) = \boldsymbol{E}^{\mathrm{GO}}(\sigma_0)\, \mathrm{e}^{-jk(\sigma_0-\sigma)} \left[\frac{(R_1+\sigma_0)(R_2+\sigma_0)}{(R_1+\sigma)(R_2+\sigma)}\right]^{1/2} \tag{7.12}$$

These results are derived by expanding the field in the integer power of k^{-1} in (7.4). Expansion in the fractional power also leads to similar results for the problems discussed in this chapter.

7.1.4 Canonical Problems

Exact analytical solutions are limited to a few simple problems. In GTD, high-frequency approximate solutions for general scatterers are derived from those for the simple ones with exact solutions. The latter are called *canonical problems,* which belong to two classes, diffraction by a conducting wedge and that by a smooth surface. Typical exact solutions for respective cases as well as its high-frequency approximation are presented here. One is the diffraction by a conducting half sheet, and the other is by that with a circular cylinder at its apex. We find from these solutions that diffracted fields behave in the same manner as geometrical optics rays. The term *diffracted field* is defined as the difference between the total field and the geometrical optics field.

(1) Diffraction by a Conducting Half Sheet

Figure 7.3(a) shows a two-dimensional diffraction from a conducting half sheet where a plane wave is incidence normally to the edge. For an E-wave (E_z^i) or H-wave (H_z^i) incidence, diffracted fields are given by

$$\begin{bmatrix} E_z^d \\ H_z^d \end{bmatrix} = \begin{bmatrix} E_z^i \\ H_z^i \end{bmatrix} \frac{-\mathrm{e}^{-j(\pi/4)-jk\rho}}{2\sqrt{\pi}} \left\{ \frac{F_-(\xi^{i^2})}{\xi^i} [\mp] \frac{F_-(\xi^{r^2})}{\xi^r} \right\} \tag{7.13}$$

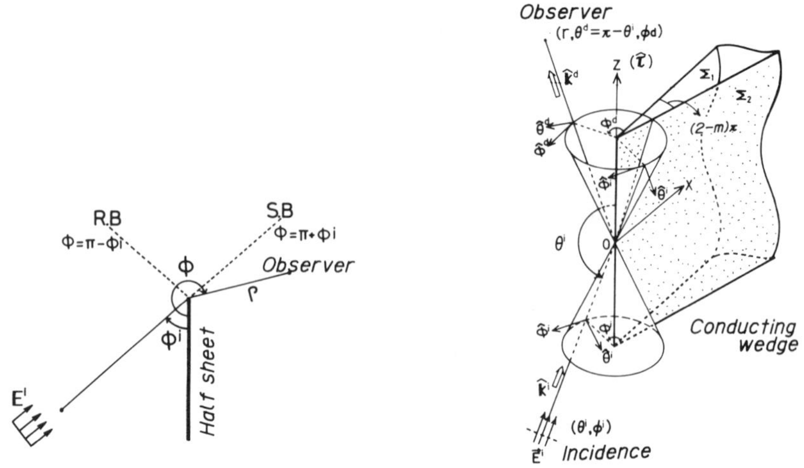

(a) Normal incidence

(b)Oblique incidence

Figure 7.3 Plane wave diffraction by a wedge: (a) normal incidence; (b) oblique incidence.

$$\xi^{[{i \atop r}]} = \sqrt{2k\rho}\cos\frac{\phi^d\,[\mp]\,\phi^i}{2} \tag{7.14}$$

$$F_{\pm}(\xi) = 2\sqrt{\xi}\,\mathrm{e}^{\mp j\xi \mp j(\pi/2)}\int_{\sqrt{\xi}}^{\infty}\mathrm{e}^{\pm j\chi^2}d\chi \tag{7.15}$$

where $F_{\pm}(\xi)$ is the Fresnel integral and for $|\xi|$ more than 10, $F_{\pm}(\xi)$ can be replaced by 1 and (7.13) is rewritten as

$$\begin{bmatrix} E_z^d \\ H_z^d \end{bmatrix} = \begin{bmatrix} E_z^i \\ H_z^i \end{bmatrix}\frac{-\mathrm{e}^{-j(\pi/4)}}{2\sqrt{2\pi k}}\frac{\mathrm{e}^{-jk\rho}}{\sqrt{\rho}}\left(\sec\frac{\phi^d-\phi^i}{2}\,[\mp]\,\sec\frac{\phi^d+\phi^i}{2}\right) \tag{7.16}$$

The ρ dependence of this result indicates the cylindrical wave. Hence, the diffracted field propagates in a manner similar to that of geometrical optics ray in (7.12) for $R_1 = \infty$. For small $|\xi|$, however, the Fresnel integral cannot be approximated by 1, and (7.13) must be used. The observation angles of $\phi^d = \pi + \phi^i$ and $\phi^d = \pi - \phi^i$ are optical boundaries that correspond to $|\xi| = 0$ and are called *shadow* (SB) and the *reflection* (RB) *boundaries*. As a more general problem, Figure 7.3(b) shows the diffraction by a wedge for an obliquely incident plane wave. The wedge angle is $(2 - m)\pi$. The unit tangent to the edge is $\hat{\tau}$ and the local spherical coordinate system as well as the unit vectors $\hat{\boldsymbol{\theta}}^i$ and $\hat{\boldsymbol{\phi}}^i$ is defined in the figure. First, the incident electric field is decomposed in terms of these vectors. For a half sheet ($m = 2$), the fields for the oblique incidence are derived from (7.13), (7.14), and (7.16) by changing the variables as

$$\begin{bmatrix} E_\theta^d \\ E_\phi^d \end{bmatrix} = \begin{bmatrix} E_\theta^i \\ -E_\phi^i \end{bmatrix} \frac{-e^{-j(\pi/4)-jkr}}{2\sqrt{\pi}} \left\{ \frac{F_-(\xi^{i^2})}{\xi^i} [\mp] \frac{F_-(\xi^{r^2})}{\xi^r} \right\} \tag{7.17}$$

$$\xi^{[{i \atop r}]} = \sqrt{2kr}\, \sin\beta^i \cos\frac{\phi^d[\mp]\phi^i}{2} \tag{7.18}$$

For sufficiently large $\xi^{i,r}$, we have simpler expressions:

$$\begin{bmatrix} E_\theta^d \\ E_\phi^d \end{bmatrix} = \begin{bmatrix} E_\theta^i \\ -E_\phi^i \end{bmatrix} \frac{e^{-j(\pi/4)-jkr}}{2\sqrt{2\pi kr}/\sin\beta^i} (\chi^i[\mp]\chi^r) \tag{7.19}$$

$$\chi^{[{i \atop r}]} = -\sec\frac{\phi^d[\mp]\phi^i}{2} \tag{7.20}$$

For a thick wedge with $m \neq 2$, Pauli gave the asymptotic solutions, and (7.20) should be replaced by

$$\chi^{[{i \atop r}]} = \frac{\dfrac{2}{m}\sin\dfrac{\pi}{m}}{\cos\dfrac{\pi}{m} - \cos\dfrac{\pi + \Psi[{i \atop r}]}{m}} \tag{7.21}$$

where $\Psi^{i,(r)}$ are angles between the projection of $\boldsymbol{k}^d$ and $\boldsymbol{k}^{i,(r)}$ onto the plane perpendicular to $\hat{\boldsymbol{r}}$ and their signs are defined as minus and plus for the illuminated and the shadow regions, respectively. In measuring Ψ, the wedge should not be crossed. The results of (7.19) are valid only for the observer far from optical boundaries, SB and RB, whereas (7.17) is valid uniformly.

(2) Diffraction by a Half Sheet with a Cylinder at Its Apex [6]

A diffraction problem discussed here is presented in Figure 7.4. Using the notation of $\rho_>$ and $\rho_<$ for the larger and the smaller distance of either ρ or ρ_s, respectively, we get diffracted fields in the form of an eigenfunction expansion:

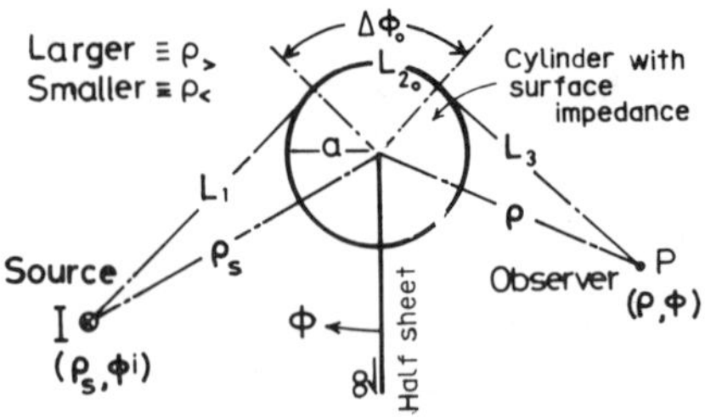

Figure 7.4 Diffraction by a half sheet with a cylinder.

$$E_z = -\frac{\omega\mu I}{2}\sum_{n=0}^{\infty} H^{(2)}_{n/2}(k\rho_>)\{J_{n/2}(k\rho_<) - B_{n/2}H^{(2)}_{n/2}(k\rho_<)\}$$
$$\cdot \sin\frac{n\phi^d}{2}\sin\frac{n\phi^i}{2} \tag{7.22}$$

$$B_v = \frac{J'_v(ka) - jCJ_v(ka)}{H^{(2)\prime}_v(ka) - jCH^{(2)}_v(ka)} \tag{7.23}$$

where C is the surface admittance of the cylinder and $C = \infty$ and $C = 0$ correspond to the electrically and magnetically conducting cylinder. The series in (7.22) is slowly convergent at high frequencies, and the physical meaning of the phenomena is not clear. The high-frequency asymptotics is derived by Watson transform. The final result in series converging only for the shadow region reads

$$E_z = -\frac{\omega\mu I}{2} j\pi \sum_n \sum_{l=1}^{\infty} e^{-jv_l|\phi^d - \phi^i_n|} H^{(2)}_{v_l}(k\rho_<)\, H^{(2)}_{v_l}(k\rho_>)$$
$$\cdot \frac{J'_{v_l}(ka) - jCJ_{v_l}(ka)}{\frac{\partial}{\partial v}\{H^{(2)\prime}_v(ka) - jCH^{(2)}_v(ka)\}_{v=v_l}} \tag{7.24}$$

The sum with respect to l accounts for propagation modes along the ϕ direction and vl's are poles of Bv in quadrant IV of the complex v-plane, in order of decreasing imaginary parts. The values of vl for various size of cylinders are depicted in Figure 7.5. The real parts of vl's are almost equal to ka, and their imaginary parts decrease as the radius a and the admittance C increase. For a large ka, they are approximated as

$$v_l \simeq ka - \left(\frac{ka}{2}\right)^{1/3} e^{-j(4\pi/3)} \begin{bmatrix} q_{sl} \\ q_{hl} \end{bmatrix} \begin{cases} C = \infty \\ C = 0 \end{cases} \tag{7.25}$$

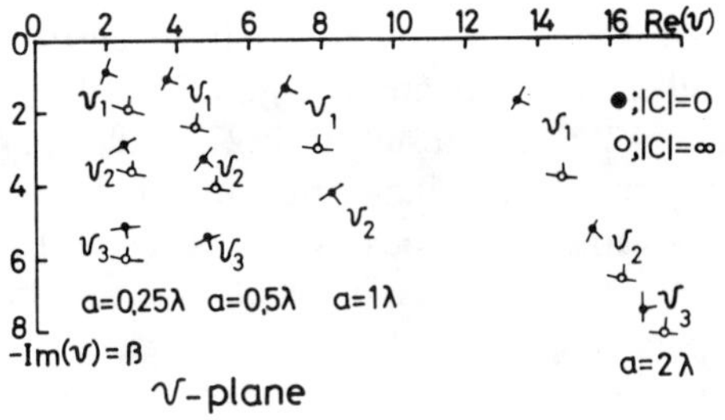

Figure 7.5 Location of the eigenvalues ν_l as functions of the radius a (for $C = 0, \infty$).

where q_s and q_h are zeros of Airy function and its derivative listed in Table 7.1 [4, 5].

The summation with respect to n corresponds to the contribution of images due to reflection by the half sheet, where ϕ_n^i is given by

$$\phi_n^i = 4n\pi \pm \phi^i \tag{7.26}$$

At high frequencies, creeping wave representation is available. For $\rho_< >> a$, Debye's asymptotic expansion of the Hankel functions is applied to (7.24) and then fields are rewritten in the form

$$E_z = E^i 2\sqrt{2\pi}\, e^{j(3/4)\pi} \sum_n \sum_{l=1}^{\infty} \frac{e^{-j\nu_l \Delta\phi_n - jkL_3}}{\sqrt{kL_3}} \cdot \frac{J'_{\nu_l}(ka) - jCJ_{\nu_l}(ka)}{\dfrac{\partial}{\partial u}\{H_\nu^{(2)\prime}(ka) - jCH_\nu^{(2)}(ka)\}_{\nu=\nu_l}} \tag{7.27}$$

Table 7.1
Airy Function Zeros and Associated Values
(after [4], p. 23, Table 2.1)

	$A_i(-q_s)=0$	$A_i'(-q_h)=0$		
l	q_{sl}	q_{lh}	$A_i(-q_{hl})$	$A_i'(-q_{sl})$
1	2.338	1.019	0.536	0.701
2	4.088	3.248	−0.419	−0.803
3	5.521	4.820	0.380	0.865
4	6.787	6.163	−0.358	−0.911
5	7.944	7.372	0.342	0.947

where E^i is the incident field at Q_1 on the cylinder surface $[-\omega\mu I \cdot \exp(-jkL_3 + j\pi/4)/\sqrt{8\pi kL_3}]$ and other notations are in Figure 7.4. Ray paths and $\Delta\phi_n$ for each n are presented in Figure 7.6. The result in (7.27) indicates that radiation from the source reaches the cylinder at a glancing angle; and, after traveling the distance $a\Delta\phi_n$ along the surface, it goes toward the observer. Because v_l has a negative imaginary part, the field on the surface decays exponentially. Such a ray is termed a *creeping wave*. Note that the path lengths of the rays in Figure 7.6 are the minimal ones among those that connect the source, the cylinder, the half sheet, and the observer, which satisfies extended Fermat's principle. Furthermore, diffracted fields propagate in cylindrical rays after leaving the surface. Figure 7.7 demonstrates the rapid convergence of the series with respect to n and l. Only a first term gives excellent approximation, though the cylinder is small and the radius is only a half-wavelength, provided the observer is in the shadow region ($\Delta\phi > 0$). Evaluation of diffracted fields in the lit region is slightly cumbersome but discussed in the references [4, 5, 7].

The two examples discussed here demonstrate that the behavior of diffracted fields is similar to that of geometrical optics fields and that the diffracted ray satisfies the extended Fermat's principle. In addition to these qualitative suggestions, the concrete expressions in (7.19) and (7.27) play important roles in GTD.

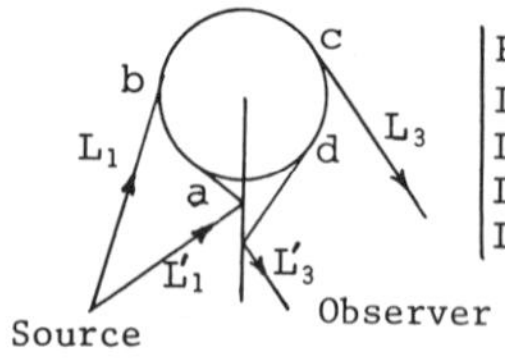

Ray	n	$\Delta\phi_n$	L_{2n}
L_1L_3	0	bc	$\overline{bc}$
L'_1L_3	1	ac	$\overline{ac}$
$L_1L'_3$	−1	bd	$\overline{bd}$
$L'_1L'_3$	2	ad	$\overline{ad}$

Figure 7.6 Ray paths for index n.

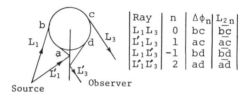

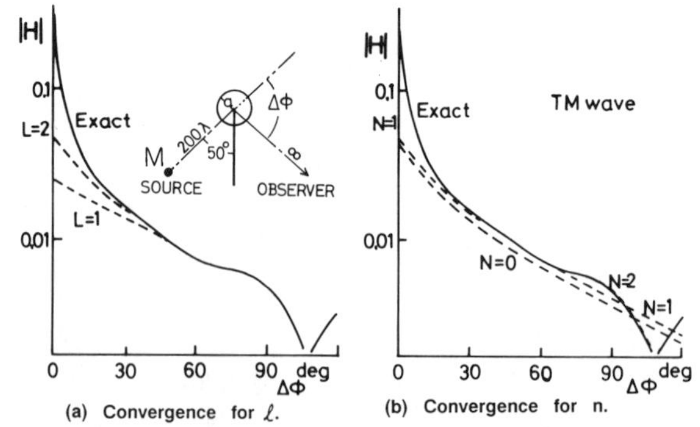

Figure 7.7 Convergence of a creeping wave for a sphere ($a = 0.5\lambda$): (a) convergence for l; (b) convergence for n.

7.1.5 Keller's GTD

(1) Fermat's Principle and Diffraction Points

Electromagnetic fields consist of geometrical optics fields and diffracted fields. The objective of GTD is to find diffracted fields simply and accurately. Diffraction occurs at diffraction points on the scatterer, which are determined applying Fermat's principle as used in optics. Fermat's principle states that reflected or refracted rays are those rays for which the optical path length between the source and the observer with one point on the interface between two media is stationary with respect to variations in path [4]. As is stated in Section 7.1.4, diffracted rays also are those for which the path length with one point on the diffracting edge or surface is also stationary. Hence, diffraction points are determined by obtaining the minimum or maximum rays among those connecting the source, the scatterer, and the observer. An example of such procedure is shown in Figure 7.8, where various types of diffracted rays exist in accordance with the shape of the scatterer. Geometrical optics rays, such as reflected and direct rays, are not included.

Edge-Diffracted Ray (D_1)

Rays with a stationary path length with one point on the edge of the scatterer are called *edge diffracted rays*. This condition requires the following relation. Diffraction law:

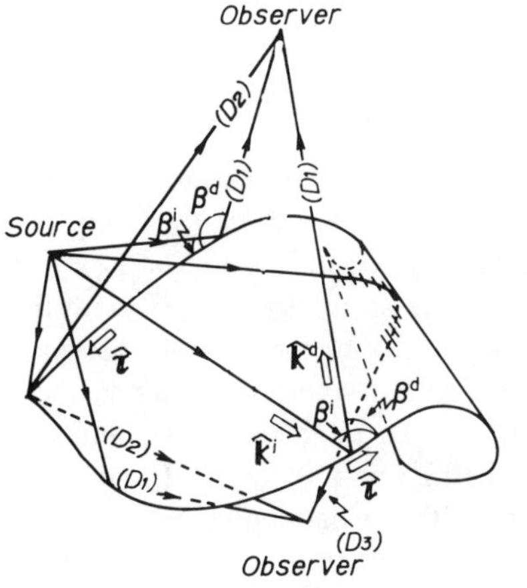

igure 7.8 Various types of diffracted rays.

$$(\hat{\boldsymbol{k}}^i - \hat{\boldsymbol{k}}^d) \cdot \hat{\boldsymbol{\tau}} = 0 \quad (\beta^i = \pi - \beta^d) \tag{7.28}$$

where $\hat{\boldsymbol{k}}^i$ and $\hat{\boldsymbol{k}}^d$ are the unit vectors indicating the directions of the incidence and the observer, respectively. The tangential unit to the edge is $\hat{\boldsymbol{\tau}}$.

Corner-Diffracted Ray (D_2)

Rays with a path with one point at the corner of the scatterer are called *corner diffracted rays*.

Creeping Wave (D_3)

Rays reaching and launching the surface at glancing angles and traveling along the surface correspond to creeping waves.

The preceding rays belong to the diffracted class and do not appear in geometrical optics. In GTD, every ray path (D_1) to (D_3) just stated brings about a diffracted ray. For example, if some point satisfies (7.28), the edge diffraction occurs at this point. The relation (7.28) reflects only the local shape of the scatterer and suggests the local property of the diffraction phenomenon. Total field is given by the sum of all the diffracted rays and geometrical optics fields. The magnitude of edge, corner diffraction, and creeping wave is $O(k^{-0.5})$, $O(k^{-1.0})$ and $O(\exp(-\alpha k))$, respectively, where $k(= \omega\sqrt{\varepsilon\mu} = 2\pi/\lambda)$ is the wavenumber and $O(k^0)$ of geometrical optics field is assumed. Hence, at high frequencies, edge diffracted rays for the scatterer with edges or creeping waves for the smooth object are the dominant contributions and discussed intensively in GTD.

(2) Diffracted Rays

Expressions for diffracted rays in GTD are summarized. After leaving from the scatterer, the field amplitude and the phase of the ray vary in a manner similar to geometrical optics fields, stated in Section 7.1.3. The key features are as follows: The diffracted field propagates in the straight ray tubes with the free space wavenumber k. The amplitude decays according to the law of energy conservation. The electric and magnetic fields are in the plane perpendicular to the direction of propagation. Figure 7.9 shows ray tubes of an edge diffracted and a creeping wave. For edge diffraction, one of the caustics ($\sigma = 0$) coincides with the edge, whereas for the creeping wave it coincides with the launching point on the surface. The term R is the distance to another caustic. Substituting $R_1 = 0$ and $R_2 = R$ into (7.12), we get the relation between the diffracted field at an arbitrary point (σ) and that at the reference point (σ_0):

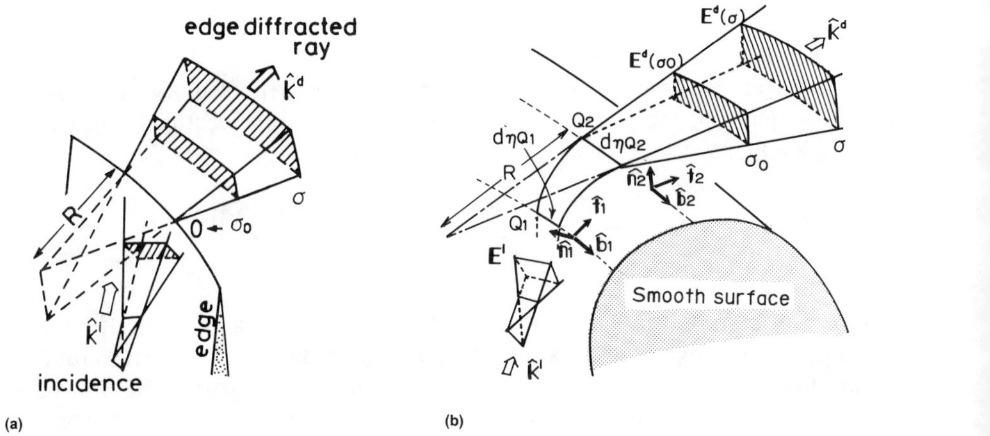

Figure 7.9 Diffracted rays: (a) an edge-diffracted ray; (b) a creeping wave.

$$\boldsymbol{E}^d(\sigma) = \boldsymbol{E}^d(\sigma_0)\, e^{-jk(\sigma-\sigma_0)} \left\{\frac{\sigma_0(R+\sigma_0)}{\sigma(R+\sigma)}\right\}^{1/2} \tag{7.29}$$

A unique feature of GTD consists in determining the diffracted field at the reference point. Consider a limiting case of $\sigma_0 \to 0$. Since $\boldsymbol{E}^d(\sigma)$ is independent of σ_0, $\boldsymbol{E}^d(\sigma_0)\sqrt{\sigma_0}$ approaches some finite value. Invoking the locality of diffraction, Keller determined this value as a function of only the incidence $\boldsymbol{E}^i$ and the scatterer shape at the diffraction point. $\boldsymbol{E}^d(\sigma_0 \to 0)$ and $\boldsymbol{E}^i$ are related as

$$\lim_{\sigma_0 \to 0} \boldsymbol{E}^d(\sigma_0)\sqrt{\sigma_0} = \begin{cases} \boldsymbol{E}^i \cdot \bar{\bar{D}}; & \text{edge diffraction} \\ \boldsymbol{E}^i \cdot \bar{\bar{T}}; & \text{creeping wave} \end{cases} \tag{7.30}$$

where $\bar{\bar{D}}$ and $\bar{\bar{T}}$ are dyadic diffraction coefficients and their explicit forms are derived directly from the solutions for canonical problems. Final results for diffracted fields are then given as

$$\boldsymbol{E}^d(\sigma) = \boldsymbol{E}^i \cdot \begin{bmatrix} \bar{\bar{D}} \\ \bar{\bar{T}} \end{bmatrix} e^{-jk\sigma} \left\{\frac{R}{\sigma(R+\sigma)}\right\}^{1/2} \begin{cases} \text{edge diffraction} \\ \text{creeping wave} \end{cases} \tag{7.31}$$

For a negative value of R in (7.31), the factor in $\sqrt{\ }$ may be negative depending upon the values of σ. By defining $\sqrt{-1} = j$, the appropriate phase jump of $\pi/4$ occurring at the caustic is automatically accounted for. In GTD analyses of diffraction, such as the edge diffraction and the creeping wave, only diffraction coefficients $\bar{\bar{D}}$ or $\bar{\bar{T}}$ and the distance to the caustic R are required.

(3) *Calculation of the Radius of Curvature R*

In edge diffraction, the radius of curvature R of the diffracted wavefront is determined by imposing the phase-matching condition of the incident and diffracted rays along the edge:

$$\frac{1}{R} = \frac{\cos^2 \Omega}{R_{ai}} + \frac{\sin^2 \Omega}{R_{bi}} + \frac{\hat{\boldsymbol{n}}_\rho \cdot (\hat{\boldsymbol{k}}^i - \hat{\boldsymbol{k}}^d)}{\sin^2 \beta^i \rho} \tag{7.32}$$

The parameters used in this calculation are presented in Figure 7.10. The radius of curvature of the edge is ρ; $\hat{\boldsymbol{n}}_\rho$ is the unit pointing to the center of curvature; R_{ai} and R_{bi} are the principal radii of curvature of the incident wavefront. The projection of edge tangential unit $\hat{\boldsymbol{\tau}}$ onto the wavefront is making the angle Ω with the line of curvature for R_{ai}. For a creeping wave from smooth surface, a general form of R is rather complicated and an example for a sphere is derived later by evaluating the spreading of the geodesic.

(4) *Diffraction Coefficients* [4]

Edge-Diffracted Wave

Diffraction coefficients for an edge are derived from the canonical problem of a wedge diffraction for an oblique incidence, shown in Figure 7.3(b). Substituting $R \to \infty$ in (7.31) and comparing it with (7.19), we arrive at the dyadic diffraction coefficient in the form [8]

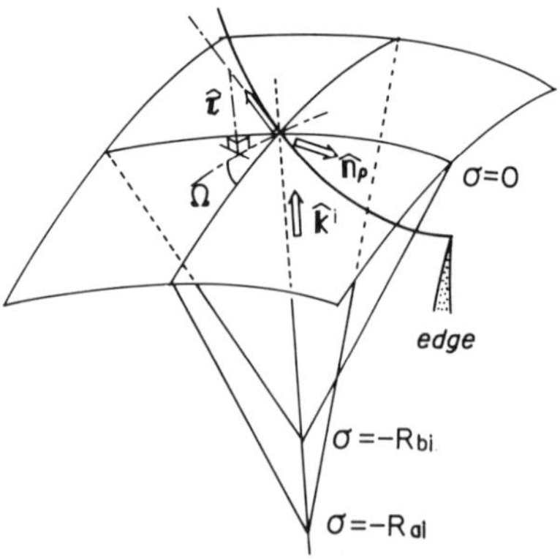

Figure 7.10 Calculation of the radius of curvature *R*.

$$\bar{\bar{D}} = D_s\hat{\boldsymbol{\theta}}^i\hat{\boldsymbol{\theta}}^d - D_h\hat{\boldsymbol{\phi}}^i\hat{\boldsymbol{\phi}}^d \tag{7.33}$$

$$D_{\left[\substack{s\\h}\right]} = \frac{e^{-j(\pi/4)}}{2\sqrt{2\pi k}\,\sin\beta^i}(\chi^i[\mp]\chi^r) \tag{7.34}$$

where $\chi^{i,r}$ is as defined in (7.21).

Creeping Wave [7]

In Figure 7.9(b), a creeping wave and some parameters are depicted. The terms $\hat{\boldsymbol{b}}$ and $\hat{\boldsymbol{n}}$ are unit vectors perpendicular to $\hat{\boldsymbol{\tau}}$, which is in the propagation direction, and are parallel and perpendicular to the surface, respectively. The subscripts 1 and 2 are for values at the incident and launching points, respectively. $\bar{\bar{T}}$ relates the incident field $\boldsymbol{E}^i$ and $\lim_{\sigma_0\to 0}\boldsymbol{E}(\sigma_0)\sqrt{\sigma_0}$ at the launching point. This coefficient includes the coupling factors at the incident and the launching points, exponential decay, and phase variation of the field during propagation along the surface. Equating (7.31) for $R = \infty$ and (7.27), we can extend $\bar{\bar{T}}$ to a general form:

$$\bar{\bar{T}} = T_s\hat{\boldsymbol{b}}_1\hat{\boldsymbol{b}}_2 + T_h\hat{\boldsymbol{n}}_1\hat{\boldsymbol{n}}_2 \tag{7.35}$$

$$T_{\left[\substack{s\\h}\right]} = \sum_{l=1}^{\infty} D_{\left[\substack{s\\h}\right]1}^{l}\left\{\exp\left[-jkt - \int_{t_1}^{t_2}\alpha_{\left[\substack{s\\h}\right]l}dt\right]\right\}\sqrt{\frac{d\eta_{Q1}}{d\eta_{Q2}}}\,D_{\left[\substack{s\\h}\right]2}^{l} \tag{7.36}$$

$$(D_{\left[\substack{s\\h}\right]}^{l})^2 = \sqrt{\frac{1}{2\pi k}}\left(\frac{k\rho_g}{2}\right)^{1/3} e^{-j(\pi/12)}\begin{bmatrix}\{A_i'(-q_{sl})\}^{-2}\\ q_{hl}^{-1}\{Ai(-q_{hl})\}^{-2}\end{bmatrix} \tag{7.37}$$

$$\alpha_{\left[\substack{s\\h}\right]}^{l} = \frac{1}{\rho_g}\left(\frac{k\rho_g}{2}\right)^{1/3} e^{j(\pi/6)}\begin{bmatrix}q_{sl}\\ q_{hl}\end{bmatrix} \tag{7.38}$$

where ρ_g is the radius of surface curvature measured in incident plane. The denominator in (7.27) is rewritten using Airy functions.

7.2 APPLICATION OF GTD AND ITS DEFECTS

7.2.1 GTD for Diffraction from a Circular Disk

GTD is applied to diffraction by a circular disk, as is shown in Figure 7.11. The disk is in the plane $z = -d$ and has a radius a. An x-directed small dipole source

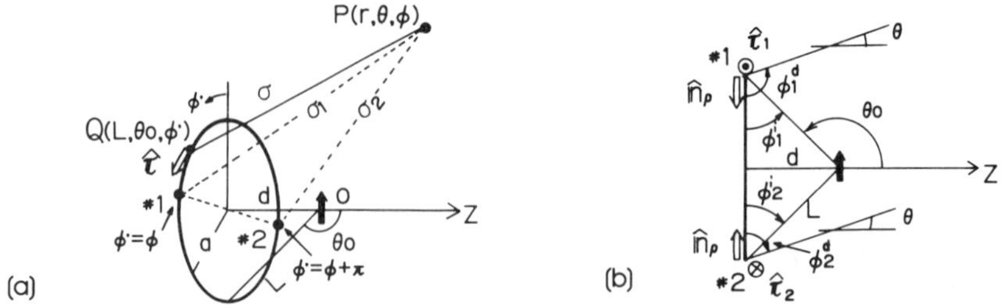

Figure 7.11 Local spherical coordinates for diffraction by a disk: (a) diffraction points; (b) local coordinates.

is at a distance d from the center of the disk. A direct field at the observer $P(r, \theta, \phi)$ from this source is given by

$$\boldsymbol{E}^i = \frac{e^{-jkr}}{r}(\cos\theta\cos\phi\hat{\boldsymbol{\theta}} - \sin\phi\hat{\boldsymbol{\phi}}) \tag{7.39}$$

$Q(r' = L, \theta_0, \phi')$ is a point on the edge and $\hat{\tau}$ is a unit tangent to the edge. The distance between Q and P is σ. Only a diffracted far field ($k\sigma \gg 1$) is calculated.

(1) Diffraction Points

The path length from the source to the observer with one point Q on the edge is given by

$$L + \sigma = L + \sigma_0 - a\sin\theta\cos(\phi - \phi') \tag{7.40}$$

Differentiating (7.40) with respect to ϕ', we can find two diffraction points for which the path length is stationary.

$$\#1\text{: } \phi_1' = \phi \quad \#2\text{: } \phi_2' = \phi + \pi \tag{7.41}$$

(2) Local Spherical Coordinate Systems [9]

At two diffraction points, local spherical coordinates are defined so that the z-axis may coincide with $\hat{\tau}$ at the points. According to the definition in Figure 7.3(b), the units $\hat{\boldsymbol{\theta}}^i$, $\hat{\boldsymbol{\phi}}^i$ are defined. Other parameters are derived in (7.32) to (7.34) and listed in Table 7.2.

Diffraction coefficients $D_s(h)_1$ are singular at $\theta = \theta_0$ and $\pi - \theta_0$. These angles correspond to SB and RB, and GTD is not applicable there, as it is here.

Table 7.2
Parameters in GTD Analysis of Diffraction by a Disk

# 1	# 2
$R_{ai}=R_{bi}=L,\ \rho=a,\ \beta=\pi/2$	
$\phi_1{}^i=\theta_0-\pi/2$	$\phi_2{}^i=\theta_0-\pi/2$
$\phi_1{}^d=\theta+\pi/2$	$\phi_2{}^d=\begin{cases}5\pi/2-\theta \\ \pi/2-\theta\end{cases}\quad \theta\gtrless\pi/2$
$\hat{\theta}_1=-\hat{\phi}$	$\hat{\theta}_2=\hat{\phi}$
$\hat{\phi}_1=\hat{\theta}$	$\hat{\phi}_2=-\hat{\theta}$
$\sigma_1=\sigma_0-a\sin\theta$	$\sigma_2=\sigma_0+a\sin\theta$
$R_1=a/\sin\theta$	$R_2=-a/\sin\theta$
$\chi_1{}^i=\text{cosec}\dfrac{\theta-\theta_0}{2}$	$\chi_2{}^i=\pm\text{cosec}\dfrac{\theta+\theta_0}{2}\quad \theta\gtrless\dfrac{\pi}{2}$
$\chi_1{}^r=-\sec\dfrac{\theta+\theta_0}{2}$	$\chi_2{}^r=\pm\sec\dfrac{\theta_0-\theta}{2}\quad \theta\gtrless\dfrac{\pi}{2}$
$D_{\left[{s \atop h}\right]1}=\dfrac{e^{-j(\pi/4)}}{2\sqrt{2\pi k}}(\chi_1{}^i\mp\chi_1{}^r)$	$D_{\left[{s \atop h}\right]2}=\dfrac{e^{-j(\pi/4)}}{2\sqrt{2\pi k}}(\chi_2{}^i\mp\chi_2{}^r)$

(3) Diffracted Fields

Using parameters in Table 7.2, contributions from two diffraction points are expressed as follows:

$$\boldsymbol{E}_{\left[{1 \atop 2}\right]^d}=\frac{e^{-jkL}}{L}\frac{e^{-jk\sigma_0}}{\sigma_0}\sqrt{\frac{[\pm]a}{\sin\theta}}\left(-\sin\phi D_{s\left[{1 \atop 2}\right]}\hat{\boldsymbol{\phi}}-\frac{d}{L}\cos\phi D_{h\left[{1 \atop 2}\right]}\hat{\boldsymbol{\theta}}\right)\cdot e^{[\pm]jka\sin\theta} \qquad (7.42)$$

The diffracted field is the sum of these; and to get the total field, geometrical optics fields should also be added. Figure 7.12 shows the diffracted fields in the E ($E\theta$: $\phi=0°$) and H ($E\phi$: $\phi=90°$) planes, calculated by (7.42) for $a=1.5$ wavelength and $d=2.75$ wavelength. Due to the symmetry of diffracted fields with respect to $\theta=90°$, only the patterns for $0°<\theta<90°$ are presented. An exact series solution is available for this problem [10] and given by dashed lines for comparison. In the

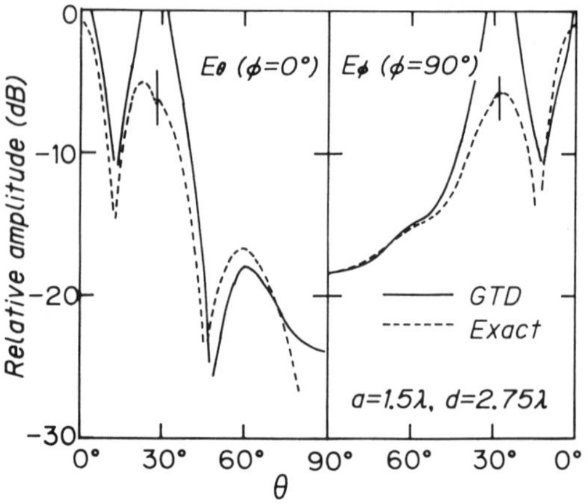

Figure 7.12 Application of GTD to diffraction by a disk.

figure, qualitative agreement between the exact and GTD solutions is observed. However, infinities at caustic ($\theta = 0°$), RB($\theta = 28°$) and SB($\theta = 152°$) affect the results in nearby angular regions; and GTD approximation is not satisfactory as a whole. At higher frequencies, degradation due to these infinities is localized on close vicinities of the caustic, RB, and SB; and in the limit of $k \to \infty$, higher accuracy is expected except near these angles.

7.2.2 Diffraction from a Sphere

Diffraction of a dipole field by a conducting sphere is discussed as an example o smooth-surface diffraction. The diffracted fields are observed in the far field, and the z-directed dipole is located at a distance of r_0 from the center of the sphere. Th incident electric field has only $\hat{\boldsymbol{b}}$ and $\hat{\boldsymbol{n}}$ components for the electric and magneti dipoles, respectively. Propagation of a creeping wave as well as parameters used i the analysis are depicted in Figure 7.13. The line ϕ = constant coincides with th geodesic, and the z-axis is a caustic. Hence, the spreading factor of the creepin wave and the distance R of the caustic are obtained as

$$\sqrt{\frac{d\eta_{Q1}}{d\eta_{Q2}}} = \sqrt{\frac{\cos\theta_0}{\cos\theta}} \qquad (7.4\text{:}$$

$$R = \frac{a}{-\tan\theta} \qquad (7.4$$

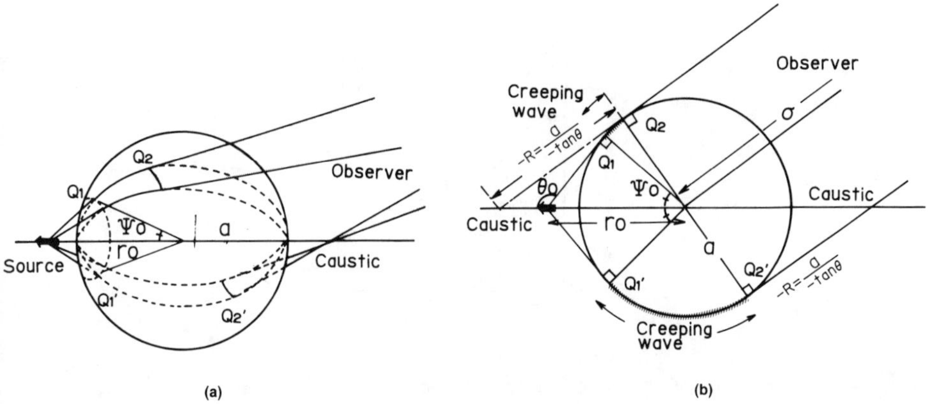

Figure 7.13 Diffraction by a sphere: (a) creeping wave; (b) parameters in the analysis.

Because the radius of curvature of the surface is a constant, the GTD creeping wave in the shadow region is expressed as follows, for the electric and magnetic dipoles:

$$\begin{bmatrix} E_\theta^d \\ H_\theta^d \end{bmatrix} = \begin{bmatrix} E_n^i \\ H_n^i \end{bmatrix} \sqrt{\frac{-a\cos\theta_0}{\sin\theta}} \{U_{[^h_s]}(\theta) - jU_{[^h_s]}(2\pi - \theta)\} \frac{e^{-jk\sigma}}{\sigma} \tag{7.45}$$

$$U_{[^h_s]}(\theta) = \sum_{l=1}^{\infty} \frac{\sqrt{\frac{2\pi}{k}} \left(\frac{ka}{2}\right)^{1/3}}{1 + e^{-j2\pi v_1}} \exp -j(\pi/12) - jv_l(\theta - \pi/2$$

$$- \text{arc}\cos(-a/r_0)) \begin{bmatrix} q^{-1}h_l & \{A_i(-q_{h_l})\}^{-2} \\ \{A_i' & (-q_{s_l})\}^{-2} \end{bmatrix} \tag{7.46}$$

which converges only in a shadow region. The first and second terms are creeping waves that travel around the sphere clockwise and counterclockwise, respectively. The denominator of (7.46) results from the sum of the geometric series with the ratio of $-\exp(-jv_l 2\pi)$, which accounts for higher-order rays traveling several times around the surface. In Figure 7.14, the creeping wave solution in the shadow region is compared with the exact eigenfunction expansion for a surface impedance sphere with the radius of 5 wavelengths. Though only a few terms with respect to l are included, excellent accuracy of GTD is achieved. Note that the diffraction coefficients used in (7.46) are not for a sphere but a cylinder; the local property of diffraction phenomenon assures the wide applicability of the coefficients.

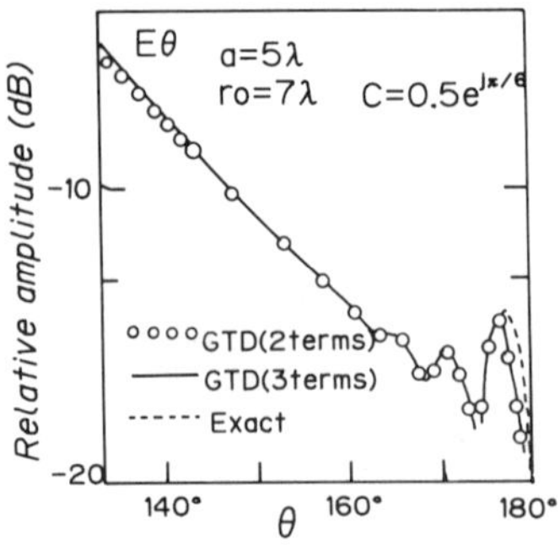

Figure 7.14 Application of GTD to diffraction by a smooth surface.

7.2.3 Defects of GTD

The field expression in GTD for the existence of N diffracted rays is

$$\boldsymbol{E} = \boldsymbol{E}^{\text{GO}} + \boldsymbol{E}^{d} = \boldsymbol{E}^{\text{GO}} + \sum_{n=1}^{N} \boldsymbol{E}_n^i \cdot \begin{bmatrix} \bar{\bar{D}} \\ \bar{\bar{T}} \end{bmatrix} \sqrt{\frac{R_n}{\sigma_n(R_n + \sigma_n)}}\, \mathrm{e}^{-jk\sigma_n} \tag{7.47}$$

The following difficulties are inherent in this representation:

1. Infinities of diffraction coefficients $\bar{\bar{D}}$, $\bar{\bar{T}}$ at SB and RB.
2. Infinities of the space-spreading factor at caustics.
3. Null-field prediction for null incidence at diffraction points.
4. Infinities of geometrical optics fields in focusing systems, such as reflecto antennas.

Among these, difficulties 1 and 2 were already indicated in Figure 7.12. Lo cation of these difficulties are illustrated in Figure 7.15 for a flat plate and a parabola SB is determined solely by the angle of incidence, whereas RB depends on the di rection of reflection at the diffraction points as well. The latter falls in the mainbean of focusing systems and cause serious difficulties in antenna analysis.

Various ideas are proposed to solve one or all of the difficulties (1) to (4). Th roots of difficulties and the basic strategy for improvements, for respective items are summarized:

1. Asymptotic evaluation of Fresnel integral in deriving the diffraction coeff cients from canonical problems produces infinity. Hence, we must use unifor coefficients, with Fresnel integral as it is (UTD [8], UAT [11]) or with infi ities cancelled using PO coefficients (PTD [12]).
2. This is a common limitation to ray techniques, where only a finite number diffracted rays are assumed. Continuous distribution of equivalent currents shou be integrated (PTD, EEC [13], (UAT)) to give finite fields near caustic.

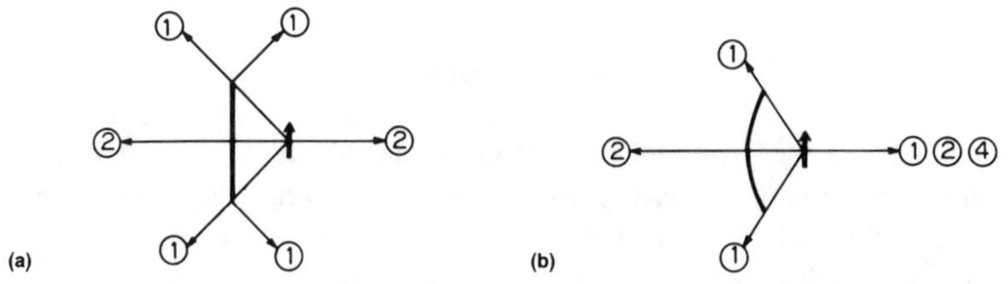

Figure 7.15 Singularities in GTD analysis: (a) a dipole and a disk; (b) a parabola.

3. As a first approximation, the diffracted field actually vanishes. To improve the accuracy, higher-order terms must be included (UAT, Slope wave [14], (EEC)).
4. This is a common limitation in optics, where only a finite number of reflection points are assumed. Reflection must be evaluated by integrating surface currents like PO (PTD).

7.3 IMPROVEMENTS FOR KELLER'S GTD

7.3.1 Improvements to Diffraction Coefficients

The argument of the Fresnel function vanishes at SB and RB. Hence, approximation of Fresnel integral in the transform from (7.13) to (7.16) causes the infinities of diffraction coefficients: the singularities disappear by recalling the Fresnel integral. Note that the diffracted fields behave like geometrical rays only when an integral is asymptotically approximated and the dependence of $\exp(-jk\sigma)/\sqrt{\sigma}$ is extracted. The behavior of diffracted fields at SB and RB is actually different from that of geometrical ray. Two methods for improving diffraction coefficients by recalling Fresnel integrals are explained.

UTD [8]

Fresnel integral is recalled by multiplication and infinities are cancelled. For an edge in a plane sheet ($m = 2$), uniform diffraction coefficients are written as

$$D_{\left[\substack{s \\ h}\right]} = \frac{-e^{-j(\pi/4)}}{2\sqrt{2\pi k}\,\sin\beta^i}$$

$$\left\{ \frac{F_-\left(2k\bar{L}\cos^2\dfrac{\phi^d - \phi^i}{2}\right)}{\cos\dfrac{\phi^d - \phi^i}{2}} [\mp] \frac{F_-\left(2k\bar{L}\cos^2\dfrac{\phi^d + \phi^i}{2}\right)}{\cos\dfrac{\phi^d + \phi^i}{2}} \right\} \qquad (7.48)$$

$$\bar{L} = \frac{\sigma(R_e + \sigma)R_a R_b}{R_e(R_a + \sigma)(R_b + \sigma)} \sin^2\beta^i \tag{7.49}$$

where R_a and R_b are the principal radii of curvature of incident wavefront, R_e is that taken in the plane containing $\hat{\tau}$, the unit tangent to the edge. R_e is given by the sum of the first two terms in (7.32). The original argument of Fresnel integral is given in (7.18) for a two-dimensional plane wave diffraction, whereas (7.49) is the generalized argument for arbitrary incidence. For a large argument of the integral, UTD coefficients in (7.48) asymptotically approach those in GTD (7.33).

UAT [11]

Subtraction of (7.17) from (7.19) gives the correction term that cancels singularities of GTD:

$$\begin{aligned}
\Delta &= \begin{bmatrix} E_\theta^{d(\mathrm{EXACT})} - E_\theta^d \\ E_\phi^{d(\mathrm{EXACT})} - E_\phi^d \end{bmatrix} \\
&= \begin{bmatrix} E^i_{\theta^i} \\ -E^i_{\phi^i} \end{bmatrix} \frac{-e^{-j(\pi/4)-jk\sigma}}{2\sqrt{\pi}} \left\{ \frac{F_-(\xi^{i^2}) - 1}{\xi^i} [\mp] \frac{F_-(\xi^{r^2}) - 1}{\xi^r} \right\} \qquad (7.50) \\
&= \begin{bmatrix} E_\theta^{i(\mathrm{GO})} \\ -E_\phi^{i(\mathrm{GO})} \end{bmatrix} G(\xi^i)[\mp] \begin{bmatrix} E_\theta^{i(\mathrm{GO})} \\ -E_\phi^{r(\mathrm{GO})} \end{bmatrix} G(\xi^r)
\end{aligned}$$

$$G(\chi) = \frac{e^{j(\pi/4)}}{2\sqrt{\pi}\chi\, e^{j\chi^2}} \{F_-(\chi^2) - 1\} \tag{7.51}$$

The argument $\xi^{i(r)}$ is defined so that ξ^2 indicates the detour parameter; that is, the difference between the diffracted ray path and the direct (reflected) ray path reaching the observer. The sign of $\xi^{i(r)}$ is plus and minus for the observer in the lit and shadow regions of the incidence (reflection), respectively. $\boldsymbol{E}^{i(\mathrm{GO})}$ and $\boldsymbol{E}^{r(\mathrm{GO})}$ are geometrical optics fields for incident and reflected waves. The correction terms in (7.50) tend to zero as the argument ξ becomes larger; UAT results also approach GTD in this case.

We now apply UTD and UAT to diffraction by a circular disk. In UTD, χ in Table 7.2 is replaced by $\bar{\chi}$ given by

$$\bar{\chi}^i_1 = \chi^i_1 F_-\left(2kL \sin^2\frac{\theta - \theta_0}{2}\right), \quad \bar{\chi}^i_2 = \chi^i_2 F_-\left(2kL \sin^2\frac{\theta + \theta_0}{2}\right) \tag{7.52}$$

$$\bar{\chi}^r_1 = \chi^r_1 F_-\left(2kL \cos^2\frac{\theta + \theta_0}{2}\right), \quad \bar{\chi}^r_2 = \chi^r_2 F_-\left(2kL \cos^2\frac{\theta - \theta_0}{2}\right) \tag{7.53}$$

In UAT, on the other hand, the following correction term should be added to E_1 of GTD results in (7.42):

$$\Delta = \frac{e^{-jk\sigma_0}}{\sigma_0} e^{-jkd\cos\theta}(-\sin\phi\hat{\boldsymbol{\phi}} + \cos\theta\cos\phi\hat{\boldsymbol{\theta}})G(\xi^i) \tag{7.54}$$

$$\xi^i = \sqrt{2kL}\cos\frac{\phi^d - \phi^i}{2} \tag{7.55}$$

where E_2 is not corrected because ξ_2 is large on RB associated with diffraction point #1. Numerical results by UTD and UAT are compared with the exact one in Figure 7.16(a). Infinity at RB ($\theta = 28°$) disappears and reasonable accuracy is obtained. Infinity at the caustic ($\theta = 0°$) remains unchanged after these corrections.

7.3.2 Method of Equivalent Edge Currents

1) Principle and Features [5, 13]

Diffracted ray paths obey the extended Fermat's principle, which requires that the path length is stationary with respect to variations in path; that is, its first derivative vanishes. A caustic appears at the observation point for which the second derivative of the path length becomes zero. For a caustic, an infinite number of diffraction points exist along the edge. By extending GTD, we can assume that the diffracted fields are radiated from a "line source" along the edge. The radiation of each current segment is integrated over the edge and produces the resultant diffracted fields. At high frequencies, the line integral can be evaluated asymptotically and the principal contributions from the stationary phase points give identical results with GTD. If the second derivative vanishes at the stationary phase point, the asymptotic evaluation of the integral becomes infinite, producing a caustic. Hence, these singularities disappear if the integral is evaluated numerically but not asymptotically. This is the principle of *equivalent edge currents* (EEC) proposed to improve GTD singularity at a caustic. The equivalent currents along the edge are derived from GTD solutions and then potential theory is used to find the diffracted fields:

$$\boldsymbol{E}_d = j\omega(\boldsymbol{A} + \eta\boldsymbol{B}\times\hat{\boldsymbol{r}})\times\hat{\boldsymbol{r}}\times\hat{\boldsymbol{r}} \tag{7.56}$$

$$\begin{bmatrix}\boldsymbol{A}\\ \boldsymbol{B}\end{bmatrix} = \frac{1}{4\pi}\begin{bmatrix}\mu_0\\ \varepsilon_0\end{bmatrix}\int\begin{bmatrix}I_{\mathrm{GTD}}\\ M_{\mathrm{GTD}}\end{bmatrix}\frac{e^{-jkr}}{r}\,d\boldsymbol{\tau} \tag{7.57}$$

where I_{GTD} and M_{GTD} are the equivalent electric and magnetic edge currents, respectively. Note that these currents are determined by GTD only for points where

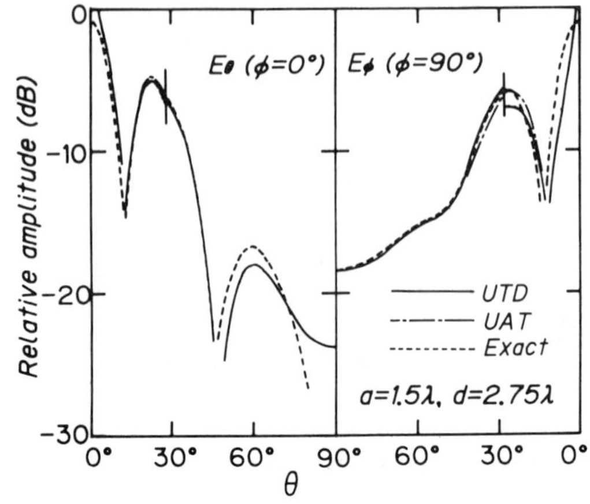

(a)

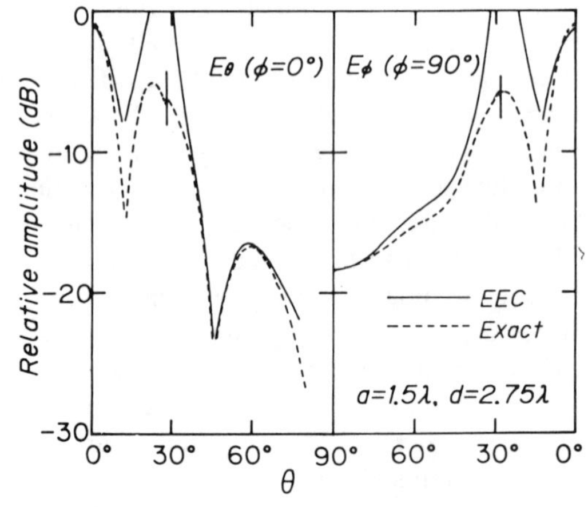

(b)

Figure 7.16 Accuracy of various methods in diffraction by a disk: (a) methods improving diffractio coefficients; (b) methods eliminating caustic; (c) physical theory of diffraction.

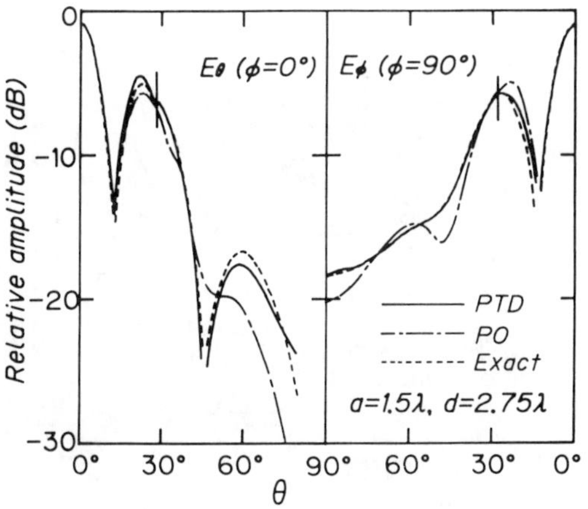

(c)

Figure 7.16 Continued

the law of diffraction is satisfied. In EEC, the currents for general points as well are used in the integration; this causes the ambiguity of EEC. Conversely, EEC automatically accounts for the space-spreading factor of the ray, and calculation of the radius R of curvature is needless.

(2) Equivalent Edge Currents

Equivalent edge currents I_{GTD} and M_{GTD} at some point are derived from GTD in a following procedure: (a) For a given direction of incidence and a given position on edge, the position of observer is assumed to satisfy the law of diffraction. (b) Instead of the practical scatterer, a conducting half sheet that is tangent to the surface forming the edge at the points of interest is considered. On the bases of (a) and (b), the equivalent currents are perfectly determined by GTD. Figure 7.17 shows the half sheet assumed at point P and a local spherical coordinate system used there. In a manner similar to Figure 7.3, the incidence is expressed in $\hat{\boldsymbol{\theta}}^i$ and $\hat{\boldsymbol{\phi}}^i$ components and then equivalent currents have the following form:

$$\begin{bmatrix} I_{\text{GTD}} \\ M_{\text{GTD}} \end{bmatrix} = \frac{\chi^i \mp \chi^r}{j\begin{bmatrix} \omega\mu \\ k \end{bmatrix} \sin\beta^i} \begin{bmatrix} E_{\theta^i} \\ E_{\phi^i} \end{bmatrix} = \frac{2\sqrt{2\pi k}}{j\begin{bmatrix} \omega\mu \\ k \end{bmatrix}} \begin{bmatrix} D_s E_{\theta^i} \\ D_h E_{\phi^i} \end{bmatrix} \times e^{j(\pi/4)} \tag{7.58}$$

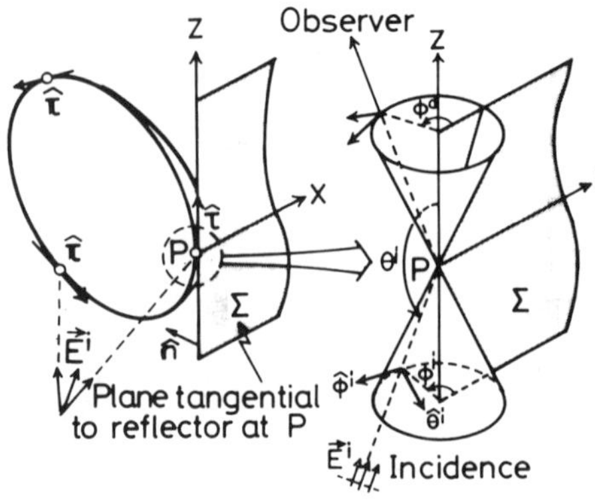

Figure 7.17 A local spherical coordinate at edge points.

where χ^i and χ^r are the coefficients in (7.21), and D_s and D_h are the diffraction coefficients of GTD in (7.34). Hence, the equivalent currents are the products of the incident-field component and the diffraction coefficient. This current is meaningful only for the observer on the Keller cone ($\theta^d = \pi - \theta^i$), satisfying the law of diffraction in the strict sense. Here, in this section, (7.58) is used to obtain the currents at all the points on the edge [9], using D_s and D_h at diffraction points. Because the incident-field components are easily defined at all points, the diffraction coefficients are approximated in this extension. Various expressions of the currents for the points except diffraction points have been proposed.

(3) Application of EEC

EEC is now applied to the diffraction of a dipole wave by a disk, shown in Figure 7.13. The equivalent currents are

$$\begin{bmatrix} I_{\mathrm{GTD}} \\ M_{\mathrm{GTD}} \end{bmatrix} = \frac{\mathrm{e}^{-jkL}}{j\begin{bmatrix} \omega\mu \\ k \end{bmatrix} L^2} \begin{bmatrix} \sin\phi' \\ \cos\phi' \end{bmatrix} \begin{bmatrix} L \\ d \end{bmatrix} (\chi^i \mp \chi^r) \tag{7.59}$$

The total diffracted fields are obtained by integrating these currents and presented in Figure 7.16(b). It is clear that the GTD infinity at caustic ($\theta = 0°$) is removed in EEC, and fine accuracy is observed in this area. In the region far away from the caustic, the results of EEC approach to those by GTD.

7.3.3 Modified Physical Theory of Diffraction

(1) Nonuniform Edge Currents

The *physical theory of diffraction* (PTD) eliminates the infinities both in diffraction coefficients of GTD and the geometrical optics field. The idea of PTD was proposed by Ufimtsev [12].

1. Scattered fields consist of the contributions of PO currents and nonuniform edge currents.
2. Nonuniform edge currents are constructed so that the field given by PTD approaches that given by GTD in the region where GTD is accurate.

In reference [12], the fields due to nonuniform currents were derived from the difference between GTD and asymptotically evaluated PO for special problems. It includes no explicit forms of the currents for general use; its contribution is conceptual rather than practical [15]. The infinities at SB and RB are eliminated by subtraction of the two methods. To remove the infinity at caustic for general problems, it is necessary to obtain the explicit forms for nonuniform edge currents and integrate them along the edge as in EEC. The procedure is illustrated in Table 7.3. PTD furnished with the explicit nonuniform edge currents are applicable uniformly throughout the region. Nonuniform edge currents are defined as the difference between the equivalent edge currents in PO and those in GTD in the form [9]

$$I_{\mathrm{PTD}} = \frac{2}{-j\omega\mu \sin\beta^i}\left(\frac{\sin\dfrac{\phi^i}{2}}{\cos\dfrac{\phi^i}{2} + \sin\dfrac{\phi^d}{2}} E_{\theta^i} + \cos\beta^i E_{\phi^i}\right) \tag{7.60}$$

$$M_{\mathrm{PTD}} = \frac{2}{-jk \sin\beta^i} \frac{\cos\dfrac{\phi^d}{2}}{\cos\dfrac{\phi^i}{2} + \sin\dfrac{\phi^d}{2}} E_{\phi^i} \tag{7.61}$$

where the local spherical coordinates in Figure 7.17 are used. The correction terms to be added to PO are obtained by substituting these currents into (7.56) and (7.57). Different expressions for nonuniform currents have also been proposed [15, 16].

The behavior of PTD is surveyed. In the mainbeam region of the focusing system, the PO term predominates the correction term, and PTD results approach PO. In the region far away from the mainbeams, PTD approaches EEC (~ GTD)

Table 7.3
Derivation of Physical Theory of Diffraction (from [9])

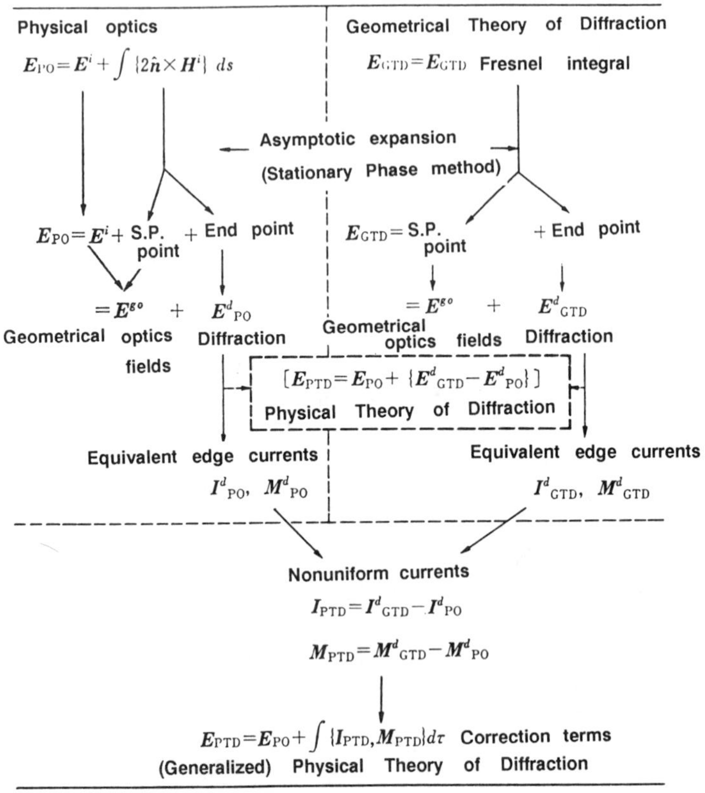

because the PO integration tends to its asymptotic value, which cancels the contribution of PO edge currents in the correction term, leaving GTD edge currents alone.

(2) Application of PTD

To demonstrate the accuracy of PTD, diffraction by a flat disk is calculated. PTD results are indicated by ○ in Figure 7.16c. Almost perfect agreement with the exact one is observed in all the angular regions. PO results also are compared, and the error is notable in the wide-angle region. The effect of the correction term in PTD is evident from these data. PTD, thus constructed, surely eliminates the difficulties (1), (2), and (4) in GTD.

.4 RADIATION PATTERN ANALYSIS OF REFLECTOR ANTENNAS

O has been widely used in the design of reflector antennas. It is known that PO ıtegration predicts the accurate fields in the mainbeam region. In the wide-angle ɛgion, however, numerical integration becomes time consuming and the accuracy greatly degraded. This is because PO currents ($2\hat{n} \times \boldsymbol{H}^i$) are the most erroneous t the edge, and the end point contribution predominates in this region.

In contrast GTD needs negligibly little computing time and still maintains high ccuracy. Currently, GTD is an important method in reflector antenna pattern analsis. For practical use, the difficulties cited in Section 7.2.3 should be removed. mong others, the infinities of the geometrical optics fields (4) fall on the mainbeam nd are the serious defects in antenna analysis. Patterns of a parabola fed by a small ipole are presented in Figure 7.18 to demonstrate the defects (1)–(4) in 7.2.3. In ocusing systems, the infinities at RB and caustics also appear in the mainbeam. To btain the full angle pattern, the hybrid use of PO and GTD has been popular for ıe mainbeam and the wide-angle regions, respectively. In contrast, PTD can predict atterns uniformly in all angular regions. The results by PTD are presented in Figure .18 by ○. The results by PTD shift automatically from PO to GTD at about $\theta =$ 0°. This indicates that PTD approaches best approximation in each angular region. igure 7.19 compares the patterns for an offset parabola with an offset (defocused) ɛed. In this case, the results by PO and GTD differ greatly with each other, but TD connects them smoothly.

Thus, the potential of PTD in antenna pattern analysis has been demonstrated. he computation time of PTD is comparable or a bit longer than that of PO, and ıe maximum use of GTD beyond the angle at which GTD approaches PTD (not

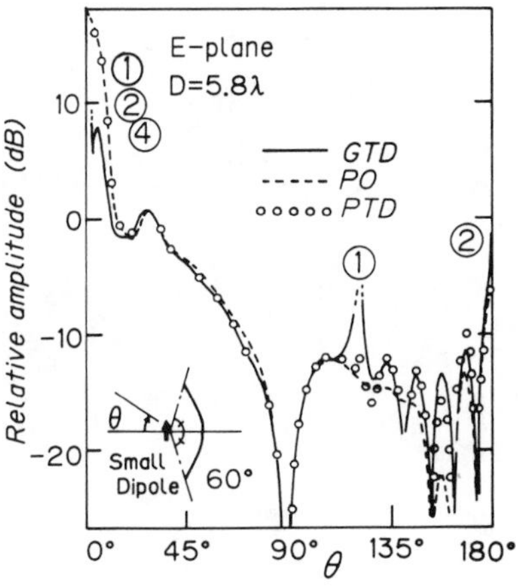

igure 7.18 Radiation patterns of a dipole-fed parabola.

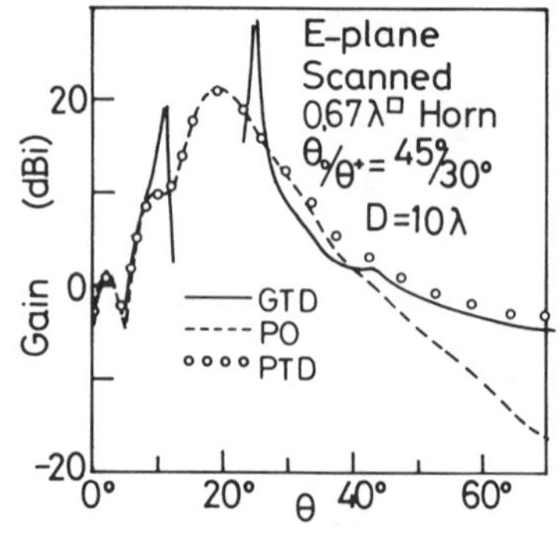

Figure 7.19 Radiation patterns of an offset parabola.

PO) is still recommended. An accuracy check of PTD for a wider class of canonical problems is important and is now being investigated. The criteria for the alternative use of PO or GTD is also discussed on the basis of PTD.

7.5 SOME PRECAUTIONS FOR PROGRAMMING

GTD is an approximation method, but it is still analytical; its physical meaning is quite clear, in contrast with other numerical methods. No special techniques are required in GTD calculation. The expressions for the unit vectors and angles used in the local spherical coordinates in Figure 7.3(b) are listed for practical use.

$$\hat{\boldsymbol{\phi}}^i = \frac{\hat{\boldsymbol{k}}^i \times \hat{\boldsymbol{\tau}}}{|\hat{\boldsymbol{k}}^i \times \hat{\boldsymbol{\tau}}|} \tag{7.62}$$

$$\hat{\boldsymbol{\theta}}^i = \hat{\boldsymbol{k}}^i \times \hat{\boldsymbol{\phi}}^i \tag{7.63}$$

$$\phi^i = \text{arc}\cos(\hat{\boldsymbol{n}} \cdot \hat{\boldsymbol{\phi}}^i), \quad 0 < \phi^i < \pi \tag{7.64}$$

$$\theta^i = \frac{\pi}{2} + \text{arc}\sin(\hat{\boldsymbol{\tau}} \cdot \hat{\boldsymbol{k}}^i), \quad 0 < \theta^i < \pi \tag{7.65}$$

$$\hat{\boldsymbol{\phi}}^d = \frac{-\hat{\boldsymbol{k}}^d \times \hat{\boldsymbol{\tau}}}{|\hat{\boldsymbol{k}}^d \times \hat{\boldsymbol{\tau}}|} \tag{7.66}$$

$$\hat{\boldsymbol{\theta}}^d = -\hat{\boldsymbol{k}}^d \times \hat{\boldsymbol{\phi}}^d \tag{7.67}$$

$$\phi^d = -\text{arc}\tan\left(\frac{\hat{\boldsymbol{\phi}}^d \cdot (\hat{\boldsymbol{n}} \times \hat{\boldsymbol{\tau}})}{\hat{\boldsymbol{\phi}}^d \cdot \hat{\boldsymbol{n}}}\right) \tag{7.68}$$

REFERENCES

[1] Keller, J.B., "Geometrical Theory of Diffraction," *J. Opt. Soc. Amer.*, Vol. 52, 1962, p. 116–130.

[2] Bowman, J.J., T.B.A. Senior, and P.L.E. Uslenghi, ed., *Electromagnetic and Acoustic Scattering by Simple Shapes,* North-Holland, Amsterdam, 1969.

[3] Kouyoumjian, R.G., "Asymptotic High-Frequency Methods," *Proc. IEEE,* Vol. 53, 1965, pp. 864–876.

[4] Games, G.L., *Geometrical Theory of Diffraction,* Peter Peregrinus, Stevenage, 1976.

[5] Ito, K., T. Katagi, K. Hongo, and M. Ando, "Geometrical Theory of Diffraction and Its Application," *Proc. IEICE,* Vol. 70, 1987, pp. 607–612, 745–749, 839–845, 945–950, 1059–1064 [in Japanese].

[6] Ando, M., "Diffracted Fields from Semi-Infinite Conducting Plane and Methods for Their Suppression," *Trans. IECE of Japan,* Vol. 60-B, No. 11, 1977, pp. 812–819 [in Japanese].

[7] Pathak, P.H., W.D. Burnside, and R.J. Marhefka, "A Uniform GTD Analysis of the Diffraction of Electromagnetic Waves by a Smooth Convex Surface," *IEEE Trans.,* Vol. AP-28, No. 5, Sept. 1980, pp. 631–642.

[8] Kouyoumjian, R.G., and P.H. Pathak, "A Uniform Geometrical Theory of Diffraction for an Edge in a Perfectly Conducting Surface," *Proc. IEEE,* Vol. 62, No. 11, 1974, pp. 1448–1461.

[9] Ando, M., "Radiation Pattern Analysis of Reflector Antennas," *Electronics & Communication in Japan,* Part 1, Vol. 68, No. 4, 1985, p. 93–102; translated from *Trans. IECE of Japan,* Vol. 67-B, No. 8, 1985, pp. 853–860.

[10] Inawashiro, S., "Diffraction of Electromagnetic Waves from an Electric Dipole by a Conducting Circular Disk," *J. Phys. Soc. Japan,* Vol. 18, No. 2, 1963, pp. 273–287.

[11] Lee, S.W., and G.A. Deschamps, "A Uniform Asymptotic Theory of Electromagnetic Diffraction by a Curved Edge," *IEEE Trans.,* Vol. AP-24, 1976, pp. 25–34.

[12] Ufimtsev, P.Y., "Method of Edge Waves in the Physical Theory of Diffraction," trans. U.S. Air Force Foreign Technology Division Wright Patterson AFB, Dayton, Ohio, 1971.

[13] Millar, R.F., "An Approximate Theory of the Diffraction of an Electromagnetic Wave by an Aperture in a Plane Screen," *Proc. IEEE,* Vol. 103c, 1956, pp. 177–185.

[14] Rudduck, R.C., and D.C. Wu, "Slope Diffraction Analysis of TEM Parallel-Plate Guide Radiation Patterns," *IEEE Trans.,* Vol. AP-17, 1969, pp. 797–799.

[15] Michaeli, A., "Elimination of Infinities in Equivalent Currents, Part 1: Fringe Current Components," *IEEE Trans.,* Vol. AP-34, No. 7, July 1986, pp. 912–918.

[16] Hongo, K., and H. Kume, "Diffraction of a Plane Electromagnetic Wave by a Convex Aperture," *Trans. IECE of Japan,* Vol. 61-B, No. 11, Nov. 1978, pp. 927–934 [in Japanese].

Chapter 8
Wiener-Hopf and Modified Residue Calculus Techniques

Kazuya Kobayashi

8.1 INTRODUCTION

8.1.1 Historical Background

Theoretical analysis of wave scattering and diffraction problems has been carried out extensively with regard to a number of scatterers. In 1896, Sommerfeld [1] introduced the idea of multiple-valued functions that are single-valued, bounded, and continuous on the appropriate Riemann surface, and he applied this idea to obtain the exact solution to the diffraction by a wedge, which was expressed in the form of complex integral representations. He further considered a semiinfinite plate as a special case and showed that the solution could be considerably simplified to yield the Fresnel integral representations. His method of solution for scattering and diffraction problems, known as the *Sommerfeld theory of diffraction,* is famous as a classic result for diffraction by wedge-shaped obstacles [2].

Because the scattering and diffraction theory requires knowledge of applied mathematics, the theory was closely tied to the progress of mathematics. As is well known in general, boundary value problems concerning the wave equation can be reduced to the integral equation using the Green's formula together with the boundary conditions. Therefore, Fredholm's theory of integral equations had a great impact on the progress of the diffraction theory. On the other hand, in 1931, Wiener and Hopf [3] showed that a certain singular integral equation could be solved exactly using the theory of complex Fourier transforms and functions of a complex variable. This integral equation and their method of solution are known as the *Wiener-Hopf integral equation* and the *Wiener-Hopf technique,* respectively. In 1941, Magnus [4] reduced the problem of diffraction by a semiinfinite plate to the solution of a singular integral equation. Shortly thereafter, Schwinger [5] and Copson [6] independently, for the first time, applied the Wiener-Hopf technique to solve this integral equation and derived the same solution as Sommerfeld. Since then, the importance of the Wiener-Hopf technique has been recognized, and a number of radiation and diffraction problems concerning semiinfinite waveguides have been solved using this method by many authors such as Carlson and Heins [7], Heins [8], Levine and Schwinger [9], and Iijima [10]. Great progress on the wave scattering and diffraction theory followed their pioneering work.

Vajnshtejn (Weinstein) [11], Karp [12], and Clemmow [13] noted that if the problems regarding semiinfinite boundaries are formulated in terms of certain *dual* integral equations, then the exact solution could be obtained by applying a simpler method than the original Wiener-Hopf technique. In particular, the method developed by Clemmow is based on formulating the problems after representing the fields in terms of a superposition of plane waves with arbitrary angular spectrum; it is known as the *plane-wave spectrum method*. This method was later applied by Clemmow [14] and Horiuchi [15] to solve the complicated diffraction problem involving a semiinfinite plate. The method of solution based on dual integral equations [16, 17] proposed by Vajnshtejn, Karp, and Clemmow is equivalent to the Wiener-Hopf technique. Most of the earlier results related to the Wiener-Hopf technique during the 1940s and 1950s are discussed in detail by Noble [18].

The Wiener-Hopf technique was originally based on using the complex Fourier transform of the Wiener-Hopf integral equation to derive a certain functional equation in the complex domain and then solving the resultant equation with the aid of the function-theoretic method. This functional equation is known as the *Wiener-Hopf functional equation,* or simply the *Wiener-Hopf equation*. In 1952, Jones [19] developed a more direct, simpler method of formulation for obtaining the Wiener-Hopf equation. This procedure was to take the complex Fourier transform of the reduced wave equation and derive the Wiener-Hopf equation by applying the boundary conditions in the complex domain. His formulation, called *Jones's method,* has the advantage of bypassing the derivation of the integral equation in the real space.

Rigorously speaking, the Wiener-Hopf technique is applicable only to a class of geometries with semiinfinite boundaries and no thickness. However, Jones [20, 21] developed a method of solving diffraction problems for which exact Wiener-Hopf analysis is difficult and showed that a certain class of problems could be formally treated with the aid of the Wiener-Hopf technique and solved approximately. Since Jones, the range of problems solvable by the Wiener-Hopf technique has been widely extended [18, 22–29].

Berz [30] and Whitehead [31] developed a method of exact solution for boundary value problems called the *residue-calculus technique* in 1951, and applied this method to the scattering by a semiinfinite parallel plate waveguide array. It is well known that, if the scattered field is represented by an eigenfunction expansion, then the problem in general is formulated in terms of an infinite set of equations satisfied by unknown expansion coefficients. The residue-calculus technique obtains the exact solution by finding certain regular functions that fit the structure of this set of equations. Since the work by Berz and Whitehead, a number of waveguide boundary value problems have been treated by the residue calculus technique [32, 33]. In 1968, an extension called the *modified residue-calculus technique* was proposed by Mittra, Lee, and Van Blaricum [34] and applied to a wide range of boundary value problems [35–37]. This method was developed for a class of boundary value problems that could not be treated by the residue-calculus technique, and it was based on constructing regular functions corresponding to the solution of the infinite set of equations approximately. The modified residue-calculus technique recently has been extended to obtain the approximate solution to the Wiener-Hopf equations [37–39].

The Wiener-Hopf technique garnered much attention after the middle of 1970s, particularly since various function-theoretic methods were developed and applied to several complicated problems. In 1976, Hurd [40] established the *Wiener-Hopf-Hilbert method* to rigorously analyze the diffraction by a semiinfinite plate with special boundary conditions. Following this work, authors such as Daniele [41], Hurd and Lüneburg [42], Rawlins and Williams [43], and Jones [44] have investigated a method of solving the vector Wiener-Hopf equations, and many important results have been obtained. The diffraction problems concerned with complicated geometries have also been treated by Aoki *et al.* [45–48], Kobayashi *et al.* [49–53], and Rojas [54, 55], and several approximate methods of solution related to the Wiener-Hopf technique have been developed.

8.1.2 Range of Applications

The Wiener-Hopf technique was invented to solve the following integral equation of the first kind:

$$\int_0^\infty f(z')k(z - z')dz' = g(z), \quad 0 < z < \infty \tag{8.1}$$

where $k(z)$ and $g(z)$ are known, and $f(z)$ is the unknown function to be determined. The basic step for the Wiener-Hopf technique is to take the complex Fourier transform of equation (8.1) to derive the functional equation in the complex domain (Wiener-Hopf equation), which is then solved via a procedure called *factorization*. This procedure represents a function $K(\alpha)$ of a complex variable α $(= \sigma + i\tau)$ regular in the strip $\tau_- < \tau < \tau_+$ as the multiplication of the two functions, as in

$$K(\alpha) = K_+(\alpha)K_-(\alpha) \tag{8.2}$$

where $K_+(\alpha)$ and $K_-(\alpha)$ are regular and nonzero in the half-planes $\tau > \tau_-$ and $\tau < \tau_+$, respectively, and show algebraic behavior at infinity.

When we treat scattering and diffraction problems using the Wiener-Hopf technique, the following conditions are required for an exact solution:

(WH1) The obstacle has a semiinfinite straight boundary.

(WH2) The complex Fourier transform is taken such that the transform axis does not traverse the surface of the obstacle.

The geometries satisfying these conditions are called the *Wiener-Hopf geometry*. Therefore, problems belonging to this class in general can be solved exactly by the Wiener-Hopf technique if factorization of the complex Fourier transform $K(\alpha)$ of $k(z)$ appearing in (8.1) is possible. Because $k(z)$ is the kernel of the Wiener-Hopf integral equation, $K(\alpha)$ is called the *kernel function*. The problem discussed in Section 8.3 belongs to a class of the Wiener-Hopf geometry.

Next, as an example of the problems that cannot be reduced to the Wiener-Hopf integral equation, let us consider the following two cases:

(WH1′) The obstacle has a finite straight boundary.

(WH2′) The complex Fourier transform axis chosen arbitrarily traverses the surface of the obstacle.

Here, (WH1′) and (WH2′) are the modifications of (WH1) and (WH2), respectively. It is known that if factorization of the kernel function is possible, then the Wiener-Hopf technique is formally applicable and an approximate solution is obtained. The geometries that belong to the class of (WH1′) and (WH2′) are called the *modified Wiener-Hopf geometry of the first kind* and *the second kind,* respectively. The scatterers treated in Sections 8.4 and 8.5 are of the modified Wiener-Hopf geometry of the first and second kinds, respectively. Furthermore, the problems regarding the scatterers with hybrid modification of (WH1′) and (WH2′) can also be treated by the Wiener-Hopf technique and approximate solutions are obtained [27, 45, 47, 49–51, 53], but we will not discuss the details here. The geometries belonging to a class of this hybrid modification are called the *modified Wiener-Hopf geometry of the third kind*.

The residue-calculus technique and the modified residue-calculus technique were both developed for solving a certain infinite set of equations derived via an application of boundary conditions; they are the exact and approximate methods, respectively. Most of the problems exactly solvable by the residue-calculus technique can also be treated using the Wiener-Hopf technique, and there is a certain equivalence between these two methods [37]. Therefore, the scatterers must be of the Wiener-Hopf geometry even when we apply the residue-calculus technique to scattering and diffraction problems. However, the residue-calculus technique is not appropriate for analyzing *open-region* problems. As stated in the previous subsection, the modified residue-calculus technique has recently been extended so as to obtain the approximate solution to the Wiener-Hopf equations. This extension gave us a scheme for increasing the accuracy of the approximate solution from a point of view of numerical computations. The modified residue-calculus technique can be applied effectively to *closed-region* problems concerned with the modified Wiener-Hopf geometries of all kinds, but for open-region problems, geometries to which the modified residue-calculus technique is applicable in general are limited to a class of the modified Wiener-Hopf geometry of the second kind.

In subsequent sections, after some mathematical preliminaries necessary for development of the Wiener-Hopf technique, we shall treat the three typical diffraction problems and develop a solution based on Jones's method. The residue-calculus technique will be omitted and the modified residue-calculus technique will be focused only on the applications to the Wiener-Hopf technique. Readers who are interested in the method of solution for an infinite set of equations may refer to the related references [30–37].

8.2 MATHEMATICAL PRELIMINARIES FROM THE THEORY OF FOURIER INTEGRALS AND FUNCTIONS OF A COMPLEX VARIABLE

The Wiener-Hopf technique is one of the powerful methods of solving wave scattering and diffraction problems and requires knowledge of the theory of Fourier integrals and functions of a complex variable. In this section, we shall first summarize the main theorems and formulas necessary for the developments in subsequent sections and then describe a method of solution based on the Wiener-Hopf technique. Due to spatial limitations, proof of all the theorems and formulas will be omitted.

8.2.1 Complex Fourier Integrals

Theorems 8.1 to 8.3 are known for analytical properties of the functions defined by certain integral transforms [18, 37].

Theorem 8.1. Let $f(z)$ be a function of a real variable satisfying the following conditions over the finite interval $a < z < b$:

(i) $f(z)$ has at most a finite number of discontinuities and is continuous over any subinterval in $a < z < b$, not containing these discontinuities.
(ii) $f(z)$ is absolutely integrable over $a \leqslant z \leqslant b$.

Then the function defined by

$$F_1(\alpha) = (2\pi)^{-1/2} \int_a^b f(z)\, e^{i\alpha z} dz \tag{8.3}$$

is an entire function of $\alpha\ (= \sigma + i\tau)$.

Theorem 8.2. Let $f(z)$ be a function of a real variable satisfying the conditions (i) and (ii) stated in Theorem 8.1 over any finite subinterval $a < z < b$ involved in $0 < z < \infty$. Also suppose that $f(z)$ satisfies

$$f(z) = 0[\exp(\tau_- z)] \tag{8.4}$$

for some real constant τ_- as $z \to \infty$. Then the function defined by

$$F_+(\alpha) = (2\pi)^{-1/2} \int_0^\infty f(z)\, e^{i\alpha z} dz \tag{8.5}$$

is regular in the half-plane $\tau > \tau_-$.

Theorem 8.3. Let $f(z)$ be a function of a real variable satisfying the conditions (i) and (ii) stated in Theorem 8.1 over any finite subinterval $a < z < b$ involved in $-\infty < z < 0$. Also suppose that $f(z)$ satisfies

$$f(z) = 0[\exp(\tau_+ z)] \tag{8.6}$$

for some real constant τ_+ as $z \to -\infty$. Then the function defined by

$$F_-(\alpha) = (2\pi)^{-1/2} \int_{-\infty}^0 f(z)\, e^{i\alpha z} dz \tag{8.7}$$

is regular in the half-plane $\tau < \tau_+$.

The integral of the type given by (8.3), (8.5), and (8.7) is called the *complex Fourier integral*. From Theorems 8.2 and 8.3, we can show the following theorem.

Theorem 8.4. Let $f(z)$ be a function of a real variable satisfying the conditions (i) and (ii) stated in Theorem 8.1 over any finite subinterval $a < z < b$ involved in $-\infty < z < \infty$. Also suppose that $f(z)$ satisfies

$$\begin{aligned} f(z) &= 0[\exp(\tau_+ z)], \quad z \to -\infty \\ &= 0[\exp(\tau_- z)], \quad z \to \infty \end{aligned} \tag{8.8}$$

for some real constants $\tau_\pm$ such that $\tau_- < \tau_+$. Then the function defined by

$$F(\alpha) = (2\pi)^{-1/2} \int_{-\infty}^\infty f(z)\, e^{i\alpha z} dz \tag{8.9}$$

is regular in the strip $\tau_- < \tau < \tau_+$ and the following relation holds for arbitrary c such that $\tau_- < c < \tau_+$:

$$f(z) = (2\pi)^{-1/2} \lim_{\lambda\to\infty} \int_{-\lambda+ic}^{\lambda+ic} F(\alpha)\, e^{-i\alpha z} d\alpha \tag{8.10}$$

The integrals defined by (8.9) and (8.10) are called the *complex Fourier transform* and the *inverse complex Fourier transform,* respectively. The preceding theorem is usually called the *inversion formula for complex Fourier transform.*

8.2.2 Asymptotic Behavior of the Complex Fourier Integrals

There are important formulas necessary for investigating the asymptotic behavior of the complex Fourier integrals. These formulas are closely related to *Watson's lemma* [56], established in connection with the asymptotic expansion of the Laplace integrals. Here, we shall give the important theorems [57] for deriving the asymptotic expansion of the complex Fourier integrals, which yields significant extension and the Fourier integral version of Watson's lemma. The following theorem is used for the asymptotic behavior of the complex Fourier integrals.

Theorem 8.5. Let $f(z)$ be a function of a complex variable satisfying the following conditions:

(i) $f(z)$ is regular except for the possible branch singularity at the origin.
(ii) $f(z)$ can be expanded asymptotically as

$$f(z) \sim \sum_{n=0}^{\infty} c_n z^{a_n} \tag{8.11}$$

in the neighborhood of the origin, where c_n is an arbitrary complex sequence, and $-1 < \mathrm{Re}\, a_n < \mathrm{Re}\, a_{n+1}$ for $n = 0, 1, 2, \ldots$.
(iii) $f(z)$ satisfies

$$f(z) = 0[\exp(\tau_- z)] \tag{8.12}$$

for some real constant τ_- as $z \to +\infty$.

Then $F_+(\alpha)$ defined by (8.5) has the asymptotic expansion

$$F_+(\alpha) \sim (2\pi)^{-1/2} \sum_{n=0}^{\infty} c_n \Gamma(1 + a_n)(-i\alpha)^{-1-a_n} \tag{8.13}$$

uniformly in arg α as $\alpha \to \infty$ in the sector $\varepsilon \leqslant \arg(\alpha - i\tau_-) \leqslant \pi - \varepsilon$ for any ε such that $0 < \varepsilon < \pi/2$, where $\Gamma(\cdot)$ is the Gamma function.

This theorem similarly holds when the integration range is finite, and we have the following theorem.

Theorem 8.6. Let $f(z)$ be a function of a complex variable satisfying the conditions (i) and (ii) stated in Theorem 8.5. Then the function defined by

$$F_1(\alpha) = (2\pi)^{-1/2} \int_0^d f(z)\, e^{i\alpha z} dz \tag{8.14}$$

for arbitrary positive d has the same asymptotic expansion as that given by (8.13) uniformly in arg α when $\alpha \to \infty$ in the sector $\varepsilon \leqslant \arg \alpha \leqslant \pi - \varepsilon$ for any ε such that $0 < \varepsilon < \pi/2$.

Theorems 8.5 and 8.6 are both concerned with the asymptotic behavior of the complex Fourier integrals defined along the positive real axis in the complex plane. Similar theorems can be proved for the asymptotic expansion of the complex Fourier integrals along the negative real axis, and we are led to the following theorems.

Theorem 8.7. Let $f(z)$ be a function of a complex variable satisfying the following conditions:

(i) $f(z)$ is regular except for the possible branch singularity at the origin.
(ii) $f(z)$ can be expanded asymptotically as

$$f(z) \sim \sum_{n=0}^{\infty} c_n(-z)^{a_n} \tag{8.15}$$

in the neighborhood of the origin, where c_n is an arbitrary complex sequence, and $-1 < \text{Re } a_n < \text{Re } a_{n+1}$ for $n = 0, 1, 2, \ldots$.
(iii) $f(z)$ satisfies

$$f(z) = O[\exp(\tau_+ z)] \tag{8.16}$$

for some real constant τ_+ as $z \to -\infty$.

Then $F_-(\alpha)$ defined by (8.7) has the asymptotic expansion

$$F_-(\alpha) \sim (2\pi)^{-1/2} \sum_{n=0}^{\infty} c_n \Gamma(1 + a_n)(i\alpha)^{-1-a_n} \tag{8.17}$$

uniformly in arg α as $\alpha \to \infty$ in the sector $-\pi + \varepsilon \leqslant \arg(\alpha - i\tau_+) \leqslant -\varepsilon$ for any ε such that $0 < \varepsilon < \pi/2$.

Theorem 8.8. Let $f(z)$ be a function of a complex variable satisfying the conditions (i) and (ii) stated in Theorem 8.7. Then the function defined by

$$F_1(\alpha) = (2\pi)^{-1/2} \int_{-d}^{0} f(z)\, e^{i\alpha z} dz \tag{8.18}$$

for arbitrary positive d has the same asymptotic expansion as that given by (8.17) uniformly in arg α when $\alpha \to \infty$ in the sector $-\pi + \varepsilon \leqslant \arg \alpha \leqslant -\varepsilon$ for any ε such that $0 < \varepsilon < \pi/2$.

8.2.3 Decomposition and Factorization of Functions

Using the Cauchy's integral formula, we can prove the following theorems [18, 37].

Theorem 8.9. Let $K(\alpha)$ be regular in the strip $\tau_- < \tau < \tau_+$, satisfying

$$K(\alpha) = 0(|\sigma|^{-\nu}) \tag{8.19}$$

uniformly in τ for some positive ν as $|\sigma| \to \infty$ in $\tau_- + \varepsilon \leqslant \tau \leqslant \tau_+ - \varepsilon$ for arbitrary ε such that $\varepsilon < (\tau_+ - \tau_-)/2$. Then $K(\alpha)$ can be additively split as in

$$K(\alpha) = K_+(\alpha) + K_-(\alpha) \tag{8.20}$$

where for $\tau_- < c < \tau < d < \tau_+$,

$$K_+(\alpha) = \frac{1}{2\pi i} \lim_{\lambda \to \infty} \int_{-\lambda + ic}^{\lambda + ic} \frac{K(\beta)}{\beta - \alpha} d\beta \tag{8.21a}$$

$$K_-(\alpha) = -\frac{1}{2\pi i} \lim_{\lambda \to \infty} \int_{-\lambda + id}^{\lambda + id} \frac{K(\beta)}{\beta - \alpha} d\beta \tag{8.21b}$$

$K_+(\alpha)$ and $K_-(\alpha)$ are then regular and bounded in $\tau > \tau_-$ and $\tau < \tau_+$, respectively.

Theorem 8.10. Let $K(\alpha)$ be regular and nonzero in the strip $\tau_- < \tau < \tau_+$, satisfying

$$K(\alpha) = 1 + 0(|\sigma|^{-\nu}) \tag{8.22}$$

uniformly in τ for some positive ν as $|\sigma| \to \infty$ in $\tau_- + \varepsilon \leqslant \tau \leqslant \tau_+ - \varepsilon$ for arbitrary ε such that $\varepsilon < (\tau_+ - \tau_-)/2$. Then $K(\alpha)$ can be multiplicatively split as in

$$K(\alpha) = K_+(\alpha) K_-(\alpha) \tag{8.23}$$

where for $\tau_- < c < \tau < d < \tau_+$,

$$K_+(\alpha) = \exp\left[\frac{1}{2\pi i}\lim_{\lambda\to\infty}\int_{-\lambda+ic}^{\lambda+ic}\frac{\ln K(\beta)}{\beta-\alpha}\,d\beta\right] \tag{8.24a}$$

$$K_-(\alpha) = \exp\left[-\frac{1}{2\pi i}\lim_{\lambda\to\infty}\int_{-\lambda+id}^{\lambda+id}\frac{\ln K(\beta)}{\beta-\alpha}\,d\beta\right] \tag{8.24b}$$

$K_+(\alpha)$ and $K_-(\alpha)$ are then regular, nonzero and bounded in $\tau > \tau_-$ and $\tau < \tau_+$, respectively.

Additive and multiplicative splitting are known as *decomposition* and *factorization*, respectively. We call the functions $K_\pm(\alpha)$ appearing in Theorems 8.9 and 8.10 the *split functions*. If $K(\alpha)$ is an entire function with an infinite number of zeros, then the following convenient factorization formula is known [18].

Theorem 8.11. Let $K(\alpha)$ be an entire and even function of α, having an infinite number of simple zeros α_n for $n = 1, 2, 3, \ldots$ such that

$$0 < |\alpha_n| \leqslant |\alpha_{n+1}|, \quad 0 < \mathrm{Im}\,\alpha_N \leqslant \mathrm{Im}\,\alpha_n, \quad n = 1, 2, 3, \ldots \tag{8.25}$$

for some integer N. Also suppose that α_n has the asymptotic behavior

$$\alpha_n = an + b + 0(n^{-1}) \tag{8.26}$$

as $n \to \infty$, where a and b are some complex constants. Then $K(\alpha)$ can be factorized as

$$K(\alpha) = K_+(\alpha)K_-(\alpha) \tag{8.27}$$

where $K_\pm(\alpha)$ are regular and nonzero in $\tau \gtrless \mp\,\mathrm{Im}\,\alpha_N$, defined by

$$K_\pm(\alpha) = [K(0)]^{1/2}\,\mathrm{e}^{\mp\chi(\alpha)}\prod_{n=1}^{\infty}(1 \pm \alpha/\alpha_n)\,\mathrm{e}^{\mp\alpha/an} \tag{8.28}$$

In (8.28), $\chi(\alpha)$ is an arbitrary entire function.

In the Wiener-Hopf analysis, we usually choose $\chi(\alpha)$ in (8.28) such that the split functions $K_\pm(\alpha)$ show an algebraic behavior at infinity. The following theorem holds for the asymptotic expansion of the infinite product appearing in (8.28) as $\alpha \to \infty$ [18].

Theorem 8.12. Let α_n be a complex sequence satisfying the asymptotic behavior as given by (8.26). Then the function defined by

$$P(\alpha) = \prod_{n=1}^{\infty} (1 + \alpha/\alpha_n)\, e^{-\alpha/an} \tag{8.29}$$

has the asymptotic expansion

$$\begin{aligned} P(\alpha) &\sim 2^{-1/2} K_p \alpha^{-1/2-b/a} \exp\{(\alpha/a)[1 - C - \ln(\alpha/a)]\} \text{ for } |\arg \alpha| < \pi, \\ &\sim -2^{1/2} K_p (-\alpha)^{-1/2-b/a} \sin[(\pi/a)(\alpha + b)] \\ &\quad \times \exp\{(\alpha/a)\,[1 - C - \ln(-\alpha/a)]\} \text{ for } \arg \alpha = \pi \end{aligned} \tag{8.30}$$

as $\alpha \to \infty$, where C (= 0.57721566...) is Euler's constant, and

$$K_p = \frac{a^{1/2+b/a}\Gamma(1 + b/a)}{\pi^{1/2}} \prod_{n=1}^{\infty} \frac{an + b}{\alpha_n} \tag{8.31}$$

Theorem 8.12 can be conveniently used to derive the asymptotic expansion of the split functions $K_{\pm}(\alpha)$ appearing in Theorem 8.11 for large $|\alpha|$.

8.2.4 Saddle-Point Method

There are a number of asymptotic methods for evaluation of the infinite integrals. In this section, we shall introduce the typical infinite branch-cut integral and discuss the *saddle-point method* known to be a powerful tool for deriving the asymptotic expansion of functions defined by integrals.

First, let us introduce the function

$$\gamma = (\alpha^2 - k^2)^{1/2} \equiv (\alpha + k)^{1/2}(\alpha - k)^{1/2} \tag{8.32}$$

where

$$k = k_1 + ik_2, \quad k_1 > 0, \quad k_2 > 0 \tag{8.33}$$

We see that γ is a double-valued function of α and has the branch points at $\alpha = \pm k$. We now choose the branch cuts for γ as a portion of hyperbola defined by $\sigma\tau = k_1 k_2$, which is shown in Figure 8.1. Then, we verify that Re $\gamma > 0$ for any α in the strip $-k_2 < \tau < k_2$. We henceforth adopt this choice of branch cut unless otherwise specified.

Let $\Phi(\alpha)$ be regular in the strip $\tau_- < \tau < \tau_+$, where $\tau_{\pm}$ are some constants such that $-k_2 \leqslant \tau_- < \tau_+ \leqslant k_2$. We now introduce the integral

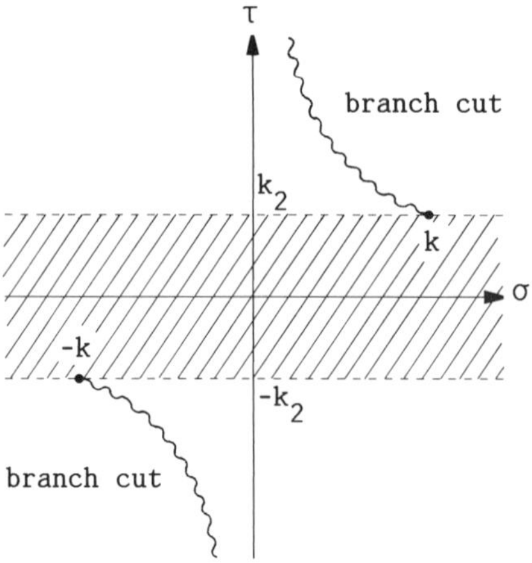

Figure 8.1 Branch cuts for γ.

$$\phi(x, z) = (2\pi)^{-1/2} \int_{-\infty+ic}^{\infty+ic} \Phi(\alpha)\, e^{-\gamma|x| - i\alpha z} d\alpha \tag{8.34}$$

for real x and z, where c is an arbitrary constant such that $\tau_- < c < \tau_+$. Because the integrand possesses branch points at $\alpha = \pm k$ due to the presence of γ, evaluation of this integral in a closed form is generally difficult. However, we can derive the asymptotic representation based on the saddle-point method of integration as $k(x^2 + z^2)^{1/2} \to \infty$ if the integrand has no singularities other than the branch points at $\alpha = \pm k$. The fundamental theorem for the asymptotic expansion is stated as follows [58].

Theorem 8.13. Let $\Phi(\alpha)$ be regular except for possible singularities at $\alpha = \pm k$, where these singularities are branch points due to the presence of γ in $\Phi(\alpha)$. Then the function $\phi(x, z)$ defined by (8.34) has the asymptotic expansion

$$\phi(\rho, \theta) \sim \frac{e^{ik\rho}}{(2k\rho)^{1/2}} \sum_{n=0}^{\infty} \frac{G^{(2n)}(0)}{n!2^{2n}} (k\rho)^{-n} \tag{8.35}$$

as $k\rho \to \infty$, where

$$G^{(2n)}(0) = \left. \frac{d^{2n}}{dt^{2n}} G(t) \right|_{t=0} \tag{8.36}$$

$$G(t) = \left. \frac{2^{1/2} e^{-i\pi/4}}{(1 + it^2/2)^{1/2}} \Phi(-k \cos w) k \sin w \right|_{w=g(t)} \tag{8.37}$$

$$g(t) = |\theta| + \cos^{-1}(1 + it^2) \tag{8.38}$$

In (8.35)–(8.37), (ρ, θ) is the polar coordinate as defined by $x = \rho \sin\theta$ and $z = \rho \cos\theta$ for $-\pi < \theta < \pi$, and the arc cosine function in (8.38) is interpreted as the principal value.

This theorem gives a complete asymptotic expansion of $\phi(x, z)$ as $k(x^2 + z^2)^{1/2} \rightarrow \infty$; however, by extracting the dominant term from the asymptotic series, we have the following simple corollary.

Corollary. Let $\Phi(\alpha)$ satisfy the hypotheses stated in Theorem 8.13. Then $\phi(x, z)$ defined by (8.34) has the asymptotic expansion

$$\phi(\rho, \theta) \sim \Phi(-k \cos\theta)k \sin|\theta| \frac{e^{i(k\rho - \pi/4)}}{(k\rho)^{1/2}} \tag{8.39}$$

as $k\rho \rightarrow \infty$ under introduction of the polar coordinate (ρ, θ) given in Theorem 8.13.

We have so far treated the case of complex k, but Theorem 8.13 and its corollary hold as well for real k by taking the limit $k_2 \rightarrow +0$.

8.2.5 Wiener-Hopf Technique

In this subsection, we provide the outline of a solution method based on the Wiener-Hopf technique [18]. The Wiener-Hopf equation discussed in this chapter is typically of the following form:

$$K(\alpha)\Phi_-(\alpha) + \Psi_+(\alpha) + F(\alpha) = 0 \tag{8.40}$$

Equation (8.40) is the complex Fourier transform of the Wiener-Hopf integral equation (8.1) and holds in an appropriate strip $\tau_- < \tau < \tau_+$ of the complex α-plane. Here, $K(\alpha)$ and $F(\alpha)$ are known functions regular in $\tau_- < \tau < \tau_+$, and $\Phi_-(\alpha)$ and $\Psi_+(\alpha)$ are the unknown functions that are regular in $\tau < \tau_+$ and $\tau > \tau_-$, respectively. $K(\alpha)$ is the kernel function that provides geometrical properties of the boundary under consideration and corresponds to the complex Fourier transform of the Green's function.

Basic procedure required for application of the Wiener-Hopf technique is first to factorize $K(\alpha)$ in the form

$$K(\alpha) = K_+(\alpha)K_-(\alpha) \tag{8.41}$$

where $K_+(\alpha)$ and $K_-(\alpha)$ are regular and nonzero in $\tau > \tau_-$ and $\tau < \tau_+$, respectively. Bearing (8.41) in mind and dividing both sides of (8.40) by $K_+(\alpha)$, we obtain

$$K_-(\alpha)\Phi_-(\alpha) + \frac{\Psi_+(\alpha)}{K_+(\alpha)} + \frac{F(\alpha)}{K_+(\alpha)} = 0 \tag{8.42}$$

Next we must decompose the third term on the left-hand side of (8.42) in the form

$$\frac{F(\alpha)}{K_+(\alpha)} = F_+(\alpha) + F_-(\alpha) \tag{8.43}$$

where $F_+(\alpha)$ and $F_-(\alpha)$ are regular in $\tau > \tau_-$ and $\tau < \tau_+$, respectively. Substituting (8.43) into (8.42) and making some arrangements, we can define a function $P(\alpha)$ by

$$P(\alpha) = Q_-(\alpha) = R_+(\alpha) \tag{8.44}$$

where

$$Q_-(\alpha) = K_-(\alpha)\Phi_-(\alpha) + F_-(\alpha) \tag{8.45a}$$

$$R_+(\alpha) = -\frac{\Psi_+(\alpha)}{K_+(\alpha)} - F_+(\alpha) \tag{8.45b}$$

Because (8.44) has been derived in the strip $\tau_- < \tau < \tau_+$, $P(\alpha)$ is regular there. On the other hand, we find from (8.45a) and (8.45b) that $Q_-(\alpha)$ and $R_+(\alpha)$ are regular in the half-planes $\tau < \tau_+$ and $\tau > \tau_-$, respectively; and they are equal in the overlapping strip $\tau_- < \tau < \tau_+$ of these two half-planes. Therefore, by employing the argument of analytic continuation, we see that $P(\alpha)$ is equal to a certain entire function. We now assume that $Q_-(\alpha)$ and $R_+(\alpha)$ have the asymptotic behavior

$$Q_-(\alpha) = 0(\alpha^q) \quad \text{for} \quad \tau < \tau_+ \tag{8.46a}$$

$$R_+(\alpha) = 0(\alpha^r) \quad \text{for} \quad \tau > \tau_- \tag{8.46b}$$

as $\alpha \to \infty$, where q and r are some real constants. Then, by the extension of Liouville's theorem, we see that $P(\alpha)$ must be a polynomial $p(\alpha)$ of a degree no greater than the integral part of min(q, r). Therefore, substituting this result into (8.45a) and (8.45b) and solving for $\Phi_-(\alpha)$ and $\Psi_+(\alpha)$, we derive that

$$\Phi_-(\alpha) = \frac{p(\alpha) - F_-(\alpha)}{K_-(\alpha)} \tag{8.47a}$$

$$\Psi_+(\alpha) = -K_+(\alpha)[p(\alpha) + F_+(\alpha)] \tag{8.47b}$$

This shows that the Wiener-Hopf equation (8.40) has been solved exactly within the arbitrary polynomial $p(\alpha)$. The remaining step for the solution in real space is to take the inverse complex Fourier transform of (8.47a) and (8.47b) according to Theorem 8.4. The constants q and r in (8.46a) and (8.46b) in general are known to be given explicitly for the scattering and diffraction problems; and hence, the polynomial $p(\alpha)$ can be determined uniquely. This guarantees the uniqueness of the solution of the Wiener-Hopf integral equation.

The procedure just explained is the outline of a method based on the Wiener-Hopf technique, and factorization and decomposition are the most important steps for the solution. With the aid of Theorems 8.9 and 8.10, we can carry out these procedures for general functions.

8.3 DIFFRACTION BY A SEMIINFINITE PLATE

In dealing with electromagnetic boundary value problems, boundary conditions are necessary to the solution, which governs the behavior of tangential electromagnetic fields at the interface of different media. However, we often encounter a situation in which the region of interest involves either boundaries at infinity or geometrical singularities. Under such circumstances, we must impose certain additional constraints to ensure the uniqueness of the solution. These constraints are stated using the radiation condition and the edge condition.

First, in this section, we shall describe the radiation condition and the edge condition briefly. Then, as one of the typical problems that can be solved exactly using the Wiener-Hopf technique, we treat the diffraction of a plane electromagnetic wave by a semiinfinite plate and develop a solution in detail. As mentioned in Section 8.1.1, this problem was solved using the Wiener-Hopf technique by Schwinger [5] and, independently, by Copson [6]. We assume that all the field quantities have a time variation $\exp(-i\omega t)$, where ω is the angular frequency in radians. We will omit the time factor throughout the remaining part of this chapter.

8.3.1 Radiation Condition and Edge Condition

Consider the situation in which all sources in an unbounded space are contained in a finite region and the surface of the obstacle is smooth. Then the additional constraint that ensures the uniqueness of the solution to the reduced wave equation is stated in terms of the *radiation condition*. This condition is expressed in two ways as follows:

1. For a lossy medium, the field must vanish at infinity.
2. For a lossless medium, the field behavior is governed by the *Sommerfeld radiation condition*. In particular, when treating the two-dimensional problems, this condition is stated as

$$\lim_{\rho\to\infty} \rho^{1/2}\left(\frac{\partial\psi}{\partial\rho} - ik\psi\right) = 0 \tag{8.48}$$

using the cylindrical coordinate (ρ, θ, z), where the limit must hold uniformly in θ, and ψ is a nonzero component of electromagnetic fields. The term k (= $\omega\sqrt{\mu\varepsilon}$) is the wavenumber of the medium under consideration, where ε and μ are the permittivity and permeability of the medium, respectively.

If the configuration of the obstacle involves geometrical singularities, such as sharp edges, an additional constraint, known as the *edge condition,* is required to guarantee the uniqueness of the solution. This condition is stated as follows [37, 59]:

$$\lim_{V\to 0} \int_V (\varepsilon|\boldsymbol{E}|^2 + \mu|\boldsymbol{H}|^2)dV = 0 \tag{8.49}$$

where the limit is taken such that the volume V contracts to the edge. Equation (8.49) implies that the electromagnetic energy stored in any neighborhood of the edge must be finite. For the problems treated in this chapter, it is convenient to consider a two-dimensional perfectly conducting wedge as shown in Figure 8.2, where the cylindrical coordinate (ρ, θ, z) with its origin at the edge has been introduced. We see that the electromagnetic fields must behave like

$$E_\rho, E_\theta, H_\rho, H_\theta = 0(\rho^{-1+\nu}) \tag{8.50a}$$

$$E_z, H_z = 0(\rho^\nu) \tag{8.50b}$$

as $\rho \to 0$, where ρ is the distance from the edge, and

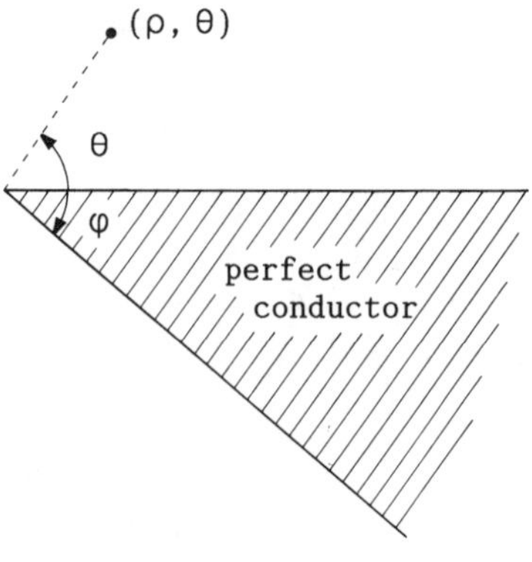

Figure 8.2 Two-dimensional perfectly conducting wedge.

$$\nu = \pi/(2\pi - \varphi) \tag{8.51}$$

In (8.51), φ is the angle between two surfaces of the wedge.

8.3.2 Formulation of the Problem

We consider the diffraction of a plane electromagnetic wave by a perfectly conducting and infinitely thin semiinfinite plate located on the plane $x = 0$ for $z \leqslant 0$ as shown in Figure 8.3. The incident field is assumed to be of E-polarization, which means that the electric field is parallel to the edge. In view of the geometry and characteristics of the incident field, this is a two-dimensional problem. We also find that the boundary under consideration satisfies the conditions (WH1) and (WH2) stated in Section 8.1.2 and hence is of the Wiener-Hopf geometry, because the complex Fourier transform can be taken along the z-axis without crossing the surface of the plate.

Let the total electric field $\phi^t(x, z)$ [$\equiv E_y^t(x, z)$] be

$$\phi^t(x, z) = \phi^i(x, z) + \phi(x, z) \tag{8.52}$$

where $\phi^i(x, z)$ is the incident field given by

$$\phi^i(x, z) = \exp[-ik(x \sin\theta_0 + z \cos\theta_0)] \tag{8.53}$$

for $0 < \theta_0 < \pi/2$. In (8.53), k ($= \omega\sqrt{\mu_0\varepsilon_0}$) is the free-space wavenumber, where ε_0 and μ_0 are the permittivity and permeability of vacuum, respectively. The term $\phi(x, z)$ in (8.52) is the unknown scattered field and satisfies the two-dimensional reduced wave equation:

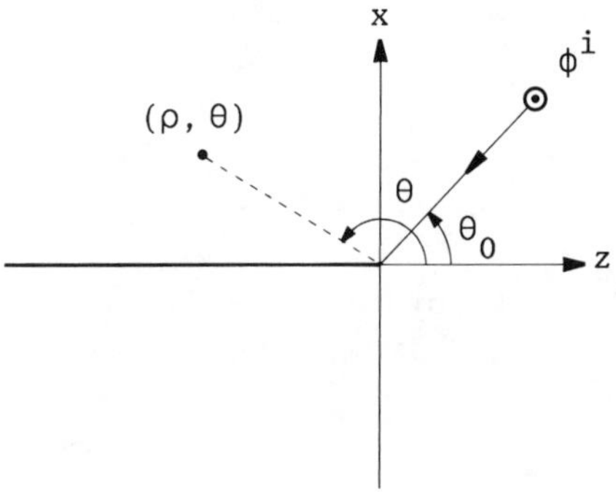

Figure 8.3 Semiinfinite plate.

$$(\partial^2/\partial x^2 + \partial^2/\partial z^2 + k^2)\phi(x, z) = 0 \tag{8.54}$$

We want to solve (8.54) subject to the appropriate boundary conditions. The nonzero components of electromagnetic fields are derived from the relation

$$(E_y, H_x, H_z) = \left(\phi, \frac{i}{\omega\mu_0}\frac{\partial\phi}{\partial z}, \frac{1}{i\omega\mu_0}\frac{\partial\phi}{\partial x}\right) \tag{8.55}$$

We now investigate the asymptotic behavior of the scattered far field. Let us introduce the polar coordinate (ρ, θ) with its center at the origin as shown in Figure 8.3 and divide a space into the three regions as in $-\pi < \theta < -\pi + \theta_0$, $|\theta| < \pi - \theta_0$, and $\pi - \theta_0 < \theta < \pi$, which we denote regions I, II, and III, respectively. Then the scattered far field behaves like

$$\begin{aligned}\phi(x, z) &\sim -\phi^i(x, z) + \phi^d(x, z) \quad \text{in region I}\\ &\sim \phi^d(x, z) \quad \text{in region II}\\ &\sim -\phi^r(x, z) + \phi^d(x, z) \quad \text{in region III}\end{aligned} \tag{8.56}$$

where $\phi^r(x, z)$ is the reflected field from a semiinfinite plate as given by

$$\phi^r(x, z) = \exp[ik(x \sin\theta_0 - z \cos\theta_0)] \tag{8.57}$$

In (8.56), $\phi^d(x, z)$ is the field diffracted by the edge and behaves like

$$\phi^d(\rho, \theta) \sim F(\theta)\frac{e^{ik\rho}}{(k\rho)^{1/2}} \tag{8.58}$$

as $k\rho \to \infty$ according to the Sommerfeld radiation condition given by (8.48).

It is convenient to introduce a slight loss in the medium so that wavenumber k becomes complex, as in

$$k = k_1 + ik_2, \quad 0 < k_2 \ll k_1 \tag{8.59}$$

We can recover the case for real k by letting $k_2 \to 0$ at the end of analysis. It follows from (8.53) and (8.56) to (8.58) that the asymptotic behavior of the scattered field for large $|z|$ is given by

$$\begin{aligned}\phi(x, z) &= 0[\exp(k_2 z \cos\theta_0)], \quad z \to -\infty\\ &= 0[\exp(-k_2 z)], \quad z \to \infty\end{aligned} \tag{8.60}$$

We define the complex Fourier transform of $\phi(x, z)$ with respect to z as

$$\Phi(x, \alpha) = (2\pi)^{-1/2} \int_{-\infty}^{\infty} \phi(x, z)\, e^{i\alpha z} dz \tag{8.61}$$

where $\alpha = \sigma + i\tau$. Then we see with the aid of (8.60) and Theorem 8.4 that $\Phi(x, \alpha)$ is regular in the strip $-k_2 < \tau < k_2 \cos\theta_0$. If we further introduce the complex Fourier integrals

$$\Phi_{\pm}(x, \alpha) = \pm(2\pi)^{-1/2} \int_{0}^{\pm\infty} \phi(x, z)\, e^{i\alpha z} dz \tag{8.62}$$

then we find from (8.60) and Theorems 8.2 and 8.3 that $\Phi_+(x, \alpha)$ and $\Phi_-(x, \alpha)$ are regular in $\tau > -k_2$ and $\tau < k_2 \cos\theta_0$, respectively. Using (8.62), $\Phi(x, \alpha)$ is written as

$$\Phi(x, \alpha) = \Phi_-(x, \alpha) + \Phi_+(x, \alpha) \tag{8.63}$$

We can also see with the aid of (8.60) that $\Phi(x, \alpha)$ is bounded as $|x| \to \infty$.

Taking the complex Fourier transform of (8.54) and making use of (8.60), we derive that

$$(d^2/dx^2 - \gamma^2)\Phi(x, \alpha) = 0 \tag{8.64}$$

for any α in the strip $-k_2 < \tau < k_2 \cos\theta_0$, where

$$\gamma = (\alpha^2 - k^2)^{1/2} \tag{8.65}$$

Equation (8.64) is called the *transformed-wave equation*. We now choose branch cuts for γ as the simple curves discussed in Section 8.2.4. Because Re $\gamma > 0$ for any α in $|\tau| < k_2$ according to the choice of branch cuts, the solution of (8.64) can be written as

$$\begin{aligned} \Phi(x, \alpha) &= A(\alpha)\, e^{-\gamma x} \quad \text{for} \quad x > 0 \\ &= B(\alpha)\, e^{\gamma x} \quad \text{for} \quad x < 0 \end{aligned} \tag{8.66}$$

by imposing the condition that $\Phi(x, \alpha)$ be bounded as $|x| \to \infty$.

From the boundary conditions for tangential electric fields, it follows that

$$\phi(+0, z) = \phi(-0, z)\ [\equiv \phi(0, z)] \tag{8.67}$$

for $0 < z < \infty$ and

$$\phi(0, z) = -\phi^i(0, z) \tag{8.68}$$

for $-\infty < z < 0$. Setting $x = \pm 0$ in (8.66) and using (8.67) and (8.68), we derive that

$$\begin{aligned} \Phi(x, \alpha) &= U_{(+)}(\alpha)\, e^{-\gamma x} \quad \text{for} \quad x \geqslant 0 \\ &= U_{(+)}(\alpha)\, e^{\gamma x} \quad \text{for} \quad x \leqslant 0 \end{aligned} \tag{8.69}$$

where

$$U_{(+)}(\alpha) = \Phi_+(0, \alpha) - \frac{A}{\alpha - k\cos\theta_0} \tag{8.70}$$

$$A = \frac{1}{(2\pi)^{1/2} i} \tag{8.71}$$

The parentheses in the subscript of $U_{(+)}(\alpha)$ imply that $U_{(+)}(\alpha)$ is regular in $\tau > -k_2$ except for a simple pole at $\alpha = k\cos\theta_0$. We now differentiate (8.69) with respect to x and set $x = \pm 0$ in the resultant equation. This yields

$$\Phi'(+0, \alpha) = -\gamma U_{(+)}(\alpha) \tag{8.72a}$$

$$\Phi'(-0, \alpha) = \gamma U_{(+)}(\alpha) \tag{8.72b}$$

where the prime implies differentiation with respect to x. Subtracting (8.72b) from (8.72a) and taking into account the boundary condition for tangential magnetic fields

$$\frac{\partial \phi(+0, z)}{\partial x} = \frac{\partial \phi(-0, z)}{\partial x} \tag{8.73}$$

for $0 < z < \infty$, we obtain that

$$K(\alpha) J_-(0, \alpha) + U_{(+)}(\alpha) = 0, \quad -k_2 < \tau < k_2 \cos\theta_0 \tag{8.74}$$

where

$$J_-(0, \alpha) = \Phi'_-(+0, \alpha) - \Phi'_-(-0, \alpha) \tag{8.75}$$

$$K(\alpha) = (2\gamma)^{-1} \tag{8.76}$$

Here, $K(\alpha)$ is the kernel function, and $J_-(0, \alpha)$ corresponds to the surface current induced on the semiinfinite plate. Equation (8.74) is the desired Wiener-Hopf equation and can be solved exactly by following the procedure outlined in Section 8.2.5.

8.3.3 Exact Solution of the Wiener-Hopf Equation

By taking into account the choice of branch cuts for γ, we find that γ reduces to $-ik$ when $\alpha = 0$. Keeping this in mind, the kernel function $K(\alpha)$ given by (8.76) is found to be factorized as follows:

$$K(\alpha) = K_+(\alpha)K_-(\alpha) \tag{8.77}$$

where

$$K_\pm(\alpha) = 2^{-1/2}\, e^{i\pi/4}(k \pm \alpha)^{-1/2} \tag{8.78}$$

We can see from (8.78) that $K_+(\alpha)$ and $K_-(\alpha)$ are regular and nonzero in $\tau > -k_2$ and $\tau < k_2$, respectively. We divide (8.74) by $K_+(\alpha)$ and decompose the resultant equation as follows:

$$\frac{U_{(+)}(\alpha)}{K_+(\alpha)} + \frac{A}{K_+(k\cos\theta_0)(\alpha - k\cos\theta_0)} = -K_-(\alpha)J_-(0, \alpha) + \frac{A}{K_+(k\cos\theta_0)(\alpha - k\cos\theta_0)} \tag{8.79}$$

We see that the left-hand and right-hand sides of (8.79) are regular in $\tau > -k_2$ and $\tau < k_2 \cos\theta_0$, respectively; and they have a common strip of regularity $-k_2 < \tau < k_2 \cos\theta_0$. Therefore, the argument of analytic continuation shows that both sides of (8.79) must be equal to a certain entire function, which we denote by $P(\alpha)$. Now, for the solution, we must determine $P(\alpha)$.

Setting $\varphi = 0$ in (8.51), we find from (8.50a) and (8.50b), (8.52), and (8.55) that the edge condition to this problem is given by

$$\phi(0, z) = -1 + 0(z^{1/2}), \quad z \to +0 \tag{8.80a}$$

$$\frac{\partial \phi(\pm 0, z)}{\partial x} = 0(z^{-1/2}), \quad z \to -0 \tag{8.80b}$$

Hence, application of Theorems 8.5 and 8.7 with the aid of (8.70) and (8.75) shows that

$$U_{(+)}(\alpha) = 0(\alpha^{-3/2}), \quad \tau > -k_2 \tag{8.81a}$$

$$J_{-}(0, \alpha) = 0(\alpha^{-1/2}), \quad \tau < k_2 \cos\theta_0 \tag{8.81b}$$

as $\alpha \to \infty$. We also see from (8.78) that the split functions $K_{\pm}(\alpha)$ are $0(\alpha^{-1/2})$ as $\alpha \to \infty$ in $\tau \gtrless \mp k_2$. These considerations show that

$$P(\alpha) = 0(\alpha^{-1}), \quad \alpha \to \infty \tag{8.82}$$

Thus, we can conclude with the aid of Liouville's theorem that $P(\alpha)$ must be identically zero. Equating both sides of (8.79) to zero and solving for $U_{(+)}(\alpha)$ and $J_{-}(0, \alpha)$, we obtain

$$U_{(+)}(\alpha) = -\frac{A}{K_{+}(k\cos\theta_0)} \frac{K_{+}(\alpha)}{\alpha - k\cos\theta_0} \tag{8.83a}$$

$$J_{-}(0, \alpha) = \frac{A}{K_{+}(k\cos\theta_0)} \frac{1}{K_{-}(\alpha)(\alpha - k\cos\theta_0)} \tag{8.83b}$$

Equations (8.83a) and (8.83b) form the exact solution of the Wiener-Hopf equation (8.74).

8.3.4 Scattered-Field Representations

Substituting (8.83a) into (8.69) and taking the inverse complex Fourier transform of the result according to Theorem 8.4, we obtain the scattered-field representation as follows:

$$\phi(x, z) = -\frac{(k\cos\theta_0 + k)^{1/2}}{2\pi i} \int_{-\infty+ic}^{\infty+ic} \frac{e^{-\gamma|x| - i\alpha z}}{(\alpha + k)^{1/2}(\alpha - k\cos\theta_0)}\, d\alpha, \tag{8.84}$$

$$d\alpha - k_2 < c < k_2 \cos\theta_0$$

Letting the medium loss k_2 tend to zero and taking into account that the branch cuts discussed in Section 8.2.4 are then degenerated into the real and imaginary axes we have

$$\phi(x, z) = -\frac{(k\cos\theta_0 + k)^{1/2}}{2\pi i}\int_C \frac{e^{-\gamma|x| - i\alpha z}}{(\alpha + k)^{1/2}(\alpha - k\cos\theta_0)}\,d\alpha \tag{8.85}$$

where C is the integration path composed of semicircles around the singularities at $\alpha = \pm k$, $k\cos\theta_0$, and a portion of the real axis as shown in Figure 8.4. Equation (8.85) is the exact integral representation of the solution to the original problem for real k.

We introduce the polar coordinate $x = \rho\sin\theta$ and $z = \rho\cos\theta$ for $-\pi < \theta < \pi$ and make a change of variable $\alpha = -k\cos w$ [$\equiv -k\cos(u + iv)$], which is also written as $w = i\ln[(\gamma - \alpha)/k]$ by taking into account the choice of branch cuts. Then, (8.85) can be written as

$$\phi(\rho, \theta) = \frac{1}{4\pi i}[\phi_+(\rho, \theta) + \phi_-(\rho, \theta)] \tag{8.86}$$

where

$$\phi_\pm(\rho, \theta) = \int_{C_\pm} \sec\frac{w \pm \theta_0}{2}\, e^{k\rho g(w)}dw \tag{8.87}$$

$$g(w) = i\cos(w - |\theta|) \tag{8.88}$$

In (8.87), C_+ and C_- are the infinite integration paths as shown in Figures 8.5(a) and 8.5(b), respectively.

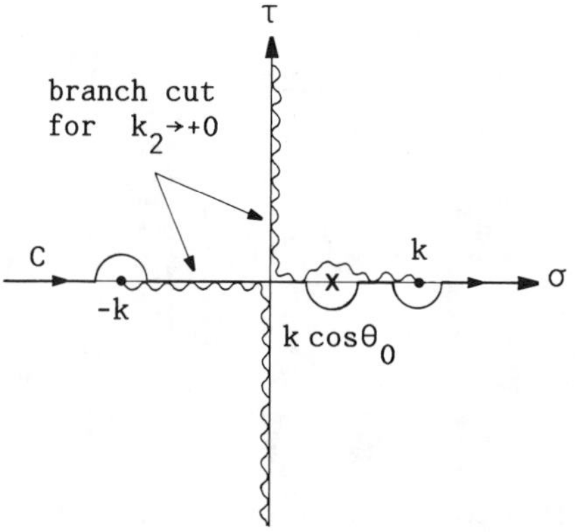

Figure 8.4 Integration path C in equation (8.85).

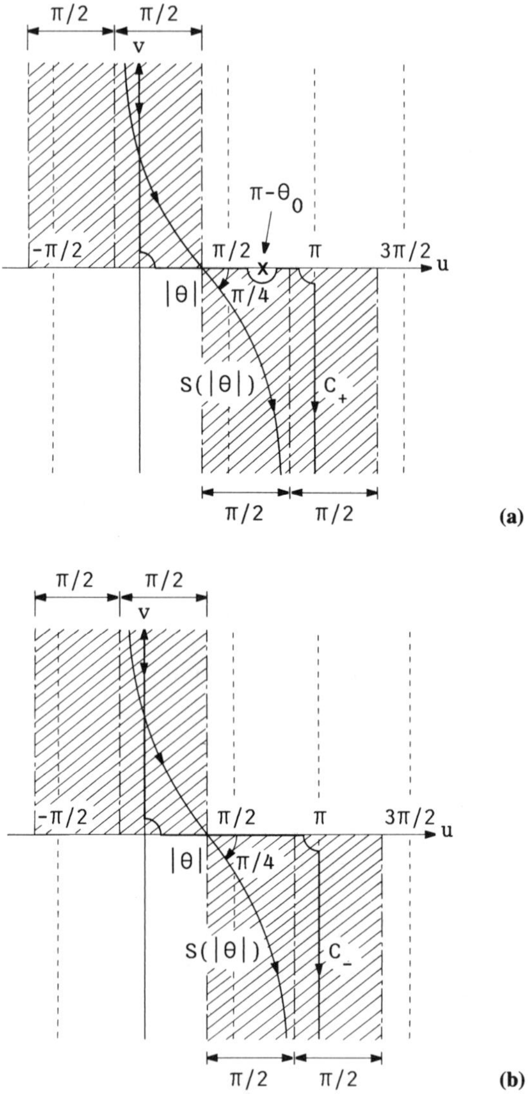

Figure 8.5 Integration paths $C_{\pm}$ in equation (8.87) and $S(|\theta|)$ defined by equation (8.90).

Let us consider the following equation:

$$\left.\frac{d}{dw} g(w)\right|_{w=w_s} = 0 \tag{8.89}$$

Solving (8.89) for w_s, we find that $w_s = |\theta|$. The point w_s defined by (8.89), known as a *saddle point,* plays an important role in the evaluation of integrals. We now introduce the new integration path $S(|\theta|)$ as defined by

$$\text{Im } g(w) = 1, \quad \text{Re } g(w) \leqslant 0 \tag{8.90}$$

The path $S(|\theta|)$ also is sketched in Figure 8.5. We find that if we let $|v| \equiv |\text{Im } w| \rightarrow \infty$ in the shaded region of Figure 8.5, then the magnitude of the integrand of (8.87) decays exponentially. We also find that the saddle point $w_s = |\theta|$ is located on the path $S(|\theta|)$, and the magnitude of the integrand decays most rapidly away from the saddle point along $S(|\theta|)$. The integration path satisfying these properties in general is called the *steepest-descent path.*

From these considerations, the paths $C_\pm$ in (8.87) can be deformed into the steepest-descent path $S(|\theta|)$; but, for $\phi_+(\rho, \theta)$, the residue contribution due to a simple pole at $w = \pi - \theta_0$ must be included in the process of deformation of the contour when $|\theta| > \pi - \theta_0$. Therefore, we finally arrive at the result

$$\phi_+(\rho, \theta) = -4\pi i \exp[-ik\rho \cos(|\theta| + \theta_0)]H(|\theta| - \pi + \theta_0) + \phi_+^s(\rho, \theta) \tag{8.91a}$$

$$\phi_-(\rho, \theta) = \phi_-^s(\rho, \theta) \tag{8.91b}$$

where

$$\phi_\pm^s(\rho, \theta) = \int_{S(|\theta|)} \sec \frac{w \pm \theta_0}{2} e^{k\rho g(w)} dw \tag{8.92}$$

$$H(\xi) = 1 \quad \text{for} \quad \xi > 0$$

$$= 0 \quad \text{for} \quad \xi < 0 \tag{8.93}$$

ntroducing a change of variable $w - |\theta| \rightarrow w$ in (8.92) and rearranging the result, ve derive that

$$\phi_{\pm}^{s}(\rho, \theta) = \frac{1}{2}\int_{S(0)} \left(\sec\frac{w + |\theta| \pm \theta_0}{2} + \sec\frac{w - |\theta| \mp \theta_0}{2}\right) e^{ik\rho\cos w}dw$$

$$= 2\int_{S(0)} \frac{\cos(w/2)\cos[(|\theta| \pm \theta_0)/2]}{\cos w + \cos(|\theta| \pm \theta_0)} e^{ik\rho\cos w}dw \tag{8.94}$$

We make a further change of variable $t = 2^{1/2}e^{i\pi/4}\sin(w/2)$ in (8.94). Taking into account that the path $S(0)$ in the w-plane is mapped to the real axis in the t-plane, this yields

$$\phi_{\pm}^{s}(\rho, \theta) = -2\, e^{i(k\rho + \pi/4)}\eta_{\pm}J(k\rho, \eta_{\pm}) \tag{8.95}$$

where

$$J(\xi, \eta) = \int_{-\infty}^{\infty} \frac{\exp(-\xi t^2)}{t^2 - i\eta^2}dt, \quad \xi > 0, \quad -\infty < \eta < \infty \tag{8.96}$$

$$\eta_{\pm} = 2^{1/2}\cos[(|\theta| \pm \theta_0)/2] \tag{8.97}$$

Replacing the variable ξ in (8.96) by ζ and carrying out some manipulations, we obtain

$$\frac{d}{d\zeta}[J(\zeta, \eta)\exp(i\zeta\eta^2)] = -\left(\frac{\pi}{\zeta}\right)^{1/2}\exp(i\zeta\eta^2) \tag{8.98}$$

Integrating both sides of (8.98) from ξ to infinity and making use of the relation

$$\lim_{\zeta\to\infty} J(\zeta, \eta) = 0 \tag{8.99}$$

we derive

$$J(\xi, \eta) = \frac{2\pi}{|\eta|}\exp[-i(\xi\eta^2 - \pi/4)]F(\xi^{1/2}|\eta|) \tag{8.100}$$

where $F(\cdot)$ is the Fresnel integral defined by

$$F(x) = \frac{e^{-i\pi/4}}{\pi^{1/2}}\int_0^{\infty} \exp(it^2)dt \tag{8.101}$$

Substituting (8.100) into (8.95) and taking into account (8.91a) and (8.91b), after some manipulations, we have

$$\begin{aligned} \phi_{\pm}(\rho, \theta) = & -4\pi i \exp[-ik\rho \cos(|\theta| \pm \theta_0)] \\ & \times F\{(2k\rho)^{1/2} \cos[(|\theta| \pm \theta_0)/2]\} \end{aligned} \tag{8.102}$$

In the derivation of (8.102), we used the following *reflection formula* for the Fresnel integral:

$$F(x) + F(-x) = 1 \tag{8.103}$$

Further substitution of (8.102) into (8.86) yields

$$\begin{aligned} \phi(\rho, \theta) = & -\exp[-ik\rho \cos(\theta - \theta_0)]F\{(2k\rho)^{1/2} \cos[(\theta - \theta_0)/2]\} \\ & - \exp[-ik\rho \cos(\theta + \theta_0)]F\{(2k\rho)^{1/2} \cos[(\theta + \theta_0)/2]\} \end{aligned} \tag{8.104}$$

Although (8.104) has been derived on the assumption that $0 < \theta_0 < \pi/2$, we can obtain the same scattered-field representation as given by (8.104) even for $\pi/2 < \theta_0 < \pi$ after slight modifications of the analysis. Therefore, (8.104) is the exact solution to this problem, which holds for the arbitrary incident angle θ_0.

The radiation and diffraction problems concerned with a semiinfinite parallel plate waveguide [7, 8, 11] and a semiinfinite circular waveguide [9–11] can also be treated by using the method developed in this section.

8.4 DIFFRACTION BY A STRIP

Diffraction problems can be solved exactly by the Wiener-Hopf technique only if the obstacle has a semiinfinite boundary and the complex Fourier transform is taken such that the transform axis does not traverse the surface of the obstacle. In other words, the obstacle must be of the Wiener-Hopf geometry for the exact solution. However, as pointed out in Section 8.1.2, there is a class of problems that can be treated by formal application of the Wiener-Hopf technique and solved approximately. In this section, as an example of such problems, we analyze the diffraction of a plane electromagnetic wave by a strip and develop a solution method in detail [60]. Because the strip is of finite width, this problem belongs to a class of the modified Wiener-Hopf geometry of the first kind. In the following text, we will show how the Wiener-Hopf technique can be applied. As in the previous section, we shall introduce a slight loss into the medium so that wavenumber k has a small positive imaginary part as shown in (8.59).

8.4.1 Formulation of the Problem

We consider the diffraction of an E-polarized plane wave by a perfectly conducting and infinitely thin strip located on the plane $x = 0$ for $|z| \leqslant a$, as shown in Figure 8.6. The incident field $\phi^i(x, z)$ [$\equiv E_y^i(x, z)$] is given by

$$\phi^i(x, z) = \exp[-ik(x \sin\theta_0 + z \cos\theta_0)] \tag{8.105}$$

for $0 < \theta_0 < \pi/2$. From the geometry of the strip and the polarization of the incident field, this is a two-dimensional problem. If we define the total field $\phi^t(x, z)$ [$\equiv E_y^t(x, z)$] by (8.52), then the scattered field $\phi(x, z)$ satisfies the two-dimensional reduced-wave equation (8.54) and the nonzero components of electromagnetic fields can be deduced from (8.55).

Dividing a space into appropriate regions as in Section 8.3.2 and investigating the asymptotic behavior of $\phi(x, z)$ for large $|z|$, we see that $\phi(x, z)$ behaves like the diffracted field $\phi^d(\rho, \theta)$ given by (8.58) as $|k|\rho \to \infty$. Therefore, it follows that

$$\begin{aligned} \phi(x, z) &= 0[\exp(-k_2|z|)] \\ &= 0[\exp(-k_2|z| \cos\theta_0)], \quad |z| \to \infty \end{aligned} \tag{8.106}$$

If we define the complex Fourier transform of $\phi(x, z)$ with respect to z as (8.61), then we can see with the aid of (8.106) and Theorem 8.4 that $\Phi(x, \alpha)$ is regular in the strip $|\tau| < k_2 \cos\theta_0$. Let us further introduce the complex Fourier integrals as in

$$\Phi_\pm(x, \alpha) = \pm(2\pi)^{-1/2} \int_{\pm a}^{\pm\infty} \phi(x, z)\, e^{i\alpha(z \mp a)} dz \tag{8.107a}$$

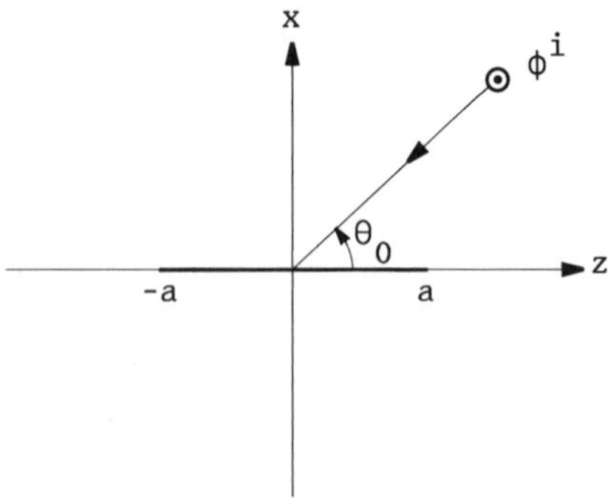

Figure 8.6 Strip.

$$\Phi_1(x, \alpha) = (2\pi)^{-1/2} \int_{-a}^{a} \phi(x, z)\, e^{i\alpha z} dz \tag{8.107b}$$

ıen $\Phi(x, \alpha)$ can be written as

$$\Phi(x, \alpha) = e^{-i\alpha a}\Phi_-(x, \alpha) + \Phi_1(x, \alpha) + e^{i\alpha a}\Phi_+(x, \alpha) \tag{8.108}$$

sing Theorems 8.1 to 8.3, we find that $\Phi_\pm(x, \alpha)$ are regular in $\tau \gtrless \mp k_2 \cos\theta_0$ and $_1(x, \alpha)$ is an entire function.

Taking the complex Fourier transform of (8.54) with the aid of (8.106), we ɔtain the transformed-wave equation (8.64) in $|\tau| < k_2 \cos\theta_0$. Therefore, the solution (8.64) for this problem can be written as

$$\Phi(x, \alpha) = [e^{-i\alpha a} U_-(\alpha) + e^{i\alpha a} U_{(+)}(\alpha)]\, e^{-\gamma|x|} \tag{8.109}$$

y taking into account the boundary conditions that

$$\phi(+0, z) = \phi(-0, z)\ [\equiv \phi(0, z)] \tag{8.110}$$

ɔr $|z| > a$ and

$$\phi(0, z) = -\phi^i(0, z) \tag{8.111}$$

ɔr $|z| < a$, where

$$U_-(\alpha) = \Phi_-(0, \alpha) + \frac{A}{\alpha - k\cos\theta_0} \tag{8.112a}$$

$$U_{(+)}(\alpha) = \Phi_+(0, \alpha) - \frac{B}{\alpha - k\cos\theta_0} \tag{8.112b}$$

$$A = \frac{\exp(ika\cos\theta_0)}{(2\pi)^{1/2} i}, \quad B = \frac{\exp(-ika\cos\theta_0)}{(2\pi)^{1/2} i} \tag{8.113}$$

ifferentiating (8.109) with respect to x and setting $x = \pm 0$ in the resultant equation, fter some arrangements, we derive

$$e^{-i\alpha a} U_-(\alpha) + K(\alpha) J_1(0, \alpha) + e^{i\alpha a} U_{(+)}(\alpha) = 0 \tag{8.114}$$

ɔr any α in $|\tau| < k_2 \cos\theta_0$, where

$$J_1(0, \alpha) = \Phi_1'(+0, \alpha) - \Phi_1'(-0, \alpha) \quad (8.115)$$

$$K(\alpha) = (2\gamma)^{-1} \quad (8.116)$$

In the derivation of (8.114), we have used the boundary condition for tangential magnetic fields:

$$\frac{\partial\phi(+0, z)}{\partial x} = \frac{\partial\phi(-0, z)}{\partial x}, \quad |z| > a \quad (8.117)$$

Equation (8.114) is the Wiener-Hopf equation satisfied by the unknown functions. This equation is not of the form exactly solvable by the Wiener-Hopf technique, but we can apply the Wiener-Hopf technique formally to obtain the approximate solution.

8.4.2 Formal Solution of the Wiener-Hopf Equation

Factorization of the kernel function $K(\alpha)$ defined by (8.116) has already been discussed in Section 8.3.3 as given by (8.77) and (8.78). Multiplying both sides of (8.114) by $\exp(i\alpha a)/K_-(\alpha)$, we obtain

$$e^{i\alpha a} J_1(0, \alpha) K_+(\alpha) = -\frac{U_-(\alpha)}{K_-(\alpha)} - e^{2i\alpha a}\frac{U_{(+)}(\alpha)}{K_-(\alpha)} \quad (8.118)$$

Because $\exp(i\alpha a)J_1(0, \alpha)$ is an entire function bounded for $\tau > 0$, the left-hand side of (8.118) is found to be regular and bounded in $\tau > -k_2 \cos\theta_0$. On the other hand the first and second terms on the right-hand side of (8.118) are regular in $\tau < k_2 \cos\theta_0$ and $|\tau| < k_2 \cos\theta_0$, respectively. Therefore, we must decompose the second term on the right-hand side of (8.118) into the sum of two functions regular in $\tau \gtrless \mp k_2 \cos\theta_0$.

According to the edge condition, we have

$$\phi(0, z) = -\exp(\mp ika \cos\theta_0) + 0[(z \mp a)^{1/2}], \quad z \rightarrow \pm a \pm 0 \quad (8.119a)$$

$$\frac{\partial\phi(+0, z)}{\partial x}, \ \frac{\partial\phi(-0, z)}{\partial x} = 0[(z \mp a)^{-1/2}], \quad z \rightarrow \pm a \mp 0 \quad (8.119b)$$

Therefore, we find, using Theorems 8.5 to 8.7, that

$$U_{(\pm)}(\alpha) = 0(\alpha^{-3/2}), \quad \tau \gtrless \mp k_2 \cos\theta_0 \quad (8.120a)$$

$$e^{\pm i\alpha a}J_1(0, \alpha) = 0(\alpha^{-1/2}), \quad \tau \gtrless 0 \tag{8.120b}$$

; $\alpha \to \infty$. It follows from (8.120a) that the second term on the right-hand side of ;.118) is $0(|\sigma|^{-1})$ uniformly in τ for $|\tau| < k_2 \cos\theta_0$, because the split functions $K_\pm(\alpha)$ 'e $0(\alpha^{-1/2})$ as $\alpha \to \infty$ in $\tau \gtrless \mp k_2$. Therefore, application of Theorem 8.9 yields

$$e^{i\alpha a}J_1(0, \alpha)K_+(\alpha) + \frac{1}{2\pi i}\int_{C_1} \frac{e^{2i\beta a}U_{(+)}(\beta)}{K_-(\beta)(\beta - \alpha)}\, d\beta = -\frac{U_-(\alpha)}{K_-(\alpha)} + \frac{1}{2\pi i}\int_{C_2} \frac{e^{2i\beta a}U_{(+)}(\beta)}{K_-(\beta)(\beta - \alpha)}\, d\beta \tag{8.121}$$

'here C_1 and C_2 are the infinite integration paths parallel to the real axis, as shown ι Figure 8.7. Note that the left-hand and right-hand sides of (8.121) are regular in $> -k_2 \cos\theta_0$ and $\tau < k_2 \cos\theta_0$, respectively, and both sides have a common strip f regularity $|\tau| < k_2 \cos\theta_0$. Therefore, the argument of analytic continuation shows ιat both sides of (8.121) must be equal to a certain entire function, which we denote y $P(\alpha)$. Taking into account (8.120a) and the asymptotic behavior of the split func-ons, we also find that the two integrals appearing in (8.121) are $o(\alpha^{-1/2})$ as $\alpha \to$ '. These considerations show with the aid of (8.120a) and (8.120b) that the entire ınction $P(\alpha)$ derived earlier is $o(\alpha^{-1/2})$ as $\alpha \to \infty$. Therefore, application of Liou-ille's theorem yields $P(\alpha) \equiv 0$. Equating the right-hand side of (8.121) to zero, it ɔllows that

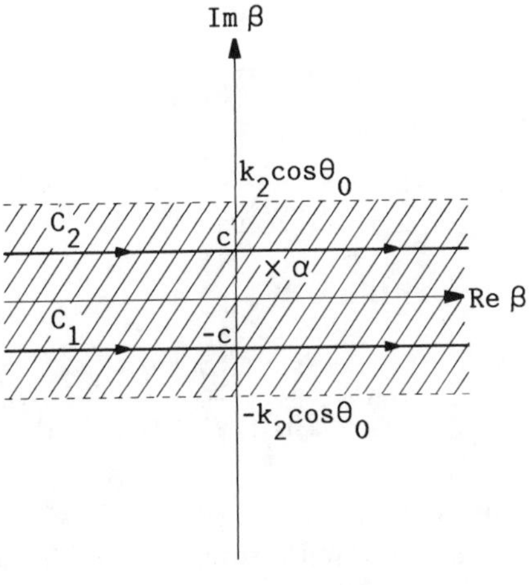

'igure 8.7 Integration paths C_1 and C_2 for decomposition ($0 < |\tau| < c < k_2 \cos\theta_0$).

$$\frac{U_-(\alpha)}{K_-(\alpha)} - \frac{1}{2\pi i}\int_{C_2} \frac{e^{2i\beta a}U_{(+)}(\beta)}{K_-(\beta)(\beta - \alpha)}\, d\beta = 0 \tag{8.122}$$

Next we multiply both sides of (8.114) by $\exp(-i\alpha a)/K_+(\alpha)$ and decompose the resultant equation by following the same procedure as earlier with the aid of (8.120a) and (8.120b). This yields, after some manipulations,

$$\frac{U_{(+)}(\alpha)}{K_+(\alpha)} + \frac{B}{K_+(k\cos\theta_0)(\alpha - k\cos\theta_0)} + \frac{1}{2\pi i}\int_{C_1} \frac{e^{-2i\beta a}U_-(\beta)}{K_+(\beta)(\beta - \alpha)}\, d\beta = 0 \tag{8.123}$$

Setting $\alpha \to -\alpha$ in (8.122) and making a change of variable $\beta \to -\beta$ in (8.123) after taking the sum and difference of the resultant equations, we derive

$$\frac{U^{s,d}_{(+)}(\alpha)}{K_+(\alpha)} + \frac{B}{K_+(k\cos\theta_0)(\alpha - k\cos\theta_0)} \mp \frac{1}{2\pi i}\int_{C_2} \frac{e^{2i\beta a}U^{s,d}_{(+)}(\beta)}{K_-(\beta)(\beta + \alpha)}\, d\beta = 0 \tag{8.124}$$

where

$$U^{s,d}_{(+)}(\alpha) = U_{(+)}(\alpha) \pm U_-(-\alpha) \tag{8.125}$$

The integrand of the integral involved in (8.124) has a branch point at $\beta = k$, and we choose the branch cut emanating from $\beta = k$ as a straight line parallel to the imaginary axis and extending to infinity in the upper half-plane, as shown in Figure 8.8. Evaluating the integral from (8.124), yields

$$U^{s,d}_{(+)}(\alpha) = -K_+(\alpha)\left[\frac{B}{K_+(k\cos\theta_0)(\alpha - k\cos\theta_0)} \pm \frac{A}{K_-(k\cos\theta_0)(\alpha + k\cos\theta_0)} \mp u_{s,d}(\alpha)\right] \tag{8.126}$$

where

$$u_{s,d}(\alpha) = \frac{1}{2\pi i}\int_C \frac{e^{2i\beta a}U^{s,d}_{(+)}(\beta)}{K_-(\beta)(\beta + \alpha)}\, d\beta \tag{8.127}$$

In (8.127), C is the contour composed of a portion C_ε of a circle of a small radius ε centered at $\beta = k$ and the semiinfinite straight paths $C_\pm$ along the branch cut as shown in Figure 8.8. We find that the contribution from C_ε tends to zero by letting $\varepsilon \to 0$. On the other hand, the contributions from $C_\pm$ can be combined to yield the single branch-cut integral by noting that

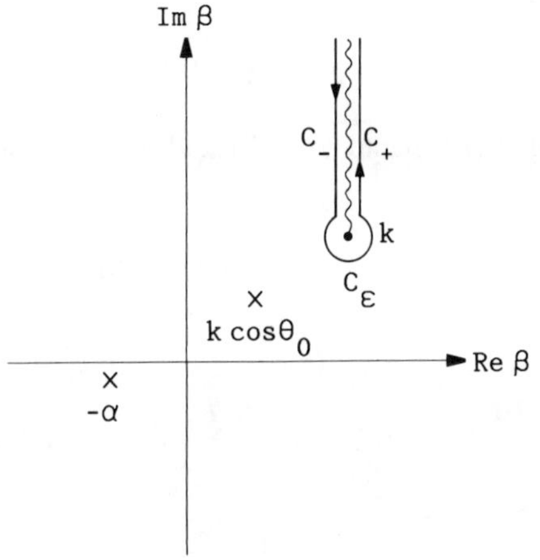

Figure 8.8 Integration path C (= $C_- + C_\varepsilon + C_+$) in equation (8.127).

$$(\beta^2 - k^2)^{1/2}|_{\beta \in C_+} = -(\beta^2 - k^2)^{1/2}|_{\beta \in C_-} \tag{8.128}$$

After letting $\varepsilon \to 0$ and making some arrangements, (8.127) can be reduced to

$$u_{s,d}(\alpha) = \frac{1}{\pi i}\int_k^{k+i\infty} \frac{e^{2i\beta a}U^{s,d}_{(+)}(\beta)}{K_-(\beta)(\beta+\alpha)}\,d\beta \tag{8.129}$$

where the integration path is the one running on the right of the branch cut.

From (8.125) and (8.126), we can obtain the explicit representations of $U_-(\alpha)$ and $U_{(+)}(\alpha)$, which give the exact solution to the Wiener-Hopf equation (8.114). However, they are formal in the sense that there are infinite branch-cut integrals $u_{s,d}(\alpha)$ with the unknown integrands $U_-(\alpha)$ and $U_{(+)}(\alpha)$ as defined by (8.112a) and (8.112b). Therefore, we must carry out approximate evaluation of the integrals for practical numerical computations. From this reason, the type of solution given by (8.126) is called the *formal solution*. In the next subsection, we develop a method of asymptotic evaluation of the integrals and derive the approximate solution to the Wiener-Hopf equation as $ka \to \infty$.

8.4.3 High-Frequency Asymptotic Solution

Let us introduce the functions

$$\Phi^{s,d}_{+}(\alpha) = \Phi_+(0, \alpha) \pm \Phi_-(0, -\alpha) \tag{8.130}$$

Then it follows from (8.112a), (8.112b) and (8.125) that

$$U^{s,d}_{(+)}(\alpha) = \Phi^{s,d}_{+}(\alpha) - \frac{B}{\alpha - k\cos\theta_0} \mp \frac{A}{\alpha + k\cos\theta_0} \tag{8.131}$$

Therefore, substituting (8.131) into (8.129) and carrying out some manipulations, we are led to

$$u_{s,d}(\alpha) = v_{s,d}(\alpha) + B\eta_b(\alpha) \pm A\eta_a(\alpha) \tag{8.132}$$

where

$$v_{s,d}(\alpha) = \frac{2^{1/2}\,e^{-i\pi/4}}{\pi}\int_k^{k+i\infty} \Phi^{s,d}_{+}(\beta)\,\frac{e^{2i\beta a}(\beta - k)^{1/2}}{\beta + \alpha}\,d\beta \tag{8.133}$$

$$\eta_{b,a}(\alpha) = \frac{\xi(\alpha) - \xi(\mp k\cos\theta_0)}{\alpha \pm k\cos\theta_0} \tag{8.134}$$

$$\xi(\alpha) = \frac{e^{2ika}}{\pi a^{1/2}}\,\zeta_0[-2i(\alpha + k)a] \tag{8.135}$$

$$\zeta_n(z) = \int_0^{\infty} \frac{t^{n+1/2}e^{-t}}{t + z}\,dt, \quad n > -3/2, \quad |\arg z| < \pi \tag{8.136}$$

Because the unknown functions $\Phi^{s,d}_{+}(\beta)$ are involved in the integrand of (8.133), exact evaluation of the integrals is difficult. In the following, we derive the asymptotic expansion of $u_{s,d}(\alpha)$ given by (8.132) as $ka \to \infty$. For this purpose, we use the following lemma [61].

Lemma. Let $H(\beta)$ be regular in the neighborhood of $\beta = k$ and bounded for Im $\beta > 0$, and consider the integral

$$I(\nu, \alpha) = \frac{1}{\pi i}\int_k^{k+i\infty} H(\beta)\,\frac{e^{i\beta\nu}(\beta - k)^{1/2}}{\beta + \alpha}\,d\beta \tag{8.137}$$

for $\nu > 0$, $\alpha \neq -k$ and $\arg(\alpha + k) \neq -\pi/2, 3\pi/2$, where $k = k_1 + ik_2$, ($k_1 > 0$, $k_2 > 0$) and the distance between k and the closest possible singularity of $H(\beta)$ is assumed not too small. Then $I(\nu, \alpha)$ has the asymptotic expansion

$$I(\nu, \alpha) \sim \frac{e^{i(k\nu - \pi/4)}}{\pi}\sum_{n=0}^{\infty} i^n \frac{h_n}{\nu^{n+1/2}}\,\zeta_n[-i(\alpha + k)\nu] \tag{8.138}$$

as $\nu \to \infty$, where $\zeta_n(\cdot)$ is defined by (8.136), and

$$h_n = \frac{1}{n!}\frac{d^n}{d\beta^n}H(\beta)\bigg|_{\beta=k} \tag{8.139}$$

From (8.130), we find that $\Phi_+^{s,d}(\beta)$ are regular and bounded for Im $\beta > -k_2 \cos\theta_0$, and hence the above Lemma can be applied directly to obtain the asymptotic expansion of $v_{s,d}(\alpha)$ as defined by (8.133) for large ka. Therefore, it follows from substituting this result into (8.132) that

$$u_{s,d}(\alpha) \sim \Phi_+^{s,d}(k)\xi(\alpha) + B\eta_b(\alpha) \pm A\eta_a(\alpha) \tag{8.140}$$

as $ka \to \infty$. Substituting (8.131) into (8.126) and solving for $\Phi_+^{s,d}(\alpha)$, we have

$$\Phi_+^{s,d}(\alpha) = K_+(\alpha)[BP_b(\alpha) \pm AP_a(\alpha) \pm u_{s,d}(\alpha)] \tag{8.141}$$

where

$$P_{b,a}(\alpha) = \frac{1}{\alpha \mp k\cos\theta_0}\left[\frac{1}{K_+(\alpha)} - \frac{1}{K_\pm(k\cos\theta_0)}\right] \tag{8.142}$$

Application of (8.140) to (8.141) yields the asymptotic expansion of $\Phi_+^{s,d}(\alpha)$ with the result

$$\Phi_+^{s,d}(\alpha) \sim K_+(\alpha)[\pm\chi_b(\alpha) + \chi_a(\alpha) \pm \Phi_+^{s,d}(k)\xi(\alpha)] \tag{8.143}$$

as $ka \to \infty$, where

$$\chi_b(\alpha) = B\eta_b(\alpha) + AP_a(\alpha) \tag{8.144a}$$

$$\chi_a(\alpha) = A\eta_a(\alpha) + BP_b(\alpha) \tag{8.144b}$$

We can derive the simple equation for $\Phi_+^{s,d}(k)$ by setting $\alpha = k$ in (8.143). Thus, solving this equation and substituting the result into (8.140) yields

$$u_{s,d}(\alpha) \sim \frac{K_+(k)[\pm\chi_b(k) + \chi_a(k)]}{1 \mp K_+(k)\xi(k)}\xi(\alpha) + B\eta_b(\alpha) \pm A\eta_a(\alpha) \tag{8.145}$$

which is the asymptotic representation of $u_{s,d}(\alpha)$ as $ka \to \infty$. Substituting (8.145) into (8.126) and taking into account (8.125), we can obtain the asymptotic solution of the Wiener-Hopf equation (8.114) with the result

$$U_{-}(\alpha) \sim K_{-}(\alpha)\left[\frac{A}{K_{-}(k\cos\theta_0)(\alpha - k\cos\theta_0)} + B\eta_b(-\alpha) + C_1\xi(-\alpha)\right] \quad (8.146a)$$

$$U_{(+)}(\alpha) \sim K_{+}(\alpha)\left[-\frac{B}{K_{+}(k\cos\theta_0)(\alpha - k\cos\theta_0)} + A\eta_a(\alpha) + C_2\xi(\alpha)\right] \quad (8.146b)$$

as $ka \to \infty$, where

$$C_{1,2} = \frac{K_{+}(k)[\chi_{a,b}(k) + K_{+}(k)\xi(k)\chi_{b,a}(k)]}{1 - [K_{+}(k)\xi(k)]^2} \quad (8.147)$$

8.4.4 Scattered Far Field

Integral representation for the scattered field is derived by taking the inverse complex Fourier transform of (8.109) with the aid of Theorem 8.4 as follows:

$$\phi(x, z) = (2\pi)^{-1/2}\int_{-\infty+ic}^{\infty+ic} [e^{-i\alpha a}U_{-}(\alpha) + e^{i\alpha a}U_{(+)}(\alpha)]\, e^{-\gamma|x| - i\alpha z}\, d\alpha \quad (8.148)$$

where c is a constant such that $|c| < k_2 \cos\theta_0$. Substituting (8.146a) and (8.146b) into (8.148) and rearranging the result, after letting $k_2 \to 0$, we obtain

$$\phi(x, z) \sim \phi_1(x, z) + \phi_2(x, z) \quad (8.149)$$

as $ka \to \infty$, where

$$\phi_1(x, z) = (2\pi)^{-1/2}\int_C \Psi_{(+)}(0, \alpha)\, e^{-\gamma|x| - i\alpha(z-a)}\, d\alpha + (2\pi)^{-1/2}\int_C \Psi_{-}(0, \alpha)\, e^{-\gamma|x| - i\alpha(z+a)}\, d\alpha \quad (8.150)$$

$$\phi_2(x, z) = (2\pi)^{-1/2}\int_{C'} \{e^{-i\alpha a}K_{-}(\alpha)[B\eta_b(-\alpha) + C_1\xi(-\alpha)] + e^{i\alpha a}K_{+}(\alpha)[A\eta_a(\alpha) + C_2\xi(\alpha)]\}\, e^{-\gamma|x| - i\alpha z}\, d\alpha \quad (8.151)$$

$$\Psi_{(\pm)}(0, \alpha) = \pm\frac{\exp(\mp ika\cos\theta_0)i(k\cos\theta_0 \pm k)^{1/2}}{(2\pi)^{1/2}(\alpha \pm k)^{1/2}(\alpha - k\cos\theta_0)} \quad (8.152)$$

The integration paths C and C' in (8.150) and (8.151) are those ones composed of semicircles centered at the singularities of the integrands and a portion of the real axis, as shown in Figures 8.9(a) and 8.9(b), respectively.

Let us introduce the polar coordinates (ρ, θ), (ρ_1, θ_1), and (ρ_2, θ_2) as follows:

$$x = \rho \sin\theta, \quad z = \rho \cos\theta \quad \text{for} \quad -\pi < \theta < \pi \tag{8.153a}$$

$$x = \rho_1 \sin\theta_1, \quad z - a = \rho_1 \cos\theta_1 \quad \text{for} \quad -\pi < \theta_1 < \pi \tag{8.153b}$$

$$x = \rho_2 \sin\theta_2, \quad z + a = \rho_2 \cos\theta_2 \quad \text{for} \quad -\pi < \theta_2 < \pi \tag{8.153c}$$

Then $\phi_1(x, z)$ defined by (8.150) can be evaluated following a procedure similar to that employed in Section 8.3.4 with the result that

$$\begin{aligned}\phi_1(\rho, \theta; \rho_1, \theta_1; \rho_2, \theta_2) = &- \exp[-ik\rho \cos(\theta - \theta_0)]F\{(2k\rho_1)^{1/2} \\ &\cdot \cos[(\theta_1 - \theta_0)/2]\} - \exp[-ik\rho \\ &\cdot \cos(\theta + \theta_0)]F\{(2k\rho_1)^{1/2} \cos[(\theta_1 + \theta_0)/2]\} \\ &\mp (\exp[-ik\rho \cos(\theta - \theta_0)]F\{(2k\rho_2)^{1/2} \\ &\cdot \cos[(\theta_2 - \theta_0)/2]\} - \exp[-ik\rho \\ &\cdot \cos(\theta + \theta_0)]F\{(2k\rho_2)^{1/2} \cos[(\theta_2 + \theta_0)/2]\})\end{aligned} \tag{8.154}$$

for $\theta, \theta_1, \theta_2 \gtrless 0$, where $F(\cdot)$ is the Fresnel integral defined by (8.101). For $\phi_2(x, z)$ defined by (8.151), evaluation of the integral in a closed form is difficult, because the integrand involves the function $\xi(\alpha)$. However, if we observe that the

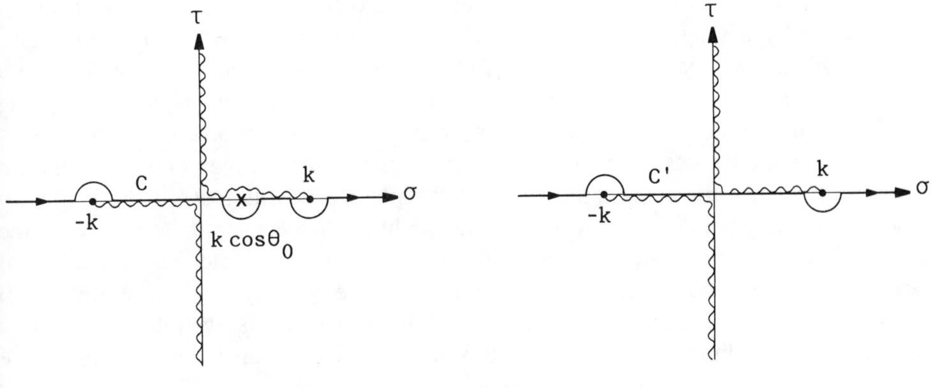

Figure 8.9 Integration paths C and C' in equations (8.150) and (8.151).

integrand has no singularities other than the branch points at $\alpha = \pm k$, then the simple saddle-point method can be used for derivation of the scattered far field. Using the polar coordinate (ρ, θ) as defined by (8.153a) and applying a corollary of Theorem 8.13, we arrive at

$$\phi_2(\rho, \theta) \sim [f(\theta, \theta_0)K_-(k\cos\theta) + f(\pi - \theta, \pi - \theta_0)K_+(k\cos\theta)]\, k\sin|\theta| \frac{e^{i(k\rho - \pi/4)}}{(k\rho)^{1/2}} \tag{8.155}$$

as $k\rho \to \infty$, where

$$f(\theta, \theta_0) = e^{-ika\cos\theta}[C_2\xi(-k\cos\theta) + A\eta_a(-k\cos\theta)] \tag{8.156}$$

Substitution of (8.154) and (8.155) into (8.149) yields the scattered far field expression as $ka \to \infty$. Further discussions regarding the high-frequency asymptotic expansion of the scattered far field can be found in Kobayashi [60]. The preceding analysis has been restricted to the case for $0 < \theta_0 < \pi/2$, but this restriction to the angle of incidence does not lose the generality of the result.

The diffraction by a slit [62], a finite parallel plate waveguide [20], a finite circular waveguide [24], and a parallel plate grating [39, 52] also belongs to a class of the problems that can be treated by the method developed in this section.

8.5 DIFFRACTION BY A THICK SEMIINFINITE PLATE

In this section, as another example of the problems that can be treated approximately using the Wiener-Hopf technique, we analyze the diffraction of a plane electromagnetic wave by a thick semiinfinite plate and develop a solution method in detail. This problem was analyzed by Jones [21] for the first time, but here we develop a method somewhat different from his. Because the semiinfinite plate has a finite thickness, this problem is considered a generalization to the problem treated in Section 8.3. Unlike the case of no thickness, the axis for the complex Fourier transform chosen arbitrarily traverses the surface of the plate. For this reason, the geometry considered here belongs to a class of the modified Wiener-Hopf geometry of the second kind, and the exact solution to this problem is not available. However, formal analysis with the aid of the Wiener-Hopf technique is possible, and therefore an approximate solution can be obtained. In the following analysis, we introduce a slight loss into the medium, as in Sections 8.3 and 8.4, and assume that the free-space wavenumber k has a small positive imaginary part, as shown in (8.59).

8.5.1 Transformed-Wave Equations

We consider the diffraction of an E-polarized plane wave by a perfectly conducting thick semiinfinite plate, as shown in Figure 8.10. As in Section 8.3, let us define the total electric field $\phi^t(x, z)$ and the incident field $\phi^i(x, z)$ by (8.52) and (8.53), respectively. Then the scattered field $\phi(x, z)$ satisfies the reduced-wave equation (8.54) and the nonzero components of electromagnetic fields can be derived using (8.55).

Dividing a space into appropriate regions and investigating the scattered far field, we see that $\phi(x, z)$ has the same asymptotic behavior as that given by (8.60). Therefore, defining the complex Fourier transform and the complex Fourier integrals by (8.61) and (8.62), respectively, we find that $\Phi_+(x, \alpha)$, $\Phi_-(x, \alpha)$, and $\Phi(x, \alpha)$ are regular in $\tau > -k_2$, $\tau < k_2 \cos\theta_0$, and $-k_2 < \tau < k_2 \cos\theta_0$, respectively. The term $\Phi(x, \alpha)$ can now be represented as

$$\Phi(x, \alpha) = \Phi_-(x, \alpha) + \Phi_+(x, \alpha) \tag{8.157}$$

Because the plate is perfectly conducting, (8.157) is further rewritten as

$$\Phi(x, \alpha) = \Psi_{(+)}(x, \alpha) \tag{8.158}$$

for $|x| \leqslant b$, where

$$\Psi_{(+)}(x, \alpha) = \Phi_+(x, \alpha) - A \frac{\exp(-ikx \sin\theta_0)}{\alpha - k \cos\theta_0} \tag{8.159}$$

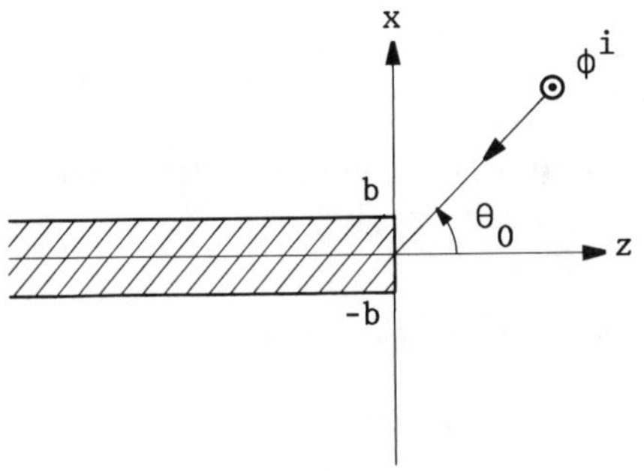

Figure 8.10 Thick semiinfinite plate.

$$A = \frac{1}{(2\pi)^{1/2} i} \tag{8.160}$$

The transformed wave equation satisfied by $\Phi(x, \alpha)$ for the region $|x| > b$ can be derived by taking the complex Fourier transform of the reduced wave equation (8.54), and it is of the same form as that given by (8.64); that is,

$$(d^2/dx^2 - \gamma^2)\Phi(x, \alpha) = 0, \quad -k_2 < \tau < k_2 \cos\theta_0 \tag{8.161}$$

On the other hand, for the region $|x| < b$, the derivation of the transformed wave equation is complicated, as there is a medium discontinuity across the surface $z = 0$. We multiply both sides of (8.54) by $(2\pi)^{-1/2} \exp(i\alpha z)$ and integrate with respect to z over the range $0 < z < \infty$. This leads to

$$(d^2/dx^2 - \gamma^2)\Phi_+(x, \alpha) = (2\pi)^{-1/2}\left[\frac{\partial\phi(x, +0)}{\partial z} - i\alpha\phi(x, +0)\right] \tag{8.162}$$

for any α in $\tau > -k_2$ taking into account (8.60). Simple manipulation shows that

$$(d^2/dx^2 - \gamma^2)A\frac{\exp(-ikx\sin\theta_0)}{\alpha - k\cos\theta_0} = -(2\pi)^{-1/2}\left[\frac{\partial\phi^i(x, 0)}{\partial z} - i\alpha\phi^i(x, 0)\right] \tag{8.163}$$

for all α except $\alpha = k\cos\theta_0$. Therefore, subtracting (8.163) from (8.162) using (8.158) and (8.159) and taking into account the boundary condition $\phi^t(x, 0) = 0$ for $|x| < b$, we obtain

$$(d^2/dx^2 - \gamma^2)\Phi(x, \alpha) = f(x) \tag{8.164}$$

for $\tau > -k_2$ with $\alpha \neq k\cos\theta_0$, where

$$f(x) = (2\pi)^{-1/2}\frac{\partial\phi^t(x, +0)}{\partial z} \tag{8.165}$$

Equations (8.161) and (8.164) are the desired transformed-wave equations for the regions $|x| > b$ and $|x| < b$, respectively.

8.5.2 Simultaneous Wiener-Hopf Equations

The solution of (8.161) can be written as

$$\begin{aligned}\Phi(x, \alpha) &= \Psi_{(+)}(b, \alpha)\, e^{-\gamma(x-b)} \quad \text{for} \quad x \geqslant b \\ &= \Psi_{(+)}(-b, \alpha)\, e^{\gamma(x+b)} \quad \text{for} \quad x \leqslant -b\end{aligned} \tag{8.166}$$

by taking into account (8.158) and the continuity of tangential electric fields across the surface $x = \pm b$. Because (8.164) is the ordinary differential equation with the inhomogeneous term $f(x)$, we can express the solution as

$$\Phi(x, \alpha) = \Psi_{(+)}(b, \alpha) \frac{\sinh \gamma(x + b)}{\sinh 2\gamma b} - \Psi_{(+)}(-b, \alpha) \frac{\sinh \gamma(x - b)}{\sinh 2\gamma b} + F(x, \alpha) - F(b, \alpha) \frac{\sinh \gamma(x + b)}{\sinh 2\gamma b} \tag{8.167}$$

where $F(x, \alpha)$ is the particular solution given by

$$F(x, \alpha) = \frac{1}{\gamma} \int_{-b}^{x} f(t) \sinh \gamma(x - t) dt \tag{8.168}$$

We now expand the inhomogeneous term $f(x)$ given by (8.165) in terms of the Fourier sine series, as in

$$f(x) \sim \frac{1}{b} \sum_{n=1}^{\infty} f_n \sin \frac{n\pi}{2b} (x + b) \tag{8.169}$$

Substituting (8.169) into (8.168) and integrating term by term, we obtain

$$F(x, \alpha) = -\frac{1}{b} \sum_{n=1}^{\infty} \frac{f_n}{\alpha^2 + \gamma_n^2} \sin \frac{n\pi}{2b} (x + b) + \frac{\sinh \gamma(x + b)}{\gamma b} \sum_{n=1}^{\infty} \frac{n\pi}{2b} \frac{f_n}{\alpha^2 + \gamma_n^2} \tag{8.170}$$

where

$$\gamma_n = [(n\pi/2b)^2 - k^2]^{1/2} \tag{8.171}$$

Because we find with the aid of the edge condition that $f(x)$ is absolutely integrable over $-b \leqslant x \leqslant b$, convergence of the infinite series contained in (8.170) is guaranteed from the theory of Fourier series. The series can be differentiated term by term with respect to x and the resultant series will also converge for any x in $-b \leqslant x \leqslant b$. We can also prove that γ_n as given by (8.171) satisfies

$$\operatorname{Re} \gamma_n \geqslant k_2, \quad -k_1 \leqslant \operatorname{Im} \gamma_n \leqslant 0 \tag{8.172}$$

for all n. Substituting (8.170) into (8.167) yields

$$\Phi(x, \alpha) = \Psi_{(+)}(b, \alpha)\frac{\sinh \gamma(x+b)}{\sinh 2\gamma b} - \Psi_{(+)}(-b, \alpha)\frac{\sinh \gamma(x-b)}{\sinh 2\gamma b}$$
$$-\frac{1}{b}\sum_{n=1}^{\infty}\frac{f_n}{\alpha^2+\gamma_n^2}\sin\frac{n\pi}{2b}(x+b) \tag{8.173}$$

Let us introduce the functions

$$J_-^{s,d}(\alpha) = J_-(b, \alpha) \pm J_-(-b, \alpha) \tag{8.174}$$

where

$$J_-(\pm b, \alpha) = \Phi'(\pm b \pm 0, \alpha) - \Phi'(\pm b \mp 0, \alpha) \tag{8.175}$$

Taking into account the continuity of tangential magnetic fields across the surface $x = \pm b$ for $z > 0$, we can show that $J_-(\pm b, \alpha)$ defined by (8.175) is expressed by $\Phi'_-(\pm b \pm 0, \alpha)$ and, hence, is regular in $\tau < k_2 \cos\theta_0$. Using (8.166) and (8.173) to derive the explicit expressions of $J_-(\pm b, \alpha)$ and substituting the results into (8.174), after some arrangements, we obtain

$$J_-^d(\alpha) = -\frac{U_{(+)}(\alpha)}{L(\alpha)} - \sum_{n=1,\mathrm{odd}}^{\infty}\frac{n\pi}{b^2}\frac{f_n}{\alpha^2+\gamma_n^2} \tag{8.176a}$$

and

$$J_-^s(\alpha) = -\frac{V_{(+)}(\alpha)}{N(\alpha)} + \sum_{n=2,\mathrm{even}}^{\infty}\frac{n\pi}{b^2}\frac{f_n}{\alpha^2+\gamma_n^2} \tag{8.176b}$$

for $-k_2 < \tau < k_2 \cos\theta_0$, where

$$U_{(+)}(\alpha) = \Psi_{(+)}(b, \alpha) + \Psi_{(+)}(-b, \alpha) \tag{8.177a}$$

$$V_{(+)}(\alpha) = \Psi_{(+)}(b, \alpha) - \Psi_{(+)}(-b, \alpha) \tag{8.177b}$$

$$L(\alpha) = \frac{1+e^{-2\gamma b}}{2\gamma}, \quad N(\alpha) = \frac{1-e^{-2\gamma b}}{2\gamma} \tag{8.178}$$

Equations (8.176a) and (8.176b) are the desired simultaneous Wiener-Hopf equations satisfied by the unknown functions. $L(\alpha)$ and $N(\alpha)$ as defined by (8.178) are kernel functions.

8.5.3 Factorization of Kernel Functions

We express the kernel functions $L(\alpha)$ and $N(\alpha)$ defined by (8.178) as follows:

$$L(\alpha) = K(\alpha)G_u(\alpha), \quad N(\alpha) = K(\alpha)G_v(\alpha) \tag{8.179}$$

where

$$K(\alpha) = (2\gamma)^{-1}, \quad G_{u,v}(\alpha) = 1 \pm \mathrm{e}^{-2\gamma b} \tag{8.180}$$

In (8.180), $K(\alpha)$ is the same function as that introduced by (8.76) in connection with the diffraction by a semiinfinite plate, and the factorization has been carried out in (8.77) and (8.78). Because Re $\gamma > 0$ for any α in the strip $|\tau| < k_2$ as has been shown in Section 8.2.4, we have $\gamma \sim |\sigma|$ as $|\sigma| \to \infty$ in this strip. Therefore, it follows that

$$G_{u,v}(\alpha) = 1 + 0\,(\mathrm{e}^{-2|\sigma|b}) \tag{8.181}$$

uniformly in τ as $|\sigma| \to \infty$ with $|\tau| < k_2$. Because $G_{u,v}(\alpha)$ are regular and nonzero in $|\tau| < k_2$, simple application of Theorem 8.10 yields the factorization of $G_{u,v}(\alpha)$ with the result that

$$G_{u,v}(\alpha) = G_+^{u,v}(\alpha)G_-^{u,v}(\alpha) \tag{8.182}$$

where for $0 < |\tau| < c < k_2$,

$$G_\pm^{u,v}(\alpha) = \exp\left[\frac{1}{2\pi i}\int_{-\infty\mp ic}^{\infty\mp ic} \frac{\ln G_{u,v}(\beta)}{\beta - \alpha}\,d\beta\right] \tag{8.183}$$

Here, $G_\pm^{u,v}(\alpha)$ is regular, nonzero, and bounded in $\tau \gtrless \mp k_2$.

From preceding discussions, we can factorize $L(\alpha)$ and $N(\alpha)$ as follows:

$$L(\alpha) = L_+(\alpha)L_-(\alpha), \quad N(\alpha) = N_+(\alpha)N_-(\alpha) \tag{8.184}$$

where

$$L_\pm(\alpha) = K_\pm(\alpha)G_\pm^u(\alpha), \quad N_\pm(\alpha) = K_\pm(\alpha)G_\pm^v(\alpha) \tag{8.185}$$

In (8.185), $K_\pm(\alpha)$ is defined by (8.78). It follows from (8.183) that

$$G_\pm^{u,v}(\alpha) \sim 1 \tag{8.186}$$

as $\alpha \to \infty$ in $\tau \gtrless \mp k_2$. Therefore, the split functions $L_\pm(\alpha)$ and $N_\pm(\alpha)$ behave like

$$L_\pm(\alpha), N_\pm(\alpha) \sim (\mp 2i\alpha)^{-1/2} \tag{8.187}$$

as $\alpha \to \infty$ in $\tau \gtrless \mp k_2$. Further details of the factorization of $L(\alpha)$ and $N(\alpha)$ can be found in Noble [18], Mittra and Lee [37], and Kobayashi [63].

8.5.4 Formal Solutions

Taking into account (8.172), we find that $\pm i\gamma_n$ for all n lie in $\tau \gtrless \pm k_2$. Bearing this in mind, we multiply both sides of (8.176a) by $L_-(\alpha)$ and decompose the resultant equation in the usual manner. This yields

$$J_-^d(\alpha)L_-(\alpha) - \frac{2A\cos(kb\sin\theta_0)}{L_+(k\cos\theta_0)(\alpha - k\cos\theta_0)}$$

$$+ \sum_{n=1,\text{odd}}^{\infty} \frac{n\pi}{b^2}\frac{f_n}{\alpha + i\gamma_n}\left[\frac{L_-(\alpha)}{\alpha - i\gamma_n} + \frac{L_+(i\gamma_n)}{2i\gamma_n}\right]$$

$$= -\frac{U_{(+)}(\alpha)}{L_+(\alpha)} - \frac{2A\cos(kb\sin\theta_0)}{L_+(k\cos\theta_0)(\alpha - k\cos\theta_0)} + \sum_{n=1,\text{odd}}^{\infty} \frac{n\pi}{2b}\frac{f_nL_+(i\gamma_n)}{bi\gamma_n(\alpha + i\gamma_n)} \tag{8.188}$$

The right-hand and left-hand sides of (8.188) are regular in $\tau > -k_2$ and $\tau < k_2 \cos\theta_0$, respectively; and they have a common strip of regularity $-k_2 < \tau < k_2\cos\theta_0$. Therefore, the argument of analytic continuation shows that both sides of (8.188) are equal to a certain entire function $P(\alpha)$.

Setting $\varphi = \pi/2$ in (8.51), we see that the edge condition to this problem is given by

$$\phi(\pm b, z) = -\exp(\mp ikb\sin\theta_0) + 0(z^{2/3}), \quad z \to +0 \tag{8.189a}$$

$$\frac{\partial\phi(\pm b \pm 0, z)}{\partial x} = 0(z^{-1/3}), \quad z \to -0 \tag{8.189b}$$

Therefore, application of Theorems 8.5 and 8.7 with the aid of (8.159), (8.174), (8.175), and (8.177a) yields

$$U_{(+)}(\alpha) = 0(\alpha^{-5/3}), \quad \tau > -k_2 \tag{8.190a}$$

$$J_-^d(\alpha) = 0(\alpha^{-1/3}), \quad \tau < k_2\cos\theta_0 \tag{8.190b}$$

as $\alpha \to \infty$. Thus, we can show by taking into account (8.187), (8.190a), and (8.190b and Liouville's theorem that the entire function $P(\alpha)$ derived earlier must be identically zero. Equating the right-hand side of (8.188) to zero, yields

$$\frac{U_{(+)}(\alpha)}{L_+(\alpha)} + \frac{2A\cos(kb\sin\theta_0)}{L_+(k\cos\theta_0)(\alpha - k\cos\theta_0)} - \sum_{n=1,\text{odd}}^{\infty} \frac{n\pi}{2b} \frac{f_n L_+(i\gamma_n)}{bi\gamma_n(\alpha + i\gamma_n)} = 0 \quad (8.191)$$

A similar procedure can be applied to decompose another Wiener-Hopf equation (8.176b), and we arrive at

$$\frac{V_{(+)}(\alpha)}{N_+(\alpha)} - \frac{2iA\sin(kb\sin\theta_0)}{N_+(k\cos\theta_0)(\alpha - k\cos\theta_0)} + \sum_{n=2,\text{even}}^{\infty} \frac{n\pi}{2b} \frac{f_n N_+(i\gamma_n)}{bi\gamma_n(\alpha + i\gamma_n)} = 0 \quad (8.192)$$

The unknown Fourier coefficients f_n for $n = 1, 2, 3, \ldots$ are involved in (8.191) and (8.192). Therefore, we must investigate the relation between these Fourier coefficients and the unknown functions $U_{(+)}(\alpha)$ and $V_{(+)}(\alpha)$ to derive the explicit representations of the solution.

We can see from (8.158) and (8.159) that $\Phi(x, \alpha)$ for the region $|x| \leqslant b$ is regular in $\tau > -k_2$ except for a simple pole at $\alpha = k\cos\theta_0$. Thus, we have

$$\lim_{\alpha \to i\gamma_n} (\alpha - i\gamma_n)\Phi(x, \alpha) = 0, \quad n = 1, 2, 3, \ldots \quad (8.193)$$

Substituting (8.173) into (8.193) and carrying out some manipulations, it follows that

$$\begin{aligned} f_n &= (n\pi/2b)U_{(+)}(i\gamma_n), \quad n = 1, 3, 5, \ldots \\ &= -(n\pi/2b)V_{(+)}(i\gamma_n), \quad n = 2, 4, 6, \ldots \end{aligned} \quad (8.194)$$

Therefore, substituting (8.194) into (8.191) and (8.192) and solving for $U_{(+)}(\alpha)$ and $V_{(+)}(\alpha)$, we are led to

$$\begin{aligned} U_{(+)}(\alpha) = -L_+(\alpha)\Bigg[&\frac{2A\cos(kb\sin\theta_0)}{L_+(k\cos\theta_0)(\alpha - k\cos\theta_0)} \\ &- \sum_{n=1,\text{odd}}^{\infty} \left(\frac{n\pi}{2b}\right)^2 \frac{L_+(i\gamma_n)U_{(+)}(i\gamma_n)}{bi\gamma_n(\alpha + i\gamma_n)}\Bigg] \end{aligned} \quad (8.195a)$$

$$\begin{aligned} V_{(+)}(\alpha) = N_+(\alpha)\Bigg[&\frac{2iA\sin(kb\sin\theta_0)}{N_+(k\cos\theta_0)(\alpha - k\cos\theta_0)} \\ &+ \sum_{n=2,\text{even}}^{\infty} \left(\frac{n\pi}{2b}\right)^2 \frac{N_+(i\gamma_n)V_{(+)}(i\gamma_n)}{bi\gamma_n(\alpha + i\gamma_n)}\Bigg] \end{aligned} \quad (8.195b)$$

Equations (8.195a) and (8.195b) are the formal solutions to the simultaneous Wiener-Hopf equations (8.176a) and (8.176b); and an infinite number of the unknowns $U_{(+)}(i\gamma_n)$ for $n = 1, 3, 5, \ldots$ and $V_{(+)}(i\gamma_n)$ for $n = 2, 4, 6, \ldots$ are involved there. These unknowns are shown to satisfy the infinite set of equations obtained by setting $\alpha = i\gamma_n$ for $n = 1, 3, 5, \ldots$ and $\alpha = i\gamma_n$ for $n = 2, 4, 6, \ldots$ in (8.195a) and (8.195b), respectively.

8.5.5 Application of the Modified Residue-Calculus Technique

In this subsection, we develop a method of finding the highly convergent approximate solutions based on the modified residue-calculus technique. The approach used here is essentially similar to that employed by Mittra and Lee [38]. We see from (8.178), (8.184), (8.190a), and (8.195a) that $U_{(+)}(\alpha)$ satisfies the following four conditions:

(U1) $U_{(+)}(-i\gamma_n) - U_{(+)}(i\gamma_n) = 0, \quad \text{for} \quad n = 1, 3, 5, \ldots$

(U2) $U_{(+)}(\alpha)$ has a branch point at $\alpha = -k$

(U3) $U_{(+)}(\alpha)$ has a simple pole at $\alpha = k\cos\theta_0$ and satisfies Res $U_{(+)}(k\cos\theta_0)$ $= -2A\cos(kb\sin\theta_0)$, where Res $U_{(+)}(k\cos\theta_0)$ implies the residue of $U_{(+)}(\alpha)$ at $\alpha = k\cos\theta_0$

(U4) $U_{(+)}(\alpha) = 0(\alpha^{-5/3})$ as $\alpha \to \infty$ in $\tau > -k_2$

For convenience, we introduce the function

$$\tilde{U}_{(+)}(\alpha) = -\frac{2A\cos(kb\sin\theta_0)}{L_+(k\cos\theta_0)} \frac{L_+(\alpha)}{\alpha - k\cos\theta_0} \tag{8.196}$$

which can be obtained by setting $U_{(+)}(i\gamma_n) \equiv 0$ for all n in (8.195a). We can show with the aid of (8.178), (8.184), and (8.187) that $\tilde{U}_{(+)}(\alpha)$ satisfies the following conditions:

($\tilde{\text{U}}$1) $\tilde{U}_{(+)}(\alpha)$ has an infinite number of simple zeros at $\alpha = -i\gamma_n$ for $n = 1, 3, 5, \ldots$

($\tilde{\text{U}}$2) $\tilde{U}_{(+)}(\alpha)$ has a branch point at $\alpha = -k$

($\tilde{\text{U}}$3) $\tilde{U}_{(+)}(\alpha)$ has a simple pole at $\alpha = k\cos\theta_0$ and satisfies Res $\tilde{U}_{(+)}(k\cos\theta_0)$ $= -2A\cos(kb\sin\theta_0)$

($\tilde{\text{U}}$4) $\tilde{U}_{(+)}(\alpha) = 0(\alpha^{-3/2})$ as $\alpha \to \infty$ in $\tau > -k_2$

Basic differences between the analytical properties of $U_{(+)}(\alpha)$ and $\tilde{U}_{(+)}(\alpha)$ are the zeros and the asymptotic behavior for large $|\alpha|$, we can find by comparing the conditions (U1) and (U4) with ($\tilde{\text{U}}$1) and ($\tilde{\text{U}}$4), respectively. We now construct the function $U_{(+)}(\alpha)$ in the form

$$U_{(+)}(\alpha) = \tilde{U}_{(+)}(\alpha)p(\alpha)q_u(\alpha) \tag{8.197}$$

where $p(\alpha)$ is an arbitrary entire function with no zeros. In (8.197), $q_u(\alpha)$ was introduced to cancel the zeros of $\tilde{U}_{(+)}(\alpha)$ at $\alpha = i\gamma_{2n-1}$ for $n = 1, 2, 3, \ldots$ and is defined by

$$q_u(\alpha) = \prod_{n=1}^{\infty} \frac{1 + \alpha/i\Gamma^u_{2n-1}}{1 + \alpha/i\gamma_{2n-1}} \tag{8.198}$$

where

$$\Gamma^u_{2n-1} = \gamma_{2n-1} + \Delta^u_n, \quad n = 1, 2, 3, \ldots \tag{8.199}$$

Δ^u_n for $n = 1, 2, 3, \ldots$ appearing in (8.199) is the shift of zeros from $i\gamma_{2n-1}$ and is unknown at this stage, which is to be determined from the condition (U1). These considerations show that our remaining task is to develop a method of finding the entire function $p(\alpha)$ in (8.197) and the shifts Δ^u_n for $n = 1, 2, 3, \ldots$ so that $U_{(+)}(\alpha)$ satisfies the asymptotic behavior for large $|\alpha|$ as given by the condition (U4).

Let us assume that Δ^u_n has the following limit:

$$\lim_{n\to\infty} \Delta^u_n = \pi\Delta_u/b \tag{8.200}$$

Then, by applying Theorem 8.12, $q_u(\alpha)$ defined by (8.198) can be expanded asymptotically with the result

$$\begin{aligned} q_u(\alpha) &\sim K_q(-i\alpha)^{-\Delta_u} \quad \text{for} \quad -\pi/2 < \arg\alpha < 3\pi/2 \\ &\sim K_q \frac{\cos(i\alpha b - \pi\Delta_u)}{\cos(i\alpha b)} (i\alpha)^{-\Delta_u} \quad \text{for} \quad \arg\alpha = -\pi/2 \end{aligned} \tag{8.201}$$

as $\alpha \to \infty$, where

$$K_q = \pi^{-1/2}(\pi/b)^{\Delta_u}\Gamma(1/2 + \Delta_u) \prod_{n=1}^{\infty} \frac{(2n - 1 + 2\Delta_u)\gamma_{2n-1}}{(2n - 1)\Gamma^u_{2n-1}} \tag{8.202}$$

Applying ($\tilde{\text{U}}$4) and (8.201) to (8.197), we derive

$$U_{(+)}(\alpha) = 0[p(\alpha)\alpha^{-3/2-\Delta_u}] \tag{8.203}$$

as $\alpha \to \infty$ in $\tau > -k_2$. Because $U_{(+)}(\alpha)$ must satisfy the condition (U4), it follows from (8.203) that

$$p(\alpha) = 0(\alpha^{-1/6+\Delta_u}), \quad \alpha \to \infty \tag{8.204}$$

Noting that $p(\alpha)$ is an entire function with no zeros, Liouville's theorem and its extension show that

$$p(\alpha) \equiv p_0 \text{ (constant)} \tag{8.205}$$

Therefore, we can determine the value of Δ_u simultaneously from the condition (U4) as in

$$\Delta_u = 1/6 \tag{8.206}$$

Substituting (8.205) into (8.197) and taking into account the conditions (U3) and ($\tilde{\text{U}}$3), we can derive the explicit representation of $U_{(+)}(\alpha)$ with the result

$$U_{(+)}(\alpha) = \tilde{U}_{(+)}(\alpha)\,\frac{q_u(\alpha)}{q_u(k\cos\theta_0)} \tag{8.207}$$

The shifted zeros $-i\Gamma^u_{2n-1}$ for $n = 1, 2, 3, \ldots$ of $U_{(+)}(\alpha)$ can be determined by solving the following infinite set of equations:

$$\lim_{\alpha \to -i\gamma_n} \tilde{U}_{(+)}(\alpha) q_u(\alpha) - \tilde{U}_{(+)}(i\gamma_n) q_u(i\gamma_n) = 0, \quad n = 1, 3, 5, \ldots \tag{8.208}$$

The function $U_{(+)}(\alpha)$ as derived by (8.207) satisfies the conditions (U1) to (U4).

By following a procedure similar to that just employed, we can construct $V_{(+)}(\alpha)$ in the form convenient for numerical computations as follows:

$$V_{(+)}(\alpha) = \tilde{V}_{(+)}(\alpha)\,\frac{q_v(\alpha)}{q_v(k\cos\theta_0)} \tag{8.209}$$

where

$$\tilde{V}_{(+)}(\alpha) = \frac{2iA\sin(kb\sin\theta_0)}{N_+(k\cos\theta_0)}\,\frac{N_+(\alpha)}{\alpha - k\cos\theta_0} \tag{8.210}$$

$$q_v(\alpha) = \prod_{n=1}^{\infty} \frac{1 + \alpha/i\Gamma^v_{2n}}{1 + \alpha/i\gamma_{2n}} \tag{8.211}$$

$$\Gamma^v_{2n} = \gamma_{2n} + \Delta^v_n, \quad n = 1, 2, 3, \ldots \tag{8.212}$$

In (8.211) and (8.212), $-i\Gamma_{2n}^{v}$ for $n = 1, 2, 3, \ldots$ are the zeros of $V_{(+)}(\alpha)$, and Δ_n^v are the shifted amounts of zeros from $-i\gamma_{2n}$. Investigating the asymptotic behavior of (8.211) for large $|\alpha|$ with the aid of the edge condition, we can show that Δ_n^v has the following limit:

$$\lim_{n \to \infty} \Delta_n^v = \pi/6b \tag{8.213}$$

We also find that the shifted zeros $-i\Gamma_{2n}^{v}$ for $n = 1, 2, 3, \ldots$ are determined by solving the following infinite set of equations:

$$\lim_{\alpha \to -i\gamma_n} \tilde{V}_{(+)}(\alpha)q_v(\alpha) - \tilde{V}_{(+)}(i\gamma_n)q_v(i\gamma_n) = 0, \quad n = 2, 4, 6, \ldots \tag{8.214}$$

We have constructed the solutions to the simultaneous Wiener-Hopf equations (8.176a) and (8.176b) in the form convenient for numerical computations. We can see that $\tilde{U}_{(+)}(\alpha)$ and $\tilde{V}_{(+)}(\alpha)$ defined by (8.196) and (8.209) are the exact solution to the diffraction by a semiinfinite parallel plate waveguide [8, 11, 13], as shown in Figure 8.11. Therefore, $q_u(\alpha)$ and $q_v(\alpha)$ as defined by (8.198) and (8.211) are interpreted as a correction term due to filling a perfect conductor in the waveguide region. As we found in the discussion of this subsection, the modified residue-calculus technique is the one for constructing the desired solution approximately by introducing the idea of the shift of zeros appearing in the exact solution to the canonical problem.

8.5.6 Determination of Zeros

As shown in the previous subsection, the zeros of $U_{(+)}(\alpha)$ and $V_{(+)}(\alpha)$ can be determined by solving the two infinite sets of equations (8.208) and (8.214) numerically.

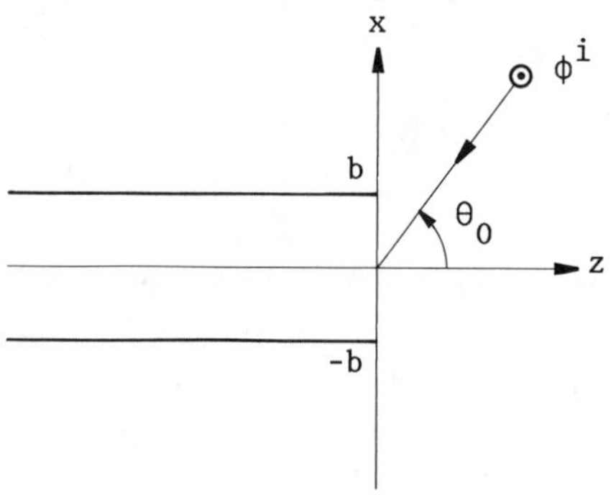

igure 8.11 Semiinfinite parallel plate waveguide.

However, direct numerical computations based on these infinite equations are somewhat difficult, because an infinite number of unknowns Δ_n^u and Δ_n^v for $n = 1, 2, 3, \ldots$ are involved in the infinite products, as can be found from (8.198) and (8.211). In this subsection, we develop a method of finding the zeros systematically based on the iterative scheme.

Substituting (8.196) and (8.198) into (8.208) and rearranging the result with the aid of (8.199), we derive

$$\Delta_n^u = \frac{\pi^2}{8b^3} \frac{(2n-1)^2(2\gamma_{2n-1} + \Delta_n^u)(\gamma_{2n-1} - ik\cos\theta_0)}{\gamma_{2n-1}^2(\gamma_{2n-1} + ik\cos\theta_0)} [L_+(i\gamma_{2n-1})]^2$$
$$\times \prod_{m=1}^{\infty}{}^{(n)} \frac{(1 - \gamma_{2n-1}/\gamma_{2m-1})[1 + \gamma_{2n-1}/(\gamma_{2m-1} + \Delta_m^u)]}{(1 + \gamma_{2n-1}/\gamma_{2m-1})[1 - \gamma_{2n-1}/(\gamma_{2m-1} + \Delta_m^u)]} \tag{8.215}$$

for $n = 1, 2, 3, \ldots$, where the symbol (n) attached to the product sign implies that the multiplication for $m = n$ is to be omitted. Because the shifted amount Δ_n^u of zeros satisfies (8.200) and (8.206), we may take a large integer N and set approximately $\Delta_m^u \approx \pi/6b$ for all $m > N$ in the infinite product involved in (8.215). Under this approximation, (8.215) can be written

$$\Delta_n^u \approx \frac{2\gamma_{2n-1}u_n P_n[\Delta_m^u]}{1 - u_n P_n[\Delta_m^u]}, \quad n = 1, 2, 3, \ldots, N \tag{8.216}$$

after some manipulation, where

$$u_n = \frac{\pi^2}{8b^3} \frac{(2n-1)^2(\gamma_{2n-1} - ik\cos\theta_0)}{\gamma_{2n-1}^2(\gamma_{2n-1} + ik\cos\theta_0)} [L_+(i\gamma_{2n-1})]^2$$
$$\times \prod_{m=1}^{\infty}{}^{(n)} \frac{1 - \gamma_{2n-1}/\gamma_{2m-1}}{1 + \gamma_{2n-1}/\gamma_{2m-1}} \prod_{m=N+1}^{\infty} \frac{1 + \gamma_{2n-1}/(\gamma_{2m-1} + \pi/6b)}{1 - \gamma_{2n-1}/(\gamma_{2m-1} + \pi/6b)} \tag{8.217}$$

$$P_n[\Delta_m^u] = \prod_{m=1}^{N}{}^{(n)} \frac{1 + \gamma_{2n-1}/(\gamma_{2m-1} + \Delta_m^u)}{1 - \gamma_{2n-1}/(\gamma_{2m-1} + \Delta_m^u)} \tag{8.218}$$

Basic procedures according to the iterative scheme are as follows:

1. Replace Δ_m^u for $m = 1, 2, 3, \ldots, N$ appearing in the product $P_n[\Delta_m^u]$ on the right-hand side of (8.216) by its limit $\pi/6b$ for $m \to \infty$. Set Δ_n^u for $n = 1, 2, 3, \ldots, N$ then derived as the first-order iterative solution and write $\Delta_{nu}^{(1)}$.
2. Repeat the procedure just described. In other words, set Δ_m^u for $m = 1, 2, 3, \ldots, N$ appearing on the right-hand side of (8.216) equal to the $(k-1)$th-order iterative solution $\Delta_{nu}^{(k-1)}$ for $k = 2, 3, 4, \ldots$ and write the value of the left-hand side then derived as the kth-order iterative solution $\Delta_{nu}^{(k)}$.

3. Terminate the iteration when the sufficient convergence is recognized.

For convenience, we set $\Delta_n^u = \pi/6b$ for $n = 1, 2, 3, \ldots, N$ and call this value the zeroth-order iterative solution, which we write as $\Delta_{nu}^{(0)}$ for $n = 1, 2, 3, \ldots, N$. Then the formula for determining the successive iterative solution $\Delta_{nu}^{(k)}$ for $n = 1, 2, 3, \ldots, N$ is given as follows:

$$\begin{aligned} \Delta_{nu}^{(k)} &= \pi/6b, \quad k = 0, \\ &= \frac{2\gamma_{2n-1} u_n P_n[\Delta_{mu}^{(k-1)}]}{1 - u_n P_n[\Delta_{mu}^{(k-1)}]}, \quad k = 1, 2, 3, \ldots \end{aligned} \tag{8.219}$$

Substituting (8.210) and (8.211) into (8.214) and carrying out the same procedures as earlier, we have the formula for determining a finite number of the shifted amounts of zeros iteratively with the result

$$\begin{aligned} \Delta_{nv}^{(k)} &= \pi/6b, \quad k = 0, \\ &= \frac{2\gamma_{2n} v_n Q_n[\Delta_{mv}^{(k-1)}]}{1 - v_n Q_n[\Delta_{mv}^{(k-1)}]}, \quad k = 1, 2, 3, \ldots \end{aligned} \tag{8.220}$$

for $n = 1, 2, 3, \ldots, N$, where

$$\begin{aligned} v_n &= \frac{\pi^2}{2b^3} \frac{n^2(\gamma_{2n} - ik\cos\theta_0)}{\gamma_{2n}^2(\gamma_{2n} + ik\cos\theta_0)} [N_+(i\gamma_{2n})]^2 \\ &\quad \times \prod_{m=1}^{\infty}{}^{(n)} \frac{1 - \gamma_{2n}/\gamma_{2m}}{1 + \gamma_{2n}/\gamma_{2m}} \prod_{m=N+1}^{\infty} \frac{1 + \gamma_{2n}/(\gamma_{2m} + \pi/6b)}{1 - \gamma_{2n}/(\gamma_{2m} + \pi/6b)} \end{aligned} \tag{8.221}$$

$$Q_n[\Delta_m^v] = \prod_{m=1}^{N}{}^{(n)} \frac{1 + \gamma_{2n}/(\gamma_{2m} + \Delta_m^v)}{1 - \gamma_{2n}/(\gamma_{2m} + \Delta_m^v)} \tag{8.222}$$

Because the iterative scheme described here takes into account the edge condition rigorously in setting the zeroth-order iterative solution, we can determine the shifted amounts of zeros with high accuracy by carrying out a few iterations. Incidentally, in the low-frequency range $kb \to 0$, the sufficient accuracy is achieved by using only the zeroth-order iterative solution. Therefore, the method of approximate solutions developed in this subsection is an efficient method from the point of numerical computations.

8.5.7 Scattered Far Field

Substituting (8.207) and (8.209) into (8.177a) and (8.178b), respectively, and taking the sum and difference of the resulting equations, we obtain

$$\Psi_{(+)}(\pm b, \alpha) = \frac{1}{2}\left[\tilde{U}_{(+)}(\alpha)\frac{q_u(\alpha)}{q_u(k\cos\theta_0)} \pm \tilde{V}_{(+)}(\alpha)\frac{q_v(\alpha)}{q_v(k\cos\theta_0)}\right] \tag{8.223}$$

Therefore, substituting (8.223) into (8.166) and (8.173) and taking the inverse complex Fourier transform according to Theorem 8.4, we can derive the integral representation of the scattered field in real space. In particular, we find the scattered field for the region $x \gtrless \pm b$ as

$$\phi(x, z) = (2\pi)^{-1/2}\int_{-\infty+ic}^{\infty+ic} \Psi_{(+)}(\pm b, \alpha)e^{\mp\gamma(x\mp b)-i\alpha z}d\alpha \tag{8.224}$$

where c is a constant such that $-k_2 < c < k_2\cos\theta_0$. Because the integrand of (8.224) contains the branch points at $\alpha = \pm k$ and the split functions $L_\pm(\alpha)$ and $N_\pm(\alpha)$, exact evaluation of the integral is complicated. In this subsection, we derive the asymptotic expression of $\phi(x, z)$ based on the saddle-point method as $k(x^2 + z^2)^{1/2} \to \infty$. Because the region $|x| \leqslant b$ is of less interest in the far field, we consider the far field only in the region $|x| \geqslant b$.

We see from (8.196), (8.210), and (8.223) that $\Psi_{(+)}(\pm b, \alpha)$ has a simple pole at $\alpha = k\cos\theta_0$ and hence the saddle-point method discussed in Section 8.2.4 is not applicable directly. We now express $\phi(x, z)$ as in

$$\phi(x, z) = \phi_1(x, z) + \phi_2(x, z) \tag{8.225}$$

where

$$\phi_1(x, z) = (2\pi)^{-1/2}\int_{-\infty+ic}^{\infty+ic} [\Psi_{(+)}(\pm b, \alpha) - \tilde{\Phi}(\pm b, \alpha)] \times e^{\mp\gamma(x\mp b)-i\alpha z}d\alpha, \quad x \gtrless \pm b \tag{8.226}$$

$$\phi_2(x, z) = (2\pi)^{-1/2}\int_{-\infty+ic}^{\infty+ic} \tilde{\Phi}(\pm b, \alpha)\, e^{\mp\gamma(x\mp b)-i\alpha z}d\alpha, \quad x \gtrless \pm b \tag{8.227}$$

$$\tilde{\Phi}(\pm b, \alpha) = \frac{\exp(\mp ikb\sin\theta_0)i(k\cos\theta_0 + k)^{1/2}}{(2\pi)^{1/2}(\alpha + k)^{1/2}(\alpha - k\cos\theta_0)} \tag{8.228}$$

Because we can show from (8.223) that

$$\mathrm{Res}\ \Psi_{(+)}(\pm b, k\cos\theta_0) = (2\pi)^{-1/2}i\exp(\mp ikb\sin\theta_0) \tag{8.229}$$

the integrand of (8.226) is regular in the neighborhood of $\alpha = k\cos\theta_0$. Let us introduce the polar coordinates (ρ, θ), (ρ_1, θ_1) and (ρ_2, θ_2) as

$$x = \rho \sin\theta, \quad z = \rho \cos\theta \quad \text{for} \quad -\pi < \theta < \pi \tag{8.230a}$$

$$x - b = \rho_1 \sin\theta_1, \quad z = \rho_1 \cos\theta_1 \quad \text{for} \quad 0 < \theta_1 < \pi \tag{8.230b}$$

$$x + b = \rho_2 \sin\theta_2, \quad z = \rho_2 \cos\theta_2 \quad \text{for} \quad -\pi < \theta_2 < 0 \tag{8.230c}$$

Then, applying a corollary of Theorem 8.13, $\phi_1(x, z)$ defined by (8.226) can be expanded asymptotically as $k\rho_{1,2} \to \infty$ with the result

$$\phi_1(\rho_{1,2}, \theta_{1,2}) \sim \pm[\Psi_{(+)}(\pm b, -k\cos\theta_{1,2}) - \tilde{\Phi}(\pm b, -k\cos\theta_{1,2})] \times k\sin\theta_{1,2} \frac{\exp[i(k\rho_{1,2} - \pi/4)]}{(k\rho_{1,2})^{1/2}}, \quad x \gtrless \pm b \tag{8.231}$$

The term $\phi_2(x, z)$ defined by (8.227) gives the field scattered by two semiinfinite plates located on $x = \pm b$ for $z \leqslant 0$, and is evaluated exactly using the Fresnel integral, as in

$$\begin{aligned}\phi_2(\rho_{1,2}, \theta_{1,2}) = &- \exp[-ik\rho\cos(\theta - \theta_0)]F\{(2k\rho_{1,2})^{1/2}\cos[(\theta_{1,2} - \theta_0)/2]\} \\ &- \exp[-ik\rho\cos(\theta + \theta_0)]F\{(2k\rho_{1,2})^{1/2}\cos[(\theta_{1,2} + \theta_0)/2]\}\end{aligned} \tag{8.232}$$

for $x \gtrless \pm b$. Therefore, substituting (8.231) and (8.232) into (8.225) yields the scattered far-field expression that holds uniformly in observation angle θ and $\theta_{1,2}$.

The scattering by a semiinfinite solid circular cylinder [22], a step discontinuity in a parallel plate waveguide [23], and semiinfinite dielectric slabs [46, 48] can also be treated using the method developed in this section.

8.6 CONCLUDING REMARKS

The Wiener-Hopf technique and the modified residue-calculus technique for analyzing wave scattering and diffraction problems have been discussed in this chapter. If the boundary under consideration is described by typical orthogonal coordinate systems, such as the Cartesian and the cylindrical coordinates, these methods are known to be very useful and elegant solutions can be obtained for a number of problems.

The residue-calculus technique has been omitted and the modified residue-calculus technique focused on only the applications to the approximate solutions of the Wiener-Hopf equations. Topics that have not been covered fully in this chapter, such as the residue and the modified residue-calculus techniques, can be found in many related references [30–37].

We have also omitted the methods of solution for the Wiener-Hopf equations of vector type and for problems belonging to a third class of the modified Wiener-Hopf geometry. Readers interested in the recent progress of this field may refer to the related papers [27, 40–45, 47, 49–51, 53]. General methods for analyzing the problems of this kind still remain to be developed. Most of the research carried out so far is restricted to two-dimensional scattering and diffraction problems. Therefore, applications of the method to three-dimensional problems must also be worked out.

ACKNOWLEDGMENT

The author would like to thank Professor K. Horiuchi of Waseda University for many helpful discussions.

REFERENCES

[1] Sommerfeld, A., "Mathematische Theorie der Diffraction," *Math. Ann.*, Vol. 47, 1896, pp. 317–374.

[2] Baker, B.B., and E.T. Copson, *The Mathematical Theory of Huygens' Principle,* second edition, Oxford University Press, London, 1953, pp. 124–152.

[3] Wiener, N., and E. Hopf, "Über eine Klasse singulärer Integralgleichungen," *Sitz. Ber. Preuss. Akad. Wiss., Phys.-Math. Kl.,* Verlag der Akademie der Wissenschaften, Berlin, 1931, pp. 696–706. See also, R.E.A.C. Paley and N. Wiener, *Fourier Transforms in the Complex Domain,* American Mathematical Society, New York, 1934, pp. 49–58.

[4] Magnus, W., "Über die Beugung elektromagnetische Wellen an einer Halbebene," *Z. Phys.*, Vol. 117, No. 3–4, Jan. 1941, pp. 168–179.

[5] Schwinger, J., *Seminar on the Theory of Guided Waves,* MIT Radiation Laboratory, Boston, 1944.

[6] Copson, E.T., "On an Integral Equation Arising in the Theory of Diffraction," *Quart. J. Math.*, Vol. 17, 1946, pp. 19–34.

[7] Carlson, J.F., and A.E. Heins, "The Reflection of an Electromagnetic Plane Wave by an Infinite Set of Plates, I," *Quart. Appl. Math.*, Vol. 4, No. 4, Jan. 1947, pp. 313–329.

[8] Heins, A.E., "The Radiation and Transmission Properties of a Pair of Semi-Infinite Parallel Plates—I and II," *Quart. Appl. Math.*, Vol. 6, No. 2, July 1948, pp. 157–166; Vol. 6, No. 3, Oct. 1948, pp. 215–220.

[9] Levine, H., and J. Schwinger, "On the Radiation of Sound from an Unflanged Circular Pipe," *Phys. Rev.*, Vol. 73, No. 4, Feb. 1948, pp. 383–406.

[10] Iijima, T., "On the Electromagnetic Fields in Case of Existence of a Semi-Infinite Hollow Conductive Circular Cylinder. I, II and III," *Research of the Electrotechnical Laboratory,* No. 518 Dec. 1950, No. 531, Sept. 1952; No. 541, Aug. 1954 [in Japanese].

[11] Vajnshtejn, L.A., "Propagation in Semi-Infinite Waveguides," six papers by L.A. Vajnshtejn written during 1948–1950, trans. J. Shmoys, *Res. Rep., Div. Electromagnetic Res., Inst. Math. Sci., New York Univ.*, No. EM-63, 1954.

[12] Karp, S.N., "Wiener-Hopf Techniques and Mixed Boundary Value Problems," *Comm. Pure Appl. Math.*, Vol. 3, No. 4, Dec. 1950, pp. 411–426.

[13] Clemmow, P.C., "A Method for the Exact Solution of a Class of Two-Dimensional Diffraction Problems," *Proc. Roy. Soc. London, Ser. A,* Vol. 205, No. A1081, Feb. 1951, pp. 286–308.

4] Clemmow, P.C., "Radio Propagation over a Flat Earth across a Boundary Separating Two Different Media," *Phil. Trans. Roy. Soc. London, Ser. A,* Vol. 246, No. A905, June 1953, pp. 1–55.

5] Horiuchi, K., "An Analytical Study on the Microwave Propagating System, Part 2—Boundary between Two Different Media and Diffraction of Electromagnetic Waves, Ph.D. dissertation, Waseda University, Tokyo, 1958 [in Japanese].

6] Clemmow, P.C., *The Plane Wave Spectrum Representation of Electromagnetic Fields,* Pergamon Press, London, 1966.

7] Weinstein (Vajnshtejn), L.A., *The Theory of Diffraction and the Factorization Method (Generalized Wiener-Hopf Technique),* trans. by P. Beckmann, Golem Press, Boulder: 1969.

8] Noble, B., *Methods Based on the Wiener-Hopf Technique for the Solution of Partial Differential Equations,* Pergamon Press, London, 1958.

9] Jones, D.S., "A Simplifying Technique in the Solution of a Class of Diffraction Problems," *Quart. J. Math. Oxford (2),* Vol. 3, 1952, pp. 189–196.

0] Jones, D.S., "Diffraction by a Wave-Guide of Finite Length," *Proc. Camb. Phil. Soc.,* Vol. 48, Part 1, Jan. 1952, pp. 118–134.

1] Jones, D.S., "Diffraction by a Thick Semi-Infinite Plate," *Proc. Roy. Soc. London, Ser. A,* Vol. 217, No. A1129, April 1953, pp. 153–175.

2] Jones, D.S., "The Scattering of a Scalar Wave by a Semi-Infinite Rod of Circular Cross Section," *Phil. Trans. Roy. Soc. London, Ser. A,* Vol. 247, No. A934, April 1955, pp. 499–528.

3] Williams, W.E., "Step Discontinuities in Waveguides," *Res. Rep., Div. Electromagnetic Res., Inst. Math. Sci., New York Univ.,* No. EM-77, 1955.

4] Williams, W.E., "Diffraction by a Cylinder of Finite Length," *Proc. Camb. Phil. Soc.,* Vol. 52, Part 2, April 1956, pp. 322–335.

5] Papadopoulos, V.M., "Scattering by a Semi-Infinite Resistive Strip of Dominant-Mode Propagation in an Infinite Rectangular Wave-Guide," *Proc. Camb. Phil. Soc.,* Vol. 52, Part 3, July 1956, pp. 553–563.

6] Papadopoulos, V.M., "The Scattering Effect of a Junction between Two Circular Waveguides," *Quart. J. Mech. Appl. Math.,* Vol. 10, Part 2, 1957, pp. 191–209.

7] Aoki, K., "Diffraction of a Plane Electromagnetic Wave by a Solid Circular Cylinder of Finite Length," Ph.D. dissertation, Kyushu University, Fukuoka, 1962 [in Japanese].

8] Igarashi, A., "Simultaneous Wiener-Hopf Equations and Their Applications to Diffraction Problems in Electromagnetic Theory," *J. Phys. Soc. Japan,* Vol. 19, No. 7, July 1964, pp. 1213–1221.

9] Igarashi, A., "Simultaneous Wiener-Hopf Equations and Their Applications to Diffraction Problems in Electromagnetic Theory. II," *J. Phys. Soc. Japan,* Vol. 25, No. 1, July 1968, pp. 260–271.

0] Berz, F., "Reflection and Refraction of Microwaves at a Set of Parallel Metallic Plates," *Proc. IEE, Part III,* Vol. 98, No. 51, Jan. 1951, pp. 47–55.

1] Whitehead, E.A.N., "The Theory of Parallel Plate Media for Microwave Lenses," *Proc. IEE, Part III,* Vol. 98, No. 52, March 1951, pp. 133–140.

2] Hurd, R.A., and H. Gruenberg, "H-Plane Bifurcation of Rectangular Waveguides," *Can. J. Phys.,* Vol. 32, No. 11, Nov. 1954, pp. 694–701.

3] Hurd, R.A., "The Propagation of an Electromagnetic Wave along an Infinite Corrugated Surface," *Can. J. Phys.,* Vol. 32, No. 12, Dec. 1954, pp. 727–734.

4] Mittra, R., S.W. Lee, and G.F. Van Blaricum, Jr., "A Modified Residue Calculus Technique," *Int. J. Eng. Sci.,* Vol. 6, No. 7, Aug. 1968, pp. 395–408.

5] Van Blaricum, Jr., G.F., and R. Mittra, "A Modified Residue-Calculus Technique for Solving a Class of Boundary Value Problems—Part I: Waveguide Discontinuities," *IEEE Trans. Microwave Theory and Tech.,* Vol. MTT-17, No. 6, June 1969, pp. 302–309.

[36] Itoh, T., and R. Mittra, "An Analytical Study of Echelette Grating with Application to Open Resonators," *IEEE Trans. Microwave Theory and Tech.*, Vol. MTT-17, No. 6, June 1969, pp. 319–327.

[37] Mittra, R., and S.W. Lee, *Analytical Techniques in the Theory of Guided Waves,* Macmillan Company, New York, 1971.

[38] Mittra, R., and S.W. Lee, "On the Solution to a Generalized Wiener-Hopf Equation," *J. Math. Phys.*, Vol. 11, No. 3, March 1970, pp. 775–783.

[39] Kobayashi, K., "Diffraction of a Plane Wave by the Parallel Plane Grating with Periodic Structure," *Trans. IECE Japan,* Vol. J63-B, No. 5, May 1980, pp. 444–451 [in Japanese].

[40] Hurd, R.A., "The Wiener-Hopf-Hilbert Method for Diffraction Problems," *Can. J. Phys.*, Vol. 54, No. 7, April 1976, pp. 775–780.

[41] Daniele, V.G., "On the Factorization of Wiener-Hopf Matrices in Problems Solvable with Hurd's Method," *IEEE Trans. Antennas and Propagation,* Vol. AP-26, No. 4, July 1978, pp. 614–616.

[42] Hurd, R.A., and E. Lüneburg, "Scattering by Hard and Soft Parallel Half-Planes," *Can. J. Phys.*, Vol. 59, No. 12, Dec. 1981, pp. 1879–1885.

[43] Rawlins, A.D., and W.E. Williams, "Matrix Wiener-Hopf Factorization," *Quart. J. Mech. Appl. Math.*, Vol. 34, Part 1, Feb. 1981, pp. 1–8.

[44] Jones, D.S., "Commutative Wiener-Hopf Factorization of a Matrix," *Proc. Roy. Soc. London, Ser. A,* Vol. 393, No. 1804, May 1984, pp. 185–192.

[45] Aoki, K., and K. Uchida, "Scattering of a Plane Electromagnetic Wave by a Conducting Rectangular Cylinder," *Trans. IECE Japan,* Vol. J63-B, No. 6, June 1980, pp. 596–603 [in Japanese]. See also, *Mem. Fac. Eng., Kyushu Univ.*, Vol. 38, No. 2, June 1978, pp. 153–175.

[46] Aoki, K., and K. Uchida, "Scattering of a Plane Electromagnetic Wave by Two Semi-Infinite Dielectric Slabs," *Trans. IECE Japan,* Vol. J62-B, No. 12, Dec. 1979, pp. 1132–1139 [in Japanese].

[47] Aoki, K., T. Matsunaga, and K. Uchida, "Scattering of a Plane Wave by a Lossy Dielectric Rectangular Cylinder," *Trans. IECE Japan,* Vol. J65-B, No. 11, Nov. 1982, pp. 1417–1424 [in Japanese]. See also, *Res. Rep., Technical Group on Electromagnetic Theory, IEE Japan,* No. EMT-80-45, Oct. 1980 [in Japanese].

[48] Uchida, K., and K. Aoki, "Scattering of Surface Waves on Transverse Discontinuities in Symmetrical Three-Layer Dielectric Waveguides," *IEEE Trans. Microwave Theory and Tech.*, Vol. MTT-32, No. 1, Jan. 1984, pp. 11–19.

[49] Kobayashi, K., "Diffraction of a Plane Wave by the Parallel Plate Grating with Dielectric Loading," *Trans. IECE Japan,* Vol. J64-B, No. 10, Oct. 1981, pp. 1091–1098 [in Japanese]. See also, *The 100th Anniversary Bull., Fac. Sci. Eng., Chuo Univ.*, Oct. 1985, pp. 473–524.

[50] Kobayashi, K., "Diffraction of a Plane Electromagnetic Wave by a Rectangular Dielectric Rod. I and II," *Res. Rep., Technical Group on Electromagnetic Theory, IEE Japan,* No. EMT-80-46, Oct. 1980; No. EMT-81-3, Jan. 1981 [in Japanese].

[51] Kobayashi, K., "Diffraction of a Plane Electromagnetic Wave by a Rectangular Conducting Rod. I and II," *Bull. Fac. Sci. Eng., Chuo Univ.*, Vol. 25, Dec. 1982, pp. 229–261, 263–282.

[52] Kobayashi, K., and T. Inoue, "Diffraction of a Plane Wave by an Inclined Parallel Plate Grating," *IEEE Trans. Antennas and Propagation,* Vol. AP-36, No. 10, Oct. 1988, pp. 1424–1434.

[53] Kobayashi, K., and K. Miura, "Diffraction of a Plane Wave by a Thick Strip Grating," *IEEE Trans. Antennas and Propagation,* Vol. AP-37, No. 4, April 1989, pp. 459–470.

[54] Rojas, R.G., "A Uniform GTD Analysis of the EM Diffraction by a Thin Dielectric/Ferrite Half-Plane and Related Configurations," Ph.D. dissertation, Ohio State University, Columbus, 1985

[55] Rojas, R.G., "Wiener-Hopf Analysis of the EM Diffraction by an Impedance Discontinuity in a Planar Surface and by an Impedance Half-Plane," *IEEE Trans. Antennas and Propagation,* Vol AP-36, No. 1, Jan. 1988, pp. 71–83.

[56] Copson, E.T., *Asymptotic Expansions,* Cambridge University Press, London, 1971, pp. 48–62

[57] Kobayashi, K., "Asymptotic Methods Applied to the Diffraction Theory of Electromagnetic Waves. I.—Watson's Lemma," *Bull. Fac. Sci. Eng., Chuo Univ.*, Vol. 25, Dec. 1982, pp. 217–228.

[58] Felsen, L.B., and N. Marcuvitz, *Radiation and Scattering of Waves,* Prentice-Hall, Englewood Cliffs, 1973, pp. 370–441.

[59] Meixner, J., "The Behavior of Electromagnetic Fields at Edges," *Res. Rep., Div. Electromagnetic Res., Inst. Math. Sci., New York Univ.,* No. EM-72, 1954. See also, *IEEE Trans. Antennas and Propagation,* Vol. AP-20, No. 4, July 1972, pp. 442–446.

[60] Kobayashi, K., "Diffraction of a Plane Wave by a Strip—Exact and Asymptotic Solutions," *Res. Rep., Technical Group on Electromagnetic Theory, IEE Japan,* No. EMT-89-75, Oct. 1989 [in Japanese].

[61] Kobayashi, K., "Asymptotic Expansions of Integrals with an Exponentially Decaying Kernel," *Res. Rep., Technical Group on Electromagnetic Theory, IEE Japan,* No. EMT-90-28, March 1990 [in Japanese].

[62] Seshadri, S.R., "High-Frequency Diffraction of Plane Waves by an Infinite Slit—I and II," *Proc. Nat. Inst. Sci. India, Ser. A,* Vol. 25, No. 6, Nov. 1959, pp. 301–321, 322–336.

[63] Kobayashi, K., "Wiener-Hopf Technique and Its Applications to the Scattering and Diffraction Problems," in *A Course of Applied Mathematics,* ed. by K. Horiuchi, Corona Publishing, Tokyo, 1989, pp. 344–408 [in Japanese].

Chapter 9
Asymptotic Expansion Methods

Masahiro Hashimoto

.1 HISTORICAL BACKGROUND

symptotic expansion in mathematics is defined as the expansion of a certain func-
on in terms of one of its parameters. However, because such a series expansion
overs most of the series we encounter, we often limit this mathematical terminology
) diverging series and exclude converging series. This is quite a vague statement
nd the reader may be confused, so we shall proceed with our discussions after we
orm a clear definition.

The history of the asymptotic expansions starts at the beginning of the nine-
enth century. In those days, the wave nature of light became evident and was found
) obey the same elliptical wave equation as the vibrations of a rigid body, water
aves, and sound waves. In 1836, C. Sturm discussed the general solutions of the
ne-dimensional differential equation of the second order. In the next year, J. Liou-
ille tried to solve the Sturm problem approximately, introducing a skillful technique
f variable transformation. The asymptotic expansion method appears to have orig-
nated in those days. When, in 1987, G. Green analyzed the motion of water waves
unning along a canal of variable width, he had already given the geometrical optics
olution as the first-order term in the asymptotic expansion, which may now be familiar.

After G.B. Airy, asymptotic expansions became involved in mathematical
roblems. Airy was interested in the real nature of the rainbow (primary bow), par-
icularly the presence of a dark part over the bow and the bright part below the bow.
Ie showed, from the Huygens principle and the geometrical condition imposed on
s boundary, that the wave of light in the neighborhood of the rainbow could be
escribed in terms of the famous Airy function. He computed the values of the Airy
unction by the method of quadratures and provided for it numerical tables to the
even decimal places with respect to the variable m. However, when the value of m
xceeds 2.5, we can hardly compute it. Although he always had great concern about

the validity of the method of quadratures for a large m and tried to use other methods, he failed.

In fact, the Airy function is governed by a certain differential equation of second order. G.G. Stokes, who was aware of this, suggested a way to obtain the approximate solution of the differential equation by means of asymptotic expansions, so that the obtained solution is accurate for a large m. Stokes eventually discovered (in 1864) the discontinuity phenomenon, in which arbitrary constants that appear in the series solution could take discontinuous values as the sign of m changed. This important aspect of the diverging series solution, the so-called Stokes phenomenon, was discussed by a number of mathematicians.

In 1911, A. Sommerfeld and J. Runge introduced it again to describe the wave motion of light in three dimensions. Subsequently, under favor of the Airy function in an inhomogeneous medium, L. Rayleigh succeeded in "matching" the different fields of wave in both sides of the reflection point where the sound wave was totally reflected. R. Gans and H. Jeffreys reached the same goal independently.

There has been another development in the field of physics. Three physicists—G. Wentzel, H. Kramers, and L. Brillouin—rediscovered the asymptotic solution for the atomic level of energy, which was a new solution, thus allowing a new insight into quantum physics. The method, named the *WKB method,* came to the notice of physicists around 1926. Nowadays, the name *WKB* refers to a general method of asymptotic expansions for use in wave motion systems. Therefore, unfavorable extension of its use to the substantial formalism or the individual techniques should be avoided.

9.2 MATHEMATICAL FOUNDATIONS OF ASYMPTOTIC EXPANSIONS

9.2.1 Definition of Asymptotic Expansions

Let $f(x)$ be a function of the variable x, and suppose that for any fixed (positive) integer n:

$$f(x) = a_0 + \frac{a_1}{x} + \frac{a_2}{x^2} + \cdots + \frac{a_n}{x^n} + R_n(x) \tag{9.1}$$

where $R_n(x)$ is the remainder between $f(x)$ and the partial nth sum of the series. If, then,

$$x^n R_n(x) \to 0$$

as $x \to \infty$; namely, if

$$R_n(x) = 0(x^{-n-1}) \tag{9.2}$$

say that the partial nth sum of the series is an asymptotic expansion of $f(x)$ at $= \infty$,* and we formally write

$$f(x) \sim a_0 + \frac{a_1}{x} + \frac{a_2}{x^2} + \cdots$$

$$f(x) \sim \sum_{n=0}^{\infty} \frac{a_n}{x^n} \tag{9.3}$$

e careful in taking this expression as formal at all, because it includes diverging ries. Following H. Poincaré, who investigated the details of such diverging series ter Stokes, we call equation (9.3) an asymptotic expansion of the Poincaré type. he general definition for Poincaré expansions is as follows [1].

Let $\{\varphi_n(x)\}$ be a sequence of functions such that, as $x \to x_1$,

$$\varphi_{n+1}(x)/\varphi_n(x) \to 0 \quad (x \to x_1) \tag{9.4}$$

, for any integer n,

$$f_n(x) = \sum_{n=0}^{n} a_n\varphi_n(x) + R_n(x) \tag{9.5}$$

satisfied and, in addition,

$$R_n(x)/\varphi_n(x) \to 0 \quad \text{as} \quad x \to x_1 \tag{9.6}$$

ien, we call the following expansion the *Poincaré asymptotic expansion* of $f(x)$:

$$f(x) \sim \sum_{n=0}^{\infty} a_n\varphi_n(x) \tag{9.7}$$

ike (9.2), (9.6) can be written

$$R_n(x) = 0\{\varphi_{n+1}(x)\} \tag{9.8}$$

*The term *asymptotic* was coined by Poincaré [H. Poincaré, *Acta Math.*, Vol. 8, 1896, pp. 295–44]. He called the diverging series the *asymptotic series* and the converging series the *normal series*. owever, according to the definition, we call both the expansions the *asymptotic series*.

If (9.8) is fulfilled only for $n \leqslant N$ or, more loosely, if (9.6) is fulfilled f
$n \leqslant N$, we call

$$f(x) \sim \sum_{n=0}^{N} a_n \varphi_n(x) \tag{9.9}$$

the *asymptotic expansion* to $N + 1$ terms.

9.2.2 Liouville-Green Asymptotic Expansions

The Sturm-Liouville type of differential equation is given by

$$\frac{d}{dx}\left\{p(x)\frac{dy}{dx}\right\} + \{\lambda r(x) - q(x)\}y = 0 \tag{9.10}$$

where λ is a parameter with large value. In 1837, Liouville introduced the trans
formation of variables

$$\left.\begin{aligned} z &= \int_0^x \sqrt{\frac{r}{p}}\,dx \\ y &= \theta u, \quad \theta = \frac{1}{\sqrt[4]{pr}} \end{aligned}\right\} \tag{9.11}$$

into (9.10) and tried to solve the reduced equation

$$\frac{d^2u}{dz^2} + \lambda u = su \tag{9.12}$$

$$s = -\frac{1}{r\theta}\left\{\frac{d}{dx}\left(p\frac{d\theta}{dx}\right) - q\theta\right\} \tag{9.13}$$

approximately. A little later that year, Green solved the problem of $q = 0$ by th
same technique. Accordingly, (9.11) is called the *Liouville-Green transformation*.

A general solution of (9.12), with A and B as arbitrary constants, is

$$u = A\cos(\sqrt{\lambda}z) + B\sin(\sqrt{\lambda}z) + \frac{1}{\sqrt{\lambda}}\int_0^z s(z')u(z')\sin\{\sqrt{\lambda}(z - z')\}dz' \tag{9.14}$$

›r a large λ, neglecting the third term in the right-hand side of this equation, y (=
:) becomes

$$y \approx \frac{A}{\sqrt[4]{pr}} \cos\left(\sqrt{\lambda} \int_0^x \sqrt{\frac{r}{p}}\, dx\right) + \frac{B}{\sqrt[4]{pr}} \sin\left(\sqrt{\lambda} \int_0^x \sqrt{\frac{r}{p}}\, dx\right) \tag{9.15}$$

'e call this approximation the *Liouville-Green approximation.*

2.3 The Airy Function

ıppose that light emanating from a source is reflected at a point located on a con-ıve mirror and arrives at an observation point. The point of reflection can be de-rmined from the Fermat principle: "the total path length L is invariant within the nge of first variations δL even if light is reflected at the neighboring point apart om the true point." If L is invariant for all higher-order variations, the point is ılled the *focusing point*. If, however, L is invariant within the range of second ıriations, the point of reflection is distributed on a surface, and such a surface is ılled the *caustic*. This is a definition of the *caustic* in geometrical optics. Light on ıe side of the caustic is brilliant but dark in the other side.

Therefore, the caustic is a boundary surface between the bright and dark sides. ccording to this definition, Airy determined the point of reflection and calculated ıe field of light in the neighborhood of the caustic, using the phase and amplitude f light on the mirror and the Huygens principle [2]. As a result, he showed* that ıe field of light at a point apart from the caustic by x could be given by

$$y = A_i(x) \equiv \frac{1}{\pi} \int_0^\infty \cos\left(\frac{t^3}{3} + xt\right) dt \tag{9.16}$$

$$y = \frac{1}{2\pi} \int_{-\infty}^\infty e^{j[(t^3/3)+xt]} dt \tag{9.17}$$

pplying $\partial/\partial x = jt$, $\partial^2/\partial x^2 = -t^2$, $j\partial/\partial t = -t^2 - x$ to this equation, that is, applying $/\partial x^2 - x = j\partial/\partial t$, leads to $y'' - xy = 1/2\pi[j\, e^{j[(t^3/3)+xt]}]_{-\infty}^{\infty} = 0$, where $y'' = d^2y/$ x^2. Hence, the Airy function satisfies

$$y'' - xy = 0 \tag{9.18}$$

*His original expression was $\int_0^\infty \cos[(\pi/2)(t^3 - mt)]dt$ where m was a parameter corresponding › x.

This is a special case of the Sturm-Liouville equation for $x > 0$, $p(x) = 1$, q(x
$= 0$, $\lambda = -1$, $r(x) = x$. Hence, the approximate solution of the Sturm-Liouvill
equation, which represents an evanescent wave as x tends to infinity, is

$$y \approx \frac{C}{\sqrt[4]{x}} e^{-(2/3)x^{3/2}} \tag{9.19}$$

with C constant. For $x < 0$, on the other hand, $p(x) = 1$, $q(x) = 0$, $\lambda = 1$, and r(x
$= -x$. Hence, we have

$$y \approx \frac{A}{\sqrt[4]{-x}} \cos\left\{\frac{2}{3}(-x)^{3/2}\right\} - \frac{B}{\sqrt[4]{-x}} \sin\left\{\frac{2}{3}(-x)^{3/2}\right\} \tag{9.20}$$

or

$$y \approx \frac{D}{\sqrt[4]{-x}} e^{j(2/3)(-x)^{3/2}} + \frac{E}{\sqrt[4]{-x}} e^{-j(2/3)(-x)^{3/2}} \tag{9.21}$$

where A, B, D, and E are constants. We emphasize again that (9.19) represents a
evanescent wave whereas (9.20) represents an oscillatory wave, so $x = 0$ is a locatio
of the caustic.

Does the Airy function behave just like (9.19) and (9.20) in both the sides o
$x = 0$? To answer this, we introduce the two historical methods, called the *saddle point methods*, and, by means of them, we expand the Airy function into an asymp
totic series. One is the stationary phase method used by Lord Kelvin in 1887 t
analyze a pulse wave in dispersive water. The other is the steepest descent method
The latter was developed in 1908 by P. Debije (P. Debye) who analyzed the scat
tering of waves from a dielectric rod in conjunction with the theory of rainbow.

Consider the integral of the form

$$y = \frac{1}{2\pi} \int_{-\infty}^{\infty} e^{f(t)} dt \tag{9.22}$$

Then, the point at which the first derivative of $f(t)$ vanishes with respect to t, tha
is, the point t_0 such that $f'(t_0) = 0$, is called the *saddle point*. In the neighborhoo
of t_0, $f(t)$ can be approximated by

$$f(t) \approx f(t_0) + \frac{1}{2} f''(t_0)(t - t_0)^2 \tag{9.23}$$

Figure 9.1(a) illustrates the real part of $f(x)$ on a complex t-plane, indicating a saddl
for a rider on the back of a horse. The imaginary part of $f(x)$ also produces the sam

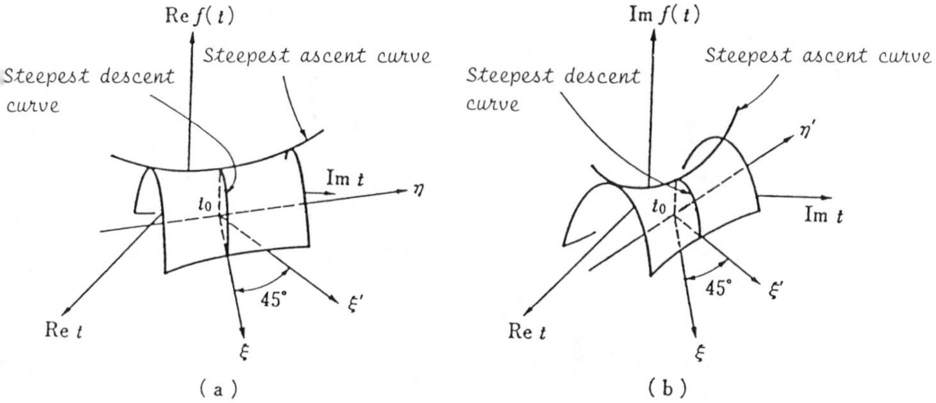

Figure 9.1 Saddle points: steepest descent curve, steepest ascent curve.

picture on a complex t-plane that rotates by 45°, as shown in Figure 9.1(b). This is why t_0 is called the *saddle point*.

When the point moves along the ξ axis (see the figure), the real part of $f(x)$ represents the steepest descent curve; and when, on the other hand, the point moves along the η axis, it represents the steepest ascent curve. On the other axes, for example, ξ' and η' making an angle of 45° to the ξ and η axes, the real part of $f(x)$ keeps the constant value Re $f(t_0)$ and, instead, the imaginary $f(x)$ represents the steepest descent curve on the ξ' axis and the steepest ascent curve on the η' axis, and its value remains constant on the ξ axis (see Figure 9.1(b)). Integrating (9.22) along the ξ' or η' axis, passing through point t_0, namely, along the path such that

$$\mathrm{Re}\,\{f(t)\} = \mathrm{Re}\,\{f(t_0)\} \tag{9.24}$$

we may be able to calculate the approximate value of y from the mathematical formula

$$\int_{-\infty}^{\infty} e^{\pm j\alpha\xi'^2} d\xi' = \sqrt{\frac{\pi}{\alpha}}\, e^{\pm j\pi/4} \tag{9.25}$$

This method is called the *stationary-phase method*. On the contrary, performing the integration along the ξ axis, namely, along the path such that

$$\mathrm{Im}\,\{f(t)\} = \mathrm{Im}\,\{f(t_0)\} \tag{9.26}$$

we may also be able to obtain the approximate value from the mathematical formula

$$\int_{-\infty}^{\infty} e^{-\alpha\xi^2} d\xi = \sqrt{\frac{\pi}{\alpha}} \tag{9.27}$$

This method is called the *steepest-descent method*. In either case if the argument of $-f''(t_0)$ exists between $-\pi$ and $+\pi$, we have

$$y \approx \frac{1}{\sqrt{-2\pi f''(t_0)}} e^{f(t_0)} \tag{9.28}$$

We shall apply the method to (9.17), noting that $f(t) = j[(t^3/3) + xt]$, $t_0 = \pm\sqrt{-x}$. Because, for $x > 0$, $t_0 = \pm j\sqrt{x}$ and $f(x)$ is approximated in the neighborhood of $t_0 = +j\sqrt{x}$ by

$$f(t) \approx -\frac{2}{3}x^{3/2} - \sqrt{x}(t - t_0)^2 \tag{9.29}$$

the steepest-descent calculation results in

$$y \approx \frac{1}{2\sqrt{\pi}} \frac{1}{\sqrt[4]{x}} e^{-(2/3)x^{3/2}} \tag{9.30}$$

Hence, we see that, in (9.19),

$$C = \frac{1}{2\sqrt{\pi}} \tag{9.31}$$

For $x < 0$, we have $t_0 = \pm\sqrt{-x}$, so that two saddle points appear on the real axis. The path of integration starting from $t = -\infty$ and terminating at $t = +\infty$ on the t-plane may necessarily pass through these two saddle points; it is impossible to draw the path to escape passing through one of the two. We therefore take account of the two contributions to the calculation of the integral. In the neighborhood of these saddle points, $f(t)$ becomes

$$f(t) \approx \mp j\frac{2}{3}(-x)^{3/2} \pm j\sqrt{-x}(t - t_0)^2 \tag{9.32}$$

and thus, by the method of stationary phase,

$$\begin{aligned} y &\approx \frac{1}{2\pi} e^{-j(2/3)(-x)^{3/2}} \int_{-\infty}^{\infty} e^{j\sqrt{-x}(t-t_0)^2} dt + \frac{1}{2\pi} e^{j(2/3)(-x)^{3/2}} \int_{-\infty}^{\infty} e^{-j\sqrt{-x}(t-t_0)^2} dt \\ &= \frac{1}{2\sqrt{\pi}} \frac{1}{\sqrt[4]{-x}} e^{-j(2/3)(-x)^{3/2}+j\pi/4} + \frac{1}{2\sqrt{\pi}} \frac{1}{\sqrt[4]{-x}} e^{j(2/3)(-x)^{3/2}-j\pi/4} \end{aligned} \tag{9.33}$$

From this, the constants D and E in (9.21) are determined:

$$D = \frac{1}{2\sqrt{\pi}} \mathrm{e}^{-j\pi/4}, \quad E = \frac{1}{2\sqrt{\pi}} \mathrm{e}^{j\pi/4} \tag{9.34}$$

The asymptotic expansions for the Airy function in the case of $|x| \gg 1$ have been completed separately. The question may now be raised as to whether the asymptotic solution (9.30) valid for $x > 0$ provides another solution (9.33) valid for $x < 0$ by the transformation $x = -x \exp(\pm j\pi)$. Unfortunately, it is clear from a straightforward substitution that such a simple transformation fails. That is to say, for the transformation $x = -x \exp(j\pi)$, we have $C = D \exp(j\pi/4)$, $E = 0$; and also for the other transformation $x = -x \exp(-j\pi)$, we have $C = E \exp(-j\pi/4)$, $D = 0$. It follows that the values of D and E change abruptly as the sign of x changes. This is a phenomenal aspect of the diverging series, called the *Stokes phenomenon* [3].

9.2.4 The Method of Matching by Gans

We have already mentioned that because the field of light behaves in different ways depending on if it is in a bright region or a dark region beyond the caustic, the asymptotic expressions obtained are necessarily different. We have found therefore that the asymptotic expansions are not uniformly valid in the neighborhood of the caustic. This region is usually called the *boundary layer,* and the method for matching two different expressions derived outside the boundary layer is called the *boundary-layer method,* the *method of matched asymptotic expansions,* or more inclusively the *singular-perturbation method*. A general idea of the method is to use the asymptotic expansions of the Airy function, which originated from the Rayleigh's analysis (in 1912) of sound waves in an inhomogeneous medium. The method that Gans used to analyze the total reflection of light in 1915 is more general and thus the base of the current boundary-layer methods. We outline the method as follows.

Let the following equation be the basic equation:

$$\varepsilon^2 y'' + Qy = 0 \tag{9.35}$$

where ε is a positive parameter with small value, and Q is a smooth function of x. We assume the value of Q to have a zero of the first order at $x = x_1$, and to be positive or negative for $x < x_1$ or $x > x_1$, respectively. We then have, from the Liouville-Green approximation,

$$\begin{aligned} y &\approx \frac{C}{\sqrt[4]{-Q}} \mathrm{e}^{-(1/\varepsilon)\int_{x_1}^{x}\sqrt{-Q}dx} \quad (x > x_1) \\ y &\approx \frac{D}{\sqrt[4]{Q}} \mathrm{e}^{-j(1/\varepsilon)\int_{x_1}^{x}\sqrt{Q}dx} + \frac{E}{\sqrt[4]{Q}} \mathrm{e}^{j(1/\varepsilon)\int_{x_1}^{x}\sqrt{Q}dx} \quad (x < x_1) \end{aligned} \tag{9.36}$$

Arbitrary constants, C, D, and E, in (9.36) are determined by the boundary-layer method.

In the boundary layer (in the neighborhood of x_1), the function Q is regarded as an adequately smooth function and thus is approximated by

$$Q(x) = Q'(x_1)(x - x_1) \tag{9.37}$$

where the prime denotes differentiation with respect to x. According to the assumption stated earlier, $Q'(x_1) < 0$. Thus, by the variable ξ

$$\xi = \sqrt[3]{-Q'(x_1)/\varepsilon^2}(x - x_1) \tag{9.38}$$

equation (9.35) is reduced to the Airy equation:

$$y''(\xi) - \xi y(\xi) = 0 \tag{9.39}$$

which again may be applied to the asymptotic solutions (9.30) and (9.33), replacing x by ξ. For $x > x_1$,

$$y \approx \frac{1}{2\sqrt{\pi}} \frac{a}{\sqrt[4]{-Q}} \mathrm{e}^{-(1/\varepsilon)\int_{x_1}^{x}\sqrt{-Q}dx} \tag{9.40}$$

and for $x < x_1$,

$$y \approx \frac{1}{2\sqrt{\pi}} \frac{a}{\sqrt[4]{-Q}} \left\{ \mathrm{e}^{-j(1/\varepsilon)\int_{x_1}^{x}\sqrt{Q}dx - j\pi/4} + \mathrm{e}^{j(1/\varepsilon)\int_{x_1}^{x}\sqrt{Q}dx + j\pi/4} \right\} \tag{9.41}$$

where $a = [-Q'(x_1)\varepsilon]^{1/6}$. We see therefore that $D = C\exp(-j\pi/4)$, and $E = C\exp(j\pi/4)$; that is, the fields on both sides of the caustic are connected to each other by

$$\frac{1}{\sqrt[4]{-Q}} \mathrm{e}^{-(1/\varepsilon)\int_{x_1}^{x}\sqrt{-Q}dx} \leftrightarrow \frac{2}{\sqrt[4]{Q}} \cos\left(\frac{1}{\varepsilon}\int_{x_1}^{x}\sqrt{Q}dx + \frac{\pi}{4}\right) \tag{9.42}$$

Although we have derived this relation using $A_i(\xi)$ as a solution of (9.39), it is possible to derive another relation that connects the growing field in a dark region to the oscillatory field in a bright region, using the associated Airy function $B_i(\xi)$. The method is quite similar, but further discussions are omitted here. The final result is

$$\frac{1}{\sqrt[4]{-Q}} \mathrm{e}^{(1/\varepsilon)\int_{x_1}^{x}\sqrt{-Q}dx} \leftrightarrow \frac{1}{\sqrt[4]{Q}} \sin\left(\frac{1}{\varepsilon}\int_{x_1}^{x}\sqrt{Q}dx + \frac{\pi}{4}\right) \tag{9.43}$$

.3 FORMALISM BY WENTZEL, KRAMERS, AND BRILLOUIN (WKB METHOD)

Again, let us define the basic equation to be

$$\varepsilon^2 y'' + Qy = 0 \tag{9.44}$$

with Q having zeros of the first order at $x = x_1$ and x_2 and being positive for $x_1 < x < x_2$. Because the wave is trapped between x_1 and x_2, the system of the wave is in some eigenstate (the so-called modal state). The problem of solving (9.44) is thus reduced to the mathematical eigenvalue problem. In 1926, Brillouin, Wentzel, and Kramers and, again, Brillouin developed asymptotic methods to determine the quantum atomic level of energy. They assumed the solution to be given by the form

$$y = e^{-j(1/\varepsilon)\int^x p dx} \tag{9.45}$$

Substituting this equation into (9.44), we obtain the equation that the unknown function p obeys.

$$j\varepsilon p' + p^2 = Q \tag{9.46}$$

This is a nonlinear differential equation of the first order, the so-called Riccati nonlinear equation. We now expand p in terms of ε:

$$p = p_0 + \varepsilon p_1 + \varepsilon^2 p_2 + \cdots \tag{9.47}$$

where p_0 equals $\sqrt{Q}$. Substituting this into (9.46) and equating the same power terms on the left- and right-hand sides, we have

$$jp'_{n-1} = -\sum_{m=0}^{n} p_m p_{n-m} \tag{9.48}$$

For example, when $n = 1$, the left-hand side yields jp'_0, which is the derivative of the given function, whereas the right-hand side yields $-p_0p_1 - p_1p_0$, so that $jp'_0 = -2p_0p_1$ from which p_1 can be determined. Similarly, the higher-order terms p_m can be determined from $p_0, p_1, \ldots, p_{m-1}$ or their derivatives. The results follow:

$$\begin{aligned}
p_0 &= \sqrt{Q} \\
p_1 &= -jp'_0/2p_0 \\
p_2 &= -jp'_1/2p_0 - p_1^2/2p_0 = -j(p_1/2p_0)' + p_1^2/2p_0 \\
p_3 &= -jp'_2/2p_0 - p_1p_2/p_0 = -j(p_2/2p_0)' \\
p_4 &= -jp'_3/2p_0 - p_2^2/2p_0 - p_1p_3/p_0 = -j(p_3/2p_0)' - p_2^2/2p_0
\end{aligned} \tag{9.49}$$

Because the values of Q are negative for $x < x_1$ and $x > x_2$, in these regions, there exist the growing solution and the evanescent solution. The physically acceptable solution, however, is the evanescent wave. The oscillatory solution, which is connected to this evanescent solution, from the relation (9.42) applied at x_2, is

$$y \approx \frac{A}{\sqrt[4]{Q}} \cos\left(\frac{1}{\varepsilon}\int_{x_2}^{x} \sqrt{Q}dx + \frac{\pi}{4}\right) \tag{9.50}$$

and, from that applied at x_1,

$$y \approx \frac{B}{\sqrt[4]{Q}} \cos\left(\frac{1}{\varepsilon}\int_{x}^{x_1} \sqrt{Q}dx + \frac{\pi}{4}\right) \tag{9.51}$$

These expressions, valid for $x_1 < x < x_2$, must be identical, so that

$$\frac{1}{\varepsilon}\int_{x_2}^{x} \sqrt{Q}dx + \frac{\pi}{4} = -\left(\frac{1}{\varepsilon}\int_{x}^{x_1} \sqrt{Q}dx + \frac{\pi}{4} + n\pi\right) \tag{9.52}$$

where n is an integer. From this, we obtain

$$\frac{1}{\varepsilon}\int_{x_1}^{x_2} \sqrt{Q}dx = \left(n + \frac{1}{2}\right)\pi \tag{9.53}$$

or

$$\frac{1}{\varepsilon}\oint \sqrt{Q}dx = (2n + 1)\pi \tag{9.54}$$

Note that the symbol $\oint$ denotes the integration over one period in which the wave propagates from x_1 to x_2, is reflected at x_2, returns to x_1, and is again reflected at x_1. In this case, it is twice the integration over the interval between x_1 and x_2. Equation (9.54) was derived in 1926 by Kramers. This equation is sometimes called the *half-odd-integer quantum condition* in contrast to the old Bohr condition:

$$\frac{1}{\varepsilon}\oint \sqrt{Q}dx = (2n + 2)\pi \tag{9.55}$$

.4 HIGHER-ORDER ASYMPTOTIC SOLUTIONS

.4.1 Langer Transformation

o generalize the method just described, we express the solution of (9.35), y, in rms of function t of x and the function $\mathfrak{z}$ of t as follows:

$$y = \frac{1}{\sqrt{t'}}\,\mathfrak{z}(t) \tag{9.56}$$

here $t = t(x)$, $t' = dt/dx$. Substituting this into (9.35), we derive a new differential juation of the second order:

$$\varepsilon^2 \mathfrak{z}''(t) + S(t)\mathfrak{z}(t) = 0 \tag{9.57}$$

here

$$\left.\begin{aligned} S(t) &= \bar{S}(x)/(t')^2 \\ \bar{S}(x) &= Q + \varepsilon^2\sqrt{t'}(1/\sqrt{t'})'' \end{aligned}\right\} \tag{9.58}$$

ecause (9.57) belongs to the same class of equations as (9.35), the asymptotic ıethods described earlier for (9.35) are applicable to the present problem. However, ı the system of (9.56), either t or $\mathfrak{z}$ is arbitrary. The best way to choose t or $\mathfrak{z}$ is ıch that (9.57) is reduced to a certain equation of which the approximate solution known and given in analytic form. Around 1930, R.E. Langer applied this idea obtain approximate solutions of various types of differential equations. Thus, (9.56) called the *Langer transformation* [4]. If we take

$$t = \int^x \sqrt{Q}dx \tag{9.59}$$

).56) becomes the Liouville-Green transformation as described in Section 9.2.2. lso, if $\mathfrak{z}(t)$ is assumed to be $\exp(-jt/\varepsilon)$, then $S(t)$ is unity; and thus we obtain a onlinear differential equation of the second order with respect to t:

$$\varepsilon^2\sqrt{t'}(1/\sqrt{t'})'' + Q = (t')^2 \tag{9.60a}$$

r alternatively,

$$\varepsilon^2 u'' + Qu = u^{-3}, \quad u = 1/\sqrt{t'} \tag{9.60b}$$

In 1930, W.E. Milne used this equation and, in the next year, E. Madelung introduced it independently to obtain the solution of the wave equation. Thus, it is named the *Milne-Madelung transformation*.

These two transformations are particular examples useful to illustrate the earlier-mentioned approaches. A more sophisticated transformation that combines advantages of the examples just stated will be discussed.

9.4.2 Fröman-Fröman Method

Swedish scientists N. Fröman and P.O. Fröman have tried extensively to improve the WKB method since the publication of their book *JWKB Approximation, Contributions to the Theory* (North-Holland Press, Amsterdam, 1965). In the 1970s, they invented the original method of asymptotic expansions using the Langer transformation successfully. The method is worth mentioning [5] because it covers most of the methods published so far.

They first divided the asymptotic series (9.47) into the even series p_{even} with respect to ε and the odd series p_{odd}:

$$p = p_{\text{even}} + p_{\text{odd}} \tag{9.61}$$

$$p_{\text{even}} = p_0 + \varepsilon^2 p_2 + \cdots, \quad p_{\text{odd}} = \varepsilon p_1 + \varepsilon^3 p_3 + \cdots$$

Because we find the following relation between them,

$$j\varepsilon p'_{\text{even}} + 2p_{\text{even}} p_{\text{odd}} = 0 \tag{9.62}$$

we can replace p_{odd}d by $-j\varepsilon p'_{\text{even}}/2p_{\text{even}}$, thereby rewriting (9.45) as

$$y = \frac{1}{\sqrt{p_{\text{even}}}} e^{-j(1/\varepsilon)\int^x p_{\text{even}} dx} \tag{9.63}$$

This implies that y can completely be described in terms of p_{even}.

We now define the new variables ξ and μ as

$$\left.\begin{aligned} \xi &\equiv \frac{1}{\varepsilon}\int^x \sqrt{Q}dx \\ \mu &\equiv \varepsilon^2\left(\frac{1}{\sqrt[4]{Q}}\right)^3 \frac{d^2}{dx^2}\left(\frac{1}{\sqrt[4]{Q}}\right) \end{aligned}\right\} \tag{9.64}$$

By these, p_{2m} can be expressed as follows (for example, see J.A. Campbell, *J. Comput. Phys.*, Vol. 10, 1972, pp. 308–315):

$$
\begin{aligned}
p_0 &= \sqrt{Q} \\
p_2 &= \frac{1}{2\varepsilon^2}\sqrt{Q}\mu \\
p_4 &= -\frac{1}{8\varepsilon^4}\sqrt{Q}\left(\mu^2 + \frac{d^2\mu}{d\xi^2}\right) \\
p_6 &= \frac{1}{32\varepsilon^6}\sqrt{Q}\left\{2\mu^3 + 6\mu\frac{d^2\mu}{d\xi^2} + 5\left(\frac{d\mu}{d\xi}\right)^2 + \frac{d^4\mu}{d\xi^4}\right\}
\end{aligned}
\tag{9.65}
$$

and so on. Applying a similar idea to (9.57), we obtain

$$
\mathfrak{z}(t) = \frac{1}{\sqrt{q_{\text{even}}}}\, e^{-j(1/\varepsilon)\int^t q_{\text{even}} dt} \tag{9.66}
$$

$$
q_{\text{even}} = q_0 + \varepsilon^2 q_2 + \varepsilon^4 q_4 + \cdots \tag{9.67}
$$

$$
\left.
\begin{aligned}
q_0 &= \sqrt{S} \\
q_2 &= \frac{1}{2\varepsilon^2}\sqrt{S}\tau \\
q_4 &= -\frac{1}{8\varepsilon^4}\sqrt{S}\left(\tau^2 + \frac{d^2\tau}{d\zeta^2}\right) \\
&\cdots
\end{aligned}
\right\}
\tag{9.68}
$$

$$
\zeta \equiv \frac{1}{\varepsilon}\int^t \sqrt{S}dt = \frac{1}{\varepsilon}\int^x \sqrt{\bar{S}(x)}dx \tag{9.69}
$$

$$
\begin{aligned}
\tau &\equiv \varepsilon^2\left(\frac{1}{\sqrt[4]{S}}\right)^3 \frac{d^2}{dt^2}\left(\frac{1}{\sqrt[4]{S}}\right) \\
&= \frac{Q(x) - \bar{S}(x)}{\bar{S}(x)} + \varepsilon^2\left(\frac{1}{\sqrt[4]{\bar{S}(x)}}\right)^3 \frac{d^2}{dx^2}\left(\frac{1}{\sqrt[4]{\bar{S}(x)}}\right)
\end{aligned}
\tag{9.70}
$$

From (9.56),

$$y = \frac{1}{\sqrt{t' q_{\text{even}}}} e^{-j(1/\varepsilon)\int^t q_{\text{even}} dt} = \frac{1}{\sqrt{t' q_{\text{even}}}} e^{-j(1/\varepsilon)\int^x t' q_{\text{even}} dx} \tag{9.71}$$

We notice immediately that, from (9.58),

$$t' = \sqrt{\bar{S}(x)/S(t)} \tag{9.72}$$

thus, $t' q_{\text{even}}$ involved in (9.71) can be given as a function of x such that

$$t' q_{\text{even}} = \sqrt{\bar{S}(x)} \left\{ 1 + \frac{\tau}{2} - \frac{\tau^2}{8} - \frac{1}{8}\frac{d^2\tau}{d\zeta^2} + \cdots \right\} \tag{9.73}$$

As defined in (9.69) and (9.70), the variables ζ and τ are known quantities, provided that $\bar{S}(x)$ is substantiated. The right-hand side of (9.73) also is a known quantity, so that the solution y can be described in terms of $\bar{S}(x)$. This means that a well-chosen $\bar{S}(x)$ will give an adequate approximation to y when we compromise by truncating the right-hand side of (9.73); that is, by neglecting higher-order terms. For example, putting $\bar{S}(x) = Q(x)$, the function τ takes a small value of order ε^2. Therefore, neglecting it, we arrive at the Liouville-Green approximation. Although, of course, the best choice for $\bar{S}(x)$ is such that the value of τ becomes zero identically, to do this, we have to solve (9.60) as we see from (9.70) with $\tau = 0$, and accordingly we arrive at the Milne-Madelung transformation. Thus, we conclude that there are two ways of obtaining approximate solutions: choosing a suitable $\bar{S}(x)$ so that τ becomes small enough to neglect higher-order terms, or applying a simple function for $\bar{S}(x)$ that enables us to calculate the complicated higher-order terms.

9.4.3 Uniform Asymptotic Solutions

Because most of the asymptotic expansions mentioned earlier are formal expansions expressed in the form of a power series with respect to ε, they fail in the boundary layer where $Q \approx 0$. If other coordinates are introduced so that the narrow boundary layer is enlarged enough, the asymptotic expansions associated with these coordinates may converge. In such systems, however, the asymptotic solutions are given by power series expansions containing powers of $\varepsilon^\nu (0 < \nu < 1)$. The uniformly converging series expansions over the entire region are called the *uniform asymptotic expansions*. In the following text, we show the uniform method for the case of $\nu = 2/3$ [6] by C.L. Pekeris, which was used to study the propagation of microwaves along atmospheric ducts.

Here, we again assume that Q has a zero of the first order at $x = x_1$; namely, the Taylor series starts from the leading term with a power of $x - x_1$. Let us enlarge the coordinate such that

$$t = \frac{1}{\varepsilon}\int_{x_1}^{x} \sqrt{Q}dx \tag{9.74}$$

hen, $S(t)$ in (9.57) becomes

$$\left.\begin{aligned} S(t) &= \varepsilon^2(1 + \mu) \\ \mu &= \varepsilon^2\left(\frac{1}{\sqrt[4]{Q}}\right)^3 \frac{d^2}{dx^2}\left(\frac{1}{\sqrt[4]{Q}}\right) \end{aligned}\right\} \tag{9.75}$$

y expanding Q into the Taylor series at $x = x_1$ and expressing it in terms of t, we nd

$$\mu = \frac{5}{36}t^{-2} + \varepsilon^{4/3}At^{-2/3} + \varepsilon^2 B + \varepsilon^{8/3}Ct^{2/3} + \cdots \tag{9.76}$$

Jse of (9.75) and (9.76) in (9.57) provides an asymptotic expansion of the form

$$\begin{aligned} \mathfrak{z}(t) &= t^{1/2}H_{1/3}^{(i)}(t) + \varepsilon^{4/3}\frac{3}{2}At^{5/6}H_{-2/3}^{(i)}(t) \\ &- \varepsilon^2\frac{1}{2}Bt^{3/2}H_{4/3}^{(i)}(t) + 0(\varepsilon^{8/3}) \quad (i = 1, 2) \end{aligned} \tag{9.77}$$

here $H_l^{(i)}(t)$ is the Hankel function, and A, B, and C are constants expressed in rms of higher derivatives of Q. For example,

$$\begin{aligned} A &= \frac{(3/2)^{4/3}}{315}\left[9\left(\frac{Q''}{Q'}\right)^2 - 10\left(\frac{Q'''}{Q'}\right)\right]_{x=x_1} \\ B &= \frac{1}{450}\left\{\frac{1}{Q'}\left[-25\left(\frac{Q''''}{Q'}\right) + 70\left(\frac{Q''Q'''}{Q'^2}\right) - 42\left(\frac{Q''}{Q'}\right)^3\right]\right\}_{x=x_1} \end{aligned} \tag{9.78}$$

he first term on the right-hand side of (9.77) is the Gans uniform solution obtained n 1915. In fact, Gans applied this expression to matching waves and did not use he Airy function. Use of the Airy function in Section 9.2.4, however, is only for onvenience' sake.

.5 EIGENVALUE PROBLEMS

.5.1 Wentzel-Dunham Quantum Condition

f Q has, at least, two zeros of the first order at $x = x_1$ and x_2 and, further, if the alue of Q is positive for $x_1 < x < x_2$, the wave will be localized in this interval.

If the energy of a wave is confined absolutely, the dynamic motion continues forever in its form without any external supply of energy, and we call this situation *resonance*. The resonant wave resulting from such resonance has zeros of the first order at finite points in space. Although the proof is omitted, this will be understood clearly if we think of the wave in a cavity resonator. For the resonant wave represented in the form of (9.45), the function $-jp/\varepsilon$ must have poles of the first order at the points mentioned earlier. Because the value of the integral of p/ε over a simple closed path that contains a pole is 2π, the value of the integral over a closed path that contains finite poles is some integer times 2π. It follows therefore that

$$\frac{1}{\varepsilon}\oint p dx = (2n + 2)\pi \quad (n = 0, 1, 2, \ldots) \tag{9.79}$$

where the integration is performed over a suitable closed path so that all poles are enclosed.

In general, the resonant wave is expressed as a sum of the two "conjugate waves," each of which is progressive. For example, in a uniform isotropic medium the resonant wave is decomposed into a forward wave and a backward wave with phase variations $\exp(\mp jkx)$. It is not possible to construct resonances by a single progressive wave. Therefore, (9.79), in principle, is inapplicable to the system of propagation of a progressive wave. Nevertheless, for the use of the asymptotic function p, which provides an approximate expression to the rigorous p-function, (9.79) still survives. Namely, the quantum condition (9.79) holds valid for the asymptotic series solution p that gives rise to an asymptotically progressive wave. Here, such a series solution must be a divergent series at $x = x_1$ and x_2, thus yielding a formal solution. The integration is performed over a closed path on the complex x-plane, as shown in Figure 9.2. Under these conditions, (9.79) is called the *Wentzel-Dunham quantum condition*. When, in 1926, Wentzel applied it by replacing p with $p_0(=\sqrt{Q})$, the physical significance appeared unclear because of its crudeness [7], and further interpretation of this mathematical calculation was not attempted. A beautiful account of the complex quantum condition (9.79) was given in 1932 by J.L. Dunham [8]. The actual calculations for the left-hand side of (9.79), however, use the integrals of p_m listed in (9.49). For example,

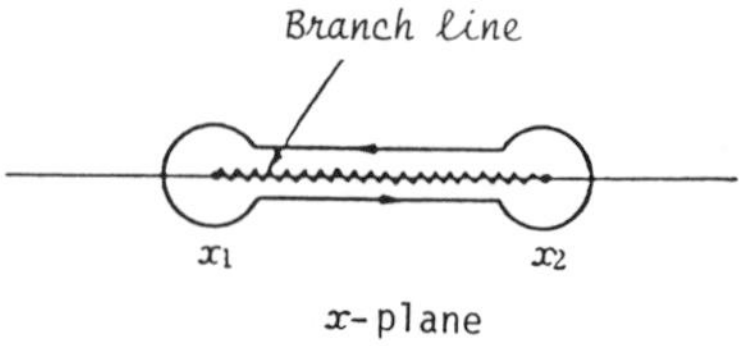

Figure 9.2 A closed path of integration over the complex x-plane.

$$
\begin{aligned}
&\oint p_0 dx = \oint \sqrt{Q} dx \\
&\oint p_1 dx = \pi \\
&\oint p_2 dx = \oint \frac{p_1^2}{2p_0} dx = -\frac{1}{32}\oint \frac{Q'^2}{Q^{5/2}} dx \\
&\oint p_3 dx = 0 \\
&\oint p_4 dx = -\oint \frac{p_2^2}{2p_0} dx \\
&\qquad = -\oint \frac{1}{Q^{11/2}}\left\{\frac{Q''^2Q^2}{128} - \frac{5Q''Q'^2Q}{256} + \frac{25Q'^4}{2048}\right\}dx \\
&\qquad = -\oint \frac{1}{Q^{11/2}}\left\{\frac{49Q'^4}{2048} - \frac{Q'''Q'Q^2}{128}\right\}dx
\end{aligned}
\tag{9.80}
$$

9.5.2 Fröman-Fröman Quantum Condition

Equation (9.80) tells us that all the integrals of the odd terms p_{2m+1} vanish except for p_1. To prove this more exactly, show that the value of the integral of $p_{\text{odd}}/\varepsilon$ is π by virtue of (9.62). Hence, from (9.79),

$$
\frac{1}{\varepsilon}\oint p_{\text{even}} dx = (2n + 1)\pi \quad (n = 0, 1, 2, \ldots) \tag{9.81}
$$

which leads to the Kramers quantum condition (9.54), provided that p_{even} is truncated and approximated by p_0.

The earlier-mentioned condition is invariant by the Langer transformation. Therefore, by applying the result obtained in 9.4.2,

$$
\begin{aligned}
\frac{1}{\varepsilon}\oint q_{\text{even}} dt &= \frac{1}{\varepsilon}\oint \sqrt{\bar{S}(x)}\left\{1 + \frac{\tau}{2} - \frac{\tau^2}{8} - \frac{1}{8}\frac{d^2\tau}{d\zeta^2} + \cdots\right\}dx \\
&= (2n + 1)\pi \quad (n = 0, 1, 2, \ldots)
\end{aligned}
\tag{9.82}
$$

If we put $\bar{S}(x) = Q(x)$ and neglect the higher-order terms in braces, we again obtain the Kramers condition. Instead, if we choose the function $\bar{S}(x)$ in such a way that the values of τ and its derivatives become smaller, the resulting expression is more precise.

9.5.3 Maslov-Argyres Regularization

In physics or technology, we often give the function $Q(x)$ of the form

$$Q(x) = \lambda - V(x) \tag{9.83}$$

with potential $V(x)$ and eigenvalue λ. Then, the curve of $V(x)$ is assumed to be smooth enough and, further, less than the value of λ between x_1 and x_2, whereby $Q(x)$ is positive there and has simple zeros at x_1 and x_2. In such a case, for example the following relation holds:

$$-\frac{1}{32}\frac{Q'^2}{Q^{5/2}} = -\frac{1}{24}\frac{\partial^2}{\partial\lambda^2}\left\{\frac{(V')^2}{\sqrt{Q}}\right\} \tag{9.84}$$

The left-hand side of this equation exhibits a strong singularity, whereas the term in braces on the right-hand side does a weak singularity of order $1/\sqrt{x - x_1}$, $1/\sqrt{x_2 - x}$, or, at most, similar singularities. The integration of such a weak function is possible, and, as a result, the integral has a finite value. In fact,

$$\oint p_2 dx = -\frac{1}{12}\frac{\partial^2}{\partial\lambda^2}\int_{x_1}^{x_2}\frac{(V')^2}{\sqrt{Q}}dx \tag{9.85}$$

Similarly, for p_4, we obtain

$$\oint p_4 dx = \frac{1}{1440}\frac{\partial^3}{\partial\lambda^3}\int_{x_1}^{x_2}\frac{7(V'')^2 - 5V'V'''}{\sqrt{Q}}dx \tag{9.86}$$

by integration by parts. Note that the upper and lower bounds of the integral, x_2 and x_1, are functions of λ and will later suffer from differentiation with respect to λ. This regularization of integration was found in 1960 by V.P. Maslov and independently in 1965 by P.N. Argyres, thus providing a beautiful formula that had never been obtained. When λ is expanded into an asymptotic series

$$\lambda = \lambda^{(1)} + \varepsilon^2\lambda^{(2)} + \varepsilon^4\lambda^{(3)} + \cdots \tag{9.87}$$

$\lambda^{(1)}$ is determined from

$$\frac{1}{\varepsilon}\int_{x_1}^{x_2}\sqrt{Q}dx = \pi\left(n+\frac{1}{2}\right) \tag{9.88}$$

nd the higher-order terms, $\lambda^{(2)}$, $\lambda^{(3)}$, and so on, are calculated from the following ormulas by using the obtained $\lambda^{(1)}$ [9]:

$$\lambda^{(2)} = \left\{\frac{1}{12}\frac{\partial^2}{\partial\lambda^2}\int_{x_1}^{x_2}\frac{(V')^2}{\sqrt{Q}}dx \bigg/ \int_{x_1}^{x_2}\frac{1}{\sqrt{Q}}dx\right\}_{\lambda=\lambda^{(1)}} \tag{9.89}$$

$$\begin{aligned}\lambda^{(3)} = &\left\{\left[\frac{\lambda^{(2)}}{12}\frac{\partial^3}{\partial\lambda^3}\int_{x_1}^{x_2}\frac{(V')^2}{\sqrt{Q}}dx - \frac{(\lambda^{(2)})^2}{2}\frac{\partial}{\partial\lambda}\int_{x_1}^{x_2}\frac{1}{\sqrt{Q}}dx\right.\right.\\ &\left.\left.-\frac{1}{1440}\frac{\partial^3}{\partial\lambda^3}\int_{x_1}^{x_2}\frac{7(V'')^2-5V'V'''}{\sqrt{Q}}dx\right] \bigg/ \int_{x_1}^{x_2}\frac{1}{\sqrt{Q}}dx\right\}_{\lambda=\lambda^{(1)}}\end{aligned} \tag{9.90}$$

lthough Maslov's original expression is slightly different in form, it is the same in ssentials. Hence, (9.89) and (9.90) are called the *Maslov formulas* for asymptotic igenvalues.

.5.4 Uniform Asymptotic-Perturbational Method

et $Q(x)$ be a smooth function whose zeros are x_1 and x_2, as defined earlier. Here, ve adopt the Langer transformation of the form

$$A\int_{-1}^{t}\sqrt{1-t^2}dt = \int_{x_1}^{x}\sqrt{Q}dx \tag{9.91}$$

vith A constant. If we let $t = \pm 1$ correspond to $x = x_1$ and x_2, respectively, then he value of A can be determined from the substitution of $t = 1 (x = x_2)$ as follows:

$$A = \frac{2}{\pi}\int_{x_1}^{x_2}\sqrt{Q}dx \tag{9.92}$$

nd also from (9.58) the following expressions can be derived:

$$\left.\begin{aligned}\bar{S}(x) &= Q(x) + \varepsilon^2\sqrt{\frac{Q}{1-t^2}}\frac{d^2}{dx^2}\sqrt[4]{\frac{1-t^2}{Q}}\\ S(t) &= A^2\left\{1-t^2+\varepsilon^2\left(\frac{1-t^2}{Q}\right)^{3/4}\frac{d^2}{dx^2}\left(\frac{1-t^2}{Q}\right)^{1/4}\right\}\end{aligned}\right\} \tag{9.93}$$

Neglecting the terms higher than the second-order term with respect to ε, we have $S(t) \approx A^2(1 - t^2)$. Hence, a possible solution of (9.57) may be given by

$$\mathfrak{z}(t) \approx W_\nu\left(2\sqrt{\nu + \frac{1}{2}}\, t\right), \quad \nu = \frac{A - \varepsilon}{2\varepsilon} \tag{9.94}$$

where $W_\nu(x)$ is the Weber function, which is a solution of the differential equation

$$W_\nu'' + \left(\nu + \frac{1}{2} - \frac{x^2}{4}\right)W_\nu = 0 \tag{9.95}$$

To confine the energy of waves to some bounded region as required, the condition such that $\mathfrak{z}(t) = 0$ at $t = \pm\infty$ is necessary. This is equivalent to that ν is a positive integer. In consequence,

$$A = 2n + 1 = \frac{2}{\pi\varepsilon}\int_{x_1}^{x_2} \sqrt{Q}dx \tag{9.96}$$

$$y \approx \sqrt[4]{\frac{1 - t^2}{Q}}\, W_n\left(2\sqrt{n + \frac{1}{2}}\, t\right) \tag{9.97}$$

Note that (9.96) is the Kramers condition (9.54). Although the method just described used the Langer transformation and thereby provided the Kramers approximate expression for the quantum condition, repeated use will provide more accurate expressions. This was suggested in 1959 by H. Moriguchi in improving the WKB method. Recently, the mathematical proof for the convergence of the repeated solution was given by A. Yata and H. Ikuno. Therefore, the quantum condition obtained by applying the approximation $A = 2n + 1$ after repeated use of the transformation is called the *Moriguchi-Yata-Ikuno condition*.

9.5.5 Felsen Series

So far we have discussed the eigenvalue problem in the case when two boundary layers exist at x_1 and x_2 and where the normal asymptotic expansions fail. In the present subsection, we show that the problem can be reduced to the one-boundary layer problem, provided that $Q(x)$ has the form (9.83) and the value of $V(x)$ is positive or identically zero. For the sake of simplicity, we assume that $V(x) = 0$ at $x = 0$. We also assume that the solution y is expressed in terms of the amplitude and the eigenvalue λ, all of which are expanded into power series of ε:

$$y = A\, e^{-(1/\varepsilon)\int_0^x \sqrt{V}dx} \tag{9.98}$$

$$\left.\begin{aligned} A &= A(x) = A_0(x) + \varepsilon A_1(x) + \varepsilon^2 A_2(x) + \cdots \\ \lambda &= \varepsilon\lambda_1 + \varepsilon^2\lambda_2 + \varepsilon^3\lambda_3 + \cdots \end{aligned}\right\} \tag{9.99}$$

Then, the differential equations of the first order for A_n are derivable as follows:

$$\begin{aligned} &A_0' + \left(\frac{V'}{4V} - \frac{\lambda_1}{2\sqrt{V}}\right)A_0 = 0 \\ &A_1' + \left(\frac{V'}{4V} - \frac{\lambda_1}{2\sqrt{V}}\right)A_1 = \frac{A_0'' + \lambda_2 A_0}{2\sqrt{V}} \\ &A_2' + \left(\frac{V'}{4V} - \frac{\lambda_1}{2\sqrt{V}}\right)A_2 = \frac{A_1'' + \lambda_2 A_1 + \lambda_3 A_0}{2\sqrt{V}}, \ \ldots \end{aligned} \tag{9.100}$$

By solving these, we obtain

$$\begin{aligned} &A_0 = \frac{1}{\sqrt[4]{V}}\, e^{\int^x (\lambda_1/2\sqrt{V})dx}, \quad A_1 = A_0 \int^x \frac{A_0'' + \lambda_2 A_0}{2\sqrt{V}A_0}\,dx \\ &A_2 = A_0 \int^x \frac{A_1'' + \lambda_2 A_1 + \lambda_3 A_0}{2\sqrt{V}A_0}\,dx, \ \ldots \end{aligned} \tag{9.101}$$

where the lower limits of integration are not yet determined. Assuming $V(x)$ to be analytic at $x = 0$, we require that the resulting asymptotic solution y is also analytic and bounded at $x = 0$. This requirement determines the lower limits of integration that remain undetermined. The reader should ascertain that for $V(x) = x^2/4$, $\lambda_1 = 2n + 1$, $\lambda_2 = \lambda_3 = \cdots = 0$, $y = W_n(x/\sqrt{\varepsilon})$ as an illustrative example.

This asymptotic expansion originated in the early theory by J. Horn (in 1899). Further advances in the theory of waveguides are due from L.B. Felsen in 1976 [10]. Of particular interest is Felsen's introduction of the imaginary phase of the form $-j\int^x \sqrt{V}dx$ to reduce the number of boundary layers in which the series expansions diverge.

9.6 MULTIPLE SCATTERING EXPANSIONS

9.6.1 Solutions of Coupled Equations

For a uniform transmission line where the line inductance and line capacitance per unit length are constant, the line voltage V and the line current I satisfy the coupled equations of the matrix form

$$\frac{d}{dx}\begin{pmatrix} V \\ I \end{pmatrix} = \begin{pmatrix} 0 & -j\omega L \\ -j\omega C & 0 \end{pmatrix}\begin{pmatrix} V \\ I \end{pmatrix} \tag{9.102}$$

where ω is the angular frequency of guided waves. Looking at the 2×2 matrix in the right-hand side, we notice that the off-diagonal elements are zero, thus showing the strongest coupling of equations. On the other hand, a single equation described for V in the equivalent form is

$$V'' + \omega^2 LCV = 0 \tag{9.103}$$

However, we may use any other expressions for V in the matrix form, introducing other quantities instead of V and I. For example, suppose that $\phi^{\pm}$ are defined by

$$\phi^{\pm} = \frac{1}{2}\left(V \pm \sqrt{\frac{L}{C}} I\right) \tag{9.104}$$

Then, we find that $\phi^{\pm}$ satisfies the following equations where the coupling is completely eliminated, and thereby the two equations become independent:

$$\frac{d}{dx}\begin{pmatrix} \phi^{+} \\ \phi^{-} \end{pmatrix} = \begin{pmatrix} -j\omega\sqrt{LC} & 0 \\ 0 & +j\omega\sqrt{LC} \end{pmatrix}\begin{pmatrix} \phi^{+} \\ \phi^{-} \end{pmatrix} \tag{9.105}$$

These uncoupled equations can be solved with respect to $\phi^{\pm}$.

How about a nonuniform transmission line where the line capacitance C is function of x and the line inductance L is constant. For this case, according to (9.104), we define V and I as

$$\left.\begin{aligned} V &= \phi^{+} + \phi^{-} \\ I &= -\frac{V'}{j\omega L} = \sqrt{\frac{C}{L}}(\phi^{+} - \phi^{-}) \end{aligned}\right\} \tag{9.106}$$

Then, instead of (9.105), we have

$$\frac{d}{dx}\begin{pmatrix} \phi^{+} \\ \phi^{-} \end{pmatrix} = \begin{pmatrix} -j\omega\sqrt{LC} - \dfrac{C'}{4C} & \dfrac{C'}{4C} \\ \dfrac{C'}{4C} & j\omega\sqrt{LC} - \dfrac{C'}{4C} \end{pmatrix}\begin{pmatrix} \phi^{+} \\ \phi^{-} \end{pmatrix} \tag{9.107}$$

Provided that $C(x)$, a slowly varying function whose derivative C' with respect to x is negligible, the off-diagonal elements are small enough to obtain the weakly cou-led equations. Neglecting these off-diagonal elements leads to the uncoupled equa-ons and again yields the Liouville-Green solution.

In fact, the relation (9.106) between V, I and ϕ^+, ϕ^- shows a transformation or diagonalization to the off-diagonal matrix given in (9.105). Another transfor-ation may diagonalize the weak matrix appearing on the right-hand side of (9.107). y means of it, we can obtain the more weakly coupled equations. Repeating this rocess, we can obtain a desired expression for equations in the coupled system. he method was found by N.G. Van Kampen in 1967, but a more general method, s will be seen later, was proposed that year by E. Bahar [11].

Now, we note that if $y = V$, $\varepsilon = 1/\omega$, and $Q = LC$ in (9.106), then,

$$y = \phi^+ + \phi^-, \quad y' = -j\frac{1}{\varepsilon}\sqrt{Q}(\phi^+ - \phi^-) \tag{9.108}$$

o generalize this expression, we write y and y' in terms of coupling elements C_{11}, C_{12}, C_{21}, C_{22}:

$$\left.\begin{aligned} y &= \phi^+ + \phi^- \\ y' &= (C_{11} + C_{21})\phi^+ + (C_{12} + C_{22})\phi^- \end{aligned}\right\} \tag{9.109}$$

here the values of $\phi^\pm$ are solutions of

$$\frac{d}{dx}\begin{pmatrix}\phi^+ \\ \phi^-\end{pmatrix} = \begin{pmatrix} C_{11} & C_{12} \\ C_{21} & C_{22}\end{pmatrix}\begin{pmatrix}\phi^+ \\ \phi^-\end{pmatrix} \tag{9.110}$$

eferring to (9.107), Bahar defined the "locally forward" wave g_+ and the "locally ackward" wave g_- as

$$\left.\begin{aligned} C_{11} + C_{21} &= g_+'/g_+ \\ C_{12} + C_{22} &= g_-'/g_- \end{aligned}\right\} \tag{9.111}$$

le also chose C_{12} and C_{21} to be

$$\left.\begin{aligned} C_{12} &= -\frac{g_-''/g_- + Q/\varepsilon^2}{g_+'/g_+ - g_-'/g_-} \\ C_{21} &= -\frac{g_+''/g_+ + Q/\varepsilon^2}{g_-'/g_- - g_+'/g_+} \end{aligned}\right\} \tag{9.112}$$

o that y could satisfy (9.35). If the values of $g_\pm$ were solutions of (9.35), then C_{12} = $C_{21} = 0$, thus making (9.110) solvable. A better choice for $\phi^\pm$ is such that the alues of C_{12} and C_{21} are adequately small. For example, using

$$g_{\pm} = e^{\mp j(1/\varepsilon)\int^{x}\sqrt{Q}dx} \quad (9.113)$$

in (9.111) and (9.112) to calculate C_{11} and C_{22}, and neglecting the off-diagonal elements in (9.110), we obtain the Liouville-Green solution:

$$\phi^{\pm} = \frac{1}{\sqrt[4]{Q}} e^{\mp j(1/\varepsilon)\int^{x}\sqrt{Q}dx} \quad (9.114)$$

Obviously, these are more refined solutions than $g_{\pm}$. Accordingly, to get even more refined solutions, they are substituted into $g_{\pm}$ again. By the repeated substitution more accurate solutions may be obtained.

We next express the obtained solutions in power series form. To this end, we write the solutions of (9.110) in integral form,

$$\left.\begin{aligned} \phi^{+}(x) &= \int_{-\infty}^{x} C_{12}(\xi)\phi^{-}(\xi)\, e^{\int_{\xi}^{x} C_{11}(t)dt} d\xi \\ \phi^{-}(x) &= -\int_{x}^{\infty} C_{21}(\xi)\phi^{+}(\xi)\, e^{\int_{\xi}^{x} C_{22}(t)dt} d\xi \end{aligned}\right\} \quad (9.115)$$

and we expand $\phi^{\pm}$ into

$$\left.\begin{aligned} \phi^{+} &= \phi_0^{+} + \phi_2^{+} + \phi_4^{+} \cdots \\ \phi^{-} &= \phi_1^{-} + \phi_3^{-} + \phi_5^{-} \cdots \end{aligned}\right\} \quad (9.116)$$

Then, we find the following recurrence formulas:

$$\left.\begin{aligned} \phi_n^{+}(x) &= \int_{-\infty}^{x} C_{12}(\xi)\phi_{n-1}^{-}(\xi)\, e^{\int_{\xi}^{x} C_{11}(t)dt} d\xi \\ \phi_n^{-}(x) &= -\int_{x}^{\infty} C_{21}(\xi)\phi_{n-1}^{+}(\xi)\, e^{\int_{\xi}^{x} C_{22}(t)dt} d\xi \end{aligned}\right\} \quad (9.117)$$

with the initial term of the form

$$\phi_0^{+}(x) = e^{\int^{x} C_{11}(t)dt} \quad (9.118)$$

If the power series y

$$y = \phi_0^{+} + \phi_1^{-} + \phi_2^{+} + \phi_3^{-} + \cdots \quad (9.119)$$

nverges uniformly, then y gives a rigorous solution of (9.103). However, except r special cases, there is no proof for general conditions. Today, the series expanons arising from the following choice of $g_{\pm}$ are named:

$$\begin{aligned} &(A) \quad g_{\pm} = e^{\mp j(1/\varepsilon)\sqrt{Q(-\infty)}x} &&\text{(Heading series)} \\ &(B) \quad g_{\pm} = e^{\mp j(1/\varepsilon)\int^x \sqrt{Q}dx} &&\text{(Bremmer series)} \qquad (9.120) \\ &(C) \quad g_{\pm} = \frac{1}{\sqrt[4]{Q}} e^{\mp j(1/\varepsilon)\int^x \sqrt{Q}dx} &&\text{(Sluijter series)} \end{aligned}$$

.6.2 Multilayer Solutions

gain, we return to (9.102) and solve the differential equation (9.35) by a transiission line analogy where we use the same notations as in the previous subsection:

$$y = V, \quad \varepsilon = 1/\omega, \quad Q(x) = LC(x), \quad I = -V'/j\omega L$$

/e assume that the line inductance per unit length, L, is constant and the line caacitance per unit length, $C(x)$, varies with x. Figure 9.3 shows an electrical ladder etwork system corresponding to the continuous system just stated. The length of iterval, Δx, is taken small so that, in the limit of $\Delta x \to 0$, the discrete system oincides with the original circuit system. An additional requirement is that, in the h interval, the normal modes (the line voltages such that $|a_{\pm}| = 1$ at $x = x_i$) repesenting the forward wave a_+ and the backward wave a_- are given and $C_i(x)$ is hosen to yield a good approximation to $C(x)$.

The difference in transmission characteristics between adjacent intervals, hower, causes the reflection of the wave to occur at the connecting terminals. The icident wave propagating from $x = -\infty$ to the right will be reflected partly at each rminal, and the remainder will be transmitted. A collection of these reflected waves ropagating to the left is denoted by ϕ_1^-, and the transmitted wave by ϕ_0^+. Also, a ollection of the reflected waves occurring by the reflection of the left-going wave

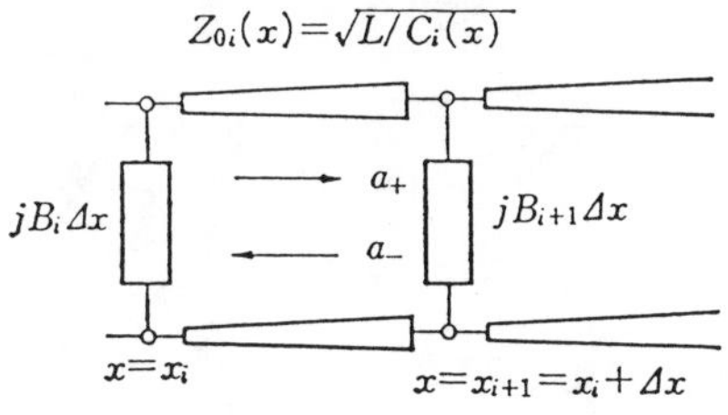

igure 9.3 General ladder network model.

ϕ_1^- is denoted by ϕ_2^+. Similarly, for higher reflections, the same designations a applied where plus or minus denotes the right- or left-going wave, respectively, an the number i denotes the degree of multiscattering. In this way, we can express th solution of (9.35) as the multiple-scattering solution. Such an expression with $\Delta x -$ 0, indeed, coincides with the integral expressions (9.115) to (9.119) derived by th coupled equation formalism. Accordingly, the series solution (9.119) obtained b solving the coupled equations is an ultimate expression for the multiple-scatterir expansions in the ladder network system.

As an example, Figure 9.4 shows several ladder circuits corresponding to (9.120 For Figure 9.4(a), the susceptance B_i mounted at the ith terminal corresponds to th increment of $C(x)$, and instead, the transmission lines in all intervals are the sam For Figure 9.4(b), on the other hand, no susceptance exists, and instead, the cha acteristic impedances of the lines are different but uniform in the intervals. Th method, sometimes called a *staircase approximation method*, already was used : 1912 by Lord Rayleigh. In 1951, H. Bremmer pointed out this relation between tw expressions for Figure 9.4(b). Figure 9.4(c) is a refined model proposed in 1969 b F.W. Sluijter, in which the capacitance $C_i(x)$ is given by the equation in the figur In this model, however, the normal modes $a_\pm(x)$ are known to be given in the Liou ville-Green form, which are not approximate solutions but absolutely exact solution We note, in general, that $a_\pm$ is the approximate solution to $g_\pm$.

(A)

$$Z_0 = \frac{L}{\sqrt{Q(-\infty)}}$$

$$B_i Z_0 = \omega \left(\frac{Q(x_i) - Q(-\infty)}{\sqrt{Q(-\infty)}} \right)$$

(B)

$$Z_{0i} = \frac{L}{\sqrt{Q(x_i)}}$$

(C)

$$Z_{0i+1}(x) = \sqrt{\frac{L}{C_{i+1}(x)}}$$

$$C_i(x) = \frac{1}{L}$$

$$\times \frac{(x_{i+1} - x_i)^4 Q(x_i) Q(x_{i+1})}{\{ \sqrt[4]{Q(x_i)}(x - x_i) + \sqrt[4]{Q(x_{i+1})}(x_{i+1} - x) \}^4}$$

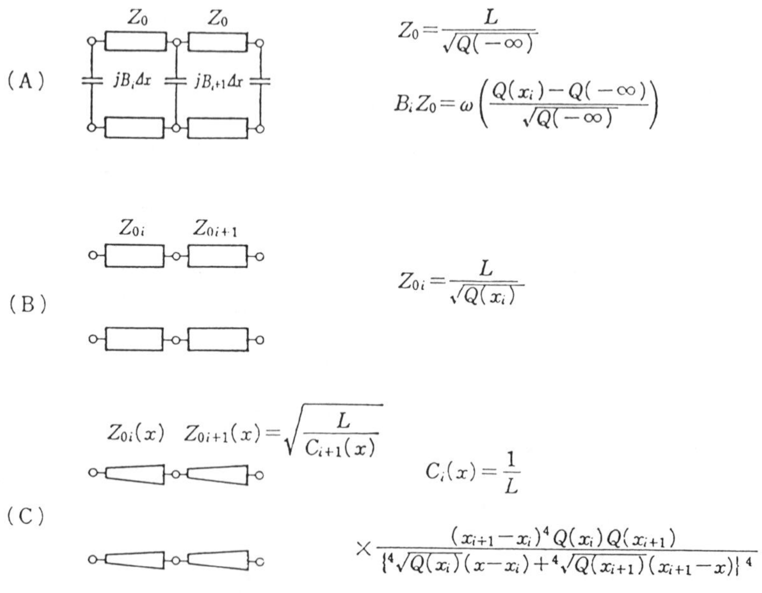

Figure 9.4 Specific examples of ladder network models.

We have so far described the equivalence between the series solution (9.119) the coupled equations and the multiple-scattering solution of the ladder network stem and have shown some examples. A unique determination of the ladder trans-ission lines that provide the same series may be impossible, however; for example, her models might provide the Sluijter series. We can prove that, if the normal odes $a_{\pm}$ in the ith interval satisfy the two relations that follow, then the ultimate lution in the limit of $\Delta x \rightarrow 0$ provides the same solution as (9.119):

$$\frac{a'_{\pm}(x_i)}{a_{\pm}(x_i)} = \frac{g'_{\pm}(x_i)}{g_{\pm}(x_i)} + \frac{1}{2}\left\{\frac{g''_{\pm}(x_i)}{g_{\pm}(x_i)} + \omega^2 LC(x_i) - \omega LB_i\right\}\Delta x + 0\{(\Delta x)^2\}$$

$$\frac{a'_{\pm}(x_{i+1})}{a_{\pm}(x_{i+1})} = \frac{g'_{\pm}(x_{i+1})}{g_{\pm}(x_{i+1})} - \frac{1}{2}\left\{\frac{g''_{\pm}(x_{i+1})}{g_{\pm}(x_{i+1})} + \omega^2 LC(x_{i+1}) - \omega LB_{i+1}\right\}\Delta x + 0\{(\Delta x)^2\} \tag{9.121}$$

ecause this statement is general, (9.121) will be advantageous for future work in-estigating a new ladder model in connection with the known solutions of the cou-led equations.

.6.3 The Invariant Imbedding Approach

s mentioned in Section 9.3, solving the linear differential equations of the second rder is equivalent to solving the nonlinear differential equation (9.46) of the first rder. For the former, two boundary conditions are necessary for the solution; whereas, or the latter, one condition is satisfactory. In such a nonlinear system, the structure f the original equation is mapped onto another system and is thus invariant. This apping is called the *imbedding*. For example, in (9.46), using the variable t

$$j\frac{p}{\varepsilon} = \tan t \tag{9.122}$$

quation (9.44) is mapped onto the following system:

$$\frac{dt}{dx} = \sin^2 t + \frac{Q}{\varepsilon^2}\cos^2 t \tag{9.123}$$

r

$$y = A\cos t, \quad y' = -A\sin t$$

his is called the *Prüfer transformation* after H. Prüfer [12]. In these two examples, ne two-point boundary value problems are mapped onto the one-point boundary value problems with the physical phenomena invariant. This idea was initially ap plied in 1942 for the reflection problem of waves by Russian astronomer V.A. Am barzumian [13] and was later developed by R. Bellman and R. Kalaba. Today, it well known as the *invariant imbedding approach.*

However, electrical engineers who studied network theory could feel that th fundamental concept of the invariant imbedding was not new. The reason is clea because the reflection coefficient R for the normal mode at each terminal has com plete information on the structure of y and y therefore can be described in terms c R, the problem is now reduced to determining R. This is the principle of invariar imbedding. Provided that the forward wave ϕ^+ and the backward wave ϕ^- are de fined by

$$\phi^+ = T, \quad \phi^- = TR \tag{9.124}$$

where T is the transmission coefficient, by substituting these into (9.110), we obtai

$$T' = C_{11}T + C_{12}TR \tag{9.125}$$

$$R' = C_{21} + (C_{22} - C_{11})R - C_{12}R^2 \tag{9.126}$$

These are the Riccati nonlinear equations and thus can be solved by machine com putation as the one-point boundary value problem. Numerical analysis can be mad more accurate when the coefficient C_{12} for the nonlinear term is much smaller.

The Riccati nonlinear equations corresponding to (9.120) are as follows:

$$\begin{aligned}
&\text{(A)} \quad R' = j\frac{1}{2\varepsilon}\frac{1}{\sqrt{Q(-\infty)}}\{[Q(x) - Q(-\infty)](1 + R^2) \\
&\qquad\qquad + 2[Q(x) + Q(-\infty)]R\} \\
&\text{(B)} \quad R' = \frac{1}{4}\frac{Q'}{Q}(1 - R^2) + j\frac{2}{\varepsilon}\sqrt{Q}R \\
&\text{(C)} \quad R' = j\frac{2}{\varepsilon}\sqrt{Q}R - j\frac{\varepsilon}{8}\frac{1}{\sqrt{Q}}\left(\frac{Q''}{Q} - \frac{5}{4}\frac{Q'^2}{Q}\right)(1 + R^2)
\end{aligned} \tag{9.127}$$

9.7 ASYMPTOTIC EXPANSIONS IN AN OPTICAL WAVEGUIDE SYSTEM

Up to now, we have described the methods of asymptotic expansions. In this section we apply these methods to the practical problem of optical waveguides. Application

) two-dimensional planar waveguides are omitted because these are described by).35). We give asymptotic solutions for vectorial waves in Maxwell equations to valuate the guided waves of optical fiber modes. Because we use notations common) the Maxwell system, which differ from those in the foregoing sections, be careful ot to be confused at new notations. In particular, ε represents the permittivity of ie medium, not the small parameter as defined previously. Also, in the coordinate ystem, we define z as the direction of propagation or the axis of the optical fiber, nd use the cylindrical coordinates r, θ, and z for geometrical convenience. The hase factor $\exp(-j\beta z)$ is assumed through this section where β is the phase constant f the guided mode. The permittivity ε and the permeability μ of the medium are unctions of r. Accordingly, all the quantities that follow are known functions of r.

$$\varepsilon = \varepsilon(r) = \text{the permittivity of the medium}$$
$$\mu = \mu(r) = \text{the permeability of the medium}$$
$$k = k(r) = \omega\sqrt{\varepsilon(r)\mu(r)} = \text{the wavenumber in the medium}$$

'or most cases where optical fibers are made of dielectric glass materials, $\mu = \mu_0$ = constant. Therefore, we treat more general cases from a theoretical point of view.

The time variation $\exp(j\omega t)$ and the angular variation $\exp(jm\theta)$ ($m = 0, 1, 2, \ldots$) are suppressed here, and the longitudinal components of the electric and magnetic fields in the Maxwell system, denoted by E_z and H_z, respectively, are written n the form

$$E_z = \frac{1}{\omega\sqrt{\mu(0)}}\left(\frac{e(r)}{\sqrt{\varepsilon(r)}}\right) e^{j\omega t + jm\theta - j\beta z} \tag{9.128}$$

$$H_z = j\frac{1}{\omega\sqrt{\mu(0)}}\left(\frac{h(r)}{\sqrt{\mu(r)}}\right) e^{j\omega t + jm\theta - j\beta z} \tag{9.129}$$

where the functions $e(r)$ and $h(r)$ obey the coupled differential equations,

$$\left.\begin{aligned} e'' + \left(\frac{1}{r} - \frac{Q'}{Q}\right)e' + Re &= -Mh \\ h'' + \left(\frac{1}{r} - \frac{Q'}{Q}\right)h' + Lh &= -Me \end{aligned}\right\} \tag{9.130}$$

with primes denoting differentiation with respect to r, and Q, R, L, and M being the following quantities:

$$
\begin{aligned}
Q &= k^2(r) - \beta^2 \\
R &= Q - \frac{m^2}{r^2} + \frac{1}{2}\left(\frac{\varepsilon'}{\varepsilon}\right)\left(\frac{Q'}{Q}\right) - \frac{1}{2r}\left(\frac{\varepsilon'}{\varepsilon}\right) - \frac{1}{2}\left(\frac{\varepsilon'}{\varepsilon}\right)' - \frac{1}{4}\left(\frac{\varepsilon'}{\varepsilon}\right)^2 \\
L &= Q - \frac{m^2}{r^2} + \frac{1}{2}\left(\frac{\mu'}{\mu}\right)\left(\frac{Q'}{Q}\right) - \frac{1}{2r}\left(\frac{\mu'}{\mu}\right) - \frac{1}{2}\left(\frac{\mu'}{\mu}\right)' - \frac{1}{4}\left(\frac{\mu'}{\mu}\right)^2 \\
M &= \frac{m}{r}\left(\frac{\beta}{k}\right)\left(\frac{Q'}{Q}\right)
\end{aligned}
\tag{9.131}
$$

Mathematically speaking, the starting point in this section is these coupled equations for e and h and the final goal is to determine the value of β by expanding it into an asymptotic series. A fuller account of calculations may be lengthy, and some manipulations for equations also may be needed. To outline the analysis, we omit them and summarize the process of calculations and the results (for details, refer to [14]).

Suppose the solutions of (9.130) are expressed as the wave progressive in the positive direction of r,

$$e = \sqrt{\frac{Q}{r}}\, \mathrm{e}^{-j\int_{-j\delta}^{r} p_e dr} \tag{9.132}$$

$$h = P\sqrt{\frac{Q}{r}}\, \mathrm{e}^{-j\int_{-j\delta}^{r} p_m dr} \tag{9.133}$$

Hence, these are progressive components of the modal wave in the optical fiber and give rise to a modal resonance together with their conjugate components of the backward wave. The progressive waves (9.128) and (9.129) do not express the true fields of the modal wave, thus diverging at $r = 0$ with the same order. To avoid difficulty in solving the mathematical problem of divergence, we choose the starting point of phase integration to be located on the imaginary axis over the complex r-plane and very near the origin such that $r = -j\delta$ (δ = a small positive parameter) (see Figure 9.5). The parameter P denotes the state of polarization of the progressive wave. Here, we define P as

$$P = \left(\frac{h}{e}\right)_{r=0} \tag{9.134}$$

However, as will be seen later, the following relation holds over the entire region of r:

$$P = -j\sqrt{\frac{\mu}{\varepsilon}\frac{H_z}{E_z}} + 0\left(\frac{1}{k}\right) \tag{9.135}$$

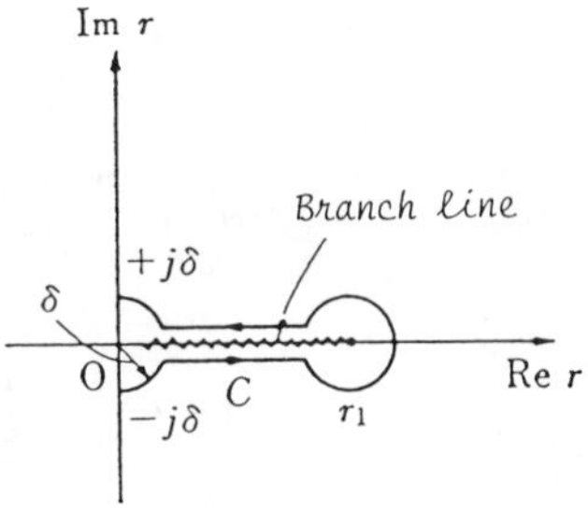

igure 9.5 The path of integration over the complex r-plane.

By deforming this somewhat, we find $P \approx j(\beta/k)E_\theta/E_r$ as the ratio of the azimuthal component E_θ of the electric field to the radial component $E_r \cdot P = +1$ shows that he progressive wave expressed by (9.128) and (9.129) is in the right-hand circularly polarized state. On the other hand, $P = -1$ shows the wave in the left-hand circularly polarized state. In this way, P is the ray parameter describing the state of polarization of the modal ray traveling along the skew path or the meridional path in the optical fiber. Because discussions on the polarization of the modal ray are beyond the scope of this book, the interested reader can refer to M. Hashimoto, *An Introduction to Electromagnetic Theory of Guided Waves* (Nikkan-Kogyo-Shinbun, Tokyo, 1985, Chapter 7).

Now, let us derive the coupled Riccati nonlinear equations for p_e and p_m by substituting (9.132) and (9.133) into (9.130) as in (9.46). Asymptotic solutions for those equations are

$$\left.\begin{matrix} p_e \\ p_m \end{matrix}\right\} = \sqrt{Q} - j\frac{1}{4}\left(\frac{Q'}{Q}\right) + \left[\frac{\dfrac{m^2 - 1/4}{r} + \dfrac{1}{4}\left(\dfrac{Q'}{Q}\right)}{2\sqrt{Q}}\right]'$$

$$+ \frac{m^2 + \dfrac{3}{4}}{4r}\left(\frac{Q'}{Q}\right)\frac{1}{\sqrt{Q}} - \frac{1}{32}\left(\frac{Q'}{Q}\right)^2\frac{1}{\sqrt{Q}} \tag{9.136}$$

$$+\begin{cases} \dfrac{1}{4}\left(\dfrac{\varepsilon'}{\varepsilon}\right)\left(\dfrac{Q'}{Q}\right)\dfrac{1}{\sqrt{Q}} + \dfrac{MP}{2\sqrt{Q}} - \dfrac{1}{4\sqrt{Q}}\left[\dfrac{1}{r}\left(\dfrac{\varepsilon'}{\varepsilon}\right) + \left(\dfrac{\varepsilon'}{\varepsilon}\right)' + \dfrac{1}{2}\left(\dfrac{\varepsilon'}{\varepsilon}\right)^2\right] \\ \dfrac{1}{4}\left(\dfrac{\mu'}{\mu}\right)\left(\dfrac{Q'}{Q}\right)\dfrac{1}{\sqrt{Q}} + \dfrac{M}{2P\sqrt{Q}} - \dfrac{1}{4\sqrt{Q}}\left[\dfrac{1}{r}\left(\dfrac{\mu'}{\mu}\right) + \left(\dfrac{\mu'}{\mu}\right)' + \dfrac{1}{2}\left(\dfrac{\mu'}{\mu}\right)^2\right] \end{cases}$$

$$+ 0(1/k^2)$$

We have given the rigorous quantum condition for determining the values of P and β in the integral form [14]

$$\int_C p_e dr = 2\pi\left(l + \frac{m}{2} + \frac{1}{4}\right)$$

$$\int_C p_m dr = 2\pi\left(l + \frac{m}{2} + \frac{1}{4}\right) \tag{9.137}$$

$$(l, m = 0, 1, 2, \ldots)$$

Substitution of the asymptotic solutions p_e and p_m into these equations gives the values of P and β in the modal state. The path of integration over the complex r-plane is taken to surround the location of the caustic, r_1, such that $Q = 0$, as indicated in Figure 9.5. We then introduce a new parameter, λ, and a new function $f(r)$ instead of β and $k(r)$, respectively:

$$k(r) = k(0)\sqrt{1 - f(r)}$$

$$\beta = k(0)\sqrt{1 - \lambda} \tag{9.138}$$

$$Q(r) = k^2(0)\{\lambda - f(r)\}$$

and expand λ into a series

$$\lambda = \lambda^{(1)} + \frac{1}{k^2(0)}\lambda^{(2)} + 0\left\{\frac{1}{k^4(0)}\right\} \tag{9.139}$$

As a result, the value of $\lambda^{(1)}$ is determined from

$$\int_0^{r_1} \sqrt{Q} dr = \pi\left(l + \frac{m}{2}\right) \tag{9.140}$$

and the value of $\lambda^{(2)}$, by the method described in Section 9.5.3, is

$$\lambda^{(2)} = \frac{1}{T}\left[\frac{1}{12}k(0)\frac{\partial^2}{\partial\lambda^2}\int_0^{r_1}\frac{(f')^2}{\sqrt{Q}}dr - k(0)\frac{\partial}{\partial\lambda}\int_0^{r_1}\frac{\dfrac{m^2 + 3/4}{r} + \dfrac{1}{2}\left(\dfrac{\varepsilon'}{\varepsilon} + \dfrac{\mu'}{\mu}\right)}{\sqrt{Q}}f' dr\right.$$

$$\left. + \frac{1}{4}k(0)\int_0^{r_1}\frac{\dfrac{1}{r}\left(\dfrac{\varepsilon'}{\varepsilon} + \dfrac{\mu'}{\mu}\right) + \left(\dfrac{\varepsilon'}{\varepsilon}\right)' + \left(\dfrac{\mu'}{\mu}\right)' + \dfrac{1}{2}\left(\dfrac{\varepsilon'}{\varepsilon}\right)^2 + \dfrac{1}{2}\left(\dfrac{\mu'}{\mu}\right)^2}{\sqrt{Q}}dr\right.$$

$$-\frac{m}{2}\left(P+\frac{1}{P}\right)D\Bigg]_{\lambda=\lambda^{(1)}} \tag{9.141}$$

where

$$T = k(0)\left[\int_0^{r_1} \frac{dr}{\sqrt{Q}}\right]_{\lambda=\lambda^{(1)}} \tag{9.142}$$

$$D = 2k(0)\beta \frac{\partial}{\partial\lambda}\int_0^{r_1} \frac{f'}{rk\sqrt{Q}}\,dr \tag{9.143}$$

and the value of P can be obtained from

$$P = \left[\frac{N \mp \sqrt{N^2 + m^2D^2}}{mD}\right]_{\lambda=\lambda^{(1)}} \tag{9.144}$$

with the upper and lower signs indicating two states of polarization, and

$$N = \frac{1}{4}k(0)\int_0^{r_1} \frac{\frac{1}{r}\left(\frac{\varepsilon'}{\varepsilon} - \frac{\mu'}{\mu}\right) - \left(\frac{\varepsilon'}{\varepsilon}\right)' + \left(\frac{\mu'}{\mu}\right)' + \frac{1}{2}\left(\frac{\varepsilon'}{\varepsilon}\right)^2 - \frac{1}{2}\left(\frac{\mu'}{\mu}\right)^2}{\sqrt{Q}}\,dr \tag{9.145}$$

Provided that $k(r)$ is given by

$$k(r) = \frac{k(0)}{1 + Ar^2}, \quad A = \text{constant} \tag{9.146}$$

and μ is constant, we have $N = 0$; namely $P = \mp 1$. Accordingly, the state of polarization is either of the left- or right-hand circular polarization. On the other hand, provided that $k(r)$ is given by

$$k(r) = \frac{k(0)}{\sqrt{1 + Ar^2}}, \quad A = \text{constant} \tag{9.147}$$

we have $D = 0$, namely, $P = 0$ or ∞, which tells us that these states are linear polarizations. When the profile of $k(r)$ deviates slightly from that of (9.147), the value of D vanishes at a certain frequency and thus the value of P becomes zero or infinity. This means that there is a sudden change of P from $P = +1$ to $P = -1$ or from $P = -1$ to $P = +1$ as the frequency varies. Such a peculiar phenomenon for the states of polarization was rediscovered by K. Morishita, T. Hinata, H. Yoshikawa, and T. Hosono (see [14]).

9.8 SIGNIFICANCE OF EXPANSIONS AND THE RANGE OF APPLICATIONS

We have thus far introduced several methods for asymptotic expansions. In general to obtain the higher terms in such asymptotic expansions is not as easy. Therefore very few works are devoted to calculating the higher terms to ascertain the divergence of the relevant expansions (refer to the review papers cited at the end of this chapter). For the practical use, however, the lower-term expansions are still efficient. The major reasons will be summarized as follows:

1. In the asymptotic expansions mentioned in this chapter, the leading term is chosen to describe the geometrical optics solution obtained from the ray configuration. Therefore, other terms also describe the physical nature of wave phenomena in the high-frequency regime, and the major parts of interest are in the lower terms.
2. The resulting expressions are analytic so that global discussions on the behavior of waves at the limit of high frequency are possible.
3. Although the obtained solutions are sometimes applied against the restrictions imposed by the theory, the results concerning the eigenvalues are better than expected. On the other hand, the results for the asymptotic fields are unsatisfactory even when applied within the range of applications.

The mathematical condition:

$$|\varepsilon^2 \sqrt{t'}(1/\sqrt{t'})''| \ll |Q| \tag{9.148}$$

arising from the theory of expansions, as indicated in Section 9.4.1, may therefore restrict us to a limited use in practical applications. However, this condition is considered a theoretical criterion of the validity for the asymptotic expansions, and it never limits the practical use, particularly in describing the physical mechanism of high-frequency operating waves.

For the Liouville-Green transformation, as an example, inequality (9.148) becomes

$$\left| \varepsilon^2 \left(\frac{1}{\sqrt[4]{Q}} \right)^3 \frac{d^2}{dx^2} \left(\frac{1}{\sqrt[4]{Q}} \right) \right| \ll 1 \tag{9.149}$$

This condition indeed fails in the neighborhood of the caustic ($Q = 0$). Nevertheless, in (9.80), the asymptotic solutions obtained under this restriction are used on the integration path, some part of which is located near the caustic and finally beyond the caustic ($x < x_1$, $x > x_2$). This is due to the analytic continuation: the original path was located in the far zone away from the caustic and mathematically deformed like the closed path shown in Figure 9.2. Accordingly, such an analytic continuation cannot be applied to the integral expression for the asymptotic field, where the integration path is not a closed path but an open arc. When we desire accurate field solutions, we can apply the uniform asymptotic solution described in Section 9.4.3, which, although complicated, may be valid over the entire region.

REFERENCES AND BIBLIOGRAPHY

Due to limited space, not all citations are given. The reader should supplement the present literature by referring to the following papers. M. Hashimoto, "Analysis Methods for Inhomogeneous Optical Waveguides (WKB Method and Multilayer Method)," *1978 Joint Conv. Rec. Four Institutes of Electrical Engineers,* Japan, pp. 40.33–36 [in Japanese]; idem., "Applications of the Method of Asymptotic Expansions in Inhomogeneous Medium Problems—Part 1," *1981 Joint Conv. Rec. of Four Institutes of Electrical Engineers,* Japan, pp. 4.115–116 [in Japanese]; idem., "Approximate Methods for Inhomogeneous Optical Waveguides," *J. Inst. Electron. Communication Engrs. Japan,* Vol. 68, No. 8, 1982, pp. 835–841 [in Japanese]; idem., "Recent Progress of Electromagnetic Field Analyses by the Perturbation Methods and the WKB Methods," National Convention Record, 1986, Institute of Electrical Engineers of Japan, S1.13–16 [in Japanese]. S. Adachi, "Electromagnetic Fields in Inhomogeneous/Anisotropic Media," in *Modern Analysis Methods for Electromagnetic Fields,* ed. T. Iijima, Inst. Electron. Communication Engrs. Japan, Tokyo, 1979, Chapter 4 [in Japanese]. S. Adachi and K. Sawaya, "Electromagnetic Field Analyses in Inhomogeneous Media," *J. Inst. Elect. Engrs. Japan,* Vol. 106, No. 9, 1986, pp. 887–894 [in Japanese].

[1] Olver, F.W.J., *Asymptotics and Special Functions,* Academic Press, New York, 1974.

[2] Airy, G.B., "On the Intensity of Light in the Neighbourhood of a Caustic," *Proc. Cambridge Philos. Soc.,* Vol. 6, 1838, pp. 379–403.

[3] Stokes, G.G., "On the Discontinuity of Arbitrary Constants Which Appear in Divergent Developments," *Proc. Cambridge Philos. Soc.,* Vol. 10, 1864, pp. 106–128.

[4] For example, see Langer, R.E., "On the Asymptotic Solutions of Ordinary Differential Equations, with an Application to the Bessel Functions of Large Order," *Trans. Am. Math. Soc.,* Vol. 33, 1931, pp. 23–64.

[5] For example, see Fröman, N., and P.O. Fröman, "On Modification of Phase Integral Approximations of Arbitrary Order," *Nuove Cimento,* Vol. B20, 1974, pp. 121–132.

[6] Pekeris, C.L., "Asymptotic Solutions for the Normal Modes in the Theory of Microwave Propagation," *J. Appl. Phys.,* Vol. 17, 1946, pp. 1108–1124.

[7] Wentzel, G., "Die mehrfach periodischen Systeme in der Quantenmechanik," *Z. Phys.,* Vol. 37, 1926, pp. 80–94; idem., "Eine Verallgemeinerung der Quantenbedingungen fur die Zwecke der Wellenmechanik," *Z. Phys.,* Vol. 38, 1926, pp. 518–529.

[8] Dunham, J.L., "The Wentzel-Brillouin-Kramers Method of Solving the Wave Equation," *Phys. Rev.*, Vol. 15, 1932, pp. 713–720.

[9] Maslov, V.P., "Asymptotic Eigenvalues of One-Dimensional and Radial Schrödinger Equations," *Usp. Math. Nauk,* Vol. 15, No. 4, 1960, pp. 220–221 [in Russian].

[10] Felsen, L.B., "Evanescent Waves," *J. Opt. Soc. Am.*, Vol. 66, No. 8, 1976, pp. 751–760.

[11] Bahar, E., "Generalized WKB Method with Application to Problems of Propagation in Inhomogeneous Media," *J. Math. Phys.*, Vol. 8, No. 9, 1967, pp. 1735–1747.

[12] Prüfer, H., "Neue Herleitung der Sturm-Liouvilleschen Reihenentwicklung stetiger Funktionen," *Math. Ann.*, Vol. 95, 1926, pp. 499–518.

[13] Ambarzumian, V.A., "Diffuse Reflection of Light by a Foggy Medium," *Dokl. Akad. Nauk SSSR,* Vol. 38, 1943, pp. 229–232.

[14] Hashimoto, M., "Asymptotic Theory of Vector Modes in Inhomogeneous Optical Fibres: Uncladded Fibres," *IEE Proc.*, Vol. 130, Part H, No. 4, 1983, pp. 261–275.

Chapter 10
The Beam Propagation Method

Takanori Okoshi and Seiko Kitazawa

10.1 INTRODUCTION

10.1.1 Historical Background

The research on optical integrated circuits and optical planar devices has become active in recent years. In this area, a principal theoretical problem is to calculate how a lightwave is propagated in an optical circuit having arbitrary refractive-index distribution. Various methods have been proposed for this purpose by researchers. One of these is the *beam-propagation method* (BPM), sometimes called the *propagating-beam method,* which was first proposed by Feit and Fleck [1–4]. The abbreviation BPM will be used throughout this chapter.

The BPM has widely been used for analyzing the performance of the light beam propagated in an optical planar circuit that has a nearly stripelike guiding structure and in which the refractive index varies smoothly compared with the wavelength. Furthermore, under specific conditions, the propagation modes as well as the radiation modes can be treated simultaneously using the BPM. The analysis of the polarization characteristics of a propagating mode is also possible.

10.1.2 Features of BPM

The BPM features the following two points: (1) that the electromagnetic fields are Fourier transformed with respect to the direction normal to that of light propagation; and (2) that an ingenious stepwise method is used for successively calculating the electromagnetic field along the axial direction.

In the BPM, we consider a small step (section) of a light propagation and perform a calculation relating the fields at the input and output planes of the section.

Such calculations are repeated for each step. In each step, the Fourier transforms and the inverse Fourier transforms are computed with respect to the transverse direction. The calculation process is rather simple in each step. However, the distance of each step cannot be long, at most one wavelength of the light beam or even much shorter in many cases, so that the number of repetitions is usually large (10^2–10^4).

10.1.3 Computing Technology

The calculating speed of computers is becoming faster. Excellent algorithms for the Fourier transform are now available. These facts make the BPM increasingly profitable. However, the number of steps cannot be excessively large because the computation error accumulates. For this reason, the BPM is not suitable for the analysis of long-distance propagation such as that in an ordinary optical fiber.

The BPM was first developed for the analysis of single-frequency light propagation in an isotropic medium, and afterward applied to the analyses of optical fibers, bent optical waveguides, optical gratings, tapered optical waveguides, and optical *Y*-junctions. Later, the application was expanded to the analyses of wave propagation with multiple frequencies and wave propagation in anisotropic media. Furthermore, even the application to the analysis of nonlinear optical waveguides recently has been reported.

10.1.4 Construction of this Chapter

In this chapter we discuss first the basic properties of the BPM, considering a two-dimensional problem as an example. In Section 10.3, the BPM in the three-dimensional media is discussed, and the properties of the Helmholtz-type and Fresnel-type BPMs are compared in three-dimensional problems. In Section 10.4, we consider the application of the BPM to light propagation in anisotropic media. Finally, examples of applications and analyses of some optical planar circuits are shown in Section 10.5. The application to wave propagation in nonlinear media is not discussed here because the comprehension of the principle of the BPM is the main purpose of this chapter. At the end of the chapter, we give the principal literature related to the BPM, including works cited in the text. Section 10.6 is the conclusion.

10.2 BASIS OF THE BEAM PROPAGATION METHOD

10.2.1 Optical Planar Circuits

We consider an optical planar circuit, a structure like that shown in Figure 10.1. Here the waveguide thickness d and permittivities ε_1, ε_2, and ε_3 are all considered

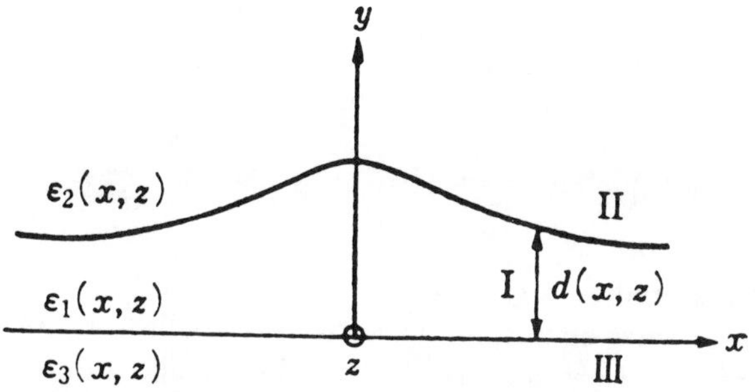

Figure 10.1 An asymmetrical, dielectric, two-dimensional waveguiding structure.

to be functions of the position (x, y), but their spatial variations are assumed gradual compared with the wavelength of light. We assume further that $\varepsilon_1 > \varepsilon_2 > \varepsilon_3$, and that all the materials are nonmagnetic. In such a case, the electric and magnetic fields in each region, $\boldsymbol{E}_i$ and $\boldsymbol{H}_i$ $(i = 1, 2, 3)$, are governed by the following equations [5]:

$$\nabla^2 \boldsymbol{E}_i + \omega^2 \varepsilon_i \mu_0 \boldsymbol{E}_i + (\nabla \ln \varepsilon_i \cdot \nabla)\boldsymbol{E}_i + (\boldsymbol{E}_i \cdot \nabla) \ln \varepsilon_i + \nabla \ln \varepsilon_i \times (\nabla \times \boldsymbol{E}_i) = 0 \quad (10.1)$$

$$\nabla^2 \boldsymbol{H}_i + \omega^2 \varepsilon_i \mu_0 \boldsymbol{H}_i + \nabla \ln \varepsilon_i \times (\nabla \times \boldsymbol{H}_i) = 0 \quad (10.2)$$

(where $\nabla = \boldsymbol{i}\dfrac{\partial}{\partial x} + \boldsymbol{j}\dfrac{\partial}{\partial y} + \boldsymbol{k}\dfrac{\partial}{\partial z}$, and $i = 1, 2, 3$).

If we approximate $\nabla \ln \varepsilon_i \approx 0$ in these equations because variations of ε_i are assumed to be gradual, simplified expressions for E_{yi} and H_{yi} can be obtained from the equations:

$$[\nabla^2 + \omega^2 \varepsilon_i(x, z)\mu_0]E_{yi} = 0 \quad (10.3)$$

$$[\nabla^2 + \omega^2 \varepsilon_i(x, z)\mu_0]H_{yi} = 0 \quad (10.4)$$

Other field components can be expressed in terms of E_{yi} and H_{yi}. These expressions are derived directly from the Maxwell equations as

$$\left(\omega^2 \varepsilon \mu_0 + \frac{\partial^2}{\partial y^2}\right)E_x = \frac{\partial^2}{\partial x \partial y}E_y + j\omega\mu_0 \frac{\partial}{\partial z}H_y \quad (10.5a)$$

$$\left(\omega^2 \varepsilon \mu_0 + \frac{\partial^2}{\partial y^2}\right)E_z = \frac{\partial^2}{\partial z \partial y}E_y - j\omega\mu_0 \frac{\partial}{\partial x}H_y \quad (10.5b)$$

$$\left(\omega^2 \varepsilon \mu_0 + \frac{\partial^2}{\partial y^2}\right) H_x = -j\omega\varepsilon \frac{\partial}{\partial z} E_y + \frac{\partial^2}{\partial x \partial y} H_y \tag{10.5c}$$

$$\left(\omega^2 \varepsilon \mu_0 + \frac{\partial^2}{\partial y^2}\right) H_z = j\omega\varepsilon \frac{\partial}{\partial x} E_y + \frac{\partial^2}{\partial x \partial y} H_y \tag{10.5d}$$

where subscript i is omitted for simplicity.

In an optical planar circuit, we have two modes that are linearly independent of each other: one with $E_{yi} = 0$ everywhere, and the other with $H_{yi} = 0$ everywhere. We next consider these modes separately [5].

(1) TE-Modes

A mode in which $E_{yi} = 0$ everywhere is called a *TE* (transverse electric) *mode* because when such a mode is propagated in the z-direction uniformly with respect to the x-axis (i.e., $\partial/\partial x = 0$), $E_{zi} = 0$ holds everywhere (see equation (10.5b)). Let H_{yi} be expressed as

$$H_{yi} = Y_i(y|\varepsilon_i(x, z), d(x, z)) \cdot X(x, z) \tag{10.6}$$

and put it into (10.4). If we assume that $\partial Y_i/\partial x \approx 0$ and $\partial Y_i/\partial z \approx 0$, we obtain

$$\frac{\dfrac{\partial^2}{\partial y^2} Y_i}{Y_i} + \frac{\left(\dfrac{\partial^2}{\partial x^2} + \dfrac{\partial^2}{\partial z^2}\right) X + k_0^2 n_i^2(x, z) X}{X} = 0 \tag{10.7}$$

where $k_0 = \omega\sqrt{\varepsilon_0 \mu_0}$, and $n_i(x, z)$ denotes the refractive index of the ith region $[n_i = (\varepsilon_i/\varepsilon_0)^{1/2}]$.

Note here that the two terms in the left-hand side are mutually independent and equate these to $\gamma_i^2(x, z)$ and $-\gamma_i^2(x, z)$, respectively. The newly introduced quantity, γ_i, is called the separation parameter. Thus we find that Y_i and X separately satisfy the following equations:

$$\frac{\partial^2}{\partial y^2} Y_i = \gamma_i^2(x, z) Y_i \tag{10.8}$$

$$\left(\frac{\partial^2}{\partial x^2} + \frac{\partial^2}{\partial z^2}\right) X + k_0^2 N^2(x, z) X = 0 \tag{10.9}$$

where

$$k_0^2 N^2(x, z) = k_0^2 n_i^2(x, z) + \gamma_i^2(x, z) \tag{10.10}$$

These equations tell us that the separation parameter γ_i directly provides the vertical (y-direction) wavenumber, whereas $N(x, z)$ determines the two-dimensional wave behavior. Because the light wave is confined to region 1, we may postulate that γ_1 is imaginary, whereas γ_2 and γ_3 are real. Therefore, to simplify the equations, we write

$$\gamma_1 = j\kappa = jk_0\sqrt{n_1^2 - N^2} \tag{10.11a}$$

$$\gamma_2 = p = k_0\sqrt{N^2 - n_2^2} \tag{10.11b}$$

$$\gamma_3 = q = k_0\sqrt{N^2 - n_3^2} \tag{10.11c}$$

and henceforth use κ, p, and q instead of γ_i's.

We first solve (10.8). The solution of this equation is given as an exponential or sinusoidal function. From the conditions that Y_2 and Y_3 must be zero at $y = \pm\infty$, we obtain

$$Y_1 = B\cos(\kappa y + \phi_1) \tag{10.12a}$$

$$Y_2 = A\exp[-p(y - d)] \tag{10.12b}$$

$$Y_3 = C\exp(qy) \tag{10.12c}$$

Parameters p, κ, q can be determined by the boundary conditions. From the conditions that E_x, E_z, H_x, and H_z must all be continuous at the lower boundary ($y = 0$) and the upper boundary ($y = d$), we find that Y and $(\partial Y/\partial y)$ also must be continuous. Hence, from (10.12), p, κ, and q must satisfy the relation

$$\kappa d = \tan^{-1}(p/\kappa) + \tan^{-1}(q/\kappa) + m\pi \quad (m = 0, 1, 2, \ldots) \tag{10.13}$$

where m denotes the mode number. We may determine p, κ, q, and also N by solving (10.11a–c) and (10.13) simultaneously, for each TE mode.

The parameter N thus determined is called the *effective refractive index* because an optical planar circuit having any distribution of ε_1, ε_2, ε_3, and d can simply be modeled as a two-dimensional waveguide having the index distribution of $N(x, z)$. Thus the analysis of an optical planar circuit is attributed to the problem of solving a two-dimensional wave equation given as (10.9). Note that $N(x, z)$ is defined for each TE-mode and satisfies $\varepsilon_3 < N^2 < \varepsilon_1$ when the mode is a propagating mode (10.11a–c).

The x and z components of the fields in TE-mode are given, from (10.5a–d), as

$$E_x = \frac{j\omega\mu_0}{k_0^2 N^2} Y \frac{\partial X}{\partial x} \tag{10.14a}$$

$$E_z = -\frac{j\omega\mu_0}{k_0^2 N^2} Y \frac{\partial X}{\partial z} \tag{10.14b}$$

$$H_x = \frac{1}{k_0^2 N^2} \frac{\partial Y}{\partial y} \frac{\partial X}{\partial x} \tag{10.14c}$$

$$H_z = \frac{1}{k_0^2 N^2} \frac{\partial Y}{\partial y} \frac{\partial X}{\partial z} \tag{10.14d}$$

(2) TM-Modes

A mode in which $H_{yi} = 0$ everywhere is called a *TM* (transverse magnetic) *mode,* because, when such a mode is propagated in the z-direction uniformly with respect to the x-axis (i.e., $\partial/\partial x = 0$), $H_{zi} = 0$ holds everywhere (see (10.5d)).

Following the preceding calculation of the TE-modes, we express E_{yi} as

$$E_{yi} = Y_i[y|\varepsilon_i(x, z), d(x, z)] \cdot X(x, z) \tag{10.15}$$

Then, after calculations similar to those for the TE-modes, we obtain Y_1, Y_2, Y_3 identical to (10.12a–c), respectively, and

$$\kappa d = \tan^{-1}\left(\frac{p}{\kappa}\frac{\varepsilon_1}{\varepsilon_2}\right) + \tan^{-1}\left(\frac{q}{\kappa}\frac{\varepsilon_1}{\varepsilon_3}\right) + m\pi \quad (m = 0, 1, 2, \ldots) \tag{10.16}$$

$$E_x = \frac{1}{k_0^2 N^2} \frac{\partial Y}{\partial y} \frac{\partial X}{\partial x} \tag{10.17a}$$

$$E_z = \frac{1}{k_0^2 N^2} \frac{\partial Y}{\partial y} \frac{\partial X}{\partial z} \tag{10.17b}$$

$$H_x = -\frac{j\omega\mu_0}{k_0^2 N^2} Y \frac{\partial X}{\partial z} \tag{10.17c}$$

$$H_z = \frac{j\omega\mu_0}{k_0^2 N^2} Y \frac{\partial X}{\partial x} \tag{10.17d}$$

where m denotes again the mode number. In the case of TM-modes, the effective refractive index is also defined similarly to TE-modes.

10.2.2 Principle of BPM

(1) Background

It has been found that when light is propagated in a waveguide composed of isotropic media, the behavior of the wave in the two-dimensional plane (x, z) is governed by the scalar Helmholtz equation about a field variable $X(x, z)$, as

$$\left(\frac{\partial^2}{\partial x^2} + \frac{\partial^2}{\partial z^2}\right)X + k_0^2 N^2(x, z)X = 0 \tag{10.18}$$

where $N(x, z)$ denotes the two-dimensional effective refractive-index distribution. For TE-modes, X denotes the (x, z) dependence of the transverse magnetic field, as

$$H_y(x, y, z) = Y(y) \cdot X(x, z)$$

For TM-modes, we may consider that $X(x, z)$ denotes the same for the transverse electric field as

$$E_y(x, y, z) = Y(y) \cdot X(x, z)$$

Thus, these two cases are unified to give the same mathematical problem of solving (10.18).

The BPM was first developed by Feit and Fleck [1, 2] as a method for analyzing a light beam having a single-frequency component and propagated in three-dimensional isotropic media. Recently, however, there has been a number of applications to two-dimensional problems. Therefore, to understand the purpose and the basic concept of the BPM, we here consider its application to a two-dimensional circuit; namely, a planar optical waveguide, but having more or less a stripline structure.

(2) Derivation of BPM Formulas

In the present case we are concerned only with a wave propagating almost parallel to the z-axis, because the optical planar circuit is assumed to have a stripelike structure.

We eliminate the dependence on time $\exp(j\omega t)$ and write the complex amplitude $X(x, z)$ as the product of a function of x (denoted by $X_1(x)$), and a component expressing the propagation in the z-direction as

$$X(x, z) = X_1(x)\exp[-j\beta(z)z] \tag{10.19}$$

In this equation, the propagation constant $\beta(z)$ is a function of z. However, if the variation of the refractive index in the z-direction is gradual compared with the light wavelength, we may consider that β is constant over a small distance in the z-direction Δz. Therefore, substituting (10.19) into (10.18), we obtain

$$\left[\frac{\partial^2}{\partial x^2} - \beta^2(z) + k_0^2 N^2(x, z)\right] X(x, z) = 0 \tag{10.20}$$

$$\frac{\partial^2}{\partial z^2} X(x, z) = -\beta^2(z) X(x, z) \tag{10.21}$$

We further write

$$\partial^2/\partial x^2 = \nabla_t^2 \tag{10.22}$$

and forget for a while that ∇_t^2 is not a mathematical quantity but an operator. Then we may "formally" write

$$\beta(z) = [\nabla_t^2 + k_0^2 N^2(x, z)]^{1/2} \tag{10.23}$$

Hence, the wave amplitude $X(x, z + \Delta z)$ at $z + \Delta z$ is expressed with $X(x, z)$ as

$$X(x, z + \Delta z) = \exp\{-j\Delta z[\nabla_t^2 + k_0^2 N^2(x, z)]^{1/2}\} X(x, z) \tag{10.24}$$

We may rewrite the square root on the right-hand side as

$$(\nabla_t^2 + k_0^2 N^2)^{1/2} = \frac{\nabla_t^2}{(\nabla_t^2 + k_0^2 N^2)^{1/2} + k_0 N} + k_0 N \tag{10.25}$$

This relation can easily be proved by simple algebraic manipulation.

The essential point in the BPM lies in the following approximation. We first consider that $N(x, z)$ may be separated into a constant N_0, typically the value in the substrate of the circuit, and a variable part $\delta N(x, z)$, as

$$N(x, z) = N_0 + \delta N(x, z) \tag{10.26}$$

If we assume here that the variable part $\delta N(x, z)$ is small compared with N_0, that is,

$$\delta N(x, z)/N_0 \ll 1 \tag{10.27}$$

then $N(x, z)$ in the denominator of the first term of (10.25) is approximated by a constant N_0 with satisfactory accuracy, so that

$$(\nabla_t^2 + k_0^2 N^2)^{1/2} \doteqdot \frac{\nabla_t^2}{(\nabla_t^2 + k_0^2 N_0^2)^{1/2} + k_0 N_0} + k_0 N_0 + k_0 \delta N(x, z) \quad (10.28)$$

The applicability of this approximation will be discussed in Section 10.3.1.

We write next the complex amplitude $X(x, z)$ as a product of another amplitude function $W(x, z)$ and a component expressing the propagation with a constant (reference) velocity; that is,

$$X(x, z) = W(x, z) \exp[-jk_0 N_0 z] \quad (10.29)$$

Putting this equation into (10.24) and using the approximation (10.28), we obtain

$$W(x, z + \Delta z) = \exp\left\{-j\Delta z\left[\frac{\nabla_t^2}{(\nabla_t^2 + k_0^2 N_0^2)^{1/2} + k_0 N_0} + \chi(x, z)\right]\right\} W(x, z) \quad (10.30)$$

where

$$\chi(x, z) = k_0 \delta N(x, z) \quad (10.31)$$

Furthermore, by dividing formally the first term of the exponential operator into two halves, (10.30) can be rewritten as

$$\begin{aligned} W(x, z + \Delta z) = {} & \exp\left\{-j\frac{\Delta z}{2}\left[\frac{\nabla_t^2}{(\nabla_t^2 + k_0^2 N_0^2)^{1/2} + k_0 N_0}\right]\right\} \\ & \times \exp[-j\Delta z \chi(x, z)] \\ & \times \exp\left\{-j\frac{\Delta z}{2}\left(\frac{\nabla_t^2}{(\nabla_t^2 + k_0^2 N_0^2)^{1/2} + k_0 N_0}\right)\right\} W(x, z) \\ & + O(\Delta z^3) \end{aligned} \quad (10.32)$$

where the final term is the remaining error caused by that we have divided the exponential operator into two parts and changed the position of one of the operators, which essentially is not exchangeable. In other words, the exponential terms in (10.32) are not algebraic coefficients but differential operators because they include ∇_t^2. Equation (10.32) is the basic formula of the BPM.

We consider here the physical model of (10.32). The first factor in (10.32),

$$\exp\left\{-j\frac{\Delta z}{2}\left[\frac{\nabla_t^2}{(\nabla_t^2 + k_0^2 N_0^2)^{1/2} + k_0 N_0}\right]\right\} W(x, z)$$

is equivalent to solving a wave equation in a homogeneous medium

$$\left(\frac{\partial^2}{\partial x^2} + \frac{\partial^2}{\partial z^2} + k_0^2 N_0^2\right)X = 0 \tag{10.33}$$

for a distance $\Delta z/2$, using $X(x, z)[= W(x, z)]$ as the initial condition. This equivalence can easily be proved by substituting $\chi = 0$ into (10.30). Therefore, the three factors (exponential operators) in (10.31) can be represented by a homogeneous medium (length $\Delta z/2$), a lens giving the phase shift $-j\Delta z\chi(x, z)$, and again a homogeneous medium (length $\Delta z/2$).

This means that advancing the solution for $W(x, z)$ by repeated application of (10.31) is equivalent to propagating the light beam through a periodic array of thin optical lenses as shown in Figure 10.2. The first lens is located at $z = \Delta z/2$. The second, third, and further lenses are separated from one another by the distance Δz. Each lens has an x-dependent phase shift given by $j\Delta z\chi(x, z)$ to the beam, whereas the beam propagation between lenses is governed by (10.33). Thus, the propagation of the beam can be traced in a step-by-step manner.

(3) Numerical Calculations

An alternative representation of (10.32), which is more suitable for numerical calculations, can be obtained by expressing $W(x, z)$ as a Fourier series having a finite number of terms:

$$W(x, z) = \sum_{n=-N/2+1}^{N/2} W_n(z) \exp(jk_{xn}x) \tag{10.34}$$

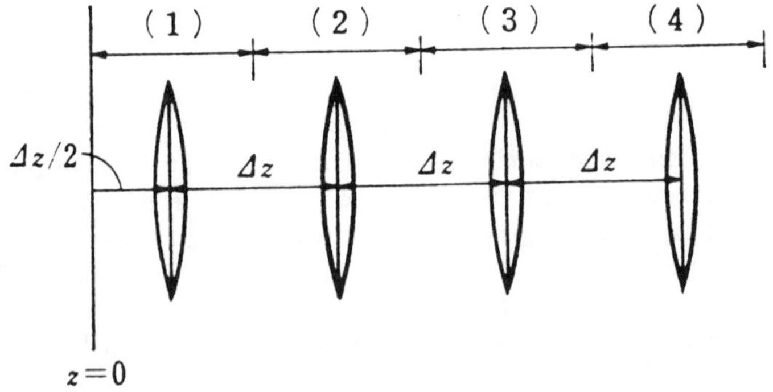

Figure 10.2 An array of lenses equivalent to the beam-shape transformation expressed by equation (10.32). One section consists of a uniform medium with a length ($\Delta z/2$), a thin lens, and a uniform medium with a length ($\Delta z/2$) (from [1]).

where k_{xn} denotes discrete transverse wavenumbers defined as

$$k_{xn} = \frac{2\pi}{L} n \tag{10.35}$$

where L is the width of the computational area.

We consider a section in Figure 10.2 that consists of a homogeneous medium having an effective index of N_0 and length $(\Delta z/2)$, a thin lens, and again a homogeneous medium of length $(\Delta z/2)$. The nth Fourier component W_n at $z + (\Delta z/2)$ can be obtained, by substituting $\chi(x, z) = 0$ into (10.30) and using $\nabla_t^2 = -k_{xn}^2$, as

$$W_n\left(z + \frac{\Delta z}{2} - 0\right) = \exp\left[-j\frac{\Delta z}{2}\frac{-k_{xn}^2}{(-k_{xn}^2 + k_0^2N_0^2)^{1/2} + k_0N_0}\right]W_n(z) \tag{10.36}$$

We may then reconstruct the real-space function W in the real space just behind the thin lens, that is, $W[x, z + (\Delta z/2) - 0]$ (symbol -0 denotes "just before the lens"), by substituting the Fourier component $W_n[z + (\Delta z/2) - 0]$ into (10.34). (The actual computation can be performed using the widely available *fast Fourier transform* (FFT) algorithm.) Then, multiplying it by the "lens" term $\exp[-j\Delta z\chi(x, z)]$ in (10.32), we obtain the real-space function just in front of the lens (at $z = z + (\Delta z/2) + 0$), as

$$W(x, z + \Delta z/2 + 0) = \exp[-j\Delta z\chi(x, z)]W(x, z + \Delta z/2 - 0) \tag{10.37}$$

Next, the Fourier components $W_n[z + (\Delta z/2) + 0]$ are calculated again using the FFT algorithm. Then, the components W_n at $z = z + \Delta z$ after the beam propagation in the following homogeneous space having length $(\Delta z/2)$ is calculated again using (10.36). The function $W(z + \Delta z)$ is finally calculated using (10.34).

Thus, the beam propagation over a single section is calculated. Of course, the final calculation (10.34) can be omitted if the beam shape (power distribution) at this point is not needed, and the calculation is to be continued to the following section. This calculating process of the BPM is illustrated in Figure 10.3.

10.3 BPM IN THREE-DIMENSIONAL MEDIA AND FRESNEL-TYPE APPROXIMATION

10.3.1 BPM Formulation of Three-Dimensional Helmholtz Equation

In this section we discuss (following Yevick [6]) the BPM in three-dimensional media, generalizing the two-dimensional analysis of the previous section. We start from a scalar Helmholtz equation describing a single-frequency three-dimensional wave:

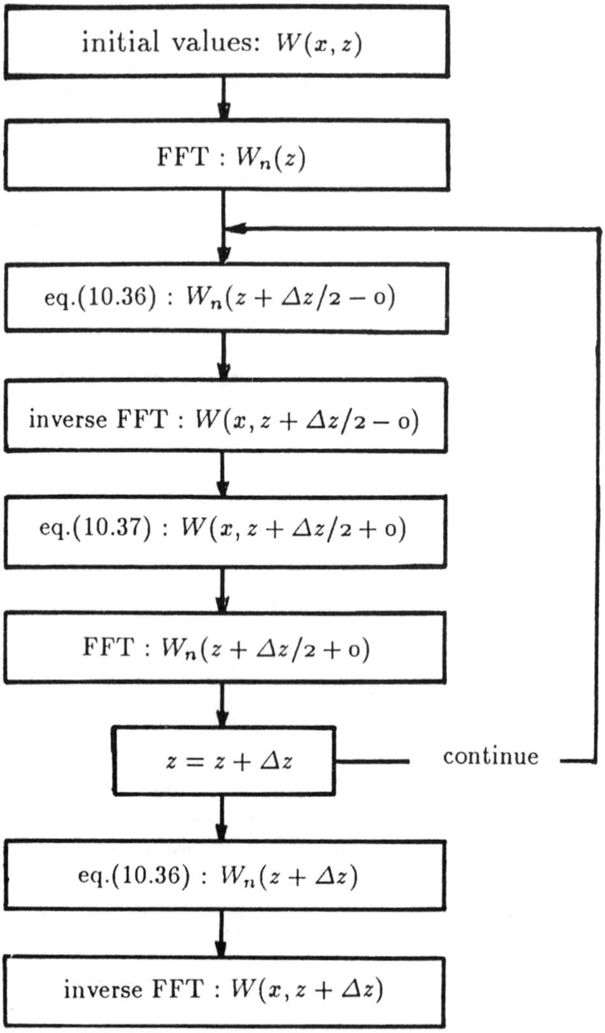

Figure 10.3 The process of calculating the BPM.

$$\frac{\partial^2 E}{\partial z^2} + \nabla_t^2 E + k_0^2 n^2(\omega, x, y, z)E = 0 \tag{10.38}$$

here $\nabla_t^2 = \partial^2/\partial x^2 + \partial^2/\partial y^2$, $E(\omega, x, y, z)\exp(j\omega t)$ denotes one component of elec-ic field at an angular frequency ω, and $n(\omega, x, y, z)$ is the refractive-index distri-ıtion. We express $n(\omega, x, y, z)$ as a sum of a constant part n_0 and a fluctuation ırt δn, as

$$n^2(x, y, z) = [n_0 + \delta n(x, y, z)]^2$$

We assume that $\partial(\delta n)/\partial z \approx 0$ over a small distance in the axial direction and ;nore that ∇_t^2 and δn are not exchangeable. In such a case, we may consider that ıe propagating-type solution of (10.38) behaves like the solution of the following rst-order differential equation:

$$\frac{\partial E}{\partial z} = -j[\nabla_t^2 + k_0^2(n_0 + \delta n)^2]^{1/2} E \tag{10.39}$$

Ve expand the square root in (10.39) in a Taylor's series and neglect the second-rder term $(\delta n/n_0)^2$ assuming that δn is small. Then the right-hand side of (10.39) an be rewritten as

$$\frac{\partial E}{\partial z} = -j\left\{\sqrt{\nabla_t^2 + k_0^2 n_0^2} + \frac{k_0^2 n_0^2}{\sqrt{\nabla_t^2 + k_0^2 n_0^2}} \frac{\delta n}{n_0}\right\}E$$

Note here that ∇_t^2 is equivalent to the transverse Fourier spectrum of the electric eld $-k_t^2$, and that usually an inequality $k_t^2 \ll k_0^2 n_0^2$ holds. Then we may simplify the enominator of the second term as $k_0 n_0$, to obtain

$$\frac{\partial E}{\partial z} = -j\left(\sqrt{\nabla_t^2 + k_0^2 n_0^2} + k_0 \delta n\right)E \tag{10.40}$$

urthermore, the first square root in (10.40) can be rewritten, as in (10.25) in the receding section, giving

$$\frac{\partial E}{\partial z} = -j\left[\frac{\nabla_t^2}{\sqrt{\nabla_t^2 + k_0^2 n_0^2} + k_0 n_0} + k_0 n_0 + k_0 \delta n\right]E \tag{10.41}$$

ɔn the other hand, (10.39) can directly be rewritten in a similar manner as

$$\frac{\partial E}{\partial z} = -j\left[\frac{\nabla_t^2}{\sqrt{\nabla_t^2 + k_0^2(n_0 + \delta n)^2} + k_0(n_0 + \delta n)} + k_0(n_0 + \delta n)\right]E \quad (10.4$$

Comparison of (10.41) and (10.42) suggests that the approximation in (10.4 is equivalent to $n_0 + \delta n$ in the first term of (10.42), which is replaced by a consta n_0. Therefore, the calculation accuracy will be improved by a more proper choic of the constant n_0 between the minimum and maximum values of $n_0 + \delta n$.

Thus, the following formula is obtained for the three-dimensional case after transformation similar to (10.32) in the two-dimensional case:

$$E(x, y, z + \Delta z) = \exp[-jk_0n_0\Delta z] \cdot \exp\left\{-j\frac{\Delta z}{2}\left[\frac{\nabla_t^2}{(\nabla_t^2 + k_0^2n_0^2)^{1/2} + k_0n_0}\right]\right\}$$

$$\cdot \exp\left(-jk_0\int_z^{z+\Delta z}\delta n\Delta z'\right)$$

$$\cdot \exp\left\{-j\frac{\Delta z}{2}\left[\frac{\nabla_t^2}{(\nabla_t^2 + k_0^2n_0^2)^{1/2} + k_0n_0}\right]\right\}E(x, y, z) + O(\Delta z^3)$$

(10.4

In this equation, the integral in the second term can usually be replaced by $\delta n\Delta z$.

10.3.2 BPM Formula Based on the Fresnel Equation

In the analysis of the characteristics of some optical waveguides including optic fibers, the so-called weak-guidance approximation $\delta(n^2)/n_0^2 \ll 1$ is often employe In this case, the Helmholtz equation can be approximated by the Fresnel equatio described in the following text, leading to a simpler BPM formula.

We again start from a scalar Helmholtz equation (10.38). For convenience, field component E is expressed as a product of the complex amplitude $\xi(\omega, x, y,$ and a carrier term expressing a wave propagated in the axial direction with a consta velocity, $\exp(-jk_0n_0z)$, so that

$$E(\omega, x, y, z) = \xi(\omega, x, y, z)\exp(-jk_0n_0z) \quad (10.4$$

where n_0 denotes the refractive index of the cladding.

We define the following variable $\delta n(x, y, z)$ expressing the refractive-inde deviation from n_0 as

$$n^2(x, y, z) = [n_0 + \delta n(x, y, z)]^2 \doteqdot n_0^2 + 2n_0\delta n(x, y, z) \tag{10.45}$$

Substituting (10.44) into (10.38), we have

$$\frac{\partial^2 \xi}{\partial z^2} + 2jk_0 n_0 \frac{\partial \xi}{\partial z} = \nabla_t^2 \xi + k_0^2[(n_0 + \delta n)^2 - n_0^2]\xi \tag{10.46}$$

If the first term of the left-hand side can be neglected because of the weak-guidance approximation, we have a parabolic differential equation:

$$2jk_0 n_0 \frac{\delta \xi'}{\delta z} = \nabla_t^2 \xi' + k_0^2 2n_0 \delta n \xi' \tag{10.47}$$

In the theory of wave propagation, an equation of this type is often referred to as the *Fresnel equation*. Equation (10.47) can further be rewritten as

$$\frac{\partial \xi'}{\partial z} = -j\left(\frac{\nabla_t^2}{2k_0 n_0} + k_0 \delta n\right)\xi' \tag{10.48}$$

By integrating (10.48), we can relate the complex amplitude $\xi'(z + \Delta z)$ after the propagation in a distance Δz to the initial amplitude $\xi'(z)$ as

$$\xi'(z + \Delta z) = \exp\left[-j\left(\frac{\nabla_t^2}{2k_0 n_0} + k_0 \delta n\right)\Delta z\right]\xi'(z) \tag{10.49}$$

If we again divide the BPM operator, we obtain the final BPM formula based on the Fresnel equation:

$$\xi'(z + \Delta z) = \exp\left(-j\frac{\nabla_t^2 \Delta z}{4k_0 n_0}\right)\exp\left(-jk_0 \int_z^{z+\Delta z} \delta n dz'\right) \cdot \exp\left(-j\frac{\nabla_t^2 \Delta z}{4k_0 n_0}\right)\xi'(z) + O(\Delta z^3) \tag{10.50}$$

The difference between the two BPM formulations, (10.50) and (10.43), lies in the first and third terms, which express the light propagation in a homogeneous medium having refractive index of n_0. Equation (10.43) is applied when the transverse wavenumber k_t^2 cannot be neglected compared with $k_0^2 n_0^2$, whereas k_t^2 in the denominator of (10.43) is neglected in (10.50).

10.3.3 Propagation Constants Obtained from Fresnel and Helmholtz Equations

In this section we investigate the relation between the propagation constants obtained from Fresnel and Helmholtz equations. We consider first the Helmholtz solution. We expand the complex amplitude $\xi(x, y, z)$ defined in (10.44) in terms of the orthogonal eigenmodes $u(x, y)$:

$$\xi(x, y, z) = \sum_n A_n u(x, y) \exp(-j\beta_n z) \tag{10.51}$$

The actual field component is then given, from (10.44) and (10.51), as

$$E(x, y, z) = \sum_n A_n u(x, y) \exp[-j(k_0 n_0 + \beta_n)z] \tag{10.52}$$

Therefore, the propagation constant in the solution of the Helmholtz equation (10.39) is expressed as

$$k_0 n_0 + \beta_n = [\nabla_t^2 + k_0^2(n_0 + \delta n)^2]^{1/2} \tag{10.53}$$

On the other hand, the solution $\xi'(x, y, z)$ for the Fresnel equation also can be expanded with the same eigenmodes $u(x, y)$ as the Helmholtz equation, to yield

$$\xi'(x, y, z) = \sum_n A_n u(x, y) \exp(-j\beta_n' z) \tag{10.54}$$

Comparing (10.54) and (10.48), we can express the propagation constant β_n' for the Fresnel equation as

$$\beta_n' = \frac{\nabla_t^2}{2k_0 n_0} + k_0 \delta n \tag{10.55}$$

From (10.55) and (10.53), we have

$$\beta_n = -k_0 n_0 (1 - \sqrt{1 + 2\beta_n'/k_0 n_0}) \tag{10.56}$$

Equation (10.56) shows that the propagation constant β_n, which has been obtained from the Helmholtz equation, can also be derived from β_n' from the Fresnel equation.

By differentiating (10.56) with respect to ω, we can write the group delay $\partial\beta_n/\partial\omega$ derived from the Helmholtz equation in terms of that derived from the Fresnel equation:

$$\frac{1}{v_n} = \frac{\partial \beta_n}{\partial \omega} = \frac{n_0}{c}\left[\frac{1 + (\beta_n'/k_0 n_0) + (\partial \beta_n'/\partial \omega)c/n_0}{\sqrt{1 + 2\beta_n'/k_0 n_0}} - 1\right] \tag{10.57}$$

The advantage of using the Fresnel equation (10.50) is that we can use a standard and more accurate algorithm in computer calculation because the formula is simpler.

10.4 BPM IN ANISOTROPIC MEDIA

The BPM as described earlier can be used in the analysis of wave propagation in isotropic media where we consider only one component of the field. However, it cannot be applied in a straightforward manner to the analysis of wave propagation in anisotropic media, which are common in optical integrated circuits where $LiNbO_3$ or GaAs is used. In such media, coupling between different electric field polarizations is induced by the off-diagonal elements in permittivity tensor ε_{ij}. In this section, the BPM is modified for application to anisotropic media, following Thylen and Yevick [7].

10.4.1 Basic Equations

The anisotropic nature of a medium is expressed by off-diagonal elements in the permittivity tensor ε_{ij}. The transverse component of the electric field $\boldsymbol{E}_t(E_x, E_y)$ of a propagating wave should satisfy the following equation derived from the Maxwell equation:

$$\nabla^2 E_t - \mu(\varepsilon_{ij})\frac{\partial^2 E_t}{\partial t^2} = 0 \tag{10.58}$$

where μ is the permeability of the medium. When the time dependence of the field is expressed as $\exp(-j\omega t)$, we have

$$\nabla^2 E_t + \mu(\varepsilon_{ij})\omega^2 E_t = 0 \tag{10.59}$$

We denote the propagation constant in the axial direction (z-direction) by k_0, so that

$$k_0^2 = \mu\varepsilon_{zz}\omega^2 \tag{10.60}$$

Equation (10.59) may then be rewritten as

$$\frac{\partial^2}{\partial z^2}\begin{bmatrix} E_x \\ E_y \end{bmatrix} = -\begin{bmatrix} \nabla_t^2 + k_0^2\varepsilon_{xx}, & k_0^2\varepsilon_{xy} \\ k_0^2\varepsilon_{xy}, & \nabla_t^2 + k_0^2\varepsilon_{yy} \end{bmatrix}\begin{bmatrix} E_x \\ E_y \end{bmatrix} \tag{10.61}$$

where we assume that $\varepsilon_{xy} = \varepsilon_{yx}$ and each element ε_{ij} is a function of the traverse coordinates x and y. Equation (10.61) is a matrix-type expression of the Helmholtz equation. We define here a matrix-type operator (M^2) as

$$(M^2) = -\begin{bmatrix} \nabla_t^2 + k_0^2\varepsilon_{xx}, & k_0^2\varepsilon_{xy} \\ k_0^2\varepsilon_{xy}, & \nabla_t^2 + k_0^2\varepsilon_{yy} \end{bmatrix} \tag{10.62}$$

10.4.2 Basic Variables and Assumptions

We consider how the transverse field components at $z = \Delta z$ [i.e., $E_x(x, y, \Delta z)$ and $E_y(x, y, \Delta z)$] can be expressed when those at $z = 0$, $E_x(x, y, 0)$ and $E_y(x, y, 0)$, are known. Here we need the following approximation similarly with the BPM analysis of scalar waves.

We assume first that the elements of the permittivity tensor ε_{xx}, ε_{yy} are expressed as the sums of constants ε_{xx}^0, ε_{yy}^0 and the variables $\delta\varepsilon_{xx}(x, y)$, $\delta\varepsilon_{yy}(x, y)$, respectively. We assume further that these variables are constant with respect to z within the small interval Δz. Therefore, we may write

$$\begin{aligned} \varepsilon_{xx} &= \varepsilon_{xx}^0 + \delta\varepsilon_{xx}(x, y) \\ \varepsilon_{yy} &= \varepsilon_{yy}^0 + \delta\varepsilon_{yy}(x, y) \end{aligned} \tag{10.63}$$

On the other hand, these variables can also be expressed in terms of the refractive-index perturbations, as

$$\begin{aligned} \varepsilon_{xx} &= [n_{xx0} + \delta n_{xx}(x, y)]^2 \\ \varepsilon_{yy} &= [n_{yy0} + \delta n_{yy}(x, y)]^2 \end{aligned} \tag{10.64}$$

We assume here that the perturbations $\delta\varepsilon_{xx}$, $\delta\varepsilon_{yy}$ are very small compared with ε_{xx}^0, ε_{yy}^0, and that $\varepsilon_{xy} \ll \varepsilon_{xx}^0$, ε_{yy}^0. We assume further for simplicity that δn_{xx} and δn_{yy} are equal, and write $\delta n_{xx}(x, y) = \delta n_{yy}(x, y) \triangleq \delta n(x, y)$. Under such assumptions, we have

$$\begin{aligned} \delta\varepsilon_{xx} &\doteqdot 2n_{xx0}\delta n(x, y) \\ \delta\varepsilon_{yy} &\doteqdot 2n_{yy0}\delta n(x, y) \end{aligned} \tag{10.65}$$

10.4.3 Formulation for BPM

In the following text, the matrix operator (M) is derived first from (M^2) in (10.62) with the approximation of (10.65). First, we rewrite (10.62) as

$$(M^2) = -\begin{bmatrix} \nabla_t^2 + k_0^2(n_{xx0}^2 + 2n_{xx0}\delta n), & k_0^2\varepsilon_{xy} \\ k_0^2\varepsilon_{xy}, & \nabla_t^2 + k_0^2(n_{yy0}^2 + 2n_{yy0}\delta n) \end{bmatrix} = \begin{bmatrix} A, & B \\ B, & C \end{bmatrix} = \begin{bmatrix} a, & b \\ b, & c \end{bmatrix}^2 = (M)^2$$

and compute the elements of the operators a, b, and c. From the assumption $\varepsilon_{xy} \ll \varepsilon_{xx}^0, \varepsilon_{yy}^0$ (see the preceding subsection), we may write $A \gg B$, and hence $a \gg b$. Consequently, element a is obtained, as in the scalar BPM analysis, as:

$$\begin{aligned} a \doteqdot \sqrt{A} &= \sqrt{\nabla_t^2 + k_0^2(n_{xx0}^2 + 2n_{xx0}\delta n)} \\ &= \frac{\nabla_t^2}{\sqrt{\nabla_t^2 + k_0^2 n_{xx0}^2} + k_0 n_{xx0}} + k_0 n_{xx0} + k_0 \delta n \end{aligned}$$

and element c is obtained in the same manner:

$$c \doteqdot \sqrt{C} = \frac{\nabla_t^2}{\sqrt{\nabla_t^2 + k_0^2 n_{yy0}^2} + k_0 n_{yy0}} + k_0 n_{yy0} + k_0 \delta n$$

Furthermore, if the traverse deviation of the ray path from z-axis is very small, we have inequalities $k_t^2 \ll k_0^2 n_{xx0}^2$ and $k_t^2 \ll k_0^2 n_{yy0}^2$, where $\nabla_t^2 = -k_t^2$. Then we have a simplified expression for element b:

$$b = B/(a + c) \doteqdot \frac{k_0 \varepsilon_{xy}}{n_{xx0} + n_{yy0}}$$

Thus, matrix (M) is expressed as

$$(M) = (M_0) + (M_\Delta) + (M_k) \tag{10.66}$$

where

$$(M_0) = (-j)\begin{bmatrix} \dfrac{\nabla_t^2}{\sqrt{\nabla_t^2 + k_0^2 n_{xx0}^2} + k_0 n_{xx0}} + k_0 n_{xx0}, & 0 \\ 0, & \dfrac{\nabla_t^2}{\sqrt{\nabla_t^2 + k_0^2 n_{yy0}^2} + k_0 n_{yy0}} + k_0 n_{yy0} \end{bmatrix}$$

$$(M_\Delta) = (-j)k_0\delta n\begin{bmatrix} 1, & 0 \\ 0, & 1 \end{bmatrix}$$

$$(M_k) = (-j)\frac{k_0 \varepsilon_{xy}}{n_{xx0} + n_{yy0}}\begin{bmatrix} 0, & 1 \\ 1, & 0 \end{bmatrix}$$

Next, we integrate (10.61) using (10.66). As a result, field components at = Δz, $E_x(x, y, \Delta z)$ and $E_y(x, y, \Delta z)$, are given in terms of $E_x(x, y, 0)$ and $E_y(x,$ 0) at $z = 0$, as

$$\begin{bmatrix} E_x(x, y, \Delta z) \\ E_y(x, y, \Delta z) \end{bmatrix} = \exp\{\Delta z[(M_0) + (M_\Delta) + (M_k)]\} \begin{bmatrix} E_x(x, y, 0) \\ E_y(x, y, 0) \end{bmatrix}$$

$$= \exp[\Delta z(M_0)] \exp[\Delta z(M_\Delta)] \exp[\Delta z(M_k)] \begin{bmatrix} E_x(x, y, 0) \\ E_y(x, y, 0) \end{bmatrix} \quad (10.6$$

Here, we split the first term of the exponential operator in (10.67) into tw halves and shift the position of the second half to the right of the third exponenti operator. The third term, on the other hand, is expressed as

$$\exp(-jq) \begin{bmatrix} 0, & 1 \\ 1, & 0 \end{bmatrix} = \begin{bmatrix} \cos(q), & -j\sin(q) \\ -j\sin(q), & \cos(q) \end{bmatrix} \quad (10.6$$

Thus, the formulation for the BPM in anisotropic media is obtained finally as

$$\begin{bmatrix} E_x(x, y, \Delta z) \\ E_y(x, y, \Delta z) \end{bmatrix}$$

$$= \begin{bmatrix} P_{xx}, & 0 \\ 0, & P_{yy} \end{bmatrix} \begin{bmatrix} Q, & 0 \\ 0, & Q \end{bmatrix} \begin{bmatrix} \cos(q), & -j\sin(q) \\ -j\sin(q), & \cos(q) \end{bmatrix} \begin{bmatrix} P_{xx}, & 0 \\ 0, & P_{yy} \end{bmatrix} \begin{bmatrix} E_x(x, y, 0) \\ E_y(x, y, 0) \end{bmatrix}$$

(10.6

In (10.69), we use the following expression

$$P_{xx} = \exp\left\{-j\frac{\Delta z}{2}\left[\frac{\nabla_t^2}{\sqrt{\nabla_t^2 + k_0^2 n_{xx0}^2} + k_0 n_{xx0}} + k_0 n_{xx0}\right]\right\}$$

Variable P_{yy} is obtained by replacing all x's in the equation by y. Variables and q in (10.69) are defined as

$$Q = \exp[-j\Delta z k_0 \delta n(x, y)]$$

$$q = \frac{\Delta z k_0 \varepsilon_{xy}}{n_{xx0} + n_{yy0}}$$

We find that the unitary nature of the exponential operator is preserved in (10.69 as in the scalar BPM.

10.4.4 Cases When Refractive-Index Variations in x- and y-Directions Are Different

In the preceding analysis we assumed that the variations of refractive index in the x-direction is the same as that in the y-direction; that is, $\delta n_{xx}(x, y) = \delta n_{yy}(x, y)$. However, actually many devices use anisotropic materials with significant difference in refractive-index variations in the x- and y-directions. Here, we consider such cases.

Substituting δn_{xx}, δn_{yy} into (10.65), we directly obtain a matrix operator (M'), instead of $(M_\Delta) + (M_k)$ in (10.66):

$$\exp[-j\Delta z(M')] = \exp\left\{ -j\Delta z k_0 \begin{bmatrix} \delta n_{xx}, & \dfrac{\varepsilon_{xy}}{n_{xx0} + n_{yy0}} \\ \dfrac{\varepsilon_{xy}}{n_{xx0} + n_{yy0}}, & \delta n_{yy} \end{bmatrix} \right\} \tag{10.70}$$

$$= \exp\begin{bmatrix} A, & B \\ B, & D \end{bmatrix} \tag{10.71}$$

We can rewrite (10.71) further using the operator-splitting method, within the accuracy in the order of (Δz^2):

$$\exp\begin{bmatrix} A, & B \\ B, & D \end{bmatrix} = \exp\begin{bmatrix} 0, & 0 \\ 0, & (D-A)/2 \end{bmatrix} \exp\begin{bmatrix} A, & B \\ B, & A \end{bmatrix} \times \exp\begin{bmatrix} 0, & 0 \\ 0, & (D-A)/2 \end{bmatrix} + O(\Delta z^3) \tag{10.72}$$

Thus, we obtain the BPM formulation for the cases when refractive-index variations in x- and y-directions are different. The unitary nature of the exponential operator also is preserved in (10.72).

10.5 EXAMPLES OF CALCULATION RESULTS

10.5.1 Cases in Isotropic Media

Some results of the BPM calculations that have appeared in [8] are shown in the following discussion. Figure 10.4 shows the propagation of a light beam in a tapered waveguide (from bottom to top), for four input modes. The height of distribution patterns gives the light amplitude. The effective refractive indices of the waveguide and substrate, N_1 and N_2, are assumed to be

$$N_2 = 1.00, \quad N_1^2 - N_2^2 = 0.02$$

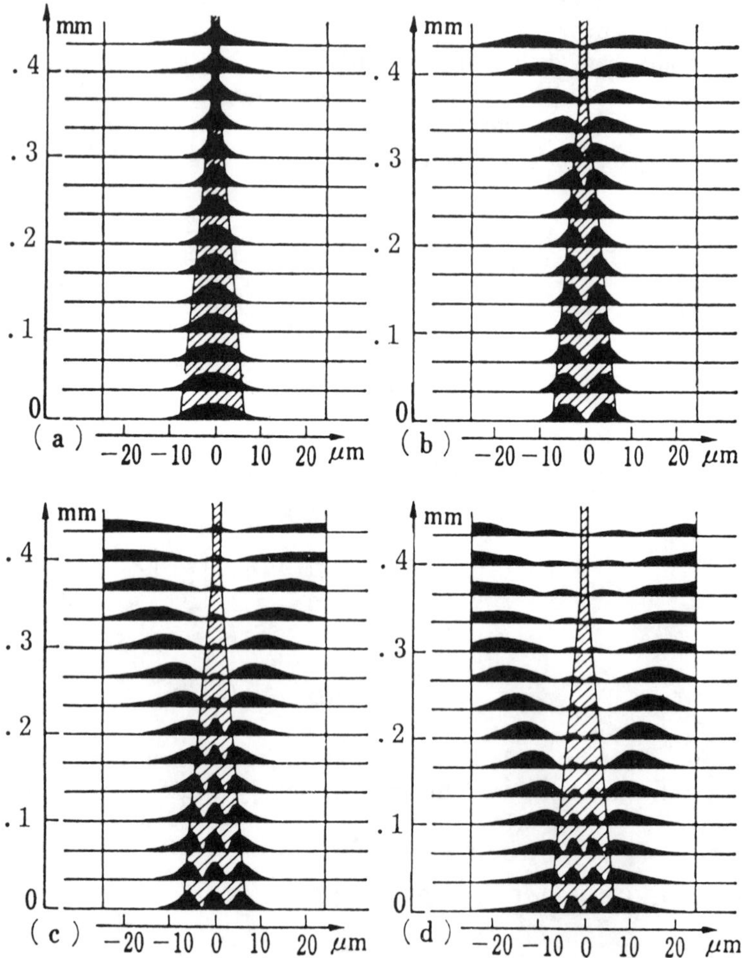

Figure 10.4 Propagation of a light beam in a tapered waveguide. The beam travels from the bottom to the top of the figure. Note that the abscissa is expanded about five times relative to the ordinate: (a) zero-order mode; (b) first-order mode; (c) second-order mode; (d) third-order mode (from [8]).

where the waveguide width at the input and output ends, d_{in} and d_{out} (Figure 10.5), are given in terms of their normalized values as

$$V_{in} = k_0 d_{in}(N_1^2 - N_2^2)^{1/2} = 10.0$$

$$V_{out} = k_0 d_{out}(N_1^2 - N_2^2)^{1/2} = 1.25$$

where k_0 is the wavenumber in vacuum. The taper angle θ (Figure 10.5) is 1°, and the wavelength of the light beam is assumed to be 1.30 μm.

Baets and Lagasse examine how the light power is transferred to the radiation mode, especially when the mode number is high. Quantitative analyses of the radiation loss and mode conversion in such tapers are also performed in [8] on the basis of the BPM analysis results.

Figure 10.6 shows the light amplitude distribution in a Y-junction [Figure 10.7(a)], where $\alpha_1 = 2°$, $\alpha_2 = 4°$, the wavelength $\lambda = 1.30$ μm, and $A = 8\lambda$. As shown in the figure, the waveguide shape is given as

$$Z = x \tan\alpha_1 + (A - L \tan\alpha_1) \sin(\pi x / 2L) \tag{10.73}$$

The waveguide width is 6 μm over the entire circuit pattern; however, the effective-refractive index is not uniform but has a smooth distribution, as shown in Figure 10.7(b), where N_1 (value at the waveguide center) and N_2 are again given as $N_2 =$ 1.00 and $N_1^2 - N_2^2 = 0.02$.

Figure 10.6 suggests that the input light power is apt to travel in a straight line; an appreciable part of the power seems to be converted to radiation modes. In the literature [8], the quantitative analyses of the radiation loss are performed for various circuit configurations on the basis of the BPM analysis results like those shown in Figure 10.6.

10.5.2 Cases in Anisotropic Media

Some results of the BPM calculations in anisotropic media reported in [7] are shown in the following text. Figure 10.8 shows a schematic of a double hetero-structure GaAs-AL$_x$Ga$_{1-x}$As[110] pn-junction waveguide electro-optic modulator. The crystal orientation, applied RF electric field E, and the direction of the light beam are shown. The index profile is assumed to be

$$n(x) = n_0 + \Delta n / \cosh^2(2x/h)$$

with the following parameters:

$$\lambda_0 = 1\ \mu\text{m},\ \Delta n = 0.001,\ h = 5.6\ \mu\text{m},\ n_0 = 3.6,\ \varepsilon_{xy} = 0.009.$$

The value of ε_{xy} corresponds to an electric RF field of 50 V/μm.

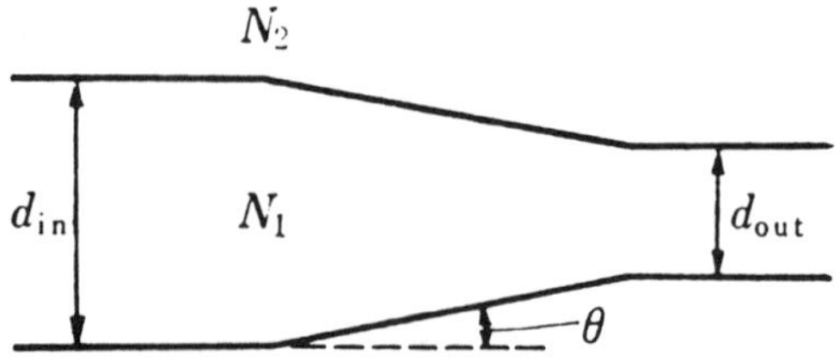

Figure 10.5 Configuration and geometry of the tapered section assumed in the analyses, whose results are shown in Figure 10.4 (from [8]).

The calculated amplitude distribution at equally spaced positions, from 25 μm to 225 μm, is shown in Figure 10.9 for when the single-mode waveguide shown in Figure 10.8 is excited by the lowest-order TE-mode. Figures 10.9(a) and 10.9(b) show the results for E_x and E_y, respectively. The computation step Δz is 5 μm. From these results, we find that the TE-mode is converted to the TM-mode at z = 200 μm with 100 percent efficiency. The values of E_x and E_y agree with those obtained by the coupled-mode theory within 0.01 percent. In this calculation, the difference between the propagation constants of TE-mode and TM-mode is neglected.

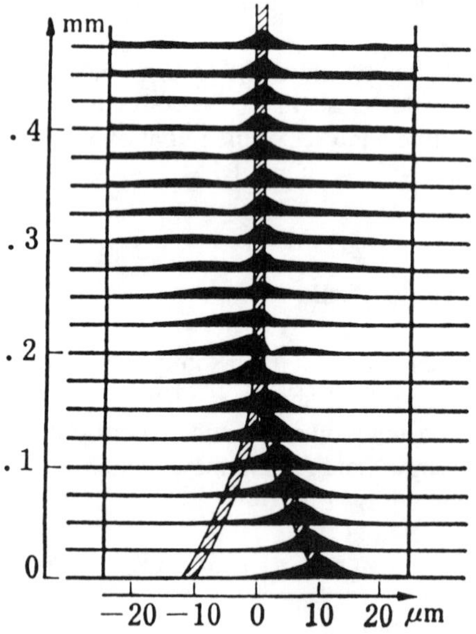

Figure 10.6 Propagation of a light beam in a waveguide Y-junction. The beam travels from the right bottom to the top of the figure. Note that the abscissa is expanded about five times relative to the ordinate (from [8]).

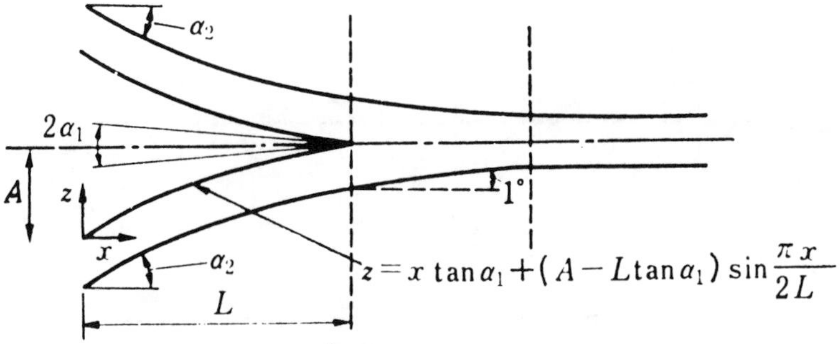

(a)

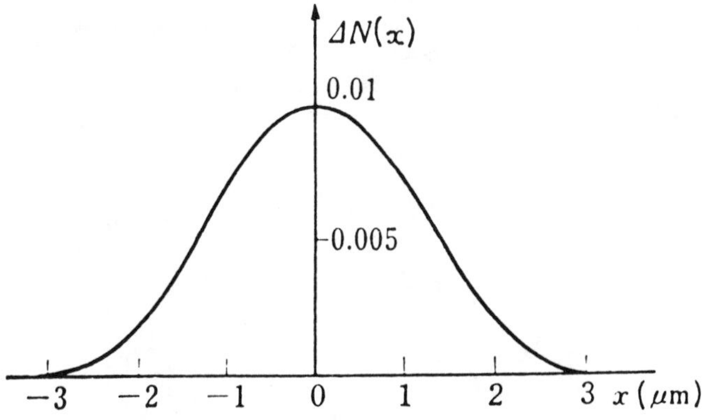

(b)

igure 10.7 (a) Configuration and geometry of the waveguide *Y*-junction; (b) effective refractive-index distribution of the waveguide used in the *Y*-junction of (a) (from [8]).

The second example is a $LiNbO_3$ waveguide mode converter in which the exted input TE-mode is coupled to the TM-mode by an electrode array that periodally modulates ε_{xy} along the axis of waveguide. Figure 10.10(a) shows the callated amplitudes of E_x and E_y at equally spaced axial positions between $z = 1.25$ m and 13.75 μm. In this case, the parameters are $\lambda = 1$ μm, $\Delta n = 0.0015$, $h =$ .6 μm, the refractive indices 2.237 and 2.157, ε_{xy} = constant, and the computation ep 1.25 μm. From the calculation results, we find that a strong mode conversion not induced by the anisotropy of the media. The results were reported to agree ith the coupled wave theory within 1.5 percent, and within 0.5 percent in the case f $\Delta z = 0.625$ μm.

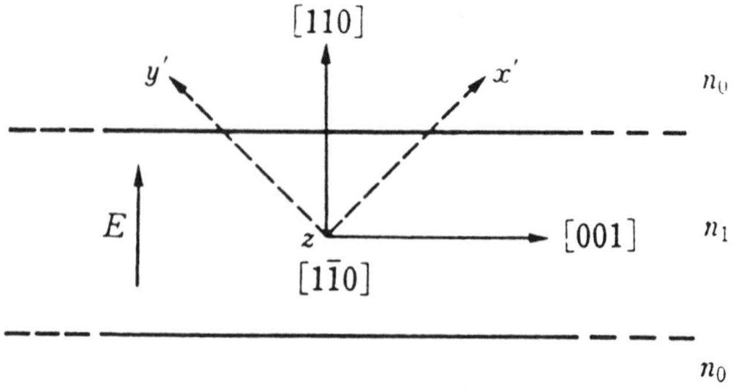

Figure 10.8 Schematic of a double hetero-structure GaAs pn-junction waveguide. The crystal orientation, RF electric field, and the direction of the light beam are shown. The waveguide material, as the result of the applied electric field, has an anisotropy for the coordinate (x', y', z'). Parameter n_1 denotes the refractive-index of the waveguide (GaAs), n_0 the refractive-index of the substrate, where $n_1 > n_0$ (from [7]).

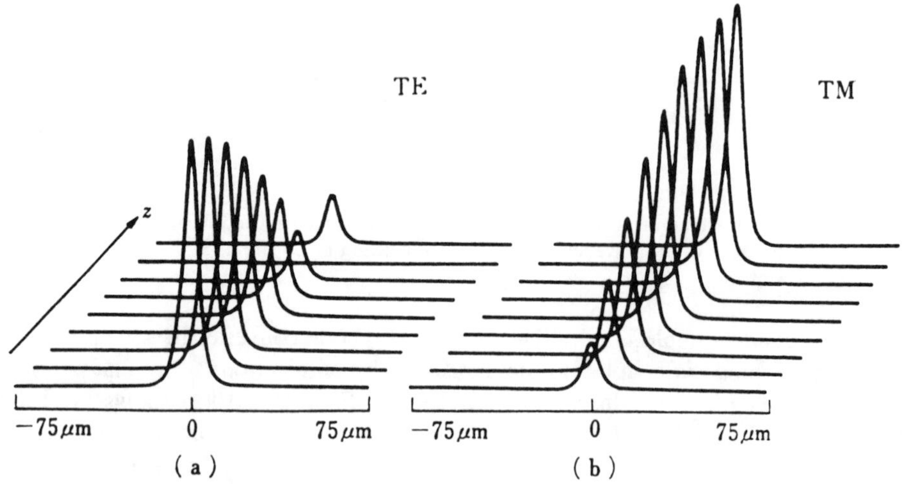

Figure 10.9 Calculation results for the amplitude distribution of the light beam in the waveguide at equally spaced axial positions: (a) is the result for E_x (TE-mode); and (b) is that for E_y (TM-mode) (from [7]).

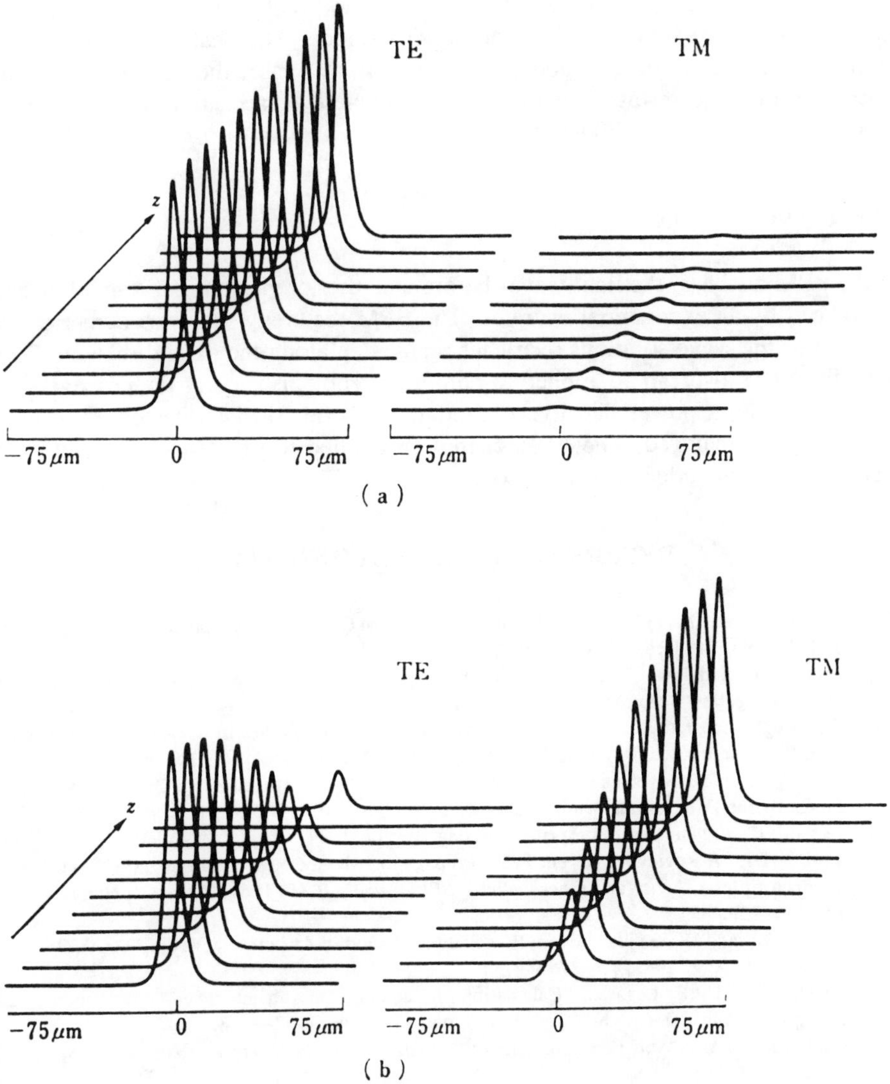

Figure 10.10 Calculation results for the amplitude distributions of a large propagation constant difference between TE- and TM-modes: (a) shows the results at equally spaced axial positions between $z = 1.25$ μm and 13.75 μm, with an interval of 1.25 μm when ε_{xy} is constant; (b) shows the results at equally spaced axial positions between $z = 20$ μm and 220 μm with an interval 20 μm when ε_{xy} varies sinusoidally (from [7]).

Figure 10.10(b) shows the results when ε_{xy} is varied sinusoidally between 0.0 and -0.01 along the waveguide axis with a period matched with the propagatio constant difference between TE-mode and TM-mode. The peak value of ε_{xy} corre sponds to an electric field of about 15 V/μm. In this case, the computation step i 1 μm. From these results, we find that a complete polarization-mode conversio occurs at the position of 200 μm.

10.6 CONCLUSION

In the BPM analysis, the light waves before and after propagation of a small distanc are related by an exponential operator. The BPM features a stepwise numerical cal culation of the wave solutions over each small step along the axial direction. There fore, the BPM also can be applied to nonlinear problems such as the propagation c light having high power density. A limitation of the BPM is that only advancin waves can be analyzed. The BPM cannot be applied to problems where an appre ciable amount of reflected waves exist.

REFERENCES AND BIBLIOGRAPHY

[1] Feit, M.D., and J.A. Fleck, Jr., "Light Propagation in Graded-Index Optical Fibers," *Appl. Opt.* Vol. 17, No. 24, Dec. 1978, pp. 3990–3998.

[2] Feit, M.D., and J.A. Fleck, Jr., "Calculation of Dispersion in Graded-Index Multimode Fibe by a Propagating Beam Method," *Appl. Opt.*, Vol. 18, No. 16, Aug. 1979, pp. 2843–2851.

[3] Feit, M.D., and J.A. Fleck, Jr., "Computation of Mode Properties in Optical Fiber Waveguide by a Propagating Beam Method," *Appl. Opt.*, Vol. 19, No. 7, April 1980, pp. 1154–1164.

[4] Feit, M.D., and J.A. Fleck, Jr., "Computation of Mode Eigenfunctions in Graded-Index Optica Fibers by the Propagating Beam Method," *Appl. Opt.*, Vol. 19, No. 13, July 1980, 2240–224

[5] Okoshi, T., *Planar Circuits,* Springer-Verlag, Berlin, 1984, Chapters 9 and 10.

[6] Yevick, D., "A Review of Recent Beam Propagation Method Results in Integrated Optics: Semi conductor Laser Theory and Fiber Optics," *SPIE Symp. OPTIKA '84,* Budapest, Hungary, 198 pp. 226–229.

[7] Thylen, L., and D. Yevick, "Beam Propagation Method in Anisotropic Media," *Appl. Opt.*, Vo 21, No. 15, Aug. 1982, pp. 2751–2754.

[8] Baets, R., and P.E. Lagasse, "Calculation of Radiation Loss in Integrated-Optic Tapers and Y Junctions," *Appl. Opt.*, Vol. 21, No. 11, June 1982, pp. 1972–1978.

[9] Marburgen, J.H., "Self-Focusing: Theory," *Prog. Quantum Electron.*, Vol. 4, 1975, pp. 35-110.

[10] Fleck, Jr., J.A., J.R. Morris, and M.D. Feit, "Time-Dependent Propagation of High Energy Lase Beams through the Atmosphere," *Appl. Phys.*, Vol. 10, 1976, pp. 129–160.

[11] Feit, M.D., and J.A. Fleck, Jr., "Propagating Beam Theory of Optical Fiber Cross Coupling," *J. Opt. Soc. Am.*, Vol. 71, No. 11, 1981, pp. 1361–1372.

[12] Thylen, L., "The Beam Propagation Method: An Analysis of Its Applicability," *Opt. and Quan tum Electron.*, Vol. 15, 1983, pp. 433–439.

[13] Roey, J.V., J.V.D. Donk, and P.E. Lagasse, "Beam-Propagation Method: Analysis and Assess ment," *J. Opt. Soc. Am.*, Vol. 71, No. 7, July 1981, pp. 803–810.

[14] Yevick, D., and P. Danielsen, "Numerical Investigation of Mode Coupling in Sinusoidally Modulated GRIN Planar Waveguides," *Appl. Opt.*, Vol. 21, No. 15, Aug. 1982, pp. 2727–2733.

[15] Yevick, D., and L. Thylen, "Analysis of Gratings by the Beam-Propagation Method," *J. Opt. Soc. Am.*, Vol. 72, No. 8, Aug. 1982, pp. 1084–1089.

[16] Yevick, D., and B. Hermansson, "A Numerical Investigation of the Mode Coupling among the Local Normal Modes of Sinusoidally Modulated Parabolic Graded-Index Planar Waveguides," *Opt. and Quantum Electron.*, Vol. 16, 1984, pp. 331–337.

[17] Hermansson, B., D. Yevick, and L. Thylen, "A Propagating Beam Method Analysis of Nonlinear Effects in Optical Waveguides," *Opt. and Quantum Electron.*, Vol. 16, 1984, pp. 525–534.

[18] Baets, R., J.-P.V.D. Capelle, and P.E. Lagasse, "Longitudinal Analysis of Semiconductor Laser with Low Reflectivity Facets," *IEEE J. Quantum Electron.*, Vol. QE-21, No. 6, June 1985, pp. 693–699.

[19] Yevick, D., and B. Hermansson, "Band Structure Calculation with the Split-Step Fast Fourier Transform Technique," *Solid State Commun.*, Vol. 54, No. 2, 1985, pp. 197–199.

[20] Feit, M.D., and J.A. Fleck, Jr., "An Analysis of Intersecting Diffused Channel Waveguides," *IEEE J. Quantum Electron.*, Vol. QE-21, No. 11, Nov. 1985, pp. 1799–1805.

[21] Yevick, D., and B. Hermansson, "New Approach to Perturbed Optical Waveguides," *Opt. Letts.*, Vol. 11, No. 2, Feb. 1986, pp. 103–105.

[22] Moloney, J.V., J. Ariyasu, C.T. Seaton, and G.I. Stegeman, "Numerical Evidence for Nonstationary, Nonlinear, Slab-Guided Waves," *Opt. Letts.*, Vol. 11, No. 5, May 1986, pp. 315–317.

[23] Setterlind, C.J., and L. Thylen, "Directional Coupler Switches with Optical Gain," *IEEE J. Quantum Electron.*, Vol. QE-22, No. 5, May 1986, pp. 595–602.

[24] Forber, R.A., and E. Marom, "Symmetric Directional Coupler Switches," *IEEE J. Quantum Electron.*, Vol. QE-22, No. 6, June 1986, pp. 911–919.

[25] Leine, L., C. Wachter, U. Langbein, and F. Lederer, "Propagation Phenomena of Nonlinear Film-Guided Waves: A Numerical Analysis," *Opt. Letts.*, Vol. 11, No. 9, Sept. 1986, pp. 590–592.

[26] Van der Donk, J., "Beam Propagation Method," *Electric Communications*, Vol. 59, No. 4, 1985, pp. 380–384.

Chapter 11
The Spectral Domain Method

Tatsuo Itoh

11.1 HISTORICAL BACKGROUND

The precursor to the spectral domain method widely used for the analysis of microstrip lines and other planar transmission lines are the papers by Yamashita published in 1968 [1, 2]. The method presented in these papers are for the solution of a quasi-TEM analysis valid at low frequencies. The method can efficiently calculate the static capacitance in the transverse cross section of a microstrip line. In general, the static capacitance can be calculated by a variational expression derived from the integral equation that relates the charge and potential. Yamashita has applied the Fourier transform to such a variational expression so that the numerical procedure is significantly simplified and made more efficient.

As the frequency is increased, the propagation in the microstrip line undergoes dispersion due to its inhomogeneous structure. Therefore, the quasi-TEM approximation is no longer applicable and hence the wave equation must be solved more rigorously. Denlinger was the first to solve the equation converted into the Fourier transform domain from the homogeneous integral equation via a rigorous electromagnetic process [3]. In his method, however, the accuracy of the solution is dictated by the way the current distribution on the strip is approximated. The spectral domain method has eliminated this deficiency by introducing Galerkin's solution so that the accuracy can be improved to any desired degree, at least in principle [4]. Subsequently, the spectral domain method has been improved by a number of researchers in terms of its algorithm, accuracy, and computation speed. The method has become one of the most popular and most frequently used solutions for eigenvalue problems in planar transmission lines and a certain class of passive components. More recently, the method has been extended to scattering problems in an *E*-plane circuit in addition to the eigenvalue problem [5].

11.2 CHARACTERISTICS OF THE METHOD AND THE RANGE OF APPLICATIONS

As a solution for eigenvalue problems, the spectral domain method can be used for calculations of the propagation constant, characteristic electromagnetic field distributions, and characteristic impedance of most planar transmission lines. For instance, the method can be applied to microstrip lines, coplanar waveguides, finlines and suspended striplines. It also can be used to analyze coupled microstrip lines consisting of more than two strips as well as for combinations of various planar transmission lines. Further, the spectral domain method is also effective for the analysis of the resonant frequency and the electromagnetic field of a resonator or a patch antenna (microstrip antenna) made of a planar transmission line of a finite extent. Even if the structure contains a multilayer substrate or a dielectric overlay, the method still can be used with a slight increase of complication. The method is equally applicable to an open or a closed structure.

The main reason for the numerical efficiency of the spectral domain method lies in its need for a significant amount of mathematical preprocessing. This in turn implies that the applicable range of the spectral domain method is substantially limited in comparison to a more numerically intensive method, such as the point-matching method. First, the most significant deficiency is that the spectral domain method can be applied only where the thickness of the strip conductor is infinitesimally thin. Also, the strip must be a perfect conductor in general, although this is not the case for the dielectric loss. Further, discontinuities of the substrate in the transverse direction are not allowed in this method. The substrate must be infinitely long in the transverse direction or terminated by two perfectly conducting side walls. If the strip or slot is discontinuous in the propagation direction—for instance, if the microstrip line is semiinfinitely long or the slot line has a step discontinuity in the line width—the spectral domain method is difficult to use.

From the numerical analysis point of view, the spectral domain method has a number of unique features:

1. It solves algebraic equations rather than integral equations.
2. Because Galerkin's method is used, the solution is equivalent to that based on the variational expression. Accuracy can be systematically improved by increasing the size of the matrix used.
3. Even if the structure becomes more complicated by such modifications as the use of a multilayered substrate, the numerical processing is not significantly affected.
4. Because the physical nature of the solution is incorporated in the solution process, no ambiguities occur in the convergence of the solution and the confirmation of the mode.

11.3 SPECTRAL DOMAIN METHOD BASED ON THE ELECTROMAGNETIC-FIELD EXPANSIONS

The spectral domain method is now applied to the microstrip line on a two-layered substrate with the structural cross section shown in Figure 11.1. This structure is fairly general in that, once a computer program is made based on the spectral domain formulation, various structures can be analyzed numerically. For instance, to analyze a microstrip line on a one-layered substrate, we may specify the structural parameters as (a) $\varepsilon_2 = \varepsilon_3 = 1$, $\varepsilon_1 \neq 1$, (b) $\varepsilon_1 = 1$, $\varepsilon_2 = \varepsilon_3 \neq 1$, (c) $d = 0$, or (d) $t = 0$. If the fundamental mode in an open structure is analyzed, we may use $h = \infty$ (which can be realized by using a numerically large number). It is also possible to analytically let $h = \infty$ before a computer program is written.

First, we assume that the transmission line is lossless and its unknown propagation constant is β. We also assume that the modal wave propagating in the z-direction (normal to the surface of the paper) takes the form of $\exp(-j\beta z)$. This factor and the time factor $\exp(j\omega t)$ are omitted in the subsequent equations. Here, ω is the angular frequency.

Because the modes in a planar transmission line are of a hybrid nature, six electromagnetic field components can be expressed in terms of two independent scalar functions. There are a number of ways to select these two scalar functions. However, it is convenient here to choose them in such a way that the hybrid modes are expressed as a superposition of the TE and TM electromagnetic fields with respect to the y direction. For instance, if we use scalar functions of E and H, $\varphi^e(x, y)$ and $\varphi^h(x, y)$, then the electromagnetic-field components are [6]

$$E_x(x, y) = \frac{1}{\hat{y}} \frac{\partial^2 \psi^e}{\partial x \partial y} - j\beta\psi^h \quad H_x(x, y) = j\beta\psi^e + \frac{1}{\hat{z}} \frac{\partial^2 \psi^h}{\partial x \partial y}$$

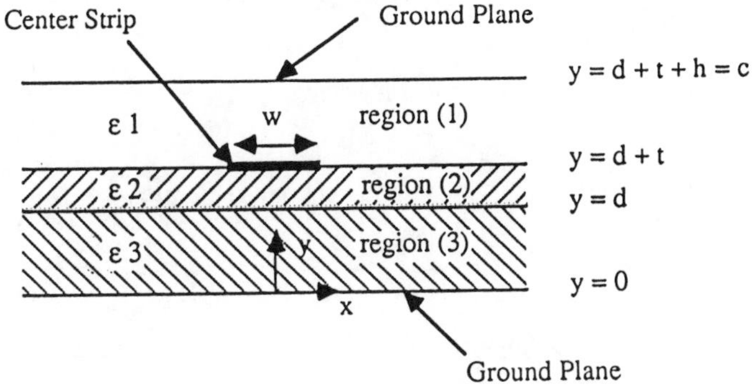

Figure 11.1 Cross-sectional view of a shielded microstrip line.

$$E_y(x, y) = \frac{1}{\hat{y}}\left(\frac{\partial^2}{\partial y^2} + k^2\right)\psi^e \quad H_y(x, y) = \frac{1}{\hat{z}}\left(\frac{\partial^2}{\partial y^2} + k^2\right)\psi^h$$

$$E_z(x, y) = -\frac{j\beta}{\hat{y}}\frac{\partial \psi^e}{\partial y} - \frac{\partial \psi^h}{\partial x} \quad H_z(x, y) = \frac{\partial \psi^e}{\partial x} - \frac{j\beta}{\hat{z}}\frac{\partial \psi^h}{\partial y}$$

where $\hat{y} = j\omega\varepsilon_0\varepsilon$, $\hat{z} = j\omega\mu_0\mu$, $k^2 = \omega^2\mu_0\varepsilon_0\mu\varepsilon$ and ε_0 and μ_0 are free space permittivity and permeability. Also, the relative permittivity ε and the relative permeability μ take the values in the region specified by the value of y. In the derivation of the equation as an eigenvalue problem, scalar functions appropriate to each region are selected and the unknown β is determined so that the electromagnetic fields derived from these scalar functions satisfy the boundary conditions. However, if these field expansions are executed as functions of x and y, the analytical labor is significant. In the spectral domain, these expansions are carried out in the Fourier transform domain. To this end, the Fourier transform in the x direction is defined as follows:

$$\tilde{\phi}(\alpha) = \int_{-\infty}^{\infty} \phi(x)e^{j\alpha x}dx \tag{11.1}$$

where $\phi(x)$ is a function with respect to x and its transform is expressed by a tilde above the symbol. If a structure with two side walls is analyzed, a finite Fourier transform is used instead of (11.1).

The Fourier transforms of the electromagnetic-field components in terms of these scalar functions can be expressed as follows in terms of the Fourier transforms of these scalar functions themselves:

$$\begin{aligned}
\tilde{E}_x(\alpha, y) &= -j\frac{\alpha}{\hat{y}}\frac{\partial \tilde{\psi}^e}{\partial y} - j\beta\tilde{\psi}^h \quad & \tilde{H}_x(\alpha, y) &= j\beta\tilde{\psi}^e - j\frac{\alpha}{\hat{z}}\frac{\partial \tilde{\psi}^h}{\partial y} \\
\tilde{E}_y(\alpha, y) &= \frac{1}{\hat{y}}\left\{\frac{\partial^2}{\partial y^2} + k^2\right\}\tilde{\psi}^e \quad & \tilde{H}_y(\alpha, y) &= \frac{1}{\hat{z}}\left\{\frac{\partial^2}{\partial y^2} + k^2\right\}\tilde{\psi}^h \\
\tilde{E}_z(\alpha, y) &= -j\frac{\beta}{\hat{y}}\frac{\partial \tilde{\psi}^e}{\partial y} + j\alpha\tilde{\psi}^h \quad & \tilde{H}_z(\alpha, y) &= -j\alpha\tilde{\psi}^e - j\frac{\beta}{\hat{z}}\frac{\partial \tilde{\psi}^h}{\partial y}
\end{aligned} \tag{11.2}$$

where

$$\tilde{\psi}^p(\alpha, y) = \int_{-\infty}^{\infty} \psi^p(x, y)\, e^{j\alpha x}dx, \quad p = e \text{ or } h$$

Further, since the scalar functions must satisfy Helmholtz equation, their Fourier transform satisfy

$$\left(-\alpha^2 + \frac{d^2}{dy^2} - \beta^2\right)\tilde{\psi} + k^2\tilde{\psi} = 0 \tag{11.3}$$

The general solution to this equation is in general given by

$$\tilde{\psi} = c_1 \cosh\gamma y + c_2 \sinh\gamma y \tag{11.4}$$

$$\gamma^2 = \alpha^2 + \beta^2 - k^2 \tag{11.5}$$

In (11.4), c_1 and c_2 are determined from the boundary conditions. Hence, the solutions in each region are as follows. Region 1:

$$\tilde{\psi}_1^e = A^e \cosh\gamma_1(c - y) \tag{11.6a}$$

$$\tilde{\psi}_1^h = A^h \sinh\gamma_1(c - y) \tag{11.6b}$$

Region 2:

$$\tilde{\psi}_2^e = B^e \sinh\gamma_2(y - d) + C^e \cosh\gamma_2(y - d) \tag{11.7a}$$

$$\tilde{\psi}_2^h = B^h \cosh\gamma_2(y - d) + C^h \sinh\gamma_2(y - d) \tag{11.7b}$$

Region 3:

$$\tilde{\psi}_3^e = D^e \cosh\gamma_3 y \tag{11.8a}$$

$$\tilde{\psi}_3^h = D^h \sinh\gamma_3 y \tag{11.8b}$$

Here, the coefficients A^e, A^h, ..., D^h are yet to be determined from the boundary conditions. Further,

$$\begin{aligned} \gamma_i^2 &= \alpha^2 + \beta^2 - k_i^2 \quad i = 1, 2, 3 \\ k_i &= \sqrt{\varepsilon_i \mu_i} k_0, \quad k_0 = \omega\sqrt{\varepsilon_0 \mu_0} \end{aligned} \tag{11.9}$$

When (11.6) through (11.8) are substituted into (11.2), the electromagnetic field expressions in each region can be obtained. In the following, we assume that $\mu_1 = \mu_2 = \mu_3 = 1$.

$$\begin{aligned} \tilde{E}_{x1} &= j\alpha\gamma_{y1} A^e \sinh\gamma_1(c - y) - j\beta A^h \sinh\gamma_1(c - y) \\ \tilde{E}_{x2} &= -j\alpha\gamma_{y2}[B^e \cosh\gamma_2(y - d) + C^e \sinh\gamma_2(y - d)] \\ &\quad - j\beta[B^h \cosh\gamma_2(y - d) + C^h \sinh\gamma_2(y - d)] \\ \tilde{E}_{x3} &= -j\alpha\gamma_{y3} D^e \sinh\gamma_3 y - j\beta D^h \sinh\gamma_3 y \end{aligned} \tag{11.10a}$$

$$\tilde{E}_{y1} = \frac{1}{\hat{y}_1}[\gamma_1^2 + k_1^2]A^e \cosh\gamma_1(c - y)$$

$$\tilde{E}_{y2} = \frac{1}{\hat{y}_2}[\gamma_2^2 + k_2^2][B^e \sinh\gamma_2(y - d) + C^e \cosh\gamma_1(y - d)] \quad (11.10b)$$

$$\tilde{E}_{y3} = \frac{1}{\hat{y}_3}[\gamma_3^2 + k_3^2]D^e \cosh\gamma_3 y$$

$$\begin{aligned} \tilde{E}_{z1} &= j\beta\gamma_{y1}A^e \sinh\gamma_1(c - y) + j\alpha A^h \sinh\gamma_1(c - y) \\ \tilde{E}_{z2} &= -j\beta\gamma_{y2}[B^e \cosh\gamma_2(y - d) + C^e \sinh\gamma_2(y - d)] \\ &\quad + j\alpha[B^h \cosh\gamma_2(y + d) + C^h \sinh\gamma_2(y - d)] \\ \tilde{E}_{z3} &= -j\beta\gamma_{y3}D^e \sinh\gamma_3 y + j\alpha D^h \sinh\gamma_3 y \end{aligned} \quad (11.10c)$$

$$\begin{aligned} \tilde{H}_{x1} &= j\beta A^e \cosh\gamma_1(c - y) + j\alpha\gamma_{z1}A^h \cosh\gamma_1(c - y) \\ \tilde{H}_{x2} &= j\beta[B^e \sinh\gamma_2(y - d) + C^e \cosh\gamma_2(y - d)] \\ &\quad - j\alpha\gamma_{z2}[B^h \sinh\gamma_2(y - d) + C^h \cosh\gamma_2(y - d)] \\ \tilde{H}_{x3} &= j\beta D^e \cosh\gamma_3 y - j\alpha\gamma_{z3}D^h \cosh\gamma_3 y \end{aligned} \quad (11.10d)$$

$$\tilde{H}_{y1} = \frac{1}{\hat{z}_1}[\gamma_1^2 + k_1^2]A^h \sinh\gamma_1(c - y)$$

$$\tilde{H}_{y2} = \frac{1}{\hat{z}_2}[\gamma_2^2 + k_2^2][B^h \cosh\gamma_2(y - d) + C^h \sinh\gamma_2(y - d)] \quad (11.10e)$$

$$\tilde{H}_{y3} = \frac{1}{\hat{z}_3}[\gamma_3^2 + k_3^2]D^h \sinh\gamma_3 y$$

$$\begin{aligned} \tilde{H}_{z1} &= -j\alpha A^e \cosh\gamma_1(c - y) + j\beta\gamma_{z1}A^h \cosh\gamma_1(c - y) \\ \tilde{H}_{z2} &= -j\alpha[B^e \sinh\gamma_2(y - d) + C^e \cosh\gamma_2(y - d)] \\ &\quad - j\beta\gamma_{z2}[B^h \sinh\gamma_2(y - d) + C^h \cosh\gamma_2(y - d)] \\ \tilde{H}_{z3} &= -j\alpha D^e \cosh\gamma_3 y - j\beta\gamma_{z3}D^h \cosh\gamma_3 y \end{aligned} \quad (11.10f)$$

$$\gamma_{yi} = \frac{\gamma_i}{\hat{y}_i}, \quad \gamma_{zi} = \frac{\gamma_i}{\hat{z}_i}, \quad i = 1, 2, 3$$

As described earlier, the boundary conditions need to be used for deriving the coefficients $A^e, \ldots, D^h$. To this end, it is necessary to study the boundary conditions in the Fourier transform domain. We start with the boundary conditions in the space domain (x = domain). At $y = d$:

$$E_{x2}(x, d) = E_{x3}(x, d), \quad |x| < \infty \tag{11.11}$$

$$E_{z2}(x, d) = E_{z3}(x, d), \quad |x| < \infty \tag{11.12}$$

$$H_{x2}(x, d) = H_{x3}(x, d), \quad |x| < \infty \tag{11.13}$$

$$H_{z2}(x, d) = H_{z3}(x, d), \quad |x| < \infty \tag{11.14}$$

At $y = d + t$:

$$E_{x2}(x, d + t) = E_{x1}(x, d + t), \quad |x| < \infty \tag{11.15}$$

$$E_{z2}(x, d + t) = E_{z1}(x, d + t), \quad |x| < \infty \tag{11.16}$$

$$H_{x2}(x, d + t) - H_{x1}(x, d + t) = \begin{cases} J_z(x), & |x| < w/2 \\ 0, & |x| > w/2 \end{cases} \tag{11.17}$$

$$H_{z2}(x, d + t) - H_{z1}(x, d + t) = \begin{cases} -J_x(x), & |x| < w/2 \\ 0, & |x| > w/2 \end{cases} \tag{11.18}$$

Further,

$$E_{z1}(x, d + t) = E_{z2}(x, d + t) = 0, \quad |x| < w/2 \tag{11.19}$$

$$E_{x1}(x, d + t) = E_{x2}(x, d + t) = 0, \quad |x| < w/2 \tag{11.20}$$

In the preceding equations, $J_z(x)$ and $J_x(x)$ are the current distributions on the strip and as yet undetermined. Because the functions need to be defined in the entire region for executing Fourier transforms, J_z and J_x are defined on the right-hand sides of (11.17) and (11.18) as the difference of the magnetic fields.

Hence, the boundary conditions in the Fourier transform domain are as follows. At $y = d$,

$$\tilde{E}_{x2}(\alpha, d) = \tilde{E}_{x3}(\alpha, d) \tag{11.21}$$

$$\tilde{E}_{z2}(\alpha, d) = \tilde{E}_{z3}(\alpha, d) \tag{11.22}$$

$$\tilde{H}_{x2}(\alpha, d) = \tilde{H}_{x3}(\alpha, d) \tag{11.23}$$

$$\tilde{H}_{z2}(\alpha, d) = \tilde{H}_{z3}(\alpha, d) \tag{11.24}$$

At $y = d + t$,

$$\tilde{E}_{x1}(\alpha, d + t) = \tilde{E}_{x2}(\alpha, d + t) \tag{11.25}$$

$$\tilde{E}_{z1}(\alpha, d + t) = \tilde{E}_{z2}(\alpha, d + t) \tag{11.26}$$

$$\tilde{H}_{x2}(\alpha, d + t) - \tilde{H}_{x1}(\alpha, d + t) = \tilde{J}_z(\alpha) \tag{11.27}$$

$$\tilde{H}_{z2}(\alpha, d + t) - \tilde{H}_{z1}(\alpha, d + t) = -\tilde{J}_x(\alpha) \tag{11.28}$$

where $\tilde{J}_z(\alpha)$ and $\tilde{J}_x(\alpha)$ are the Fourier transforms of $J_z(x)$ and $J_x(x)$, given by

$$\tilde{J}_q(\alpha) = \int_{-w/2}^{w/2} J_q(x)\, e^{j\alpha x} dx, \quad q = x \text{ or } z$$

This equation indicates that $\tilde{J}_z$ and $\tilde{J}_x$ are the Fourier transforms of the functions that are nonzero only in a finite region of x. This plays an important role in the solution process described later. Further, it should be recognized that the boundary conditions (11.19) and (11.2) have not yet been used.

Next, the electromagnetic fields expressed in (11.10) are substituted into (11.21) through (11.28). Then, all of the coefficients $A^e, \ldots, D^h$ are expressed in terms of $\tilde{J}_z$ and $\tilde{J}_x$. When these relationships are substituted once more to the representations for $\tilde{E}_{x1}$ and $\tilde{E}_{z1}$, they are expressed by $\tilde{J}_z$ and $\tilde{J}_x$ and the following simultaneous equations are obtained.

$$\tilde{E}_{z1}(\alpha, d + t) = \tilde{Z}_{zz}(\alpha, \beta)\tilde{J}_z(\alpha) + \tilde{Z}_{zx}(\alpha, \beta)\tilde{J}_x(\alpha) \tag{11.29}$$

$$\tilde{E}_{x1}(\alpha, d + t) = \tilde{Z}_{xz}(\alpha, \beta)\tilde{J}_z(\alpha) + \tilde{Z}_{xx}(\alpha, \beta)\tilde{J}_x(\alpha) \tag{11.30}$$

where

$$\tilde{Z}_{zz} = -\frac{1}{\alpha^2 + \beta^2}[\beta^2\tilde{Z}_e + \alpha^2\tilde{Z}_h] \tag{11.31}$$

$$\tilde{Z}_{zx} = -\frac{\alpha\beta}{\alpha^2 + \beta^2}[\tilde{Z}_e - \tilde{Z}_h] \tag{11.32}$$

$$\tilde{Z}_{xz} = \tilde{Z}_{zx} \tag{11.33}$$

$$\tilde{Z}_{xx} = -\frac{1}{\alpha^2 + \beta^2}[\alpha^2\tilde{Z}_e + \beta^2\tilde{Z}_h] \tag{11.34}$$

$$\tilde{Z}_e = \frac{\gamma_{y2}Ct_3 + \gamma_{y3}Ct_2}{Ct_2Ct_3 + Ct_1Ct_3\gamma_{y2}/\gamma_{y1} + Ct_1Ct_2\gamma_{y3}/\gamma_{y1} + \gamma_{y3}/\gamma_{y2}} \tag{11.35}$$

$$\tilde{Z}_h = \frac{\gamma_{z2}Ct_2 + \gamma_{z3}Ct_3}{\gamma_{z1}\gamma_{z2}Ct_1Ct_2 + \gamma_{z1}\gamma_{z3}Ct_1Ct_3 + \gamma_{z2}\gamma_{z3}Ct_2Ct_3 + \gamma_{z2}^2} \tag{11.36}$$

$$Ct_1 = \coth\gamma_1 h, \quad Ct_2 = \coth\gamma_2 t, \quad Ct_3 = \coth\gamma_3 d$$

Before moving to the solutions of the equations, let us consider the meanings of (11.29) and (11.30). If inverse Fourier transforms of these equations are taken, we obtain

$$E_{z1}(x, d + t) = \int_{-w/2}^{w/2} \{Z_{zz}(x - x', \beta)J_z(x') + Z_{zx}(x - x', \beta)J_x(x')\}dx' \tag{11.37}$$

$$E_{x1}(x, d + t) = \int_{-w/2}^{w/2} \{Z_{xz}(x - x', \beta)J_z(x') + Z_{xx}(x - x', \beta)J_x(x')\}dx' \tag{11.38}$$

where $Z_{zz}(x, \beta)$, $Z_{zx}(x, \beta)$, *et cetera* are inverse transforms of $\tilde{Z}_{zz}(\alpha, \beta)$, $\tilde{Z}_{zx}(\alpha, \beta)$, *et cetera*. From (11.29), (11.30), (11.37), and (11.38), we find that the Fourier transforms of the integrals of Faltung type result in the products of the Fourier transforms of individual functions. When the region $|x| < w/2$ is considered in (11.37) and (11.38), the left-hand sides of these equations become zero from (11.19) and (11.20) and coupled homogeneous integral equations can be obtained. From the solutions of these equations, β is found as the eigenvalue. Further, the current distributions J_x and J_z can be obtained except for their absolute amplitudes. These integral equations are the outcome of the analysis in the space (x) domain for the present problem. However, such an analysis is not necessarily simple. First, it is difficult to express Green's functions Z_{zz}, *et cetera* in simple equations. Also, if a numerical method such as the moment method is used for the solution, double integrals need to be computed. Numerical processing of double integrals is not very efficient.

On the other hand, the solution process in the spectral-domain method is much simpler. First, as shown in (11.31) through (11.34), the Fourier transformed Green's functions have closed form expressions. In addition, because (11.29) and (11.30) are simultaneous algebraic equations, they can be processed much more easily than the simultaneous integral equations. Note, however, that the left-hand sides of (11.29) and (11.30) are unknown, and hence these two equations contain a total of four unknowns. The reason for too many unknowns is that the boundary conditions corresponding to (11.19) and (11.20) have not been used as yet in the Fourier transform domain. As described in the next section, two of the unknowns, $\tilde{E}_{z1}$ and $\tilde{E}_{x1}$, can be eliminated by the use of these conditions so that the two equations can be solved for the remaining two unknowns $\tilde{J}_z$ and $\tilde{J}_x$.

11.4 GALERKIN'S METHOD OF SOLUTION

Consider solving (11.29) and (11.30) simultaneously by means of Galerkin's method. First, $\tilde{J}_z(\alpha)$ and $\tilde{J}_x(\alpha)$ are expanded in terms of sets of basis functions, $\tilde{J}_{zm}(\alpha)$ ($m = 1, 2, \ldots, N$) and $\tilde{J}_{xm}(\alpha)$ ($m = 1, 2, \ldots, M$) as follows:

$$\tilde{J}_z(\alpha) = \sum_{m=1}^{N} c_m \tilde{J}_{zm}(\alpha) \tag{11.39}$$

$$\tilde{J}_x(\alpha) = \sum_{m=1}^{M} d_m \tilde{J}_{xm}(\alpha) \tag{11.40}$$

where c_m and d_m are unknowns. Also, the basis functions must be chosen in such a way that their inverse transforms $J_{zm}(x)$ and $J_{xm}(x)$ are zero in the region $|x| > w/2$. Practical examples will be presented later in this chapter.

In the Galerkin's method, (11.39) and (11.40) are substituted into (11.29) and (11.30) and the inner products of the resultant equations with each of the basis functions $\tilde{J}_{zk}(\alpha)$ and $\tilde{J}_{xl}(\alpha)$ are taken. Then,

$$\int_{-\infty}^{\infty} \left[\tilde{J}_{zk} \tilde{Z}_{zz} \sum_{m=1}^{N} c_m \tilde{J}_{zm} + \tilde{J}_{zk} \tilde{Z}_{zx} \sum_{m=1}^{M} d_m \tilde{J}_{xm} \right] d\alpha = 0 \tag{11.41}$$

$$k = 1, 2, \ldots, N \tag{11.42}$$

$$\int_{-\infty}^{\infty} \left[\tilde{J}_{xl} \tilde{Z}_{xz} \sum_{m=1}^{N} c_m \tilde{J}_{zm} + \tilde{J}_{xl} \tilde{Z}_{xx} \sum_{m=1}^{M} d_m \tilde{J}_{xm} \right] d\alpha = 0$$

$$l = 1, 2, \ldots, M$$

The right-hand sides of (11.41) and (11.42) are zero. The reasons are easily understood if the Parseval's relationship is used. For instance, if the inner product of the left-hand side of (11.29) with $\tilde{J}_{zk}(\alpha)$ is taken

$$\int_{-\infty}^{\infty} \tilde{J}_{zk}(\alpha) \tilde{E}_{z1}(\alpha, d + t) d\alpha = 2\pi \int_{-\infty}^{\infty} J_{zk}(x) E_{z1}(x, d + t) dx = 0$$

because $J_{zk}(x) = 0$ for $|x| > w/2$ whereas $E_{z1}(x, d + t) = 0$ for $|x| < w/2$ from the boundary condition (11.19) not yet used. Hence, the integrand $J_{zk}(x)E_{z1}(x, d + t) = 0$ for any value of x.

If (11.41) and (11.42) are rewritten, we obtain

$$\sum_{m=1}^{N} K_{km}^{(1,1)} c_m + \sum_{m=1}^{M} K_{km}^{(1,2)} d_m = 0, \quad k = 1, 2, \ldots, N \tag{11.43}$$

$$\sum_{m=1}^{N} K_{lm}^{(2,1)} c_m + \sum_{m=1}^{M} K_{lm}^{(2,2)} d_m = 0, \quad l = 1, 2, \ldots, M \tag{11.44}$$

where

$$K_{km}^{(1,1)} = \int_{-\infty}^{\infty} \tilde{J}_{zk}(\alpha)\tilde{Z}_{zz}(\alpha, \beta)\tilde{J}_{zm}(\alpha)d\alpha \tag{11.45a}$$

$$K_{km}^{(1,2)} = \int_{-\infty}^{\infty} \tilde{J}_{zk}(\alpha)\tilde{Z}_{zx}(\alpha, \beta)\tilde{J}_{xm}(\alpha)d\alpha \tag{11.45b}$$

$$K_{lm}^{(2,1)} = \int_{-\infty}^{\infty} \tilde{J}_{xl}(\alpha)\tilde{Z}_{xz}(\alpha, \beta)\tilde{J}_{zm}(\alpha)d\alpha \tag{11.45c}$$

$$K_{lm}^{(2,2)} = \int_{-\infty}^{\infty} \tilde{J}_{xl}(\alpha)\tilde{Z}_{xx}(\alpha, \beta)\tilde{J}_{xm}(\alpha)d\alpha \tag{11.45d}$$

The combination of (11.43) and (11.44) makes a set of homogeneous simultaneous linear equations of order $(N + M)$. For meaningful solutions of c_m and d_m, β appearing in $K_{km}^{(1,1)}$, *et cetera* must make the determinant of the coefficient matrix of the preceding set of simultaneous equations zero for a given angular frequency ω. These values of β are the eigenvalues that provide combinations of c_m and d_m as the eigensolution. The value of β for the fundamental mode is the maximum among those satisfying $\max(\varepsilon_i) > (\beta/k_0)^2 > \min(\varepsilon_i)$.

In an actual calculation, accuracy and convergence of the solution are strongly influenced by the choice of basis functions. In the preceding, the basis functions are chosen so that their inverse transforms are zero in the region $|x| > w/2$. However, such a condition alone is insufficient, and the following points need to be considered:

1. If we choose functions that have correct singularity at the edges of the strip, then convergence is fast and a desired accuracy is reached with a small number of terms $(N + M)$.

2. Because J_z and J_x are related via the electromagnetic-field equations, their symmetries with respect to $x = 0$ are also related. For the dominant mode, J_z is an even function and J_x is an odd function.
3. In reality, of the functions of x satisfying the preceding conditions, we select functions that can be analytically Fourier transformed. Otherwise, we must calculate the basis functions by numerical calculations of Fourier transforms. Then, the numerical efficiency associated with the spectral-domain method cannot be fully utilized.

These requirements significantly restrict the types of functions to be used. Those frequently used in practice are Fourier transforms of the following functions:

$$J_{zn}(x) = \frac{\cos[2(n-1)\pi x/w]}{\sqrt{1-(2x/w)^2}}, \quad n = 1, 2, \ldots \tag{11.46a}$$

$$J_{xn}(x) = \frac{\sin[2n\pi x/w]}{\sqrt{1-(2x/w)^2}}, \quad n = 1, 2, \ldots \tag{11.46b}$$

These functional forms are applicable only in the region $|x| < w/2$ and both J_{zn} and J_{xn} are zero in the region $|x| > w/2$. The functional shapes for $n = 1$, 2, and 3 are shown in Figure 11.2. The functions just defined can be Fourier transformed analytically and the results are

$$\tilde{J}_{zn}(\alpha) = \frac{\pi w}{4}\left[J_0\left(\left|\frac{w\alpha}{2} + (n-1)\pi\right|\right) + J_0\left(\left|\frac{w\alpha}{2} - (n-1)\pi\right|\right)\right] \tag{11.47a}$$

$$\tilde{J}_{xn}(\alpha) = \frac{\pi w}{4j}\left[J_0\left(\left|\frac{w\alpha}{2} + n\pi\right|\right) - J_0\left(\left|\frac{w\alpha}{2} - n\pi\right|\right)\right] \tag{11.47b}$$

where J_0 is the Bessel function of order zero. The fact that all the basis functions can be expressed in terms J_0 alone is convenient in writing a computer program.

Once β is found, the values of c_m and d_m can be determined in relation to one of them from (11.43) and (11.44). Hence, the Fourier transforms of the current distributions can be found from (11.39) and (11.40). The current distributions in the space domain can be obtained as functions of x from J_{zm} and J_{xm}. Further, because the electromagnetic field expansion coefficients $A^e, \ldots, D^h$ in the Fourier transform domain are expressed in terms of $\tilde{J}_z$ and $\tilde{J}_x$, these coefficients are also expressed in

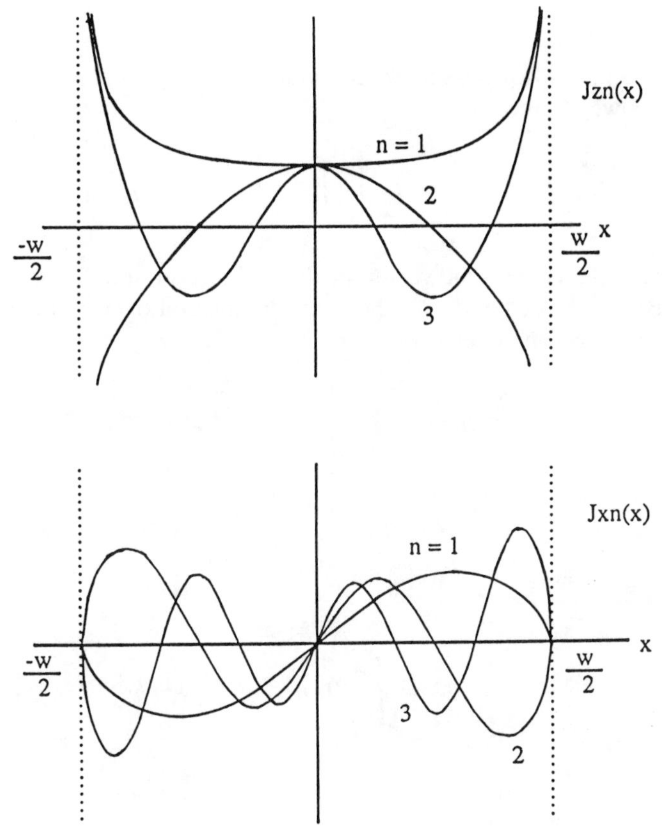

Figure 11.2 Shapes of basis functions.

terms of c_m and d_m. Hence, the Fourier transformed electromagnetic field distributions such as $\tilde{E}_{z1}$ can also be expressed in terms of c_m and d_m.

11.5 CHARACTERISTIC IMPEDANCE

Because the guided mode propagating in the microstrip line is not TEM, it is difficult to define the characteristic impedance rigorously. In general, for the planar transmission line of the strip type, the characteristic impedance defined as follows by the power transmitted and the current is commonly used:

$$Z_0 = P/I_0^2 \tag{11.48}$$

where I_0 is the current flowing on the strip in the z direction and P is the time average of Poynting power flow given by

$$P = \text{Re} \int_0^c \int_{-\infty}^{\infty} \boldsymbol{E} \times \boldsymbol{H}^* \cdot \boldsymbol{z} dxdy \tag{11.49}$$

In this equation, $\boldsymbol{z}$ is the unit vector in the z direction and * implies a complex conjugate. By means of Parseval's theorem, the integral over x in the equation can be replaced with the one over α.

$$\begin{aligned}\int_S \boldsymbol{E} \times \boldsymbol{H}^* \cdot \boldsymbol{z} dxdy &= \frac{1}{2\pi} \int_{-\infty}^{\infty} \int_0^c \tilde{\boldsymbol{E}} \times \tilde{\boldsymbol{H}} \cdot \boldsymbol{z} dy d\alpha \\ &= \frac{1}{2\pi} \int_{-\infty}^{\infty} [E_{h1} + E_{h2} + E_{h3}] d\alpha \end{aligned} \tag{11.50}$$

where

$$E_{h1} = \int_{d+t}^{c} (\tilde{E}_{x1}\tilde{H}_{y1}^* - \tilde{E}_{y1}\tilde{H}_{x1})dy$$

$$E_{h2} = \int_{c}^{d+t} (\tilde{E}_{x2}\tilde{H}_{y2}^* - \tilde{E}_{y2}\tilde{H}_{x2}^*)dy$$

$$E_{h3} = \int_{0}^{d} (\tilde{E}_{x3}\tilde{H}_{y3}^* - \tilde{E}_{y3}\tilde{H}_{x3}^*)dy$$

Here, the integrals over y can be analytically executed because the y variation of the integrands have forms like $\sinh\gamma_1 y$ and $\cosh\gamma_1 y$. Hence,

$$E_{h1} = \pm\frac{1}{2} A_1 A_4^* \left[\frac{S_1 C_1}{\gamma_1} - h\right] - \frac{1}{2} A_2 A_3^* \left[\frac{S_1 C_1}{\gamma_1} + h\right] \tag{11.51a}$$

$$\begin{aligned} E_{h2} = &\pm\frac{1}{2}(C_1 C_4^* - B_2 B_3^*)\left[\frac{S_2 C_2}{\gamma_2} - t\right] + \frac{1}{2}(B_1 B_4^* - C_2 C_3^*)\left[\frac{S_2 C_2}{\gamma_2} + t\right] \\ &+ \frac{1}{2}[C_1 B_4^* - B_2 B_3^* \pm (B_1 C_4^* - C_2 B_3^*)]\frac{S_2^2}{\gamma_2} \end{aligned} \tag{11.51b}$$

$$E_{h3} = \pm \frac{1}{2} D_1 D_4^* \left[\frac{S_3 C_3}{\gamma_3} - d \right] - \frac{1}{2} D_2 D_3^* \left[\frac{S_3 C_3}{\gamma_3} + d \right] \tag{11.51c}$$

where

$$C_1, C_2, C_3 = \cosh\gamma_1 h, \cosh\gamma_2 t, \cosh\gamma_3 d \tag{11.52a}$$

$$S_1, S_2, S_3 = \sinh\gamma_1 h, \sinh\gamma_2 t, \sinh\gamma_3 d \tag{11.52b}$$

$$\pm : + \text{ if } \gamma_i = \text{real}$$

$$- \text{ if } \gamma_i = \text{imaginary}$$

Also,

$$A_1 = j(\alpha\gamma_{y1}A^e - \beta A^h) \qquad A_2 = \frac{1}{\hat{y}_1}[\gamma_1^2 + k_1^2]A^e$$

$$B_1 = -j(\alpha\gamma_{y2}B^e + \beta B^h) \qquad B_2 = \frac{1}{\hat{y}_2}[\gamma_2^2 + k_2^2]B^e$$

$$C_1 = -j(\alpha\gamma_{y2}C^e + \beta C^h) \qquad C_2 = \frac{1}{\hat{y}_2}[\gamma_2^2 + k_2^2]C^e$$

$$D_1 = -j(\alpha\gamma_{y3}D^e + \beta D^h) \qquad D_2 = \frac{1}{\hat{y}_3}[\gamma_3^2 + k_3^2]D^e$$

$$A_3 = j(\beta A^e + \alpha\gamma_{z1}A^h) \qquad A_4 = \frac{1}{\hat{z}_1}[\gamma_1^2 + k_1^2]A^h$$

$$B_3 = j(\beta B^e - \alpha\gamma_{z2}B^h) \qquad B_4 = \frac{1}{\hat{z}_2}[\gamma_2^2 + k_2^2]B^h$$

$$C_3 = j(\beta C^e - \alpha\gamma_{z2}C^h) \qquad C_4 = \frac{1}{\hat{z}_2}[\gamma_2^2 + k_2^2]C^h$$

$$D_3 = j(\beta D^e - \alpha\gamma_{z3}D^h) \qquad D_4 = \frac{1}{\hat{z}_3}[\gamma_3^2 + k_3^2]D^h$$

Here, $A^e, \ldots, D^h$ can be expressed in terms of $\tilde{J}_z(\alpha)$ and $\tilde{J}_x(\alpha)$ and, hence, of c_m and d_m. Also, I_0 in (11.48) can be obtained by analytically integrating $J_z(x)$ over $|x| < w/2$. This completes computation of the characteristic impedance.

11.6 IMMITANCE METHOD

The method just described actually requires derivations of complicated equations. As the structural complexity is increased, such derivations become even more cumbersome and a likelihood of a mistake increases. The immitance method was invented to alleviate these deficiencies and is easily understood by electrical engineers familiar with the concept of equivalent circuits. Especially, for the derivation of the propagation constant β, the amount of effort for derivation of equations is negligible. Green's functions such as $\tilde{Z}_{zz}$ can be derived directly from the equivalent circuit [7]. Computation of the characteristic impedance is also significantly simplified.

This method can be understood if the meaning of the Fourier transform is considered. When the inverse Fourier transform is used, the quantity propagating in the z direction with the propagation constant β can be written as

$$\Phi(x, y)\, e^{-j\beta x} = \frac{1}{2\pi}\int_{-\infty}^{\infty} \tilde{\Phi}(\alpha, y)\, e^{-j(\alpha x+\beta z)} d\alpha \tag{11.53}$$

This equation implies that the quantity on the left-hand side, or the propagating electromagnetic field, is a superposition of inhomogeneous plane waves that have varying amplitudes in the y direction. Each plane wave propagates in the direction determined by the values of α and β. This direction is given by the angle $\theta = \cos^{-1}[\beta/(\alpha^2 + \beta^2)^{1/2}]$ measured from the z axis. Hence, a new (u, v) coordinate system is defined in the (x, z) plane as shown in Figure 11.3.

$$\begin{aligned} u &= z \sin\theta - x \cos\theta \\ v &= z \cos\theta + x \sin\theta \end{aligned} \tag{11.54}$$

In both coordinate systems, the y axes coincide and are in a direction perpendicular to the surface of the paper. A hybrid wave in the (x, z) coordinate system can be expressed in terms of a superposition of the $\text{TM}_y(\tilde{E}_y, \tilde{E}_v, \tilde{H}_u)$ and the $\text{TE}_y(\tilde{H}_y, \tilde{E}_u, \tilde{H}_v)$. Similarly, the current distributions $\tilde{J}_z$ and $\tilde{J}_x$ can be expressed in terms of combinations of $\tilde{J}_u$ and $\tilde{J}_v$. We can readily understand that $\tilde{J}_v$ excites only the TM_y electromagnetic fields and $\tilde{J}_u$ excites only the TE_y counterparts. Hence, Green's functions for the TM and TE can be derived independently from the equivalent circuits of transmission lines along the y direction, as shown in Figure 11.4. The current source $\tilde{J}_u$ or $\tilde{J}_v$ excites the transmission line at $y = d$ where the strip exists. The transmission line extending upward from $y = d$ corresponds to Region 1 and is short circuited at $y = h$, where the perfect conductor for upper shielding is located. For Region 2, as many transmission lines as the number of dielectric layers are cascaded. To obtain the desired Green's function, the relationship between the current and the electric

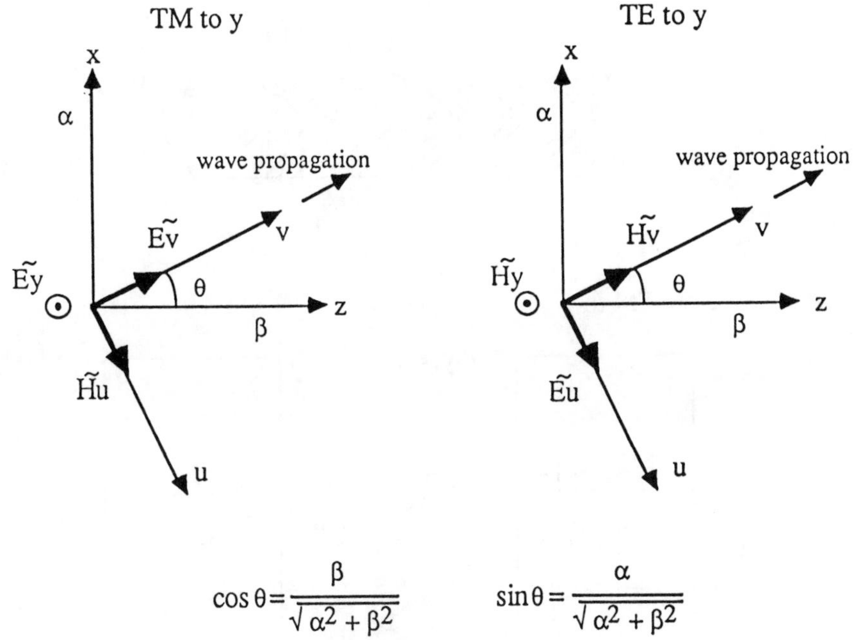

Figure 11.3 Decomposition of spectral waves in the (u, v) coordinate system.

field at $y = d$ is required. For the TM_y equivalent circuit, the circuit equations are as follows:

$$\tilde{H}_{u1} - \tilde{H}_{u2} = \tilde{J}_v \tag{11.55}$$

$$\frac{-\tilde{H}_{u1}}{\tilde{E}_{v1}} = Y_1^e \tag{11.56a}$$

$$\frac{\tilde{H}_{u2}}{\tilde{E}_{v2}} = Y_2^e \tag{11.56b}$$

$$\tilde{E}_v = -\tilde{Z}_e \tilde{J}_v \quad (\tilde{E}_{v1} = \tilde{E}_{v2} = \tilde{E}_v) \tag{11.57}$$

where

$$\tilde{Z}_e = \frac{1}{Y_1^e + Y_2^e} \tag{11.58}$$

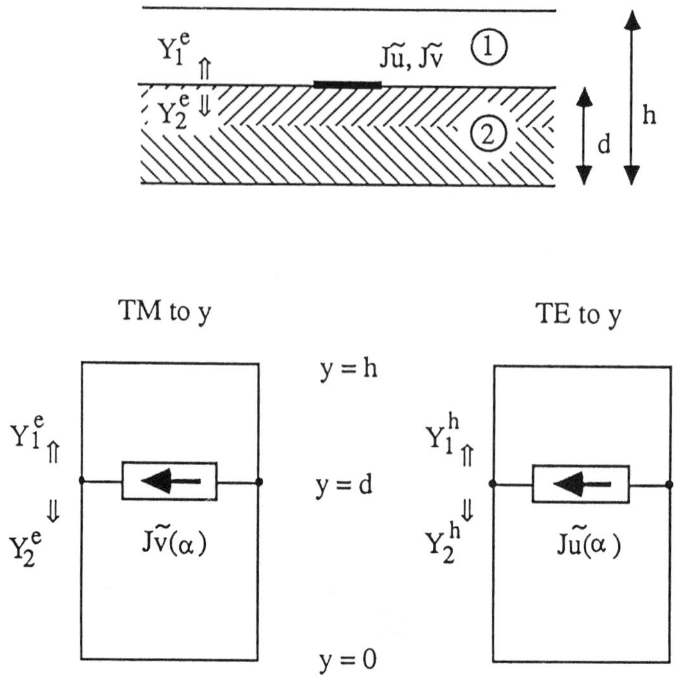

Figure 11.4 Equivalent transmission lines for the TM and the TE fields.

In the preceding, Y_1^e and Y_2^e are the input admittances looking upward and downward at $y = d$. As described later, these quantities can readily be computed once the structure is given. Similarly, for the TE_y,

$$\tilde{E}_u = -\tilde{Z}_h \tilde{J}_u \quad (\tilde{E}_{u1} = \tilde{E}_{u2} = \tilde{E}_u) \tag{11.59}$$

$$\tilde{Z}_h = \frac{1}{Y_1^h + Y_2^h} \tag{11.60}$$

where Y_1^h and Y_2^h are the input admittances for the TE wave. From (11.57) and (11.59), it is found that $\tilde{Z}_e$ and $\tilde{Z}_h$ are Green's functions for the TM and TE waves.

What is needed is the relationships between $\tilde{E}_z$, $\tilde{E}_x$ and $\tilde{J}_z$, $\tilde{J}_x$ as shown in (11.29) and (11.30). When the coordinate transformation is used in the reverse direction, these relationships are obtained.

$$\tilde{Z}_{zz} = -\frac{1}{\alpha^2 + \beta^2}(\beta^2 \tilde{Z}_e + \alpha^2 \tilde{Z}_h) \tag{11.61}$$

$$\tilde{Z}_{zx} = -\frac{\alpha\beta}{\alpha^2 + \beta^2}(\tilde{Z}_e - \tilde{Z}_h) \tag{11.62}$$

$$\tilde{Z}_{xx} = -\frac{1}{\alpha^2 + \beta^2}(\alpha^2\tilde{Z}_e + \beta^2\tilde{Z}_h) \tag{11.63}$$

These equations coincide with (11.31) through (11.34).

In practice, derivations of Y_1^e, Y_2^e, Y_1^h, and Y_2^h can be accomplished by the use of impedance transformation along the transmission line.

$$Z_{in} = Z_{0i}\frac{Z_L \coth\gamma_i h_i + Z_{0i}}{Z_{0i}\coth\gamma_i h_i + Z_L} \tag{11.64}$$

where h_i is the thickness of the ith layer, and γ_i is the complex propagation constant along the y direction and is given by (11.9). Also, for the TM and TE transmission lines, the characteristic impedance Z_{0i} of the equivalent transmission line is

$$Z_{0\mathrm{TM}i} = -\frac{\tilde{E}_v}{\tilde{H}_u} = \frac{\gamma_i}{j\omega\varepsilon_i}\ (= \gamma_{yi}) \tag{11.65a}$$

$$Z_{0\mathrm{TE}i} = \frac{\tilde{E}_u}{\tilde{H}_v} = \frac{j\epsilon\mu_i}{\gamma_i}\left(= \frac{1}{\gamma_{zi}}\right) \tag{11.65b}$$

In accordance with the process just described, $\tilde{Z}_e$ and $\tilde{Z}_h$ can be derived immediately once the structure of the planar transmission line is given. From these quantities, Green's functions $\tilde{Z}_{zz}$, $\tilde{Z}_{xz}$, and $\tilde{Z}_{xx}$ can be obtained from (11.61) through (11.63). As an example, in the case of a two-layered substrate in Figure 11.1,

$$\tilde{Z}_e = \frac{\gamma_{y2}Ct_3 + \gamma_{y3}Ct_2}{Ct_2Ct_3 + Ct_1Ct_3\gamma_{y2}/\gamma_{y1} + Ct_1Ct_2\gamma_{y3}/\gamma_{y1} + \gamma_{y3}/\gamma_{y2}}$$

$$\tilde{Z}_h = \frac{\gamma_{z2}Ct_2 + \gamma_{z3}Ct_3}{\gamma_{z1}\gamma_{z2}Ct_1Ct_2 + \gamma_{z1}\gamma_{z3}Ct_1Ct_3 + \gamma_{z2}\gamma_{z3}Ct_2Ct_3 + \gamma_{z2}^2},$$

$$Ct_1 = \coth\gamma_1 h, \quad Ct_2 = \coth\gamma_2 t, \quad Ct_3 = \coth\gamma_3 d$$

These results agree with (11.35) and (11.36). If the substrate is made of a one-layer material:

$$\tilde{Z}_e = \frac{\gamma_{y1}\gamma_{y2}}{\gamma_{y1}Ct_2 + \gamma_{y2}Ct_1} \tag{11.66a}$$

$$\tilde{Z}_h = \frac{1}{\gamma_{z1}Ct_1 + \gamma_{z2}Ct_2}, \tag{11.66b}$$

$$Ct_1 = \coth\gamma_1(h-d), \quad Ct_2 = \coth\gamma_2 d$$

More complicated structures such as the three-layered substrate shown in Figure 11.5 can also be computed in a simple manner. The results are as follows:

$$\tilde{Z}_e = \left(\frac{1}{Z_1^e} + \frac{1}{Z_2^e}\right)^{-1} \quad \tilde{Z}_h = \left(\frac{1}{Z_1^h} + \frac{1}{Z_2^h}\right)^{-1} \tag{11.67}$$

where

$$Z_1^e = \frac{\gamma_{y1}}{Ct_1} \qquad Z_1^h = \frac{1}{\gamma_{z1}Ct_1}$$

$$Z_2^e = \gamma_{y2}\frac{Z_3Ct_2 + \gamma_{y2}}{\gamma_{y2}Ct_2 + Z_3^e} \qquad Z_2^h = \frac{1}{\gamma_{z2}}\frac{\gamma_{z2}Z_3^hCt_2 + 1}{Ct_2 + \gamma_{z2}Z_3^h}$$

$$Z_3^e = \gamma_{y3}\frac{\gamma_{y4}Ct_3 + \gamma_{y3}Ct_4}{\gamma_{y3}Ct_3Ct_4 + \gamma_{y4}} \qquad Z_3^h = \frac{1}{\gamma_{z3}}\frac{\gamma_{z3}Ct_3 + \gamma_{z4}Ct_4}{\gamma_{z4}Ct_3Ct_4 + \gamma_{z3}}$$

$$Z_4^e = \frac{\gamma_{y4}}{Ct_4} \qquad Z_4^h = \frac{1}{\gamma_{z4}Ct_4},$$

$$Ct_i = \coth\gamma_i h_i, \quad i = 1, 2, 3, 4$$

Next, the concept of the immitance method is applied to the derivation of the characteristic impedance. First, in place of (11.2), the electromagnetic fields are expressed in terms of $\tilde{E}_y$ and $\tilde{H}_y$:

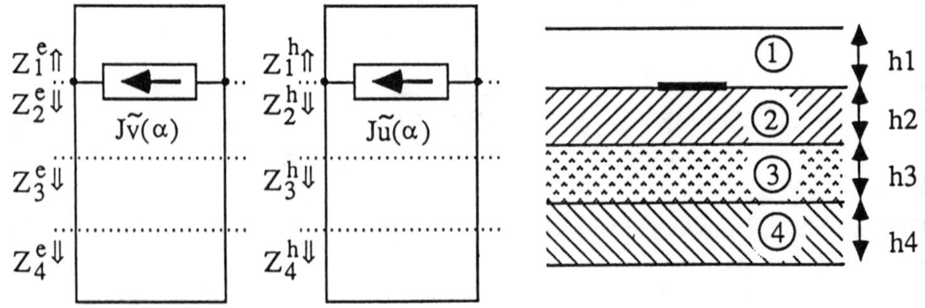

Figure 11.5 Equivalent transmission lines for a three-layer microstrip line.

TM$_y$	TE$_y$	
$\tilde{E}_x = -jR\dfrac{\partial \tilde{E}_y}{\partial y}$	$\tilde{E}_x = -j\hat{z}Q\tilde{H}_y$	(11.68)
$\tilde{E}_z = -jQ\dfrac{\partial \tilde{E}_y}{\partial y}$	$\tilde{E}_x = j\hat{z}R\tilde{H}_y$	(11.69)
$\tilde{H}_x = j\hat{y}Q\tilde{E}_y$	$\tilde{H}_x = -jR\dfrac{\partial \tilde{H}_y}{\partial y}$	(11.70)
$\tilde{H}_z = -j\hat{y}R\tilde{E}_y$	$\tilde{H}_z = -jQ\dfrac{\partial \tilde{H}_y}{\partial y}$	(11.71)

where

$$R = \frac{\alpha}{\alpha^2 + \beta^2}, \quad Q = \frac{\beta}{\alpha^2 + \beta^2} \tag{11.72}$$

In what follows, the discussion is applied to the case of a three-layered substrate shown in Figure 11.5. In each region, $\tilde{E}_y$ and $\tilde{H}_y$ are written as follows:

TM$_y$

$$\begin{aligned}
&\text{Region (4): } -j\hat{y}_4\tilde{E}_{y4} = A^e \cosh\gamma_4 y \\
&\text{Region (3): } -j\hat{y}_3\tilde{E}_{y3} = B^e \cosh\gamma_3(y - h_4) + C^e \sinh\gamma_3(y - h_4) \\
&\text{Region (2): } -j\hat{y}_2\tilde{E}_{y2} = D^e \cosh\gamma_2(y - h_3 - h_4) \\
&\qquad\qquad + E^e \sinh\gamma_2(y - h_3 - h_4) \\
&\text{Region (1): } -j\hat{y}_1\tilde{E}_{y1} = F^e \cosh\gamma_1(c - y)
\end{aligned} \tag{11.73}$$

TE$_y$

$$\begin{aligned}
&\text{Region (4): } -j\hat{z}_4\tilde{H}_{y4} = A^h \sinh\gamma_4 y \\
&\text{Region (3): } -j\hat{z}_3\tilde{H}_{y3} = B^h \sinh\gamma_3(y - h_4) + C^h \cosh\gamma_3(y - h_4) \\
&\text{Region (2): } -j\hat{z}_2\tilde{H}_{y2} = D^h \sinh\gamma_2(y - h_3 - h_4) \\
&\qquad\qquad + E^h \cosh\gamma_2(y - h_3 - h_4) \\
&\text{Region (1): } -j\hat{z}_1\tilde{H}_{y1} = F^h \sinh\gamma_1(c - y), \quad c = h_1 + h_2 + h_3 + h_4
\end{aligned} \tag{11.74}$$

The boundary conditions can be applied independently to the TM portion and the TE portion. For each portion, six equations can be derived.

At $y = h_4$,

$$\tilde{E}_{t3} = \tilde{E}_{t4} \tag{11.75}$$

$$\tilde{H}_{t3} = \tilde{H}_{t4} \tag{11.76}$$

At $y = h_3 + h_4$,

$$\tilde{E}_{t2} = \tilde{E}_{t3} \tag{11.77}$$

$$\tilde{H}_{t2} = \tilde{H}_{t3} \tag{11.78}$$

At $y = h_2 + h_3 + h_4$,

$$\tilde{E}_{t1} = \tilde{E}_{t2} \tag{11.79}$$

$$\begin{cases} \tilde{H}_{v1} - \tilde{H}_{v2} = \tilde{J}_v \quad \text{for TM}_y & (11.80\text{a}) \\ \tilde{H}_{u1} - \tilde{H}_{u2} = -\tilde{J}_u \quad \text{for TE}_y & (11.80\text{b}) \end{cases}$$

In (11.75) through (11.79), t indicates either the x or z component. Also, it is seen from (11.68) through (11.71) that both the x and z components satisfy the identical boundary conditions. Further, when (11.54) is used, the final condition (11.80) can be written by (11.70) and (11.71) as follows:

$$-j\frac{1}{\sqrt{\alpha^2 + \beta^2}}(\hat{y}_1\tilde{E}_{y1} - \hat{y}_2\tilde{E}_{y2}) = \tilde{J}_v \tag{11.81}$$

$$-j\frac{1}{\sqrt{\alpha^2 + \beta^2}}\left(\frac{\partial \tilde{H}_{y1}}{\partial y} - \frac{\partial \tilde{H}_{y2}}{\partial y}\right) = -\tilde{J}_u \tag{11.82}$$

When $\tilde{E}_y$ and $\tilde{H}_y$ expressed by (11.73) and (11.74) are substituted into the boundary conditions, we obtain for the TM and TE waves:

<u>TM_y</u>

$$\frac{1}{\hat{y}_4}A^e\gamma_4 S_4 = \frac{1}{\hat{y}_3}C^e\gamma_3$$

$$A^e C_4 = B^e$$

$$\frac{1}{\hat{y}_3} B^e \gamma_3 S_3 + \frac{1}{\hat{y}_3} C^e \gamma_3 C_3 = \frac{1}{\hat{y}_2} E^e \gamma_2$$

$$B^e C_3 + C^e S_3 = D^e$$

$$\frac{1}{\hat{y}_2} D^e \gamma_2 S_2 + \frac{1}{\hat{y}_2} E^e \gamma_2 C_2 = -\frac{1}{\hat{y}_1} F^e \gamma_1 S_1$$

$$\frac{1}{\sqrt{\alpha^2 + \beta^2}} F^e C_1 - \frac{1}{\sqrt{\alpha^2 + \beta^2}} (D^e C_2 + E^e S_2) = \tilde{J}_v$$

$\underline{\mathrm{TE}_y}$

$$A^h S_4 = C^h$$

$$A^h \gamma_4 C_4 = B^h \gamma_3$$

$$B^h S_3 + C^h C_3 = E^h$$

$$B^h \gamma_3 C_3 + C^h \gamma_3 S_3 = D^h \gamma_2$$

$$D^h S_2 + E^h C_2 = F^h S_1$$

$$-\frac{1}{\hat{z}\sqrt{\alpha^2 + \beta^2}} F^h \gamma_1 C_1 - \frac{1}{\hat{z}\sqrt{\alpha^2 + \beta^2}} (D^h \gamma_2 C_2 + E^h \gamma_2 S_2) = -\tilde{J}_u$$

where

$$C_i, S_i = \cosh\gamma_i h_i, \sinh\gamma_i h_i, \quad i = 1, 2, 3, 4$$

$$\hat{z} = j\omega\mu_0, \qquad \mu_0 = \text{permeability of free space}$$

Both of the TM and TE sets consist of 6 equations containing 6 unknown coefficients. When these equations are solved, each of these 6 coefficients can be expressed in terms of the current distribution $\tilde{J}_v$ or $\tilde{J}_u$ as follows:

$\underline{\mathrm{TM}_y}$

$$A^e = \frac{1}{C_2 C_3 C_4} H_e, \qquad B^e = \frac{1}{C_2 C_3} H_e,$$

$$C^e = \frac{1}{C_2 C_3} \frac{\gamma_{y4}}{\gamma_{y3}} \frac{1}{Ct_4} H_e, \quad D^e = \frac{1}{C_2} H_{1e} H_e, \tag{11.83}$$

$$E^e = \frac{1}{C_2} \frac{1}{\gamma_{y2}} H_{3e} H_e, \qquad F^e = -\frac{1}{S_1} \frac{1}{\gamma_{y1}} \left(\frac{\gamma_{y2}}{Ct_2} H_{1e} + H_{3e} \right) H_e$$

where

$$H_e = -\frac{\sqrt{\alpha^2 + \beta^2}\,\tilde{J}_v}{H_{1e}H_{2e} + H_{3e}H_{4e}},$$

$$H_{1e} = 1 + \frac{\gamma_{y4}}{\gamma_{y3}}\frac{1}{Ct_3Ct_4}, \qquad H_{2e} = 1 + \frac{\gamma_{y2}}{\gamma_{y1}}\frac{Ct_1}{Ct_2}, \tag{11.84}$$

$$H_{3e} = \frac{\gamma_{y3}}{Ct_3} + \frac{\gamma_{y3}}{Ct_4}, \qquad H_{4e} = \frac{Ct_1}{\gamma_{y1}} + \frac{1}{\gamma_{y2}Ct_2}$$

Also,

$\underline{\mathrm{TE}_y}$

$$A_h = \frac{1}{C_2C_3C_4}H_h \qquad B^h = \frac{1}{C_2C_3}\frac{\gamma_4}{\gamma_3}Ct_4H_h$$

$$C^h = \frac{1}{C_2C_3}H_h \qquad D^h = \frac{1}{C_2}\frac{1}{\gamma_2}H_{2h}H_h \tag{11.85}$$

$$E^h = \frac{1}{C_2}H_{4h}H_h \qquad F^h = \frac{1}{S_1}\left(\frac{1}{\gamma_2Ct_2}H_{2h} + H_{4h}\right)H_h$$

where

$$H_h = -\frac{\hat{z}\sqrt{\alpha^2 + \beta^2}\,\tilde{J}_u}{H_{1h}H_{2h} + H_{3h}H_{4h}},$$

$$H_{1h} = 1 + \frac{\gamma_1}{\gamma_2}\frac{Ct_1}{Ct_2}, \qquad H_{2h} = \frac{\gamma_3}{Ct_3} + \gamma_4Ct_4, \tag{11.86}$$

$$H_{3h} = \gamma_1Ct_1 + \frac{\gamma_2}{Ct_2}, \qquad H_{4h} = 1 + \frac{\gamma_4}{\gamma_3}\frac{Ct_4}{Ct_3}$$

Also, $C_{ti} = \coth \gamma_i h_i$, $i = 1, 2, 3, 4$.

As described in the previous section, the following integral is needed in th computation of the characteristic impedance:

$$\int_S \boldsymbol{E} \times \boldsymbol{H}^* \cdot zdxdy = \frac{1}{2\pi}\int_{-\infty}^{\infty}\int_0^c \tilde{\boldsymbol{E}} \times \tilde{\boldsymbol{H}}^* \cdot zdyd\alpha$$

$$= \frac{1}{2\pi}\int_{-\infty}^{\infty} [E_{h1} + E_{h2} + E_{h3} + E_{h4}]d\alpha$$

E_{h1}, E_{h2}, E_{h3}, and E_{h4} are given by

$$E_{h4} = \pm \frac{1}{2}\,\mathbf{A}_1\mathbf{A}_4^*\left[\pm \frac{Ct_4}{|C_2C_3|^2\gamma_4} - \frac{h_4}{|C_2C_3S_4|^2}\right]$$
$$- \frac{1}{2}\,\mathbf{A}_2\mathbf{A}_3^*\left[\pm \frac{Ct_4}{|C_2C_3|^2\gamma_4} + \frac{h_4}{|C_2C_3S_4|^2}\right]$$
$$E_{h3} = \pm \frac{1}{2}\,(\mathbf{B}_1\mathbf{B}_4^* - \mathbf{C}_2\mathbf{C}_3^*)\left[\pm \frac{1}{|C_2|^2\gamma_3Ct_3^*} - \frac{h_3}{|C_2C_3|^2}\right]$$
$$+ \frac{1}{2}\,(\mathbf{C}_1\mathbf{C}_4^* - \mathbf{B}_2\mathbf{B}_3^*)\left[\pm \frac{1}{|C_2|^2\gamma_3Ct_3^*} + \frac{h_3}{|C_2C_3|^2}\right]$$
$$+ \frac{1}{2}\,[(\mathbf{C}_1\mathbf{B}_4^* - \mathbf{B}_2\mathbf{C}_3^*) \pm (\mathbf{B}_1\mathbf{C}_4^* - \mathbf{C}_2\mathbf{B}_3^*)]\,\frac{1}{|C_2Ct_3|^2\gamma_3}$$
$$E_{h2} = \pm \frac{1}{2}\,(\mathbf{D}_1\mathbf{D}_4^* - \mathbf{E}_2\mathbf{E}_3^*)\left[\pm \frac{1}{\gamma_2Ct_2^*} - \frac{h_2}{|C_2|^2}\right]$$
$$+ \frac{1}{2}\,(\mathbf{E}_1\mathbf{E}_4^* - \mathbf{D}_2\mathbf{D}_3^*)\left[\pm \frac{1}{\gamma_2Ct_2^*} + \frac{h_2}{|C_2|^2}\right]$$
$$+ \frac{1}{2}\,[(\mathbf{E}_1\mathbf{D}_4^* - \mathbf{D}_2\mathbf{E}_3^*) \pm (\mathbf{D}_1\mathbf{E}_4^* - \mathbf{E}_2\mathbf{D}_3^*)]\,\frac{1}{|Ct_2|^2\gamma_2}$$
$$E_{h1} = \pm \frac{1}{2}\,\mathbf{F}_1\mathbf{F}_4^*\left[\pm \frac{Ct_1}{\gamma_1} - \frac{h_1}{|S_1|^2}\right] - \frac{1}{2}\,\mathbf{F}_2\mathbf{F}_3^*\left[\pm \frac{Ct_1}{\gamma_1} + \frac{h_1}{|S_1|^2}\right]$$
$$\pm : \text{for } \gamma = \text{real} \qquad - \text{ for } \gamma = \text{imaginary}$$

where

$$\mathbf{A}_i = C_2C_3S_4A_i, \quad \mathbf{B}_i = C_2C_3B_i, \quad \mathbf{C}_i = C_2C_3C_i$$
$$\mathbf{D}_i = C_2D_i, \quad \mathbf{E}_i = C_2E_i, \quad \mathbf{F}_i = S_1F_i, \quad i = 1, 2, 3, 4$$
$$A_1 = \gamma_{y4}RA^e + QA^h, \qquad A_2 = \frac{1}{-j\hat{y}_4}A^e$$
$$B_1 = \gamma_{y3}RB^e + QB^h, \qquad B_2 = \frac{1}{-j\hat{y}_3}B^e$$
$$C_1 = \gamma_{y3}RC^e + QC^h, \qquad C_2 = \frac{1}{-j\hat{y}_3}C^e$$
$$D_1 = \gamma_{y2}RD^e + QD^h, \qquad D_2 = \frac{1}{-j\hat{y}_2}D^e$$

$$E_1 = \gamma_{y2} RE^e + QE^h, \quad E_2 = \frac{1}{-j\hat{y}_2} E^e$$

$$F_1 = \gamma_{y1} RF^e - QF^h, \quad F_2 = \frac{1}{-j\hat{y}_1} F^e$$

$$A_3 = -QA^e + \gamma_{z4} RA^h, \quad A_4 = \frac{1}{-j\hat{z}} A^h$$

$$B_3 = -QB^e + \gamma_{z3} RB^h, \quad B_4 = \frac{1}{-j\hat{z}} B^h$$

$$C_3 = -QC^e + \gamma_{z3} RC^h, \quad C_4 = \frac{1}{-j\hat{z}} C^h$$

$$D_3 = -QD^e + \gamma_{z2} RD^h, \quad D_4 = \frac{1}{-j\hat{z}} D^h$$

$$E_3 = -QE^e + \gamma_{z2} RE^h, \quad E_4 = \frac{1}{-j\hat{z}} E^h$$

$$F_3 = -QF^e - \gamma_{z1} RF^h, \quad F_4 = \frac{1}{j\hat{z}} F^h$$

The expressions given by (11.83) and (11.85) are substituted into A^e, A^h, ... , F^e and F^h.

11.7 SOME PRECAUTIONS FOR PROGRAMMING

No matter how efficient the solution method is, the requirement for the computation time and computer memory depend strongly on how the computer program is written. The way the computer program should be written is also affected by the computer on which such a program is run. The following are some of the elementary precautions in writing a computer program.

11.7.1 Symmetry

Any structural symmetry can be used to reduce the amount of computational effort. For instance, the structure used in this chapter has a symmetry with respect to the y-axis that passes through the center of the strip conductor. Also the assumed current distribution has either an even or an odd symmetry. Hence, the infinite integrals need to be computed only from 0 to ∞.

1.7.2 Convergence

'he current functions used in the examples in this paper have a singularity at the dges of the strip. The asymptotic behavior of their Fourier transforms are dictated y such a singularity. In the computation of (11.45), the integrands decay at the rate f α^{-2}. Because the convergence speed at this rate is rather slow, it is important to xecute the infinite integrals efficiently. It is necessary to determine at what value f α the infinite integrals are truncated and the errors associated with the calculations aken into consideration. For the structure placed in a shield case, the finite Fourier ransform is used due to the presence of the side walls. In this case, the inverse ransforms and the inner products are given by infinite sums instead of integrals. 'he number of terms to be summed is affected by the dimensions of the shield case nd the strip width. Further, the accuracy of the solution is affected by the number f basis functions. Accuracy increases with the use of more basis functions. However, at the same time, the number of matrix elements increases at the square of that f the basis functions. The computation time, therefore, increases quickly.

1.7.3 Other Precautions

t is important not to write a computer program containing complex numbers. Even f the imaginary unit j appears or γ_i becomes purely imaginary in the formulations, he program can be written in terms of real numbers only as long as the structure reated is lossless. In finding the eigenvalue β, it is customary to compute (11.45) vith an assumed value of β, and the latter is systematically varied until the determinant of the coefficient matrix of (11.43) and (11.44) becomes numerically zero.)n such occasions, the quantities not related to β, such as $\tilde{J}_{zk}$, should not be computed every time β is varied. Also, if the computer system does not have a subroutine ackage needed for calculation then a subprogram for calculating Bessel functions, he polynomial expansions, and the asymptotic expansions presented in [8] is frequently used for developing an efficient subprogram.

In the program in which β is sought every time the frequency ω is varied by small increment, the value of β obtained in the previous value of ω can be used s the initial value to accelerate convergence of the solution. Recently, a number of omputer programs for CAD have contained an approximate formula or an empirical ormula for the microstrip line analysis. The value of β obtained from such a program an be used as the initial value for more accurate calculation by the spectral domain rogram.

11.8 CONCLUSIONS

n this chapter, the spectral domain method has been presented. In the derivation of quations, the most important steps have been included as much as possible with the least amount of omission. No explanation has been included for the spectral domain method as applied to other planar transmission lines, such as finlines. However, those readers who have followed the derivations of equations in this chapter should not face any major difficulty in deriving necessary equations for these structures. In the case of planar transmission lines of the slot type, such as finlines and slot lines, it is numerically more attractive to use equations that express the current on the conductors in terms of the electric fields in the slot than (11.29) and (11.30). In such formulations, Green's functions are of admittance type rather than impedance type ($\tilde{Z}_{zz}$ *et cetera*). Obviously, the immitance method can also be applied to such structures.

In the analysis of a three-dimensional structure, such as a microstrip resonator, double Fourier transforms in the two directions (x, z) along the surface of the substrate need to be used. The Fourier transform variables are α and β. The resonant frequency ω becomes the eigenvalue to be obtained. If the immitance method is used in the derivation of Green's functions, then the equivalent transmission line representations along the y-direction perpendicular to the substrate surface in the present three-dimensional problem are identical to those used in the two-dimensional problems.

REFERENCES

[1] Yamashita, E., and R. Mittra, "Variational Method for the Analysis of Microstrip Line," *IEEE Trans. Microwave Theory and Tech.*, April 1968, Vol. MTT-16, No. 4, pp. 251–256.

[2] Yamashita, E., "Variational Method for the Analysis of Microstrip-like Transmission Lines," *IEEE Trans. Microwave Theory and Tech.*, Aug. 1968, Vol. MTT-16, No. 8, pp. 529–535.

[3] Denlinger, E.J., "A Frequency Dependent Solution for Microstrip Transmission Lines," *IEEE Trans. Microwave Theory and Tech.*, Jan. 1971, Vol. MTT-19, No. 1, pp. 30–39.

[4] Itoh, T., and R. Mittra, "Spectral-Domain Approach for Calculating the Dispersion Characteristics of Microstrip Lines," *IEEE Trans. Microwave Theory and Tech.*, July 1973, Vol. MTT-21, No. 7, pp. 496–499.

[5] Zhang, Q., and T. Itoh, "Spectral-Domain Analysis of Scattering from E-Plane Circuit Elements," *IEEE Trans. Microwave Theory and Tech.*, Feb. 1987, Vol. MTT-35, No. 2, pp. 138–150.

[6] Harrington, R.F., *Time-Harmonic Electromagnetic Fields*, McGraw-Hill, New York, 1961.

[7] Itoh, T., "Spectral Domain Immitance Approach for Dispersion Characteristics of Generalized Printed Transmission Lines," *IEEE Trans. Microwave Theory and Tech.*, July 1980, Vol. MTT-28, No. 7, pp. 733–736.

[8] Abramowitz, M., and I.A. Stegun, *Handbook of Mathematical Functions*, Dover, New York, 1965, pp. 366–370.

Bibliography

CHAPTER ONE

Ahmed, S. and P. Daly, "Finite-element Method for Inhomogeneous Waveguides," *Proc. IEE,* Vol. 116, October 1969, pp. 1661–1664.

Angkaew, T., M. Matuhara, and N. Kumagai, "Finite-element Analysis of Waveguide Modes: A Novel Approach that Eliminates Spurious Modes," *IEEE Trans. Microwave Theory Tech.,* Vol. MTT-35, February 1987, pp. 117–123.

Chew, W.C., and M.A. Nasir, "A Variational Analysis of Anisotropic, Inhomogeneous Dielectric Waveguides," *IEEE Trans. Microwave Theory Tech.,* Vol. 37, April 1989, pp. 661–668.

Chiang, K.S., "Finite Element Analysis of Weakly Guiding Fibers with Arbitrary Refractive-index Distribution," *IEEE J. Lightwave Technol.,* Vol. LT-4, August 1986, pp. 980–990.

Chiang, K.S., "Finite Element Method for Cutoff Frequencies of Weakly Guiding Fibres of Arbitrary Cross-section," *Opt. Quantum Electron.,* Vol. 16, 1984, pp. 487–493.

Csendes, Z.J., and P. Silvester, "Numerical Solution of Dielectric Loaded Waveguides: I-Finite-element Analysis," *IEEE Trans. Microwave Theory Tech.,* Vol. MTT-18, December 1970, pp. 1124–1131.

Daly, P., "Hybrid-mode Analysis of Microstrip by Finite-element Method," *IEEE Trans. Microwave Theory Tech.,* Vol. MTT-19, January 1971, pp. 19–25.

Hano, M., "Finite-element Analysis of Dielectric-loaded Waveguides," *IEEE Trans. Microwave Theory Tech.,* Vol. MTT-32, October 1984, pp. 1275–1279.

Hayata, K., M. Eguchi, and M. Koshiba, "Finite Element Formulation for Guided-Wave Problems Using Transverse Electric Field Component," *IEEE Trans. Microwave Theory Tech.,* Vol. 37, January 1989, pp. 256–258.

Hayata, K., M. Eguchi, and M. Koshiba, "Self-consistent Finite/Infinite Element Scheme for Unbounded Guided Wave Problems," *IEEE Trans. Microwave Theory Tech.,* Vol. 36, March 1988, pp. 614–616.

Hayata, K., M. Eguchi, M. Koshiba, and M. Suzuki, "Vectorial Wave Analysis of Side-tunnel Type Polarization-maintaining Optical Fibers by Variational Finite-Elements," *IEEE J. Lightwave Technol.,* Vol. LT-4, August 1986, pp. 1090–1096.

Hayata, K., and M. Koshiba, "Characteristics of Graded-core Stress-applied Polarization-maintaining Single-mode Fibers," *J. Opt. Soc. Am. A,* Vol. 5, April 1988, p. 535–541.

Hayata, K., M. Koshiba, M. Eguchi, and M. Suzuki, "Novel Finite-element Formulation without any Spurious Solutions for Dielectric Waveguides," *Electron. Lett.,* Vol. 22, March 1986, pp. 295–296.

Hayata, K., M. Koshiba, M. Eguchi, and M. Suzuki, "Vectorial Finite-element Method without any Spurious Solutions for Dielectric Waveguiding Problem Using Transverse Magnetic-field Component," *IEEE Trans. Microwave Theory Tech.*, Vol. MTT-34, November 1986, pp. 1120–1124.

Hayata, K., K. Miura, and M. Koshiba, "Finite-element Formulation for Lossy Waveguides," *IEEE Trans. Microwave Theory Tech.*, Vol. 36, February 1988, pp. 268–276.

Hayata, K., K. Miura, and M. Koshiba, "Full Vectorical Finite Element Formulation for Lossy Anisotropic Waveguides," *IEEE Trans. Microwave Theory Tech.*, Vol. 37, May 1989, pp. 875–883.

Ikeuchi, M., H. Sawami, and H. Niki, "Analysis of Open-type Dielectric Waveguides by the Finite-element Iterative Method," *IEEE Trans. Microwave Theory Tech.*, Vol. MTT-29, March 1981, pp. 234–239.

Katz, J., "Novel Solution of 2-D Waveguides Using the Finite-element Method," *Appl. Opt.*, Vol. 21, August 1982, pp. 2747–2750.

Kobelansky, A.J., and J.B. Webb, "Eliminating Spurious Modes in Finite-element Waveguide Problems by Using Divergence-free Fields," *Electron. Lett.*, Vol. 22, May 1986, pp. 569–570.

Konrad, A., "High-order Triangular Finite Elements for Electromagnetic Waves in Anisotropic Media," *IEEE Trans. Microwave Theory Tech.*, Vol. MTT-25, May 1977, pp. 353–360.

Koshiba, M., K. Hayata, and M. Suzuki, "Approximate Scalar Finite-element Analysis of Anisotropic Optical Waveguides," *Electron. Lett.*, Vol. 18, May 1982, pp. 411–413.

Koshiba, M., K. Hayata, and M. Suzuki, "Approximate Scalar Finite-element Analysis of Anisotropic Optical Waveguides with Off-diagonal Elements in a Permittivity Tensor," *IEEE Trans. Microwave Theory Tech.*, Vol. MTT-32, June 1984, pp. 587–593.

Koshiba, M., K. Hayata, and M. Suzuki, "Finite-element Formulation in Terms of the Electric-field Vector for Electromagnetic Waveguide Problems," *IEEE Trans. Microwave Theory Tech.*, Vol. MTT-33, October 1985, pp. 900–905.

Koshiba, M., K. Hayata, and M. Suzuki, "Finite-element Solution of Anisotropic Waveguides with Arbitrary Tensor Permittivity," *IEEE J. Lightwave Technol.*, Vol. LT-4, February 1986, pp. 121–126.

Koshiba, M., K. Hayata, and M. Suzuki, "Improved Finite-element Formulation in Terms of Magnetic Field Vector for Dielectric Waveguides," *IEEE Trans. Microwave Theory Tech.*, Vol. MTT-33, March 1985, pp. 227–233.

Koshiba, M., K. Hayata, and M. Suzuki, "Vectorical Finite-element Formulation without Spurious Solutions for Dielectric Waveguide Problems," *Electron. Lett.*, Vol. 20, May 1984, pp. 409–410.

Koshiba, M., H. Kumagami, and M. Suzuki, "Finite-element Solution of Planar Arbitrarily Anisotropic Diffused Optical Waveguides," *IEEE J. Lightwave Technol.*, Vol. LT-3, August 1985, pp. 773–778.

Koshiba, M., and M. Suzuki, "Numerical Analysis of Planar Arbitrarily Anisotropic Diffused Optical Waveguides Using Finite-element Method," *Electron. Lett.*, Vol. 18, June 1982, pp. 579–581

Mabaya, N., P.E. Lagasse, and P. Vandenbulcke, "Finite Element Analysis of Optical Waveguides," *IEEE Trans. Microwave Theory Tech.*, Vol. MTT-29, June 1981, pp. 600–605.

McAalay, A.D., "Variational Finite-element Solution for Dissipative Waveguides and Transportation Application," *IEEE Trans. Microwave Theory Tech.*, Vol. MTT-25, May 1977, pp. 382–392.

Mustacich, R.V., "Scalar Finite Element Analysis of Electrooptic Modulation in Diffused Channel Waveguides and Poled Waveguides in Polymer Thin Films," *Appl. Opt.*, Vol. 27, September 1988, pp. 3732–3737.

Okamoto, K., "Comparison of Calculated and Measured Impulse Responses of Optical Fibers," *Appl. Opt.*, Vol. 18, July 1979, pp. 2199–2206.

)kamoto, K., and T. Okoshi, "Vectrial Wave Analysis of Inhomogeneous Optical Fibers Using Finite Element Method," *IEEE Trans. Microwave Theory Tech.*, Vol. MTT-26, February 1978, pp. 109–114.

)yamada, K., and T. Okoshi, "Two-dimensional Finite-element Method Calculation of Propagation Characteristics of Axially Nonsymmetrical Optical Fibers," *Radio Sci.*, Vol. 17, January-February 1982, pp. 109–116.

Rahman, B.M.A., and J.B. Davies, "Finite-element Analysis of Optical and Microwave Waveguide Problems," *IEEE Trans. Microwave Theory Tech.*, Vol. MTT-32, January 1984, pp. 20–28.

Rahman, B.M.A., and J.B. Davies, "Finite-element Solution of Integrated Optical Waveguides," *IEEE J. Lightwave Technol.*, Vol. LT-2. October 1984, pp. 682–688.

Rahman, B.M.A., and J.B. Davies, "Penalty Function Improvement of Waveguide Solution by Finite Elements," *IEEE Trans. Microwave Theory Tech.*, Vol. MTT-32, August 1984, pp. 922–928.

Strake, E., G.P. Bava, and I. Montrosset, "Guided Modes of Ti:$LiNb0_3$ Channel Waveguides: A Novel Quasi-analytical Technique in Comparison with the Scalar Finite-element Method," *IEEE J. Lightwave Technol.*, Vol. 6, June 1988, pp. 1126–1135.

Su, C.C., "A Combined Method for Dielectric Waveguides Using the Finite-element Technique and the Surface Integral Equations Method," *IEEE Trans. Microwave Theory Tech.*, Vol. MTT-34, November 1986, pp. 1140–1146.

Su, C.C., "Eigenproblems of Radially Inhomogeneous Optical Fibers from the Scalar Formulation," *IEEE J. Quantum Electron.*, Vol. QE-21, October 1985, pp. 1554–1557.

Su, C.C., "Origin of Spurious Modes in the Analysis of Optical Fiber Using the Finite-element or Finite-difference Technique," *Electron. Lett.*, Vol. 21, September 1985, pp. 858–860.

Vandenbulcke, P., and P.E. Lagasse, "Eigenmode Analysis of Anisotropic Optical Fibres or Integrated Optical Waveguides," *Electron. Lett.*, Vol. 12, March 1976, pp. 120–122.

Webb, J.P., "Finite Element Analysis of Dispersion in Waveguides with Sharp Metal Edges," *IEEE Trans. Microwave Theory Tech.*, Vol. 36, December 1988, pp. 1819–1824.

Welt, D., and J. Webb, "Finite-element Analysis of Dielectric Waveguides with Curved Boundaries," *IEEE Trans. Microwave Theory Tech.*, Vol. MTT-33, July 1985, pp. 576–586.

Wu, R.B., and C.H. Chen, "A Scalar Variational Conformal Mapping Technique for Weakly Guiding Dielectric Waveguides," *IEEE J. Quantum Electron.*, Vol. QE-22, May 1986, pp. 603–609.

Wu, R.B., and C.H. Chen, "A Variational Analysis of Dielectric Waveguides by the Conformal Mapping Technique," *IEEE Trans. Microwave Theory Tech.*, Vol. MTT-33, August 1985, pp. 681–685.

Yeh, C., S.B. Dong, and W. Oliver, "Arbitrarily Shaped Inhomogeneous Optical Fiber or Integrated Optical Waveguides," *J. Appl. Phys.*, Vol. 46, May 1975, pp. 2125–2129.

Yeh, C., K. Ha, S.B. Dong, and W.P. Brown, "Single-mode Optical Waveguides," *Appl. Opt.*, Vol. 18, May 1979, pp. 1490–1504.

Young, T.P., "Design of Integrated Optical Circuits Using Finite Elements," *IEE Proc.*, Vol. 135, Pt. A, March 1988, pp. 135–144.

CHAPTER TWO

On Integral Representations:

Courant, K., and D. Hilbert, *Methods of Mathematical Physics, Vol. II,* Chap. IV, New York: John Wiley and Sons, 1962.

Jackson, J.D., *Classical Electrodynamics,* Secs. 9.8–9.9, New York: John Wiley and Sons, 1962.

Jones, D.S., *The Theory of Electromagnetism,* Secs. 1.14–1.20, Oxford: Pergamon, 1964.

Papas, C.H., "Diffraction by a Cylindrical Obstacle," *J. Appl. Phys.*, Vol. 21, April 1950, pp. 318–325
Silver, S., *Microwave Antenna Theory and Design,* Sec. 3.8, New York: McGraw-Hill, 1949.
Van Bladel, J., *Electromagnetic Fields,* Chap. 3, New York: McGraw-Hill, 1964.

On Resonant Solutions:

Mautz, J.R., and R.F. Harrington, "A Combined-source Solution for Radiation and Scattering from a Perfectly Conducting Body," *IEEE Trans. Antennas & Propag.*, Vol. AP-27, July 1979, pp. 445–454.
Mautz, J.R., and R.F. Harrington, "Electromagnetic Scattering from a Homogeneous Material Body of Revolution," *Arch. Elektron Übertragungstech.*, Vol. 33, 1979, pp. 71–80.
Mittra, R., and C.A. Klein, "Stability and Convergence of Moment Method Solutions," in *Numerical and Asymptotic Techniques in Electromagnetics,* Mittra, R., ed., Chap. 5, New York: Springer, 1975.
Schenck, H.A., "Improved Integral Formulation for Acoustic Radiation Problems," *J. Acoust. Soc. Am.*, Vol. 44, July 1968, pp. 41–58.

On Numerical Calculation of Integral Equations:

Killes, P., "Solution of Dirichlet Problems Using a Hybrid Finite-differences and Integral-equation Method Applied to Electron Guns," *Optik,* Vol. 70, 1985, pp. 64–71.
Richmond, J.H., "Digital Computer Solutions of the Rigorous Equations for Scattering Problems," *Proc. IEEE,* Vol. 53, August 1965, pp. 796–804.
Sarkar, T.K., "A Note on the Choice Weighting Functions in the Method of Moments," *IEEE Trans. Antennas & Propag.*, Vol. AP-33, April 1985, pp. 436–441.
Spielman, B.E., and R.F. Harrington, "Waveguides of Arbitrary Cross Section by Solution of a Nonlinear Integral Eigenvalue Equation," *IEEE Trans. Microwave Theory & Tech.*, Vol. MTT-20, September 1972, pp. 575–585.

CHAPTER THREE

Bates, R.H.T., "The Rayleigh Hypothesis, the Extended Boundary Condition and Point-matching," *Electron. Lett.*, Vol. 5, 1969, pp. 654–655.
Bolle, D.M., and D.M. Fye, "Application of Point-matching Method to Scattering from Quadrilateral Cylinders," *Electron. Lett.*, Vol. 7, 1971, pp. 577–579.
Fuller, J.A., and N.F. Audeh, "The Point-matching Solution of Uniform Nonsymmetric Waveguides," *IEEE Trans. Microwave Theory Tech.*, Vol. MTT-17, No. 2, February 1969, pp. 114–115.
Kosslowski, S., F. Bogelsack, and I. Wolff, "The Application of the Point-matching Method to the Analysis of Microstrip Lines with Finite Metallization Thickness," *IEEE Trans. Microwave Theory Tech.*, Vol. MTT-36, No. 8, August 1988, pp. 1265–1271.
Lewin, L., "On the Restricted Validity of Point-matching Techniques," *IEEE Trans. Microwave Theory Tech.*, Vol. MTT-18, No. 12, December 1970, pp. 1041–1047.
Millar, R.F., and R.H.T. Bates, "On the Legitimacy of an Assumption Underlying the Point-matching Method," *IEEE Trans. Microwave Theory Tech.*, Vol. MTT-18, No. 6, June 1970, pp. 325–327.

CHAPTER FOUR

Bates, R.H.T., "Analytical Constraints on Electromagnetic Field Computations," *IEEE Trans. Microwave Theory Tech.*, Vol. MTT-23, 1975, pp. 605–623.

Hafner, C., "Computations of Electromagnetic Fields by the MMP Method," *Proc. 1989 URSI International Symposium on EM Theory,* 1989, pp. 141–143.

Hafner, C., and S. Kiener, "Parallel Computations of 3-D Electromagnetic Fields on Transputers," *Proc. 1989 International Symposium on Antennas and Propagation,* 1, 1989, pp. 105–108.

Kleev, A.I., and A.B. Manenkov, "The Convergence of Point-matching Techniques," *IEEE Trans. Antennas and Propagation,* Vol. AP-37, No. 1, 1989, pp. 50–54.

Ludwig, A.C., "A Comparison of Spherical Wave Boundary Value Matching *versus* Integral Equation Scattering Solution for a Perfectly Conducting Body," *IEEE Trans. Antennas and Propagation,* Vol. AP-34, No. 7, 1986, pp. 857–865.

Ramm, A.G., "Convergence of the T-matrix Approach to Scattering Theory," *J. Math. Phys.*, Vol. 23, 1982, pp. 1123–1125.

Ramm, A.G., *Scattering by Obstacles,* Dordrecht: D. Reidel Publishing Company, 1986.

Vekua, I.N., *New Methods for Solving Elliptic Equations,* Amsterdam: North-Holland Publishing Company, 1968.

Watermann, P.C., "Matrix Formulation of Electromagnetic Scattering," *Proc. IEEE,* Vol. 53, 1965, pp. 805–812.

CHAPTER FIVE

Spatial Network Method

Aoto, T., Y. Kakimi, N. Yoshida, and I. Fukai, "Transient Analysis of Near Field for Conductive Pillar with Thin-type Absorber," *IEEE Trans. Antennas and Propagation,* Vol. AP-37, No. 4, April 1989, pp. 471–479.

Aoto, T., N. Yoshida, and I. Fukai, "Transient Analysis of the Electromagnetic Field for a Wave Absorber in Three-Dimensional Space," *IEEE Trans. Electromagnetic Compatibility,* Vol. EMC-29, No. 1, February 1987, pp. 18–23.

Aoto, T., N. Yoshida, and I. Fukai, "Transient Analysis of Electromagnetic Field for Resonance-type Absorber in Three-Dimensional Space," *Electronics and Communication in Japan, Part 1,* Vol. 71, No. 6, June 1988, pp. 86–97.

Kakimi, Y., N. Yoshida, and I. Fukai, "Analysis of Absorbing Characteristics of Thin-type Absorber for Generalized Conditions of Incident Wave," *IEEE Trans. Electromagnetic Compatibility,* Vol. EMC-31, No. 3, August 1989, pp. 323–328.

Kashiwa, T., N. Yoshida, and I. Fukai, "Unified Analysis of Ridge-Waveguide Slot Antenna in Three-Dimensional Space and Time," *Electronics & Communications in Japan, Part I,* Vol. 70, No. 7, July 1987, pp. 88–97.

Kashiwa, T., N. Yoshida, and I. Fukai, "Analysis of Radiation Characteristics of a Planar Inverted-F Antenna on the Conductive Body of a Hand-held Transceiver by the Spatial Network Method," *Electron. Lett.*, Vol. 25, No. 16, August 1989, pp. 1044–1045.

Kashiwa, T., N. Yoshida, and I. Fukai, "Transient Analysis of a Magnetized Plasma in Three-Dimensional Space," *IEEE Trans. Antennas and Propagation,* Vol. AP-36, No. 8, August 1988, pp. 1096–1105.

Koike, S., N. Yoshida, and I. Fukai, "Transient Analysis of Coupled Microstrip-Slot-Line in Three-

Dimensional Space," *Electronics & Communications in Japan, Part I,* Vol. 69, No. 11, Nov. 1986, pp. 101–109.

Koike, S., N. Yoshida, and I. Fukai, "Transient Analysis of Coupling between Crossing Lines in Three-Dimensional Space," *IEEE Trans. Microwave Theory Tech.,* Vol. MTT-35, No. 1, January 1987, pp. 67–71.

Koike, S., N. Yoshida, and I. Fukai, "Transient Analysis of Directional Coupler Using a Coupled Microstrip Slotline in Three-Dimensional Space," *IEEE Trans. Microwave Theory Tech.,* Vol. MTT-34, No. 3, March 1986, pp. 353–357.

Koike, S., N. Yoshida, and I. Fukai, "Transient Analysis of Microstrip Gap in Three-Dimensional Space," *IEEE Trans. Microwave Theory Tech.,* Vol. MTT-33, No. 8, August 1985, pp. 726–730.

Koike, S., N. Yoshida, and I. Fukai, "Transient Analysis of Microstrip Line on Anisotropic Substrate in Three-Dimensional Space," *IEEE Trans. Microwave Theory Tech.,* Vol. MTT-36, No. 1, January 1988, pp. 34–43.

Kukutsu, N., N. Yoshida, and I. Fukai, "Transient Analysis of Ferrite in Three-Dimensional Space," *IEEE Trans. Microwave Theory Tech.,* Vol. MTT-36, No. 1, January 1988, pp. 114–125.

Kukutsu, N., N. Yoshida, and I. Fukai, "Transient Analysis of Ferrite Twin-toroidal Phase Shifter by Spatial Network Method," *Electron. Lett.,* Vol. 25, No. 9, April 1989, pp. 581–583.

Nakatsuka, H., N. Yoshida, and I. Fukai, "Three-Dimensional Analysis of a Vacuum Window Connected to Wave Guide," *IEEE Trans. Plasma Science,* Vol. PS-16, No. 4, August 1988, pp. 416–422.

Shibata, T., T. Hayashi, and T. Kimura, "Analysis of Microstrip Circuit Using Three-Dimensional Full-wave Electromagnetic Field Analysis in the Time Domain," *IEEE Trans. Microwave Theory Techn.,* Vol. MTT-36, No. 6, June 1988, pp. 1064–1070.

Soga, M., N. Yoshida, H. Homma, and I. Fukai, "Three-Dimensional Analysis of Fabry Perot Resonator," *Int. J. of Infrared and Millimeter Waves,* Vol. 4, No. 3, March 1983, pp. 345–359.

Tejika, Y., F. Komatsu, T. Masaki, N. Yoshida, and I. Fukai, "Three-Dimensional Analysis of a Microwave Heating Furnace System," *J. Microwave Power,* Vol. 22, No. 2, June 1987, pp. 107–114.

Tejika, Y., F. Komatsu, T. Masaki, N. Yoshida, and I. Fukai, "Three-Dimensional Analysis of Apertures in Furnace-type Microwave Heating System," *IEEE Trans. Electromagnetic Compatibility,* Vol. EMC-31, No. 1, January 1989, pp. 55–62.

Tejika, Y., F. Komatsu, T. Masaki, N. Yoshida, and I. Fukai, "Three-Dimensional Analysis of Microwave Heating System with Square-type Furnace," *IEEE Trans. Nuclear Science,* Vol. NS-34, No. 4, August 1988, pp. 1065–1069.

Terashima, K., N. Yoshida, and I. Fukai, "Analysis of Radiation Characteristics of NRD Guide Leaky-wave Antenna by Spatial Network Method," *Electron. Lett.,* Vol. 24, No. 18, September 1988, pp. 1164–1165.

Yoshida, N., T. Kashiwa, N. Kukutsu, and I. Fukai, "Three-Dimensional Formulation of Time-dependent Wave Propagation in a Gyroanisotropic Medium by Bergeron's Method," *Wave Motion,* Vol. 10, No. 6, December 1988, pp. 611–626.

Finite Difference Time-Domain Method

Choi, D.K., and W.J.R. Hoefer, "The Finite-difference Time-domain Method and its Application to Eigenvalue Problems," *IEEE Trans. Microwave Theory Techn.,* Vol. MTT-34, No. 12, December 1986, pp. 1464–1470.

Meriwether, D.E., R. Fisher, and F.W. Smith, "On Implementing a Numeric Huygen's Source

Scheme in a Finite-difference Program to Illuminate Scattering Bodies," *IEEE Trans. Nuclear Science,* Vol. 27, No. 6, December 1980, pp. 1819–1833.

Mur. G., "Absorbing Boundary Conditions for the Finite-difference Approximation of the Time-domain Electromagnetic-field Equations," *IEEE Trans. Electromagnetic Compatibility,* EMC-23, Vol. 6, November 1981, pp. 377–382.

Taflove, A., "Review of the Formulation and Applications of the Finite-difference Time-domain Method for Numerical Modelling of Electromagnetic Wave Interactions with Arbitrary Structure," *Wave Motion,* Vol. 10, No. 6, December 1988, pp. 547–582.

Taflove, A., and K.R. Umashankar, "The Finite-difference Time-domain (FD-TD) Method for Electromagnetic Scattering and Interaction Problems," *J. Electromagnetic Waves and Applications,* Vol. 1, No. 3, 1987, pp. 243–267.

Zhang, X., J. Fang, K.K. Mei, and Y. Liu, "Calculations of the Dispersive Characteristics of Microstrips by the Time-domain Finite Difference Method," *IEEE Trans. Microwave Theory Tech.,* Vol. MTT-36, No. 2, February 1988, pp. 263–267.

Transmission Line Matrix Method

Alexòpoulos, N.G., "Integrated-circuit Structure on Anisotropic Substrates," *IEEE Trans. Microwave Theory Tech.,* Vol. MTT-33, No. 10, October 1985, pp. 847–881.

Itoh T., *Numerical Techniques for Microwave and Millimeter-Wave Passive Structures,* New York: John Wiley and Sons, 1989.

Johns, P.B., "A Symmetrical Condensed Node for the TLM Method," *IEEE Trans. Microwave Theory Tech.,* Vol. MTT-35, No. 4, April 1987, pp. 370–377.

Mariki, G.E., and C. Yeh, "Dynamic Three-Dimensional TLM Analysis of Microstriplines on Anisotropic Substrate," *IEEE Trans. Microwave Theory Tech.,* Vol. MTT-33, No. 9, September 1985, pp. 789–799.

Saguet, P., "Le Maillage Parallelèpipèdique et le Changement de Maille dans la Mèthode TLM en Trois Dimensions," *Electron. Lett.* Vol. 20, No. 5, March 1984, pp. 222–224.

CHAPTER SIX

Hafner, C., Numerische Berechnung Elektromagnetischer Felder, Berlin: Springer, 1987.

Iskander, M.F., A. Lakhtakia, and C.H. Durney, "A New Iterative Procedure to Solve for Scattering and Absorption by Dielectric Objects," *Proc. IEEE,* Vol. 70, No. 11, November 1982, pp. 1361–1362.

Iskander, M.F., A. Lakhtakia, and C.H. Durney, "A New Procedure for Improving the Solution Stability and Extending the Frequency Range of the EBCM," *IEEE Trans. on Antennas and Propagation,* Vol. AP-31, No. 2, March 1983, pp. 317–324.

Kiener, S., "Bodies with Sharp Edges: Calculation of Near- and Far-fields," *1989 IEEE AP-S Intern'l Symp. Digest,* Vol. I, San Jose, CA, June 1989, pp. 183–186.

Klaus, G., "3D Streufeldberechnungen mit Hilfe der MMP-Methode," Diss. 7792, ETH, Zurich, 1985.

Kuster, N., "Computations of 3D Problems of High Complexity with GMT," *1989 IEEE AP-S Intern'l Symp. Digest,* Vol. I, San Jose, CA, June 1989, pp. 168–171.

Kuster, N., and R. Ballisti, "MMP Method Simulation of Antennas with Scattering Objects in the Closer Near Field," *IEEE Trans. on Magn.,* Vol. MAG-25, No. 7, July 1989, pp. 2881–2883.

Leuchtmann, P., "Automatisierung der Functionenwahl bei der MMP-Methode," Diss. 8301, ETH, Zurich, 1987.

Leuchtmann, P., "The Construction of Practically Useful Fast Converging Expansions for the GMT, *1989 IEEE AP-S Intern'l Symp. Digest,* Vol. I, San Jose, CA, June 1989, pp. 176–179.

Levensee, R.M., "The Extended Equivalent Source Technique with Undetermined Sources," *1989 IEEE AP-S Intern'l Symp. Digest,* Vol. I, San Jose, CA, June 1989, pp. 180–182.

Leviatan, Y., and A. Boag, "Analysis of Electromagnetic Scattering from Dielectric Cylinders Using a Multifilament Current Model," *IEEE Trans. on Antennas and Propagation,* Vol. AP-35, No. 10, October 1987, pp. 1119–1127.

Leviatan, Y., P.G. Li, A.T. Adams, and J. Perini, "Single-post Inductive Obstacle in Rectangular Waveguide," *IEEE Trans. Microwave Theory Tech.,* Vol. MTT-31, No. 10, October 1983, pp. 806–812.

Ludwig, A.C., "Spherical Wave Theory," in *The Handbook of Antenna Design,* Vol. I, A.W. Rudge *et al.,* Eds., Stevenage: Peter Peregrinus, 1982.

Ludwig, A.C., "A Comparison of Spherical Wave Boundary Value Matching *versus* Integral Equation Scattering Solutions for a Perfectly Conducting Body," *IEEE Trans. on Antennas and Propagation,* Vol. AP-34, No. 7, July 1986, pp. 857–865.

CHAPTER SEVEN

Akashi, T., M. Ando, and T. Kinoshita, "Effects of Multiple Diffraction in PTD Analysis of Scattered Field from a Conducting Disk," *Trans. IEICE,* Vol. E72, No. 4, April 1989, pp. 259–261.

Ando, M., "PO and PTD Analysis of Offset Reflector Antenna Patterns," *IEEE AP/S Symposium,* June 1988, pp. 112–115.

Ando, M., and T. Kinoshita, "Accuracy Comparison of PTD and PO for Plane Wave Diffraction from a Large Circular Disk," *Trans. IEICE,* Vol. E72, No. 11, November 1989, pp. 1212–1218.

Endo, N., M. Ando, and T. Sekiguchi, "Correction of Physical Optics by the Equivalent Edge Currents," *IEEE AP/S Symposium,* 1979, pp. 548–551.

Knott, E.F., "The Relationship between Mitzner's ILDC and Michaeli's Equivalent Currents," *IEEE Trans. Antennas and Propagation,* Vol. AP-33, No. 1, January 1985, pp. 112–114.

Knott, E.F., and T.B.A. Senior, "Comparison of Three High Frequency Diffraction Techniques," *Proc. IEEE,* Vol. 62, November 1974, pp. 1468–1474.

Michaeli, A., "Equivalent Edge Currents for Arbitrary Aspects of Observation," *IEEE Trans. Antennas and Propagation,* Vol. AP-32, No. 3, March 1984, pp. 252–258.

Mitzner, K.M., "Incremental Length Diffraction Coefficients," *Aircraft Northrop Corp. Tech. rep.* April 1974, AFALTR-73-296.

Shore, R.A., and A. Yaghjian; "Incremental Diffraction Coefficients for Planar Surfaces," *IEEE Trans. Antennas and Propagation,* Vol. AP-36, No. 1, January 1988, pp. 55–70.

CHAPTER EIGHT

Daniele, V.G., "On the Solution of Two Coupled Wiener-Hopf Equations," *SIAM J. Appl. Math.,* Vol. 44, No. 4, August 1984, pp. 667–680.

Dowerah, S., and A. Chakrabarti, "Extinction Cross Section of a Dielectric Strip," *IEEE Trans. Antennas and Propagation,* Vol. 36, No. 5, May 1988, pp. 696–706.

Heins, A.E., "Systems of Wiener-Hopf Integral Equations and Their Application to Some Boundary Value Problems in Electromagnetic Theory," *Proc. Symposia in Appl. Math., Vol. 2: Electromagnetic Theory,* A.H. Taub, E. Reissner and R.V. Churchill, Eds., New York: American Mathematical Society, 1950, pp. 76–81.

Hurd, R.A., and S. Przeździecki, "Diffraction by a Half-plane Perpendicular to the Distinguished Axis of a Gyrotropic Medium," *J. Math. Phys.*, Vol. 17, No. 10, October 1976, pp. 1838–1847.

Jull, E.V., "Diffraction by a Conducting Half-plane in an Anisotropic Plasma, " *Can. J. Phys.*, Vol. 42, No. 8, August 1964, pp. 1455–1468. See also for Errata: *Can. J. Phys.*, Vol. 45, No. 5, May 1967, pp. 1963–1964.

Jull, E.V., "Diffraction by a Wide Unidirectionally Conducting Strip," *Can. J. Phys.*, Vol. 46, No. 18, September 1968, pp. 2107–2117.

Lüneburg, E., and K. Westpfahl, "Diffraction of Plane Waves by an Infinite Strip Grating," *Ann. Phys.*, Vol. 27, No. 3, October 1971, pp. 257–288.

Meister, E., *Randwertaufgaben der Funktionentheorie: mit Anwendungen auf Singuläre Integralgleichungen und Schwingungsprobleme der Mathematische Physik,* Stuttgart: B.G. Teubner, 1983.

Rawlins, A.D., "The Solution of a Mixed Boundary Value Problem in the Theory of Diffraction by a Semi-infinite Plane," *Proc. Roy. Soc. London, Ser. A,* Vol. 346, No. 1647, November 1975, pp. 469–484.

Senior, T.B.A., "Diffraction by a Semi-infinite Metallic Sheet," *Proc. Roy. Soc. London, Ser. A,* Vol. 213, No. 1115, July 1952, pp. 436–458.

CHAPTER NINE

Airy, G.B., Supplement to "On the Intensity of Light in the Neighbourhood of a Caustic," *Proc. Cambridge Philos. Soc.*, Vol. 8, 1849, pp. 595–599.

Argyres, P.N., "The Bohr-Sommerfeld Quantization Rule and the Weyl Correspondence," *Physics,* Vol. 2, 1965, pp. 131–139.

Bellman, R., and R. Vasudevan, *Wave Propagation—An Invariant Imbedding Approach,* Tokyo: D. Reidel Pub., 1986.

Bremmer, H., "The W.K.B. Approximation as the First Term of a Geometric-optical Series," *Commun. Pure Appl. Math.*, Vol. 4, 1951, pp. 105–115.

Brillouin, L., "La mécanique Ondulatoire de Schrödinger; une Méthode Général de Résolution par Approximations Successives," *Compt. Rend. de L'Académie des Sciences,* Vol. 183, 1926, pp. 24–29; ---, "Remarques sur la Mécanique Ondulatoire," *J. Phys.*, Vol. 7, 1926, pp. 353–368.

Debye, P., "Näherungsformeln für die Zylinderfunktionen für groß Werte des Arguments und Unbeschränkt Veränderliche Werte des Index," *Math. Ann.*, Vol. 67, 1909, pp. 535–558.

Gans, R., "Fortpflanzung des Lichts durch ein Inhomogenes Medium," *Ann. Phys.*, Vol. 47, 1915, pp. 709–736.

Green, G., "On the Motion of Waves in a Variable Canal of Small Depth and Width," *Proc. Cambridge Philos. Soc.*, Vol. 6, 1837, pp. 457–462.

Hashimoto, M., "Asymptotic-Perturbational Eiqenvalues for the Multidimensional Wave Equations in Nonseparable Systems," *Opt. Commun.*, Vol. 47, 1983, pp. 243–247.

Hashimoto, M., "Asymptotic Vector Modes of Inhomogeneous Circular Waveguides," *Radio Sci.*, Vol. 17, No. 1, 1982, pp. 3–9.

Hashimoto, M., "Circularly Polarized Modal Skew Rays in Graded-index Optical Fibers," *J. Opt. Soc. Am.*, Vol. 72, No. 9, 1982, pp. 1147–1151.

Heading, J., "Comparison of Reflection and Transmission Formulae," *J. Res. Nat. Bureau Stand.*, Vol. 67D, No. 1, 1963, pp. 65–77.

Hinata, T., H. Yoshikawa, and T. Hosono, "Propagation Characteristics of Optical Fibers with Non-ideal Index Profiles," *Trans. Inst. Electron. Commun. Engrs. Jpn.*, Vol. J65-C, No. 8, 1982, pp. 637–644 [in Japanese].

Ikuno, H., "Vectorial Wave Analysis of Graded-index Fibers," *Radio Sci.*, Vol. 17, No. 1, 1982, pp. 37–42.

Ikuno, H., and A. Yata, "Uniform Asymptotic Solution of the Guided Modes in Inhomogeneous Sla Waveguides with an Even Polynomial Refractive-index Profile," *Memoirs of Fac. of Engineering,* Kumamoto University, Vol. 30, No. 3, 1985, pp. 17–63.

Jacobsen, G., "Evanescent Wave Theory Describing Propagation in Guiding Environments," *Rept. Tech. Univ. Denmark Lyngby,* LD43, July 1980, pp. 1–133.

Jeffreys, H., "Certain Approximate Solutions of Linear Differential Equations of the Second Order," *Proc. London Math. Soc.*, Ser. 2, Vol. 23, 1924, pp. 428–437.

Kesarwani, R.N., and Y.P. Varshni, "Five-term WKBJ Approximation," *J. Math. Phys.*, Vol. 21, 1980, pp. 90–92; ---, "The Quantum Pendulum in the WKBJ Approximation," *J. Math. Phys.,* Vol. 23, 1982, pp. 92–95.

Kramers, H.A., "Wellenmechanik und Halbzahlige Quantisierung," *Z. Phys.*, Vol. 39, 1926, pp. 828–840.

Liouville, J., "Sur le Développement des Fonctions ou Parties de Fonctions en Séries dont les Divers Terms Sont Assujettis à Satisfaire à une Même Équation Différentielle du Second Ordre Contenant un Paramètre Varable," *J. Math. Pure Appl.*, Vol. 2, 1837, pp. 16–35; ---; "Sur l'Intégration d'une Class d'Équations Différentielles du second ordre en Quantités Finies Explicites," *J. Math. Pure Appl.*, Vol. 4, 1839, pp. 423–456.

Madelung, E., "Über eine Methode zur Schnellen Numerischen Lösung von Differentialgleichungen Zweiter Ordnung," *Z. Phys.*, Vol. 67, 1931, pp. 516–518.

Milne, W. E., "The Numerical Determination of Characteristic Numbers," *Phys. Rev.*, Vol. 35, 1930, pp. 862–867.

Moriguchi, H., "An Improvement of the WKB Method in the Presence of Turning Points and the Asymptotic Solutions of a Class of Hill Equations," *J. Phys. Soc. Jpn.*, Vol. 14, No. 12, 1959, pp. 1771–1796.

Morishita, K., "Hybrid Modes in Circular Cylindrical Optical Fibers," *IEEE Trans. Microwave Theory Tech.*, Vol. MTT-31, No. 4, 1983, pp. 344–350.

Rayleigh, L., "On the Propagation of Waves through a Stratified Medium, with Special Reference to the Question of Reflection," *Proc. Roy. Soc. London,* Vol. A86, 1912, pp. 208–226; ---, "On the Reflection of Light from a Regularly Stratified Medium," *Proc. Roy. Soc. London,* Vol. A93, 1917, pp. 565–577.

Sluijter, F., "Generalization of the Bremmer Series Based on Physical Concepts," *J. Math. Anal. Appl.*, Vol. 27, 1969, pp. 282–302.

Sommerfeld, A., and J. Runge, "Anwendung der Vektorrechnung auf die Grundlagen der Geometrischen Optik," *Ann. Phys.*, Vol. 35, 1911, pp. 277–298.

Stokes, G.G., "On the Numerical Calculation of a Class of Definite Integrals and Infinite Series," *Proc. Cambridge Philos. Soc.*, Vol. 9, 1850, pp. 166–187.

Stokes, G.G., "Supplement to a Paper on the Discontinuity of Arbitrary Constants which Appear in Divergent Developments," *Proc. Cambridge Philos. Soc.*, Vol. 11, 1871, pp. 412–425.

Thomson, W. (Kelvin, L.), "On the Waves Produced by a Single Impulse in Water of any Depth, or in a Dispersive Medium," *Phil. Mag.*, Vol. 23, 1887, pp. 252–255.

Yata, A., and H. Ikuno, "Cutoff Frequencies of W-type Fibre with Polynomial Profile Core," *Electron. Lètt.*, Vol. 17, No. 1, 1981, pp. 9–11.

CHAPTER ELEVEN

Citerne, J., and W. Zieniutycz, "Spectral-domain Approach for Continuous Spectrum of Slot-like Transmission Lines," *IEEE Trans. Microwave Theory Tech.*, Vol. MTT-33, No. 9, September 1985, pp. 817–818.

Glandorf, F.J., and I. Wolff, "A Spectral-domain Analysis of Periodically Nonuniform Microstrip Lines," *IEEE Trans. Microwave Theory Tech.*, Vol. MTT-35, No. 3, March 1987, pp. 336–343.

Jansen, R.H., "The Spectral-domain Approach for Microwave Integrated Circuits," *IEEE Trans. Microwave Theory Tech.*, Vol. MTT-33, No. 10, October 1985, pp. 1043–1056.

INDEX